PROBLEME DER ASTRONOMIE

FESTSCHRIFT

FÜR

HUGO v. SEELIGER

DEM FORSCHER UND LEHRER ZUM

FÜNFUNDSIEBZIGSTEN GEBURTSTAGE

MIT 58 ABBILDUNGEN, 1 BILDNIS UND 3 TAFELN

VERLAG VON JULIUS SPRINGER · BERLIN · 1924

Additional material to this book can be downloaded from http://extras.springer.com.

ISBN 978-3-642-50455-6   ISBN 978-3-642-50764-9 (eBook)
DOI 10.1007/978-3-642-50764-9

Inhaltsverzeichnis.

The Origin of the Solar System[1].

By J. H. Jeans, London.

With 9 figures.

The astronomer of to-day has at his disposal telescopes which range in aperture from his naked eye, of aperture about one-fifth of an inch, up to the giant Mount Wilson telescope of more than 100 inches. If we lived in the midst of a uniform infinite field of stars, or in a field which was uniform as far as our telescopes could reach, the numbers of stars visible in different telescopes would be proportional to the cubes of their apertures.

In actual fact our naked eyes reveal about 5000 stars; with a one-inch telescope this number is increased to about 100,000, with a ten-inch to 5 million, and with the 100-inch telescope to perhaps 100 million. These numbers increase much less rapidly than the cubes of the apertures. We conclude that we are not surrounded by an infinite uniform field of stars. We live in a finite universe, which thins out quite perceptibly within distances reached by telescopes of very moderate size. It is estimated that the whole universe consists of some 1500 million stars, our sun being not very far from the centre of the system.

Imagine the various celestial objects in this universe arranged according to their distance from us. Disregarding altogether bodies which are much smaller than our earth, we must give first place to the planets Venus and Mars, which approach to within 26 and 35 millions of miles respectively. Next comes Mercury with a closest approach of 47 million miles, and the sun at 93 million miles. The remainder of the planets follow at distances ranging up to 2800 million miles, the radius of the orbit of Neptune.

But now comes a great gap. The first objects beyond this gap are the faint star Proxima Centauri at a distance of 24 million million miles, or more than 8000 times the distance of Neptune, and close to it, α Centauri at 25 million million miles. Next in order come the faint red star Munich 15,040 at 36 million million miles, and another faint star Lalande 21,185 at about 47 million million miles. Thus our nearest neighbours among the stars are at almost exactly a million times the distances of our nearest neighbours among the planets. After these comes Sirius,

---

[1] Discourse delivered at the Royal Institution on February 15.

the brightest star in the sky, at 50 million million miles. From here on there is a steady succession of objects until we reach distances of more than 20,000 times that of Sirius; but long before these distances are reached other objects, spiral and spheroidal nebulae, and ultimately star-clusters, are found to be mingled with the stars. The furthest object the distance of which is known with any accuracy is the star-cluster N.G.C. 7006, which Shapley estimates to be 25,000 times as distant as Sirius. This cluster is so remote that its light takes 200,000 years to reach us; even for light to cross the cluster takes hundreds of years. To all appearances the star-cloud N.G.C. 6822 is still more remote. According to Shapley its distance is about six million million million miles, a distance which light takes a million years to traverse. So far as is known at present, this brings us to the end of our universe, or perhaps I ought to say it brings us back to the beginning.

It is no easy matter to get all these different distances clearly into focus simultaneously, but let us try. The earth speeds round the sun at about twenty miles a second; in a year it describes an orbit of nearly six hundred million miles circumference. If we represent the earth's orbit by a pin-head or a full-stop of radius one-hundredth of an inch, the sun will be an invisible speck of dust, and the earth an ultra-microscopic particle one-millionth of an inch in diameter. Neptune's orbit, which encloses the whole of the solar system, will be represented by a circle the size of a threepenny-piece, while the distance to the nearest star, Proxima Centauri, will be about 75 yards and that to Sirius about 160 yards. On this same scale the distance to the remote star cluster N.G.C. 7006 is 2400 miles and that to the star-cloud N.G.C. 6822 about 12,000 miles, so that roughly speaking the whole universe may be represented by our earth.

It thus appears that we are on this occasion to discuss the origin and past history of a system which bears the same relation to the universe as a whole as does a threepenny-piece to our earth. Why are we so interested in this particular threepenny-piece? Primarily because, although a poor thing, it is our own, or at least one particle of it, one millionth of an inch in diameter, is our own. But there is a historical reason of a less sentimental kind. We have already noticed the immensity of the gap between our system and its nearest neighbours. As regards astronomical knowledge this gap has taken a great deal of crossing. Well on into last century, human knowledge of the further side of this gap was infinitesimal; the stars were scarcely more than points of light, described as "fixed stars". In those days the problem of cosmogony reduced perforce to the problem of the origin of our own system.

Recent research has changed all this, and the modern astronomer has a very extensive knowledge of the nature, structure and movements of the various bodies outside our system. The cosmogonist of a century ago could assert that the solar system had evolved in such and such a way, and need have no fear of his theories being upset by comparison with other systems. But if I put before you now a theory of the origin of our system, you will at once inquire as to the behaviour of the 1500 million or so of systems beyond the great gap. Are they following the same evolutionary course as our own system, and, if not, why not? It may be well to consider these other systems first.

Among these 1500 million or so of objects there are certain comparatively small classes the nature and interpretation of which are still enigmatical—the planetary nebulae, the Cepheid variables, the long-period variables such as Mira Ceti, and a few others. Apart from these, practically all known bodies can be arranged in one single continuous sequence. The sequence is approximately one of increasing density: it begins with nebulae of almost incredible tenuity and ends with solid stars as dense as iron. There is but little doubt that the sequence is an evolutionary one, for the laws of physics require that as a body radiates heat its density should increase, at least until it can increase no further. Let us begin our survey at the furthest point back to which we can attain on this evolutionary chain—the nebulae.

After the enigmatical "planetary" nebulae have been excluded, the remaining nebulae fall into two fairly sharply defined classes, which may be briefly described as regularly and irregularly shaped nebulae.

The irregularly shaped nebulae comprise such objects as the great nebula in Orion, and the nebulosity surrounding the Pleiades. Until quite recently these irregular nebulae were supposed to be of great evolutionary importance. It was noticed that they were usually associated with the very hottest stars: whence arose a beautifully simple cosmogony, asserting that these very hot stars were the immediate products of condensation of the nebulae, and that their after-life consisted merely of a gradual cooling until they got quite cold. This cosmogony was too simple to live for long—it was buried some ten years ago by the researches of RUSSELL, HERTZSPRUNG, and others. Thanks to these researches, we now know that the very hot stars associated with irregular nebulae, so far from being newly born, are standing at the summit of their lives awaiting their decline into old age.

A mass of hot gas isolated in space radiates heat, and this causes it to contract. If the mass radiated without contracting, it would, of course, get cooler; on the other hand, if it contracted without radiating, it would get hotter. But when radiation and contraction are proceeding together it is not obvious without mathematical investigation which of the two tendencies will take command. In 1870, HOMER LANE showed that a mass of gas of density low enough for the ordinary gas laws to be approximately obeyed, will in actual fact get hotter as it radiates heat away. Cooling does not set in until a density is reached at which

the gas laws are already beginning to fail—that is to say when lique-
faction and solidification are already within measurable distance. Thus
we see that maximum temperature is associated with middle age in a star,
the age at which the star may no longer be regarded as a perfect gas.
At this period of middle age the surface temperature of the star may
be anything up to about 25,000° C., while the temperature at its centre
will amount to millions of degrees. Its average density will probably
be something like one-tenth of that of water. It is still not known why
stars at this special maximum temperature are so commonly associated

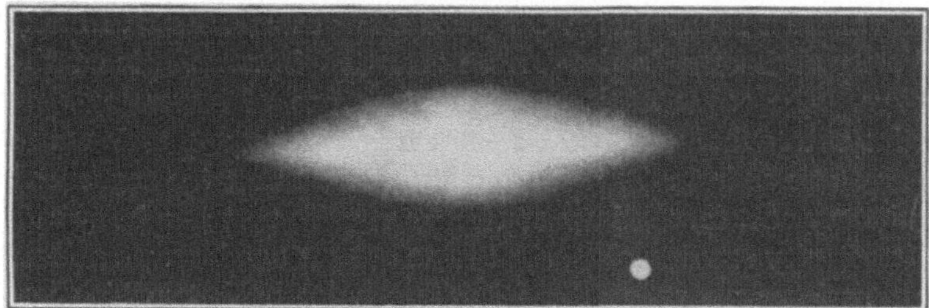
Fig. 1. Regular shaped nebula (N.G.C. 3115).

with irregular nebulae. Possibly it may be that only stars at the very
highest temperatures are capable of lighting up surrounding nebulosity
which would otherwise remain invisible. Be this as it may, it is fairly
clear that these irregular nebular masses are not an essential part of
the evolutionary chain. They are probably mere by-products, and as
such may be dismissed from further consideration.

We turn to the nebulae of regular shape. A great number of these
appear as circles or ellipses, some as ellipses drawn out at the ends of
their major-axes, sometimes almost to sharp points. An example of
this last type of figure is shown in Fig. 1 (Nebula N.G.C. 3115).

A number of these regular-shaped nebulae have been examined
spectroscopically, and in every case have been found to be rotating
with high velocities about an axis which appears in the sky as the
shortest diameter of the nebula. The mathematician can calculate what
configurations will be assumed by masses of tenuous gas in rotation.
If rotation were entirely absent the mass would, of course, assume a
spherical shape. With slow rotation its shape would be an oblate spheroid
of low ellipticity—an orange-shaped figure like our earth. At higher
rotations the spheroidal shape is departed from, the equator bulging out
more and more until finally, for quite rapid rotation, the shape is approximately that of a double convex lens having a sharp circular edge for its
equator, the shape, in fact, exhibited by the nebula shown in Fig. 1.
The whole succession of figures, if looked at along all possible lines of
sight, will exhibit precisely the series of shapes which are found to be

exhibited by the regular nebulae under discussion. There are, then, good grounds for conjecturing that these nebulae are rotating masses of gas; but we can test this conjecture further before finally accepting it.

As a mass of gas radiates its energy away it must shrink. If it is in rotation, its angular momentum will remain constant, and the shrunken

Fig. 2. Regular shaped nebula (N.G.C. 5866) with band of dark matter on equator.

mass can only carry its original dose of angular momentum by rotating more rapidly than before. This conception, which formed the cornerstone of the cosmogonies of KANT and LAPLACE, is still of fundamental importance to the cosmogonist of to-day. Thus every nebula, as it grows older, will rotate ever more and more rapidly and, barring accidents, will in due course reach the configuration shown in Fig. 1. This configuration marks a veritable landmark in the evolutionary path of a nebula. Until this configuration is reached the effect of shrinkage can be adjusted, and is adjusted, by a mere change of shape; the mass carries the same angular momentum as before, in spite of its reduced size, by the simple expedient of rotating more rapidly, and restores equilibrium by bulging out its equator. But mathematical analysis shows that this is no longer possible when once this landmark has been passed. Further shrinkage now involves an actual break-up of the nebula, the excess of the angular momentum beyond that which can be carried by the shrunken mass being thrown off into space by the ejection of matter from the equator of the nebula.

We have so far spoken of the nebular equator as being of circular shape, as it undoubtedly would be if the nebula were alone by itself in space. But an actual nebula must have neighbours, and these neighbours will raise tides on its surface, just as the sun and moon raise tides on the surface of the rotating earth. Whatever the neighbours are, there will always be two points of high tide antipodally opposite to one another, and two points of low tide intermediate between the two points of high tide. Thus the equator, instead of being strictly circular, will be slightly elliptical.

If the equator of the nebula had been a perfect circle, and if the nebula had been in all respects symmetrical about its axis of rotation,

Fig. 3.   Regular shaped nebula (N.G.C. 4594) with ring of dark matter surrounding equator.

the ejection of matter would have started from all points of the equator simultaneously. Indeed, there could be no conceivable reason why it should start at one point rather than at any other point. But in Nature we do not expect to find perfect balances of this kind; if the main factors are of exactly equal weight some quite minor factor invariably intervenes to turn the balance in one direction or another. In the present problem there could be no choice as between one point of the equator and another if the various minor factors were absent, but when these minor factors come into play, a discrimination at once takes place. Assuming, as seems likely, that the tidal irregularities are the minor factors which determine the choice of points for the ejection of matter, mathematical investigation shows that the ejection of matter will take place from the two antipodal points on the equator at which the tide is highest. The

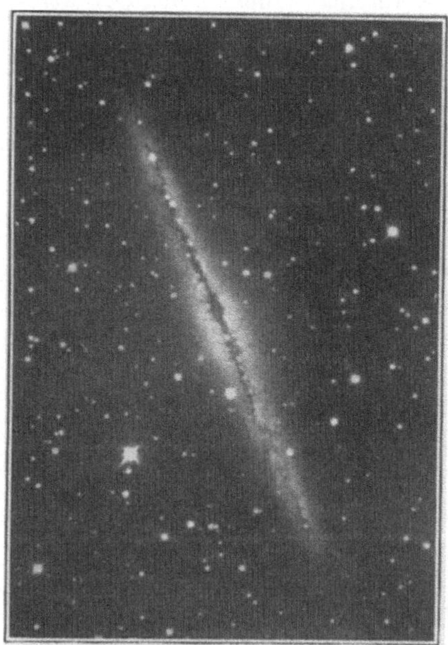

Fig. 4.   Spiral nebula (N.G.C. 891) seen edge on.

equator being slightly elliptical, these points are of course the ends of its major-axis. After the nebula has passed its critical landmark, shown in Fig. 1, its shape ought to be similar to the lenticular figure which formed the landmark, but with the additional feature of matter streaming out from two antipodal points on its equator.

This describes exactly what is observed in the spiral nebulae. Fig. 2 (N.G.C. 5866) shows a nebula in which the ejection of matter is just

Fig. 5.   Spiral nebula in Canes Venatici (M. 51).

beginning; we notice the bulge along the equator and the dark band which we may assume represents ejected matter which is already cooling. Fig. 3 (N.G.C. 4594) exhibits a more advanced state of development; and Fig. 4 (N.G.C. 891), a still later one in which the ejected matter already dwarfs the central nucleus in size, although probably not in total mass.

In all these figures we are looking at the nebulae very approximately edge-on. Fig. 5 (M. 51) shows the well-known "whirlpool" in Canes Venatici, a nebula which may be very similar physically to that shown in Fig. 4, but we see it face on: we are looking along its axis of rotation. Again the central nucleus occupies only a small part of the picture. Figs. 6 (M. 101) and 7 (M. 81) show two nebulae, the evolution of which has proceeded still further, so much so that in the last of these there is very little nucleus left, and by far the greater part of what we see is what we believe to be ejected matter.

In both of these last two nebulae it will be seen that the arms of ejected matter proceed from two antipodal points, exactly as required

Fig. 6.   Spiral nebula in Ursa Major (M. 101).

by dynamical theory. So far we have spoken of the matter in these arms as ejected matter because theory has suggested this interpretation, but we need not be satisfied with theory; there is very direct observational evidence on the point. Various astronomers, especially Van Maanen, have detected motion in the arms of many nebulae, including the three shown in Figs. 5, 6, and 7. Their observations show that the arms are in real truth jets of matter coming out of the nucleus. Fig. 8 shows the motion found by Van Maanen for about 100 points in the nebula M. 81, the arrows showing the motion in a period of 1300 years[1]), and the measures on the various other nebulae show substantially similar results; you will see that there is little room for doubting that the arms consist of matter flowing out of the nucleus. On measuring the actual velocities of flow it is found that in nebula M. 51 (Fig. 5) a particle of the jet makes a complete revolution around the nucleus in about 45,000 years; in M. 81 (Fig. 7) the corresponding figure is about 58,000 years, and in M. 101 (Fig. 6) about 85,000 years. From these figures it is possible to estimate the density of the matter in the nucleus. It

Fig. 7.   Spiral nebula in Ursa Major (M. 81).

is found that the densities must be of the order of 10^{-16} gm. per cubic centimetre, a figure representing a vacuum more perfect than any obtainable in the laboratory. The small amount of gas in an ordinary electric light bulb, if spread out through St. Paul's Cathedral, would still be something like 10,000 times as dense as the nucleus of a spiral nebula.

---

[1]) The points surrounded by small circles are stars which are believed to have no physical connexion with the nebula.

The nebula shown in Fig. 4 exhibits a lumpy or granulated appearance in its arms. In M. 51 (Fig. 5) this takes the form of pronounced condensations, and in the outer regions of M. 101 (Fig. 6) and M. 81 (Fig. 7) these condensations have further developed into detached and almost star-like points of light.

When gas is set free out of an ordinary nozzle into a vacuum it immediately spreads into the whole of the space accessible to it. Why

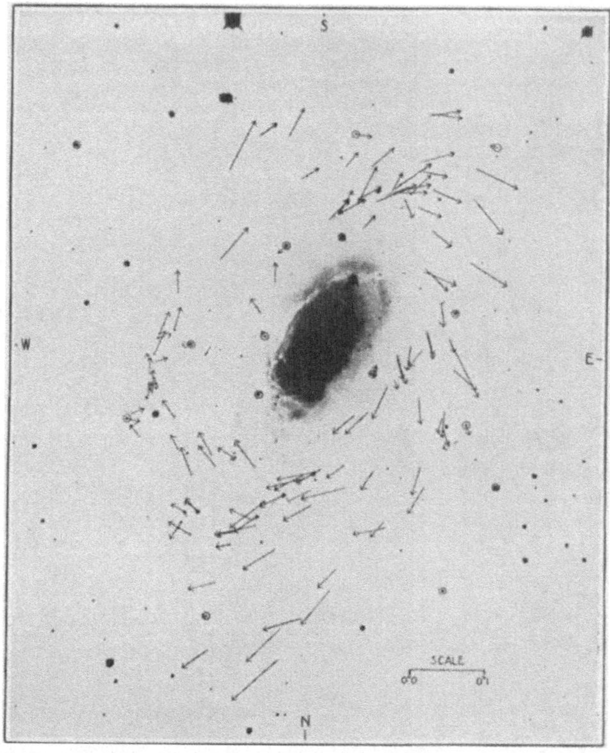

Fig. 8. Motion in the arms of the spiral nebula (M. 81).

then does not the jet of gas shot off from the equator of the nebula do the same? The explanation is to be found in the gigantic scale on which this latter process takes place. As we increase the scale of the phenomenon the mutual gravitational attraction of the particles of gas becomes of ever greater importance until finally, by the time nebular dimensions are reached, gravitation overcomes the expansive influence of gas pressure and is able to hold the jet together as a compact stream. But, as soon as this happens, dynamical theory predicts that a further phenomenon ought to appear. As regards the distribution of density along the filament, the influence of gas-pressure is in the direction of keeping the density spread out uniformly, while that of gravitation is towards making the stream condense with compact globules. When nebular dimensions are reached the latter tendency prevails, and the issuing jet of gas breaks up into drops much as a jet of water issuing from a nozzle does, although for a very different physical reason. In the photographs reproduced in Figs. 4, 5, 6, and 7 we can trace this process going on.

Dynamical theory not only predicts that these globules of gas must form, but also enables us to calculate their size, mass, and distance apart. A comparison between their distance apart, as calculated in kilometres, and their angular distance apart, as observed in the sky, leads at once to an estimate of the distance of the nebula to which they belong. It is gratifying to find that estimates of nebular distances made in this way are in good agreement with estimates made in other ways. The calculation of the masses of these condensations leads to a still more interesting and significant result. In every nebula for which the calculation can be made, the calculated mass of a single condensation proves to be approximately equal to the mass of the average star.

This gives, I believe, the key to the evolutionary process we have been considering—we have been watching the creation of the stars. In Fig. 1 we saw the raw material—a gaseous mass of extreme tenuity, already moulded, as a result of shrinkage and consequent increase of rotation, to the stage at which disintegration is about to commence. Further shrinkage takes place, and in Fig. 2 and 3 we see the ejection of jets of gas from which the future stars will in due course be made. In Figs. 4 and 5 individual stars are beginning to form, although at present only as vague condensations in what is still a continuous nebular mass. Finally, the outermost parts of Figs. 6 and 7 show us the finished product—separate masses, although still far more tenuous than ordinary stars, starting off on their independent existences. Each of these masses will go through the changes we have already briefly described. It will contract, getting hotter in doing so, until it reaches a maximum temperature just as the gas-laws are beginning to fail, after which it cools and contracts into a dead dark mass.

The family of stars born out of a single nebula may be millions in number. They may either mingle with the general mass of the stars or, if the original nebula was sufficiently remote from the main universe of stars, may form a separate colony by themselves. In illustration of the former alternative, numbers of groups of stars are known—*e.g.* the Pleiades, the stars of the Great Bear—in which all the members have a common velocity and, generally speaking, similar physical constitutions also. All the stars of any such group are voyaging through space together, and have obviously done so since they first came into being. The alternative possibility of a family of stars forming a detached colony by themselves is perhaps exemplified in the so-called "globular" star-clusters, such as the well-known cluster in Hercules (Fig. 9). These are globular only in name, for Shapley has found that they are of an elliptical structure, showing symmetry about a plane precisely as might be expected if they were the final product of a rotating nebula.

Fig. 9.   Star-cluster (M. 13) in Hercules.

Probably we ought not to regard the two possibilities just mentioned as sharply cut alternatives. It is more likely that they represent the two extreme ends of a continuous chain of possible histories for the family of stars born out of a single nebula. It seems quite possible that what we describe as "the main mass of the stars" may be nothing more than a collection of clusters of stars, each cluster having originated out of a single nebula. The clusters are by now so intermingled that it is difficult to look on them as distinct groups of stars, although we can still find some evidence that this may be the proper way of regarding them. In 1905 Kapteyn showed that the stars in the neighbourhood of the sun formed what he described as two "star-streams," each stream moving with its own velocity in space. Except that it begs the question as to the extent of these streams in space, it would have been equally accurate to describe them as forming two intermingled moving clusters. Shortly after, EDDINGTON and HALM, independently, found a third stream or moving cluster, constituted of the very hot stars which the astronomer classifies as stars of types B and O. In this case we know the extent of the cluster in space and also its approximate shape. According to Charlier, it is shaped like a round biscuit lying parallel to the Milky Way, its diameter being about 2·8 times its thickness. Any cluster of stars having a common origin, whatever shape it may assume at first, will be rapidly knocked out of shape when it begins to intermingle with other stars. Dynamical theory shows that after it has been knocked about *ad infinitum* in our universe of stars, such a cluster ought to assume the shape of a round biscuit parallel to the Milky Way, the ratio of its diameter to its thickness being about 2·5. This agrees sufficiently well with what is observed to suggest that all the stars in this stream have a common origin, and the same is true of many of the smaller known moving clusters, such as the Ursa Major cluster already mentioned Thus, although we cannot claim that anything is definitely proved, there is every justification for thinking of the main mass of the stars as a jumble of intermingled moving clusters, each cluster owing its existence to a separate nebula. This possibility has no very direct bearing on the question of the origin of our solar system; it has been mentioned merely as rounding off our knowledge of what appears to be the main evolutionary process of the stars.

In all its essentials except one, this evolutionary process is similar to, and in its earlier stages almost identical with, that which Laplace, in his famous nebular hypothesis, imagined as the origin of the solar system. We have seen before our eyes the rotating and shrinking nebula finally shedding matter from its equator; we have watched the condensation of this matter into separate masses, and have finally witnessed the start of these detached masses on their voyages into space, all precisely as pictured by Laplace.

The one essential difference is that of size. The evolutionary process we have been watching occurs on a scale such as Laplace never dreamed of. His primeval nebula was supposed to be of about the size of Neptune's orbit, a size represented on the scale I used at the beginning of this lecture by a threepenny-bit. On this same scale the nucleus alone of a good-sized spiral nebula, such as those shown in Figs. 6 and 7, would be about the size of the ALBERT HALL, while the arms would sprawl over the whole of Hyde Park and Kensington. The pictures of these nebulae that you have before you would have to be enlarged to the size of a whole country, or even possibly of a whole continent, before a body the size of our earth became visible in them at all.

Although the parent nebulae we have been considering are all incomparably greater than Laplace's imaginary nebula, yet each tiny condensation, as it starts off into space, is a gaseous nebula the mass of which is just about equal to that imagined by Laplace and the size of which is not perhaps very greatly different. If, then, this younger generation of nebulae meet with the same experiences in life as their giant parents before them, we should not have to look far for an explanation of the origin of the planets, and if the third generation again repeated the experience of their ancestors, the satellites of the planets are also accounted for. But mathematical research and observation agree in disposing of so simple an explanation of the origin of the solar system. As we have seen, it is only because the filaments in the spiral nebulae are of such huge size that gravitation is able to cause condensation in opposition to the expansive tendency of gas-pressure. A nebula of mass comparable to our sun might go through the same life-history as the bigger nebula until matter began to be thrown off from its equator, but after this the difference of scale would begin to tell, and the subsequent course of events would be widely different. The ejected matter could not condense into filaments, still less into detached globules; it would merely constitute a diffuse atmosphere surrounding the parent nebula. As such a system shrunk by the emission of radiation, the constancy of angular momentum would, at first, merely demand that more and more gas should be transferred from the centre to the atmosphere.

But mathematical investigation shows that in time, after the central star had shrunk to a certain critical density, perhaps somewhere about one-tenth of that of water, a cataclysmic period would ensue, from which the mass would emerge as a binary star—two stars of comparable masses revolving about one another nearly in contact and in approximately circular orbits. This is a formation with which the practical astronomer is very familiar. He finds that a very large proportion—perhaps about one-half—of the stars in the sky are binary, and he can detect an evolutionary sequence in these binary stars. The sequence starts with the formation just described in which the two constituent stars are almost in contact. As it progresses the stars move ever farther and farther apart, while the eccentricity of their orbits increases. Theory indicates that the process of fission which has broken up the original star into two constituents may repeat itself in either or both of these constituents, so that the final product may be a "multiple" star of either three or four constituents. Prof. H. N. RUSSELL, investigating this question theoretically, found that certain numerical relations must hold between the relative distances of the various constituents of a multiple star: he also showed that the predictions of theory are confirmed quantitatively by observation.

So far, then, theory and observation have gone hand-in-hand. We have traced the evolution of astronomical matter through stages of ever-increasing density, from the most tenuous of nebulae to the densest of multiple stars, and at almost every stage observation has confirmed the predictions of theory. Not all astronomical matter will traverse the whole length of this evolutionary course. The driving force on this course is increase of rotation consequent on the shrinkage produced by emission of radiation. When the shrinkage has proceeded a certain length solidification sets in; the rotation can increase now no further, and evolution, in the physical sense, stops. The distance along the course to which any particular system proceeds depends in effect on the amount of rotation with which it was originally endowed. Let a nebula begin its career with absolutely no rotation, and it will remain spherical in shape throughout its whole career, ending merely as a cold non-radiating, but always spherical, mass. Such a nebula never even gets away from the starting-post. It is true that this is not a likely event, but for aught we know many a nebula may freeze and die before reaching the critical configuration (Fig. 1) at which the birth of stars first commences. Similarly many of the stars may become cold and so cease to develop without ever attaining the stage at which binary systems are formed. In the same way many binary systems must fail to develop into multiple systems. Here again observation is with us: there are ten times as many purely binary systems known as there are multiple systems which have proceeded beyond the binary stage. Theory has traced out for us the whole length of the evolutionary course, but theory and observation agree that not many systems stay out the whole course.

We now come to the crux of the whole question. Nowhere on this course have we found our solar system, or anything in the least degree resembling it. If our sun had been unattended by planets we should have had no difficulty in guessing its origin. It might reasonably be supposed to have been born out of a nebula in the normal way, but to have emerged with insufficient rotation to have carried it on to the later stages of fission into a binary or a multiple system. It might, in fact, be supposed to have had the same evolutionary career as half of the stars in the sky. In support of the conjecture that our sun had been born out of a nebula in the ordinary way, we could note that its mass is about equal to what we calculate ought to be the mass of a star born out of a nebula, and that it is, apart from its planets, similar in every way to millions of other stars to which we may ascribe a nebular origin.

In support of the conjecture that it had stopped short on its evolutionary course from want of adequate rotation to carry it on further, we should merely have to note the slowness of its present rotation. A simple calculation shows that the sun has only a small fraction of the amount of angular momentum requisite for fission. Even if we add the angular momentum of all the planets, as we ought if we suppose that these at one time formed part of the sun, the result is the same—the whole system can never have had more than a fraction of the angular momentum necessary for a rotational break-up into a binary star.

Thus the sun is a quite intelligible structure. The difficulty of our problem is not the origin of the sun, but the origin of the planets and of their satellites.

Certain special types of astronomical structure have already been mentioned as not falling into place on the main line of evolutionary development. The particular examples chosen were the planetary nebulae, the Cepheid variables, and the long-period veriables. The question now arises as to whether we must add the solar system to the list. The circumstance that certain structures do not find a place in the evolutionary main line suggests that off this main line may be branch lines on to which the development of a system may in certain circumstances be turned. This indeed is only what might be anticipated. We should no more expect two stars to have precisely the same experiences in their careers than we should expect it of two humans. Our normal star has been supposed to develop in a universe of its own, where its angular momentum remained constant and where it was in every way unmolested by its neighbours. The mathematician finds it convenient to allot a whole infinite universe to each star, but Nature does not. Nevertheless, the conditions postulated by the mathematician are nearer to the truth than is often the case in his idealised problems. On the scale we have already used, on which the sun was represented by a microscopic particle $\frac{1}{10000}$ inch in diameter, the most gigantic of known giant stars may be represented by a pin-head one-thirtieth of an inch in diameter. The present spacing of the stars is such that on this scale there is less than one star to a volume equal to the interior of St. Paul's Cathedral. Space then cannot be said to be overcrowded, and although it is possible that stars may disturb one another as they move in their courses, it is clear that any serious disturbance of one star by another must be a rather exceptional event. Obviously we have been right in regarding the evolution of a star entirely undisturbed by its neighbours as the normal course of evolution, and we can now see why the vast majority of stars follow this normal course.

To all appearances, the stars which have been sidetracked off this normal course are extraordinarily few in number. The total number of stars in the sky is about equal to the total population of the earth;

the number of known exceptional systems would at most populate one small town, although, of course, we can scarcely even conjecture how many exceptional systems there may be which are still unknown to us. There is no reason for supposing that the side-tracking influence has in every case been a neighbouring star, but the systems known to be exceptional are sufficiently few to suggest that this may have been the cause in a large proportion of cases.

The immediate question before us, however, is not that of the exceptional systems in general, but of our own solar system. Was it a neighbouring star that threw it off the main line of evolutionary development? Here, for the first time, observational astronomy denies us any help. Not a single system is known outside our solar system which resembles it in the least degree. The reason is not that no such system exists, but that we could not see it if it did. An astronomer on a distant star observing our system would see Jupiter as the brightest object after our sun, but the ratio of their luminosities would be as three hundred million to one. Seen from our nearest known neighbour in space, Proxima Centauri, the sun would appear as a first magnitude star, and Jupiter as a star of magnitude 22·2, the distance between them being at most four seconds of arc. A star of magnitude 22·2 is still well beyond the range of our largest telescopes, and would be doubly invisible if it had a first magnitude star only four seconds away. We must wait for a very great increase in the power of our telescopes before there will be any hope of seeing systems similar to our own in the sky, even if they exist no further away from us than Proxima Centauri. Thus it is clear that our discussion has now left the regions in which observation can be called upon to make suggestions or to check our conclusions: henceforth we have theory alone to guide us.

Let us start on our quest by noticing that our solar system has quite clearly marked characteristics. It is no mere jumble of bodies looking as though they had fallen together by accident—had it presented this appearance the problem of its origin might reasonably be dismissed as hopeless. Not only has the principal system of the sun and its planets got clearly marked characteristics, but also these same characteristics reappear in the smaller systems formed by Jupiter and Saturn, each with its family of satellites. Each of these small systems is, to all intents and purposes, a replica in miniature of the solar system, so much so that no suggested origin for one system can be regarded as satisfactory unless at the same time it explains the origin of the other two. The principal features common to the three systems are, that the orbits in all three systems are, with few exceptions, all in or close to one plane, that these orbits are all described in the same direction, and that the masses of the secondaries, whether planets or satellites, are all small in comparison with those of the primaries around which they revolve. Thus the sun

has a mass equal to 1047 times that of his greatest planet Jupiter, while
Jupiter's mass is about 11,000 times that of his most massive satellite.
The smallest disparity in mass is found in our own Earth-Moon system
with a mass ratio of 81 to 1. In systems possessing many satellites
(those of the sun, Jupiter, and Saturn) there is a general tendency for
the masses to increase up to a maximum as we pass outwards through
the system, and then to decrease to a minimum. Thus in the main
system there is a regular progression trough Mercury, Venus, Earth,
Mars to the maximum mass of Jupiter, broken only by the anomalous
position of Mars, while on the descending side the progression through
Jupiter, Saturn, Uranus, Neptune fails in regularity only through Neptune being some few per cent. more massive than Uranus.

The main line of evolutionary progress has been supposed to be that
of a mass of shrinking, rotating matter—first gaseous, then liquid, then
solid—left to itself in space. Such a system must show one very marked
characteristic throughout its whole career, namely, a plane of symmetry.
In its earliest stage of all, when the system is a mere chaos of independent
molecules, the plane will coincide with what mathematicians describe as
the "invariable plane" of the system. Later, when the mass has assumed
the regular shape of a rotating nebula, the plane is the equatorial plane
of the nebula, the plane in which the arms subsequently appear and in
which the stellar condensations start off in their orbits. The symmetry
of spiral nebulae about their equatorial planes would of itself suggest
strongly that they have developed to their present formations as rotating
bodies practically undisturbed by external influences.

If our solar system had developed out of an undisturbed rotating
mass, it too ougth to exhibit a plane of symmetry. The orbits of nearly
all the planets and their satellites do, in actual fact, lie very nearly in one
plane, which, to this extent, is, of course, a plane of symmetry. But the
sun's axis of rotation is not perpendicular to this plane; the sun has its
own plane of symmetry in its equator, and this is inclined at an angle
of 7° to the plane of orbits.

The existence of these two distinct planes is enough in itself to suggest
that our system has not developed simply out of an undisturbed rotating
mass. Thus, in tracing our system back to its origin, we naturally look
at the effects to be expected from rotation plus some external influence.
To a first rough approximation, it is natural to suppose that the plane
of the sun's equator records the plane of rotation of the original system,
while the plane of the planetary orbits was in some way determined
by the extraneous disturbance.

Of all the interactions between two separate astronomical bodies,
gravitational attraction is likely to be by far the most potent. The moon
has been accused of exerting all kinds of influences on our earth, as, for
example, on its weather, on the destinies, the emotions, and even on the

sanity of its inhabitants; but the only influence which survives scientific examination is gravitational attraction as evidenced by the semidiurnal tides. It is true that a head-on collision between two astronomical bodies would produce more immediately dramatic results than a mere tidal pull; but we shall not consider such an event here. Head-on collisions must of necessity be exceedingly rare; systems that experience them would undoubtedly be deflected from the main line of evolutionary progress on to a branch line; but it does not seem likely that this branch line contains systems like our own. As time does not permit the exploration of all conceivable branch lines, let us turn at once to that which seems most likely to reveal the origin of our system—the branch line that diverges from the main line at the occurrence of a violent tidal encounter.

On the earth, our moon raises tides the average height of which at high tide is only a few feet. This height of high tide is only about a ten-millionth part of the earth's radius, a fraction which we may designate as the tidal fraction. If the moon were ten times as massive, the tidal fraction would be increased ten-fold; if it were brought to half its present distance, the tidal fraction would be increased eight-fold. If we agree to measure masses in terms of the body on which the tide is raised as unity and to measure lengths in terms of the radius of the same body, then the tidal fraction is equal to the mass of the tide-generating body divided by the cube of its distance, say M/R^3. Using this formula, we find that our nearest neighbour Proxima Centauri raises on the sun a tide of quite infinitesimal magnitude; the tidal fraction is about 10^{-26}, and the actual height of tide is of the order of 10^{-15} cm. or, say, one-fiftieth of the radius of an electron. This single illustration will show, and with some margin to spare, that under normal conditions the tidal influence between neighbouring stars is utterly insignificant. For tidal forces to become important to cosmogony, conditions must be abnormal.

Our sun happens at the present moment to have no especially near neighbour; but it is fairly certain that at some time, in its wanderings through the stars, it must have passed stars within a much less distance than that which now separates it from Proxima Centauri. The most trustworthy lines of evidence as to the earth's age, namely, those from geology and radioactivity, indicate an age of from 800 to 1100 million years. For precision, let us think of the sun's age as 1000 million years. Let us imagine for the moment, what is no doubt very far from the truth, that throughout all this thousand million years the sun and all the stars have moved just as they are moving now, with the same average velocities as now and keeping at the same *average* distance apart. Throughout this thousand million years the distance of our sun from its nearest neighbour will have been continually changing, and one star after another will, of course, have taken up the rôle of nearest neighbour. But there must have been some one instant in this thousand million years at which our sun was nearer than at any other instant to its nearest neighbour. A calculation based on the theory of probability indicates that this nearest distance is likely to have been of the order of 7×10^{15}cm., a distance which, although only a six-hundredth of that which now separates us from Proxima Centauri, is still equal to fifteen times the radius of Neptune's orbit. Even if the sun had filled the whole of Neptune's orbit, the tidal fraction at this closest encounter, on the supposition that the nearest star had a mass equal to the sun, would only be equal to $1/(15)^8$ or $1/3375$, giving a height of tide which is quite unimportant from the point of view of cosmogony. So long as things have been as they now are, tidal actions between separate stars must have been quite devoid of cosmogonic interest, except possibly in very special cases of quite exceptionally close approaches.

It is, of course, possible that our sun was the victim of one of those exceptionally close encounters. Nothing can be brought against the supposition of such an event, except its *a priori* improbability. The result of such a close encounter might, as we shall see, be the creation of a system in many ways resembling our solar system.

Our calculations of probabilities and improbabilities have, however, rested upon the admittedly erroneous assumption that stellar conditions have been similar to the present ones for a period of a thousand million years. On looking back trough the past history of the universe, we come to a time when conditions must have been very different from what they are now. We come to a time, which we have already considered, when our sun had not yet assumed its present stellar characteristics. It was a condensation in the arm of a spiral nebula moving with thousands of similar condensations towards a free career in space. Its density was enormously lower than it now is, and its size correspondingly greater. It was also much nearer to its neighbours than, in all probability, it has ever been since. In this early stage of its existence, the tidal effects of its neighbours may well have been enormous; we shall pass to exact figures in a moment.

In general, the passage of one star past another merely raises a tide which subsides as the tide-raising body recedes. Even when the approach is so close that the height of the tide raised is greater than the original radius of the star, the recession of the disturbing star may result in the disturbed star relapsing merely to its original spherical form. But there is a limit which must not be passed, and if the disturbing body passes this limit, all hope of the star resuming its original shape is lost. The distance of the limit depends primarily on the mass of the disturber; to a lesser degree it depends on the rotation, shape, and density-distribution of the primary star; and to some extent it depends on the velocity of the two stars relative to one another. We shall get a tolerable idea of the march of events if we suppose the primary star to be surrounded by an imaginary sphere the radius of which depends solely on the mass of the disturbing star. If this mass is equal to the mass of the primary, the radius of this imaginary sphere will be about $2\frac{1}{4}$ times the radius of the primary; if the disturbing star has eight times the mass of the primary, the radius of the imaginary sphere will be $4\frac{1}{2}$ times that of the primary, and so on. So long as the centre of the visiting star remains outside the sphere, a tide is raised which recedes as the visiting star disappears, but the moment the visiting star invades this sphere, an entirely new phenomenon appears.

As the approach of the disturber raises the tide to higher and higher levels, the highest points of the tide move ever farther away from the star's centre, into regions where the gravitational attraction of the star gets weaker and weaker. At the same time, of course, the gravitational pull of the visiting star gets stronger and stronger. Finally, just as the visiting star crosses the critical sphere, its gravitational pull just balances that of the primary—it is this condition that defines the critical sphere. If the visiting star further invades this critical sphere, the particles at high tide are shot away from the primary star, the resultant gravitational force on them now being definitely towards the visiting star; they are of course immediately replaced by others which are shot off in turn, and so on. The total effect is that a filament or jet of gas is shot out from the point of high tide. Each particle of this jet moves under the combined forces of the primary and of the visiting star, and the problem of determining its orbit is a special case of the problem of three bodies, which unfortunately is not soluble. But the general result is that the jet undergoes various contortions while moving all the time in the plane which contains the orbit of the visiting star.

If such a jet had been thrown off the sun simply by an increase of rotation consequent on shrinkage, its gravitational attraction would, as we have seen, be inadequate to resist the expansive effect of its own gaspressure, and it would have been rapidly dissipated away into space. In the present situation conditions are very different, the essential difference being that, while shrinkage from loss of radiation is a very slow process, tidal disruption may be a very rapid process. The rate of a star's rotation will alter but slightly in a thousand years, whereas ten years may suffice for a tide-raising body to come, do its work, and go away again. The filament of gas set free by increase of rotation would be of extreme tenuity; a filament set free by a tidal cataclysm might easily be of sufficient substance for its own gravitation to hold it together as a compact whole.

If gravitation is potent enough to do this, it will also be potent enough to break up the filament into condensations, just as the filaments of spiral nebulae are broken up into condensations. But here again an essential difference must be taken into account. The shrinkage of a spiral nebula is a slow secular process. Year after year, and century after century, the filament will be ejected without change of character—the process may be compared to the paying out of a coil of rope. But the tidal disruption of a star is a rapid, even cataclysmic event: within a few years the emission of the filament starts, reaches a maximum, declines, and ends. There is no steady paying out here; the process ought rather to be compared to the discharge of a torpedo, or other body which is thickest in the middle and tapers off at the two ends. When a filament of this shape breaks up into condensations it will form no long chain of similar masses, but a small number of unequal masses. It is natural to conjecture *a priori* that large masses are likely to form out of the central portions where matter is most plentiful, and smaller masses at the ends where matter is scarce. Such a question cannot of course be finally settled by *a priori* conjectures, but in the present case an exact discussion of the problem indicates that the *a priori* view is the right one, and suggests that the comparative abundance of matter in the central part of the filament may provide an explanation of the appearance of the more massive planets, Jupiter and Saturn, near the centre of the sequence of planets.

Obviously, if a tidal cataclysm can explain the existence of the planets, it can also, in general terms at least, explain the existence of the satellites of these planets. For immediately after the birth of any planet, say Jupiter, the original situation repeats itself in miniature. Jupiter now plays the part originally assigned to the sun, while either the wandering star or the sun itself, or possibly the combination of the two, acts the part of the tide-raising disturber. Again we get the emitted filament, again the formation of condensations, and again, as the ultimate result, a sequence of detached bodies with the most massive in the middle. Since Jupiter, the sun, and the disturbing star all move in the same plane, namely the plane of Jupiter's orbit, it follows that Jupiter's satellites, when formed, ought also to move in this plane, as in actual fact they are observed to do.

So long as we merely discuss the matter in general terms it looks as though the process might go on for generation after generation, each member of a family of satellites producing minor satellites to circle round itself, and so on *ad infinitum*. Common sense suggests that this cannot go on for ever: there must be a limit somewhere. Exact calculation confirms the view of common sense, with the disconcerting addition, that we are in danger of overstepping the limit if we attempt to account for the whole of the satellites in the solar system in the way just suggested.

I have already mentioned a mathematical formula which enables us to calculate the masses of the bodies formed out of the condensations in the arms of spiral nebulae. The same formula puts us in a position to calculate the masses of the planets which ought to be formed from the filament drawn out of the sun. Let us suppose that when the tidal cataclysm took place the sun had a radius equal to that of Neptune's orbit, and therefore a mean density of $5\cdot5 \times 10^{-12}$. Let us suppose that at the middle parts of the ejected filament the mean density was one-tenth of this, or $5\cdot5 \times 10^{-13}$. Let us further suppose that the temperature of the ejected matter corresponded to a molecular velocity 4×10^4, this being about the molecular velocity of hydrogen or oxygen at their ordinary boiling-points. Then our formula indicates that the masses of the planets formed out of the middle parts of the filament ought to be about 10^{30} gm., a mass intermediate between those of Jupiter and Saturn. This is satisfactory as showing that there is no numerical difficulty in supposing Jupiter and Saturn to have come into being in the way we have imagined. If we like to accept the tidal theory of their birth, we can reverse our calculation and can calculate from their present known masses what must have been the density of the matter from which they were formed.

Naturally an inverted calculation of this kind is not applicable only to Jupiter and Saturn; if the tidal hypothesis is correct, it must be applicable to all the planets and to all their satellites. For example, the first five satellites of Saturn all have masses of about 5×10^{23} gm.; our calculation shows that if these satellites came into being as gaseous condensations in a filament, the gas in this filament must have been anything from one to a million times as dense as lead. Such a conclusion is, of course, preposterous: the only proper conclusion is that these satellites cannot have originated as gaseous condensations.

This conclusion is not surprising, or even unexpected. Even now these satellites, on account of the smallness of their mass, are incapable of retaining a gaseous atmosphere, whence it follows that they can never have existed in the gaseous state. They must have been born either liquid or solid.

In this way we come upon the practical limitation to the possibility of endless generations of satellites being born. Primarily it is that after a time the satellites would be too small for their gravitation to hold them together. A brief reprieve from the operation of this law is afforded by the possibility of the matter liquefying or even solidifying before it scatters into space, and it is probably owing to the operation of this reprieve that all the satellites of the planets, and probably also the smaller planets themselves, owe their existence.

What of our Earth, which interests us above all other planets? Its present mass is rather too small to have been born out of a purely gaseous filament, but we must remember that if it were born gaseous, a large part of its mass might be immediately dissipated away into space, the present Earth representing only a remnant of a once much more massive planet. This line of investigation leads nowhere. A more promising line of attack is through a consideration of our satellite the moon. The more liquid a planet was at its birth the less likely was it to be broken up tidally by the still gaseous sun, but, in the event of this breaking up taking place, the ratio of mass between satellite and primary would be much nearer to unity than in the case of a wholly gaseous planet. Thus, as we pass from planets which were wholly gaseous at birth to planets which were wholly liquid, we ought to start from planets with large numbers of relatively small satellites and, after passing through the boundary cases of planets with a small number of relatively large satellites, reach planets having no satellites at all. This is precisely what we find in the solar system. Leaving Jupiter and Saturn each with their nine relatively small satellites we pass through Mars with its two satellites to the Earth with one relatively very large satellite, and after this come to Venus and Mercury with no satellites at all. Proceeding in the other direction from Jupiter and Saturn, we pass trough Uranus with four small satellites to Neptune with one comparatively big satellite. Looked at from this point of view, the Earth-Moon system figures as the obvious boundary case between the planets which were originally liquid and those which were originally gaseous, the corresponding boundary case on the other half of the chain being Neptune. Thus we can conjecture that Mercury and Venus were born liquid or solid, that the Earth and Neptune were born partly liquid and partly gaseous, and that Mars, Jupiter, Saturn, and Uranus were born gaseous.

We have already noticed that Mars and Uranus both have masses which are too small for their positions in the sequence of planets. If the planets were born out of a filament of continuously varying density, the mass of Mars at birth ought to have been intermediate between that of the Earth and that of Jupiter, and similarly the mass of Uranus at birth ought to have been intermediate between that of Neptune and that of Saturn. We have, however, just seen reasons for conjecturing that the two anomalous planets, Mars and Uranus, were the two smallest planets to be born in the gaseous state; they would therefore be likely to lose more mass by dissipation of their outer layers than any of the other planets. Let us introduce the supposition that Mars, and to a lesser degree Uranus, lost large parts of their mass by dissipation into space; let us suppose that they are mere fragments of what were originally much more massive planets, then all anomalies disappear, and the pieces of the puzzle begin to fit together in a very gratifying manner.

Nevertheless, and in spite of the high promise which the tidal theory seems to hold out, it is far too early to claim that it can finally explain the origin of our system; its claim to consideration at present is rather that, so far as I know, it provides the only theory of that origin which is not open to obvious and insuperable objections.

The Interior of a Star.

By **A. S. Eddington**, Cambridge (England).

> The heaven's glorious sun
> That will not be deep-searched with saucy looks;
> Small have continual plodders ever won
> Save base authority from others' books.
> SHAKESPEARE, "Love's Labour's Lost."

The first "saucy look" into the fiery regions lying deep below the visible surface of the sun and stars is contained in a paper published by HOMER LANE in 1870 "On the Theoretical Temperature of the Sun". His theory was further developed by A. RITTER, Lord KELVIN, and others; in particular the masterly work of R. EMDEN in his book *Gaskugeln* has been invaluable as a basis of recent advances. Progress in thermodynamics and atomic physics has led to modifications of the original theory and to further opportunities of advance, so that we now appear to have reached a stage at which the study of the deep-lying conditions enables us to predict to some extent the surface features accessible to observation, and we may appeal to experiment to declare that our professed knowledge of the interior is something more than "base authority from others' books".

It greatly simplifies the problem if the material of a star can be treated as a perfect gas; and most investigations refer to stars limited by this condition. In HOMER LANE's time it was doubtful if any such stars existed; the density of the sun is 1·38, and there was at that time no reason to doubt that this was typical of the densities of stars in general. It is now known, however, that there are stars ("giant stars") with mean densities comparable to air or even to the density in a vacuum tube. These can evidently be regarded as composed of perfect gas; so there will be no lack of opportunity for application to actual stars of results obtained for a perfect gas. Moreover I have very recently been led to suspect that *all* ordinary stars are in the state of a perfect gas — even though in some cases the density is as great as that of platinum. This revolutionary idea will be referred to later.

We shall take Capella as a typical gaseous star of low density since our observational knowledge of it is especially accurate and complete. The bright component of Capella is 4·2 times as massive as the Sun, and it gives out 150 times as much light. In the main the greater brilliancy is due to its diffuseness; it has a much larger radiating surface than the Sun. The temperature conditions on the Sun and Capella are very similar, as is shown by the close resemblance of their spectra. The temperature of the photosphere (i. e. of the layers from which the light is coming) is about 5500° C. That is not so very high a temperature judged even by terrestrial standards; but it is only the marginal temperature of the furnace and affords no idea of the terrific heat within. By a calculation, which now appears fairly trustworthy, we find that the temperature at the centre of Capella is about 8,000,000°, and at least 90 per cent of the mass of Capella is at a temperature above 2,000,000°. The average density is .0026, about twice the density of air; the density at the centre is probably 20 times greater.

We shall not attempt to explain in detail how these internal tempe-- ratures are calculated; but we may perhaps show that there is a clue which can be followed up by appropriate mathematical methods. Elasticity is a well-known property of a gas — familiar to everybody through its practical application in the pneumatic tyre. What gives the gas its elasticity or expansive force is its heat, that is to say the energy of motion of its molecules hastening in all directions and tending to spread apart. Now at any point inside a star a certain condition of balance must be reached; on the one hand we have the weight of all the layers above pressing down and trying to squeeze closer the gas inside; on the other hand we have the elasticity of this gas inside trying to expand and force the upper layers outwards. Since neither one thing nor the other happens, and the star remains practically unchanged for hundreds of years, we must conclude that these two tendencies just balance. At each point therefore we must attribute to the gas the heat which will give it just the elasticity needed to bear the weight of the layers above. The heat distribution has therefore to satisfy a certain condition, and it is from this condition that we are able to determine numerical values of the temperature. To perform the calculation it is necessary to know the average molecular weight of the gas constituting the star. At first widely different values of the temperature were found according to different hypotheses as to the chemical constitution of the star; it was sheer speculation whether we ought to take the molecular weight of hydrogen, carbon, iron, etc. We shall see later that this source of difficulty has recently been removed by the discovery that at stellar temperatures all the chemical elements (except hydrogen) have the same "molecular weight" within narrow limits.

These high temperatures are to be taken quite literally. Do not imagine that a million degrees indicates a temperature so vast that ordinary conceptions have broken down. Temperature is a mode of describing the speed of motion of the ultimate particles of matter. In helium at ordinary temperatures the average speed of the atoms is about 1 km. per sec.; at 4 million degrees it is 100 km. per sec. That is not a speed to feel uncomfortable over, when you reflect that in the laboratory physicists are accustomed to deal with atoms of helium moving at speeds approaching 100,000 km. per sec. (α particles). I usually find that my physical colleagues are rather disappointed with the jog-trot atoms in the stars.

Inside Capella then the atoms are rushing in all directions with speeds of the order 100 km. per sec., continually colliding and changing their courses. Each atom is pulled inwards by the gravitation of the whole mass, and as continually boosted out again by collision with the atoms below. The energy of this atomic motion constitutes a great store of heat contained in the star; but it is only part of the store. There is a store of another kind of heat, ætherial heat or æther-waves similar to those which bring us the sun's heat across 150 million kilometres of vacant space. These waves are also hastening in all directions inside the star. They are encaged by the material as in a sieve which prevents them leaking into outer space except at a slow rate. An æther-wave making for freedom is caught and absorbed by an atom, flung out again in a new direction, and passed from atom to atom; it may thread the maze for hundreds of years until by accident it finds itself at the confines of the star, free now to travel through space indefinitely, or until it reaches some distant world and perchance entering the eye of an astronomer makes known to him that a star is shining.

The possession of this double store of heat is a condition which we do not encounter in any of the hot bodies more familiar to us. It is true that a red-hot mass of iron contains a little of the ætherial heat in addition to the heat of its molecules; but the ætherial heat is less than a billionth part of the whole. Only in the stars, and more especially in the giant stars, does the ætherial portion rise to equal importance with the material portion A red-hot metal emits æther-waves but it keeps no appreciable store; it converts its material heat into this form as it is required for use The star rejects this hand-to-mouth method, and although it is continually transforming heat from one form to the other it keeps a thousand years' supply of ætherial heat always in readiness and emits its radiation by leaking ætherial heat from this store. In older theories this novel feature was not recognised; it was supposed that convection currents must exist continually bringing up hot matter from below to replace the surface-matter which had radiated and cooled. But now the problem is reversed; we have to explain how the star manages to dam back its store of æther-waves so that they do not leak from it faster than we observe.

This helps us over another difficulty which initially confronted those who tried to calculate the internal temperature. In the older theory it was necessary to know the adiabatic constant γ — the ratio of specific heats of the material of the star This could only be guessed, since the chemical nature of the material is still unknown Now in thermodynamical problems radiation behaves in most respects like a material gas with an adiabatic constant $\gamma = \frac{4}{3}$; this is because the energy of wave-motion is half kinetic and half potential; and $\frac{4}{3}$ is the value of γ for a gas whose energy is half kinetic (molecular velocity) and half concealed in internal or rotational motion of the molecules Whether we should use the value of γ for matter or æther depends on whether the material heat or the ætherial heat is the more mobile — the more rapidly conveyed from place to place seeking its equilibrium distribution, and accordingly having the first voice in deciding the temperature distribution, There is no question that the ætherial heat flows from place to place far more freely than the material heat moved by the clumsy process of transferring material. Thus the value to be used for γ is the ætherial value $\frac{4}{3}$; and the value of γ for the material gas is no longer relevant. To use the technical terms, we have substituted radiative equilibrium for convective equilibrium This amendment was first urged by R. A, Sampson in 1894, but the state of thermodynamics at that time did not admit of much development of the idea. Modern researches on radiative equilibrium originated with the work of K. Schwarzschild in 1906.

In the hot bodies of the laboratory the heat is almost entirely in the material form. In the giant stars the heat is about evenly divided between the material and ætherial forms. Can we not imagine a third condition in which the heat is almost wholly ætherial, the material portion being insignificant? We can imagine it no doubt; but the interesting, and I believe significant, thing is that we do not find it in nature We shall consider this from a slightly different point of view, paying attention to pressure instead of energy.

Light has mass and weight and momentum and exerts a minute pressure on any object which obstructs it. A beam of light or æther-waves is like a wind — a very feeb'e wind as a rule, but the intense ætherial energy rushing through the star makes a hurricane. This wind distends the star; it bears to some extent the weight of the layers of material, leaving less for the elasticity of the gas to bear. That, of course, must be taken into account in calculating the internal temperature, making them somewhat lower than the original theory supposed. The ancients imagined a giant Atlas who bore the weight of the skies on his shoulders; at a point inside the star we have two Atlas's bearing the skies above, viz. the material pressure due to the elasticity of the gas and the ætherial pressure due to the hurricane of æther-waves. How do these two Atlas's divide the burden between them We find that to a first approximation they share it in the same proportion throughout the whole interior, and further this proportion depends only on the mass of the star and not on its density or even on its chemical composition[1]). We again need to know the molecular weight, but I have already promised to explain later how that is fixed We do not make use of any astronomical knowledge in this calculation; all the constants used in the work have been determined in our terrestrial laboratories.

Let us imagine a physicist on a cloud-bound planet who has never heard tell of the stars setting to work to make these calculations for globes of gas of various dimensions Let him start with a globe containing 10 grams, then 100 grams, then 1000 grams, and so on, so that his n^{th} globe contains 10^n grams. The following table gives part of his results.

No. of Globe	Ætherial Pressure	Material Pressure
30	·00000016	·99999984
31	·000016	·999984
32	·0016	·9984
33	·106	·894
34	·570	·430
35	·850	·150
36	·951	·049
37	·984	·016
38	·9951	·0049
39	·9984	·0016
40	·99951	·00049

You will see why I omit the rest of the table; it consists of long strings of 0's and 9's: But for the 33rd, 34th and 35th globes the table gets interesting, and then lapses back into 9's and 0's again Regarded as a tussle between æther and matter to control the situation the contest is too onesided to be interesting . except from Nos. 33 to 35 where something exciting is likely to happen.

What "happens" is — the stars.

We draw aside the veil of cloud behind which our physicist has been working and let him look up into the skies. There he will find a thousand million globes of gas all of mass between the 33rd and 35th globes. The lightest known star comes just below the 33rd globe; the heaviest is just beyond the 35th. The vast majority are between Nos. 33 and 34 where the critical struggle is beginning[2]).

---

[1]) If the ratio of material to ætherial pressure is $\beta : (1 - \beta)$, the formula is
$$1 - \beta = \cdot 00309 \, M^2 \mu^4 \beta^4$$
where M is the mass (Sun = 1) and μ is the molecular weight in terms of the hydrogen atom.

[2]) The table is calculated with molecular weight 4. More recently it has been ascertained that a lower value, about 2.2, should be taken for a typical star. But we are about to apply the table to a discussion of the origin of a star, and must therefore consider the conditions when it first begins to crystallise out of primordial material; owing to the comparatively low temperature at this stage a higher molecular weight is justified.

It is remarkable that the matter of the universe has aggregated primarily into units so nearly alike in mass. The stars differ from one another widely in brightness, density, temperature, etc.; but they mostly contain about the same amount of material I think we can now form a general idea of the cause of this, although the details of the explanation may be difficult. Gravitation is the force drawing the matter together; it would if unresisted gather more and more material building globes of enormous size. Against this, radiation-pressure is the main disruptive force (doubtless assisted by the centrifugal force of the star's rotation); its function is to prevent the accumulation of very large masses. But this countervailing tendency, as we see from the table, only begins to be appreciable when the mass has already nearly reached the 33rd globe; it then rises so rapidly that before the 35th globe is reached it must perform whatever it is capable of performing. Somewhere between these two limits the balance is reached and the accumulation of mass is stopped; the precise point will depend on the rotation and possibly on other causes of disturbance affecting the particular star considered. All over the universe the masses of the stars bear witness that the gravitational aggregation proceeded just to the point at which the opposing force called into play became too strong to allow a continuance.

We have hitherto pictured the inside of a star as a hurly-burly of atoms and æther-waves; we must now introduce a third population to join in the dance. There are vast numbers of free electrons — unattached units of negative electricity. More numerous than the atoms (perhaps 20 times as numerous) the electrons dash about with a hundredfold higher speed corresponding to their small mass which is only $\frac{1}{1845}$ of a hydrogen atom. These electrons have come out of the atoms, having broken away at the high temperatures here involved. An atom has been compared to a miniature solar system: a composite central nucleus carrying a positive charge corresponds to the sun, and round it revolve in circular and elliptic orbits a number of negative electrons corresponding to the planets. We know the number of satellite electrons for each element: sodium has 11, iron 26, tin 50, uranium 92. Our own solar system with eight revolving planets would represent an atom of oxygen. There is a general law that the number of satellite electrons is roughly equal to half the atomic weight of the element.

This solves for us the difficulty in deciding what value to adopt for the molecular weight of the material of a star. Each of the free electrons has to be reckoned as a separate "molecule"; it exerts the same pressure as a gas-molecule would do. In terrestrial gases a number of atoms combine to form a molecule, but in the stellar gas an atom splits up into a number of molecules. The molecular weight is thus less than the atomic weight. If the temperature is high enough all the satellite electrons are set free and become independent molecules; and since their number is about half the atomic weight the average weight per molecule is just about 2. For example an atom of sodium seprataes into 12 molecules, viz., 11 electrons $+$ 1 nucleus; its atomic weight of 23 therefore corresponds to a weight of $\frac{23}{12}$ or 1·92 per molecule. For iron the atomic weight of 56 is divided between 27 particles, giving an average molecular weight 2·07. For tin we have 119 divided by 51; average 2·34; and so on. It is very fortunate that we are thus able to assign within narrow limits the molecular weight of stellar material although we do not know its chemical composition. I should mention, however, that hydrogen is an exceptional element; its atom breaks up into a nucleus and 1 electron giving molecular weight 0·5. I have sometimes been tempted to think that the youngest stars might consist largely of hydrogen which was gradually being converted into heavier elements, the energy set free by this process providing the main source of the heat radiated by the stars; but a hydrogen star would on account of the lower molecular weight differ so widely from stars composed of other material that this view seems to be quite untenable. I do not think we can admit more than a very moderate admixture of hydrogen in the material of a star; and the evolution of heavier elements must be far advanced before the stellar stage is reached.

Stellar temperatures are not quite high enough to detach all the satellite electrons; but except for the heaviest atoms the process is nearly complete. For example in Capella sodium has lost all its electrons, but iron retains 2 out of the 26; tin retains 3 or 4 out of 50. The possibility of calculating the condition of an atom of any element at given density and temperature is due partly to the thermodynamical theory initiated by NERNST, and partly to extensive determinations of the inner structure of the atoms revealed by x-ray experiments.

From the observational standpoint the important thing to determine is not how much heat is stored up in the star but how fast this heat leaks out. We therefore turn now to the consideration of the laws which govern the flow of heat from the interior to the outside of the star. The immediate cause of this flow is the temperature-gradient from the centre to the outside; as already explained the internal distribution of temperature can be calculated, so we know the temperature-gradient. But in order to calculate the flow of heat which results it is necessary also to know what is obstructing the flow, namely the opacity or absorption-coefficient of the material through which the radiation must force its way. If we knew the opacity we could calculate the intensity of this flow of radiation, which starts as a modest stream near the centre and gathers strength as it pursues its course to the outside; on emergence from the star it passes across space to our telescopes where its intensity can be measured directly with the bolometer or more easily deduced from its luminous quality.

The procedure may be reversed. From the observed intensity of the emitted radiation the unknown opacity of the stellar material can be found. Thus the data for Capella show that its opacity is about 120 C.G.S.units. This means that a screen of stellar material of thickness such as to contain $\frac{1}{120}$ gm. per sq.cm. would let through only 1/e (e = 2·718) of the radiation falling on it. To illustrate the significance of this let us enter Capella and find a region where the density is equal to that of the atmosphere in which we live; a slab of the material only 5 cms. thick would form a screen so opaque that only $\frac{1}{3}$ of the radiation falling on one side of it would get through to the other side, the rest being absorbed in the material. It seems at first surprising that 5 cms. of gas could stop the æther-waves so effectually, but we might have anticipated something like this from our general physical knowledge. Different names are given to æther-waves according to their wave-length. The longest are the Hertzian waves used in wireless telegraphy; then come the invisible heat waves, then light waves, then photographic or ultra-violet waves. Beyond these we have x-rays, and finally the γ-rays emitted by radioactive substances. Where in this series are we to place the æther-waves in the interior of a star? It is solely a question of temperature, and at stellar temperatures the æther-waves are x-rays — more precisely, they are very "soft" x-rays. Now x-rays, and especially soft x-rays, are strongly absorbed by all substances. The opacity which we have found for Capella has about the usual value of the opacity of substances to x-rays measured in the laboratory.

It is gratifying to be able to measure the absorption of x-rays inside a star and find a result similar to that obtained by physicists who experiment on terrestrial substances in the laboratory. It is of even greater interest if the stellar experiment turns out to be not a mere repetition but an extension of the laboratory experiments, so that the influence of conditions unattainable in terrestial surroundings may be traced. A careful examination shows that the correspondence between the stellar and the terrestrial absorption-coefficients is not so close as at first sight appeared. In the laboratory the absorption varies very rapidly according to the wave-length of the radiation that is being absorbed. By using stars of different temperatures we can also experiment over a considerable range of wave-length, and the astronomical result is that the absorption is nearly independent of wave-length; we cannot yet detect with certainty whether it increases or decreases. Here is a striking contradiction. Further the laboratory results to which I have referred apply to radiation between 1 and 2 Angström units, whereas the radiation in Capella is in the main between 5 and 10 units. If we extrapolate the results for Capella according to the stellar law —

that the wave-length does not make much difference — then we agree with the terrestrial results; but if we extrapolate the laboratory results, using the laboratory law, we arrive at exceedingly high absorption at 5 to 10 Angström units far exceeding the value found in Capella. To understand the cause of this difference of behaviour in terrestrial and stellar conditions we must plunge into the midst of those problems of atomic physics which are most exercising physicists at the present day. We started to explore the interior of a star; we find ourselves exploring the interior of an atom.

It is now generally agreed that when aether-waves fall on an atom they are not absorbed continuously. The atom lies quiet, waiting its chance, and then suddenly swallows a whole mouthful at once. The waves are done up in bundles called quanta and the atom has no option but to swallow the whole bundle or leave it alone. Generally the mouthful is too big for the atom's digestion, but the atom does not stop to consider that; it falls a victim to its own greed — in short it bursts. One of its satellite electrons shoots away at high speed, carrying off the surplus energy which the atom was unable to assimilate. This bursting could not recur continually unless there were some counter-process of repair. The ejected electrons travel about meeting other atoms; after a time a burst atom meets a loose electron under suitable conditions and induces it to stay and heal the breach. The atom is now repaired and ready to swallow another quantum as soon as it gets the chance.

From this cause a difference arises between the absorption of x-rays in the laboratory and in the stars. In the laboratory the atoms are fed very slowly; we can only produce in small quantities the bundles of x-rays which they swallow. Long before the atom has a chance of a second mouthful it is repaired and ready for it. But in the stars the intensity of the x-rays is enormous; the atoms are gorged and cannot take advantage of their abundant chances. We must allow for this saturation effect, which does not appear in terrestrial experiments. The consumption of food by the hungry hunter is limited by his skill in trapping it; the consumption by the prosperous profiteer is limited by the strength of his digestion. Laboratory experiments test the skill of the atom in catching quanta; stellar experiments test how quickly it can repair itself after the absorption of one quantum so as to be ready to deal with another. The theory of the stellar absorption coefficient thus depends on a study of the process of repair of the shattered atom, that is to say on the conditions of capture of electrons by atoms.

The most satisfactory theory of the process of electroncapture is contained in a remarkable paper by H. A. KRAMERS published in November 1923. On applying his theory to stellar material at temperatur T

and density ϱ, we find that the absorption coefficient will be proportional to $\varrho/T^{\frac{7}{2}}$; there is also a rather complicated correcting factor which may be neglected in a first approximation since it does not vary much in the range with which we are concerned in the stars. Kramers' theory not only gives the law of variation but also the absolute value of the absorption for any density and temperature, so that we have sufficient information to predict the outflow of radiant energy from a diffuse star of given mass and density. For Capella the predicted value of the absorption turns out to be about $\frac{1}{8}$ of that actually observed. It is difficult to know how much stress should be laid on the want of precise agreement. The predicted amount varies a little with the assumed chemical composition and other uncertain factors in the constitution of the star; but the limits of uncertainty are restricted and it is not easy to find an excuse for the disagreement by a factor of 8. Whether a fuller knowledge will improve the agreement or not, I have little doubt that Kramers' process of capture actually occurs in the stars and contributes to the absorption-coefficient. It is possible, however, that the major portion — the remaining $\frac{7}{8}$ — arises from some additional mode by which atoms can capture electrons, perhaps unimportant in terrestrial conditions and therefore not brought to the attention of Kramers.

The writer had previously proposed a different theory of capture which also leads to the law of absorption $\varrho/T^{\frac{7}{2}}$. When an electron meets an atom, we can trace its path through the interior of the atom under the electrical forces prevailing there. It has been shown by experiments on scattering that we have sufficient knowledge to do this correctly. In a small proportion of cases the track will be such that the electron actually hits the nucleus. We have at present no theoretical means of foretelling what will be the result of this collision, which must occur at enormous speed since the nucleus strongly attracts the electron; but if the speed of rebound is a trifling fraction less than the speed of impact the electron will have insufficient energy left to escape from the sphere of attraction of the nucleus — in other words capture has taken place. Let us assume provisionally that the great majority of the electrons which hit the nucleus suffer this loss and become captured; on calculating the absorption coefficient which results from this hypothesis we find a value which agrees as closely as possible with astronomical observation. It may be remarked that the electrons captured by Kramers' process do not hit the nucleus so that the captures here postulated are additional; further Kramers' theory does not deal with the question what happens to electrons whose tracks are interrupted by meeting the nucleus, so that we are not interfering with his scheme. At the same time there are certain difficulties in the nuclear theory of capture, which become apparent in a detailed technical discussion; and notwithstanding its perfect agreement with astronomical observation, I am not very confident that it can be maintained in its present form.

We shall shelve the difficulty attending the prediction of the absolute value of the absorption, and now concentrate attention on the law that the absorption is proportional to $\varrho/T^{\frac{7}{2}}$ which is common to both theories. Indeed very general considerations indicate that a law approximately of this form must hold. The densities and temperatures of the stars vary very widely, so that it might be thought that we should obtain widely different values of the absorption from different stars. But to a large extent the fluctuations of ϱ and of $T^{\frac{7}{2}}$ cancel one another, the less dense stars having the lower temperatures. Indeed our earlier study of the distribution of density and temperature in a star shows that for stars of equal mass ϱ is proportional to T^3, so that the absorption is constant except for the comparatively unimportant factor $T^{-\frac{1}{2}}$. Not only does this proportion hold between different stars but between different parts of the same star; this simplifies the problem, since the absorption-coefficient will then vary comparatively little throughout the main part of a star. The comparative constancy of the absorption occurs only in comparing stars of nearly the same mass but different temperatures; for stars of different mass a much wider variation appears.

By gathering together the various threads of investigation we are now in a position to predict the total radiation and hence the absolute brightness of any star which is not too dense to behave as a perfect gas. The one uncertain constant (the constant factor multiplying $\varrho/T^{\frac{7}{2}}$) may be settled by astronomical observation of a single star, preferably Capella; the formula is then complete and ready to be tested on any other star. It turns out that the absolute brightness should depend mainly on the mass; the other condition required to specify the nature of the star, viz., its spectral type or the surface-temperature corresponding to its type, has relatively small influence; but the effect of this factor is duly given by the formula and may be applied as a small correction in the comparison between theory and observation. Unfortunately Capella is the only star of low density for which the mass and absolute magnitude have been determined by direct methods. By indirect methods we can find the requisite data for certain eclipsing variables and for the Cepheid variable stars; these are found to agree very well with the formula, but perhaps the test scarcely inspires so much confidence as if the masses and luminosities had been determined in the more usual way.

If it were not necessary to limit the test to diffuse stars likely to obey the laws of a perfect gas, a great many direct determinations of mass and absolute brightness would be available to check the theoretical formula. If we boldly compare the data for dense stars with the formula, a great surprise awaits us. *The theory developed for a perfect gas predicts correctly the magnitudes of the sun and even denser stars.* In a recent investigation I have collected the masses and absolute magnitudes of about 35 stars, extending over practically the whole range of stellar mass from $\frac{1}{6}$ to 30 times the sun's mass and of absolute magnitude from -4 to $+12$; more than half of these are dense stars to which we should scarcely have expected the theory to apply; yet the average discordance between the predicted and observed magnitudes is only $\pm 0^{m}\!.56$, the greater part of which may reasonably be attributed to observational error. The test of the theory is almost too successful; the stars agree with it irrespective of whether they ought to agree or not!

When we state that matter is in the condition of a perfect gas we are ascribing to it a definite compressibility, namely, if the pressure is doubled the matter will be compressed to double the density. If the results here reached are to be taken at their face value, they indicate that stellar material retains this compressibility even in stars with density greater than water. An ordinary terrestrial gas placed under high pressure so as to reach the density of water, becomes almost incompressible; doubling the pressure will make only a trifling increase of density. If stellar material were in that stage our theory of a perfect gas would be quite inapplicable, and in fact the absolute brightness would be much reduced[1]). Hence, unless we have been greatly misled, we must accept the conclusion that in stellar conditions matter may remain a perfect gas, even when its density is as great as that of water or (in some stars) of platinum.

It is perhaps natural at first sight to dismiss this conclusion as absurd, and assume that there must have been some false step in the rather intricate investigation. But I think on the other hand that we have only been led — in a very roundabout way — to a conclusion which is almost obvious from modern physical theory though it does not seem to have occurred to anyone previously. In terrestrial gases the atoms occupy a finite bulk; they behave approximately as rigid spheres with radii of the order 10^{-8} cm. The upper limit to the compression of a gas is due to these spheres coming into contact; when they are packed as tightly as possible the maximum density is reached. But in the stars these atoms have been broken up, as already explained; what is left of the iron atom has a radius of about $2 \cdot 10^{-10}$ cm., whilst of the lighter atoms there remains only the bare nucleus having a radius less than 10^{-12} cm. Not until the density reaches 100,000 times greater value will these multilated atoms or ions begin to pack tightly. At the trifling density of water or of platinum the tiny stellar ions have plenty of room

---

[1]) For two reasons; firstly the internal temperature and temperature-gradient would be smaller because not so much heat is needed by the gaseous material to prevent it from "giving" under the weight of the outer layers; secondly the absorption is increased because $\varrho/T^{\frac{7}{2}}$ is greater than for a perfect gas.

to behave as in a perfect gas. The ideal gas is one whose molecules are of infinitesimal size, and the condition in a star which breaks up the atom into very much smaller bodies brings the material nearer to this ideal.

Thus it is quite natural in a star to have perfect gas of the density of platinum. The stars with which we have tested the theoretical formula are after all not the wrong stars. None of the ordinary stars have densities at all approaching the maximum density of stellar matter, so that there is no occasion for the formula to fail. And the test is so well fulfilled that we are tempted to hope that it is now possible to deduce the mass of a star from observation of its absolute brightness, or the absolute brightness from observation of its mass.

I do not think we need be greatly concerned as to whether these rude attempts to explore the interior of a star have brought us to any thing like the final truth. We have, I think, gained an insight into the varied factors participating in the problem. The results already attained correspond well enough with observation to encourage us to think we have begun at the right end in disentangling the problem, and we do not anywhere come across difficulties which seem likely to be insuperable. Especially do we realise that in a star matter exists in its simplest condition; and it is a much more promising subject for investigation by the mathematical physicist than the complex structure of matter at lower temperatures. And if in this article I appear to be trusting too rashly to present-day theories, pressing them to the remotest conclusions, I would plead that by doing so their possible defects may best be brought to light and remedied.

In ancient days two aviators procured to themselves wings. Daedalus flew safely through the middle air and was duly honoured on his landing. Young Icarus soared upwards towards the Sun till the wax melted which bound his wings, and his flight ended in fiasco. In weighing their achievements there is something to be said for Icarus. The classical authorities tell us he was only 'doing a stunt' but I prefer to think of him as the man who certainly brought to light a constructional defect in the flying-machines of his day. So too in Science. Cautious Daedalus will apply his theories where he feels confident they will safely go; but by his excess of caution their hidden weaknesses remain undiscovered. Icarus will strain his theories to the breaking-point, till the weak joints gape. For a spectacular stunt? Perhaps partly; that is only human. But if he is not yet destined to reach the Sun, and solve finally the riddle of its constitution, we may at least hope to learn from his journey some hints to build a better machine.

Die ruhenden Kalziumlinien.
Ein Beitrag zur Kosmogonie der O-Sterne und der Planetarischen Nebel.

Von **Hans Kienle**, Göttingen.

Mit 3 Abbildungen.

Bei der Einordnung der Planetarischen Nebel in das allgemeine Schema der Sternentwicklung ergaben sich gewisse Schwierigkeiten, die bis heute noch nicht vollkommen überwunden werden konnten. Wollte man diese Gebilde im Sinne der alten kosmogonischen Anschauungen an den Anfang der Sternentwicklung setzen, so widerstreitet dem die große Seltenheit, mit der die Planetarischen Nebel auftreten (Gesamtanzahl nach LICK XIII: 96), sowie die Tatsache, daß ihr Vorkommen am Himmel örtlich eng begrenzt ist. Die gleichen Gründe verbieten auch die Auffassung, daß die Gasnebel zusammen mit den Wolf-Rayet-Sternen (denen sie in spektraler wie auch in stellar-statistischer Beziehung nahe stehen) ein normales Durchgangsstadium, und zwar den auf der Mitte des Weges liegenden Höhepunkt der Sternentwicklung im Sinne LOCKYER-RUSSELLS darstellen. Man muß vielmehr annehmen, was schon RUSSELL und HERTZSPRUNG getan haben und was durch EDDINGTONS Arbeiten theoretisch verständlich geworden ist, daß diese Gebilde seltene Ausnahmezustände verkörpern, die zu erreichen nur wenigen Individuen beschieden ist. Eine große Schwierigkeit bleibt aber auch dann noch unbehoben, auf welche die statistische Untersuchung der Motus peculiares geführt hat. Befreit man die beobachteten Radialgeschwindigkeiten V_* der Sterne von dem Einfluß der Apexbewegung ($V_\odot$), dann erhält man nach weiterem Abzug einer für die einzelnen Spektralklassen verschiedenen Konstanten K eine von M bis B absteigende Reihe durchschnittlicher „Eigengeschwindigkeiten" $\overline{V}_*$, wie die zuletzt von GYLLENBERG angegebenen Zahlen deutlich veranschaulichen:

Typus	$\overline{V}_*$	K
B	7,0 km/sec	+ 4,3 km/sec
A	11,8 ,,	+ 0,1 ,,
F	14,5 ,,	+ 0,2 ,,
G	15,8 ,,	− 0,8 ,,
K	15,9 ,,	+ 3,6 ,,
M	17,2 ,,	+ 5,3 ,,

Aus dieser Reihe fallen die Planetarischen Nebel mit durchschnittlich 30 km/sec vollkommen heraus. Es scheint kein stetiger Übergang zu diesem Werte vorhanden zu sein, weder am Anfang noch am Ende der Spektralreihe. Im gleichen Sinne weichen, worauf wir gleich an dieser Stelle hinweisen wollen, die früher mit M d bezeichneten langperiodischen Veränderlichen von den übrigen Sternen ab.

PLASKETTS eingehende Untersuchungen über die Spektra der O-Sterne haben es sehr wahrscheinlich gemacht, daß diese Spektra sich in sinngemäßer Fortsetzung der auf die Linienintensität gegründeten Einteilung an die B-Sterne anschließen lassen als O 5, O 6, ... O 9, B 0. Wirklich lückenlos ist nach den verschiedenen Untersuchungen über die spektroskopischen Parallaxen der B- und A-Sterne der weitere Anschluß von B über A nach F. Die Frage, die uns im folgenden beschäftigen wird, ist die, ob sich die Verbindung von O und B nicht noch durch weitere Tatsachen festigen läßt und ob es nicht vielleicht möglich ist, von O auch noch eine Brücke zu schlagen zu den Planetarischen Nebeln (P). Wir knüpfen dabei an eine Erscheinung an, die man bis vor nicht allzu langer Zeit noch als Seltenheit ansah, die aber beginnt, sich als ganz allgemeines Charakteristikum dieser „frühesten" Typen zu entpuppen. Wir meinen die „ruhenden Kalziumlinien".

Die „ruhenden" oder, wie man richtiger sagte, scharfen Linien H und K des Kalziums, deren Natur seit ihrer Entdeckung durch HARTMANN[1]) im Jahre 1904 Gegenstand mehrfacher Diskussionen gewesen ist, sind, soviel läßt sich mit aller Bestimmtheit sagen, an ein bestimmtes Entwicklungsstadium der Sterne geknüpft. Sie sind bisher bei keinem Stern aufgefunden worden, dessen Spektraltypus später als B 5 ist, und zeichnen sich dadurch aus, daß sie im Gegensatz zu den breiten und diffusen Wasserstoff- und Heliumlinien, die das Spektrum dieser Sterne beherrschen, schmal und scharf sind und daß sie an den periodischen Verschiebungen der übrigen Linien, soweit solche vorhanden sind, gar nicht oder nur mit geringerer Amplitude teilnehmen. Bei einer Reihe von Sternen ist es gelungen, neben diesen besonderen Kalziumlinien auch die normalen aufzufinden, welche die gleichen Verschiebungen wie die anderen Linien des Spektrums aufweisen und die theoretisch bei keinem Typus früher als B 3 auftreten können. Es besteht kaum ein Zweifel darüber, daß da, wo die scharfen Kalziumlinien keine unveränderliche Lage aufweisen, die periodischen Schwankungen auf den Einfluß der nicht getrennt sichtbaren normalen (und schwachen) Linien zurückzuführen sind, und es läßt sich ein in den meisten Stadien bereits durch typische Sterne zu belegender kontinuierlicher Übergang schaffen von den Spektren mit nur scharfen und ruhenden Kalziumlinien zu denen mit nur normalen, an den

---

[1]) Astrophys. Journal Bd. 19, S. 268.

Schwankungen der übrigen im vollen Umfange teilnehmenden Linien, über die Spektren, in denen die Amplitude der Kalziumlinien kleiner ist als die der übrigen Linien.

Wichtig für die Deutung des Phänomens sind noch eine Reihe empirischer Feststellungen, auf die an dieser Stelle hingewiesen sei: die gleichen Eigenschaften wie die Linien H und K des Kalziums konnten von Miß HEGER[1]) und von PLASKETT[2]) in einer Reihe von Fällen auch für die beiden D-Linien des Natriums aufgezeigt werden, und es ist hier besonders interessant, wie die unmittelbar benachbarte D_3-Linie des Heliums neben den beiden Komponenten D_1 und D_2 des Natriums periodisch hin nnd her pendelt während eines Umlaufs des spektroskopischen Doppelsternpaares. Im Spektrum des aus mehrfachen Gründen oft untersuchten Systems β-Lyrae sind neben den periodisch variierenden auch anormale Linien des Wasserstoffs vorhanden, die genau wie die Kalziumlinien unveränderlich ihre Lage beibehalten. Nach neueren Mitteilungen I. S. PLASKETTS[2]) treten die scharfen Kalziumlinien nicht nur in den Spektren früher B-Sterne, sondern in praktisch allen O-Spektren und in den Spektren der Kerne der planetarischen Nebel auf, ja sie sind vielfach auch über die breiten Emissionen in den Spektren der Wolf-Rayet-Sterne als scharfe Absorptionen gelagert. Schließlich kennt man sie seit langem in den Spektren der neuen Sterne, wo sie dadurch auffallen, daß sie allein auf normale Radialgeschwindigkeiten führen (von der Größenordnung ± 10 km), während die starken Verschiebungen der übrigen Linien, als Dopplereffekte gedeutet, radiale Bewegungen bis zu $- 2000$ km ergeben (Violettverschiebung der Emissionslinien!).

Ohne uns damit auf einen bestimmten Standpunkt bezüglich der Deutung der Linienverschiebungen festzulegen, werden wir im folgenden im Anschluß an den allgemeinen astronomischen Gebrauch und mit Rücksicht auf die der Literatur zu entnehmenden Beobachtungsergebnisse alle Linienverschiebungen in km/sec angeben. Zur überschlagsweisen Orientierung sei bemerkt, daß in dem für uns im allgemeinen in Betracht kommenden Bereich ($\lambda = $ ca. 4000 Å) einer Rotverschiebung von 1 Å eine Radialgeschwindigkeit von $+ 75$ km/sec entspricht und daß die diffusen Linien in den O- und B-Spektren teilweise Breiten von mehreren Å haben.

Es darf wohl angenommen werden, daß die im vorstehenden skizzierte besondere Kalziumabsorption anderen physikalischen Bedingungen ihre Entstehung verdankt als das übrige Sternspektrum. Man braucht dabei aber keineswegs, wie dies gewöhnlich geschieht, an eine richtige „Kalziumwolke", sei sie nun interstellarer Natur oder sei es eine ausgedehnte Sternatmosphäre, zu denken. Ist doch in den letzten Jahren

---

[1]) Lick Obs. Bull. Bd. 10, S. 59 u. 141.
[2]) Monthly Notices of the Royal Astr. Soc. Bd. 84, S. 80.

oft genug darauf hingewiesen worden, welch wichtige Rolle die Anregungsbedingungen (Temperatur!) spielen, wie eng der unserer Beobachtung
bei Verwendung aller zu Gebote stehenden Hilfsmittel zugängliche
Wellenlängenbereich (3800 bis 6800) und wie begrenzt damit die Wahrscheinlichkeit für das Auftreten der charakteristischen Linien der
Elemente in den Sternspektren ist. Wenn wir daher im folgenden
gelegentlich der Kürze halber uns des Ausdrucks „Kalziumwolke" be-
dienen, so geschieht dies unter allem Vorbehalt.

Es sind bisher zwei verschiedene Hypothesen über den Sitz der
besonderen Kalziumabsorption aufgestellt worden:

A. Das von dem Stern ausgesandte Licht durchsetzt auf seinem Weg
zur Erde eine dunkle interstellare Kalziumwolke. Die Kalziumlinien
haben dann eine unveränderliche Lage, und die aus ihrer konstanten
Verschiebung abgeleitete Radialgeschwindigkeit wird im allgemeinen
nicht übereinstimmen mit der Radialgeschwindigkeit des Sternes bzw.
des Schwerpunktes des Doppelsternsystems.

B. Der Stern ist eingehüllt von einer ausgedehnten Kalzium-
atmosphäre, die im Falle eines spektroskopischen Doppelsternsystems
weit über die Bahn hinausgreift und nur an der Translation des ganzen
Systems, nicht aber an der Bahnbewegung teilnimmt. In diesem Falle
muß natürlich die aus den Kalziumlinien abgeleitete Geschwindigkeit
innerhalb der Beobachtungsgenauigkeit übereinstimmen mit der Schwerpunktsgeschwindigkeit, abgesehen von der im folgenden erwähnten
Möglichkeit.

Während HARTMANN noch die erste Hypothese als die wahrschein-
lichere betrachtete, entschieden die im Laufe der Jahre hinzukommenden
Beobachtungen immer mehr zugunsten der zweiten. Es fällt hier vor
allem die eingangs erwähnte Tatsache ins Gewicht, daß die scharfen
Kalziumlinien nur innerhalb des Bereiches der höchsten Sterntemperaturen auftreten. Die Hypothese B eröffnete eine Möglichkeit zur
Prüfung der von EINSTEIN geforderten Rotverschiebung der Spektral-
linien in starken Gravitationsfeldern; denn die Kalziumlinien entstehen
in einem Gebiet mit wesentlich geringerem Potential als die eigentlichen
Sternlinien. FREUNDLICH[1]) hat seinerzeit einen Versuch in dieser Richtung unternommen, der auch zu einem positiven Ergebnis geführt hat.
Indessen war das ihm zur Verfügung stehende Material viel zu spärlich,
als daß man dem Resultat große Beweiskraft hätte zusprechen können.
Durch eine bald darauf von YOUNG[2]) veröffentlichte umfangreiche Liste
von B-Sternen wurde denn auch das FREUNDLICHsche Ergebnis sehr
in Frage gestellt; denn KRUSE[3]) bemerkte in einem Referat, daß die
Differenzen zwischen der Systemgeschwindigkeit V_* und der aus den

---

[1]) Phys. Zeitschr. Bd. 20, S. 561.
[2]) Publ. Dominion Astrophysical Observatory Victoria Bd. 1, S. 17.
[3]) Naturwissenschaften 1921, S. 483.

Kalziumlinien abgeleiteten Geschwindigkeit V_{Ca} sich gleichmäßig über positive und negative Werte verteilen.

Es schien also, als ob auch die Hypothese B nicht in ihrer ursprünglichen Form zutreffe, vielmehr die Geschwindigkeit des Systems und die der Kalziumwolke merklich voneinander verschieden seien. Indessen glaubte Young die auftretenden Differenzen im Hinblick auf ihre Kleinheit und mit Rücksicht auf die wenig exakte Definition der Linien in diesen Spektren durch Fehler in den benutzten Wellenlängen hinreichend erklären zu können. Nach Plaskett (a. a. O.) ist dieser Ausweg nicht mehr gangbar, da bei seinen Beobachtungen die Differenzen

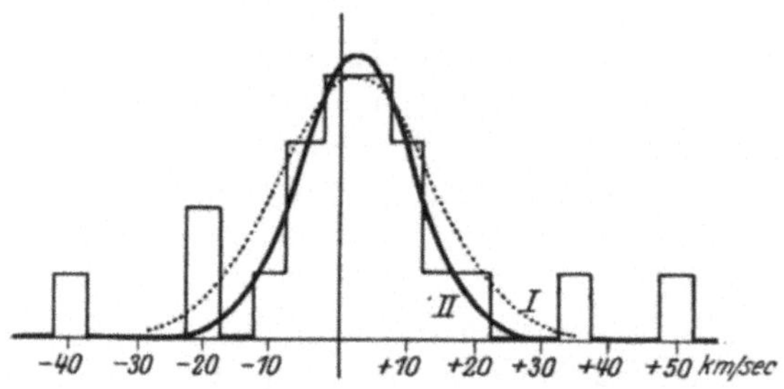

Abb. 1. Häufigkeitskurven der Differenzen $V_* - V_{Ca}$ (Young, 2 Oe-, 20 B-Sterne)

sich zwischen -59 und $+50$ km/sec bewegen. Er betrachtet sie als reelle Unterschiede in den Radialgeschwindigkeiten von Stern und Kalziumwolke und kommt so zu einer Vorstellung, welche die beiden Ausgangshypothesen vereinigt[1]): Der Stern bewegt sich durch eine von ihm unabhängige ausgedehnte dunkle Wolke und ionisiert in seiner Umgebung infolge der hohen Temperatur der von ihm ausgesandten Strahlung genügend Kalzium, um die Linien H und K in Absorption erscheinen zu lassen.

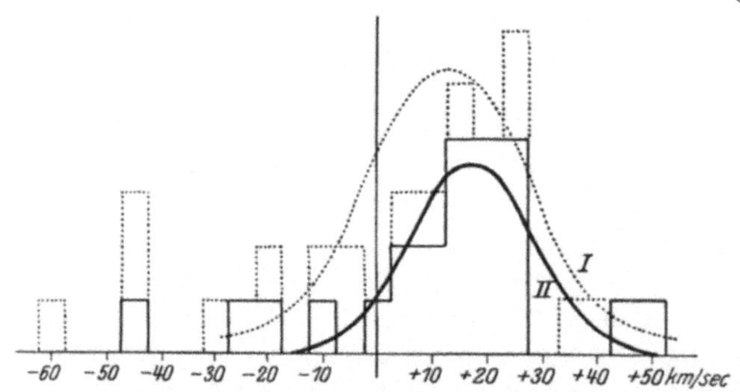

Abb. 2. Häufigkeitskurven der Differenzen $V_* - V_{Ca}$ (Plaskett, 10 Oe-, 20 O-, 8 B-Sterne)

Es ist nun von großer Wichtigkeit, zu entscheiden, ob der empirische Befund wirklich zu der Schlußfolgerung berechtigt, daß wir es hier mit reellen Geschwindigkeitsdifferenzen zu tun haben.

---

[1]) Anmerkung bei der Korrektur. Erst nach Drucklegung des vorliegenden Aufsatzes erhielt ich durch Herrn Ludendorff Kenntnis von einem Aufsatz in Astr. Nachr. Bd. 212, 3, in dem auch schon der Zusammenhang zwischen O-Sternen und planetarischen Nebeln auf Grund vorliegender Messungen von Radialgeschwindigkeiten untersucht wird. Ich möchte nicht unterlassen, darauf hinzuweisen, vor allem, weil dort auch schon der von Plaskett vorgeschlagene Ausweg der Vereinigung der beiden Hypothesen A und B diskutiert wird.

Gehen wir zunächst rein statistisch vor, indem wir die Verteilung der Differenzen durch Abzählung in Intervallen von je 5 km/sec untersuchen, so ergeben sich für die Differenzen V_*-V_{Ca} die folgenden Bilder. Die Differenzen V_*-V_{Ca} gruppieren sich nicht um den Mittelwert 0, sondern zeigen eine deutliche Bevorzugung positiver Werte, bei PLASKETT ausgesprochener als bei YOUNG. Die Hauptmasse der Sterne YOUNGS schließt sich sehr gut einer GAUSSschen Fehlerkurve an mit der Symmetrielinie bei $+ 2{,}5$ km/sec und einem mittleren Fehler von $\pm 9{,}3$ km/sec. Schließt man die vier extremen und vollkommen isoliert liegenden Werte aus, beschränkt sich also auf $- 20 < (V_*-V_{Ca}) < + 30$, dann erhält man:

$$\overline{V_*-V_{Ca}} = + 2{,}5 \text{ km/sec}, \qquad \varepsilon = \pm 7{,}0 \text{ km/sec (Kurve II)}$$

Bei PLASKETT gestaltet sich das Bild wesentlich anders. Bei einer ausgesprochenen Bevorzugung positiver Werte (25 von 38) ist die ganze Verteilung sehr viel ungleichmäßiger, und ist es kaum möglich, sie durch eine GAUSSsche Fehlerkurve wiederzugeben. Rein formal erhielte man die Symmetrielinie bei $+ 12{,}5$ km/sec, den m. F. zu $\pm 12{,}5$ km/sec. Es zeigt sich nun aber, daß die extremen Werte durchweg von Sternen geliefert werden, in deren Spektrum neben den Absorptionslinien auch Emissionslinien auftreten, und daß bei einer Reihe von Sternen die Kalziumlinien variabel sind, ihre Verschiebung daher unsicher ist. In der Figur sind diese Anzahlen punktiert eingezeichnet. Schließt man sie aus, dann wird die formal günstigste Darstellung erzielt durch die Kurve II, deren Konstante sind:

$$\overline{V_*-V_{Ca}} = + 17{,}5 \pm 2{,}1 \text{ km/sec}, \qquad \varepsilon = \pm 9{,}3 \text{ km/sec}$$

Die Streuung der Einzelwerte ist erheblich und führt auf wesentlich größere mittlere Fehler, als nach der von den Beobachtern angegebenen inneren Genauigkeit zu erwarten wäre. Indessen darf nicht vergessen werden, daß bei diesen Spektren mit ihren breiten und diffusen Linien die systematischen Fehler meist einen sehr hohen Betrag erreichen, wie aus dem Vergleich der von verschiedenen Beobachtern gemessenen Radialgeschwindigkeiten ein und desselben Sternes hervorgeht. Die meisten der obigen Messungen gehen auf nur einen Beobachter zurück; eine Elimination der systematischen Fehler hat also nicht stattgefunden.

Da ein systematischer Unterschied zwischen den Geschwindigkeiten V_* und V_{Ca} offenbar vorhanden ist, ist es wichtig, zu entscheiden, welche der beiden Geschwindigkeiten die „richtige" ist; richtig in dem Sinne verstanden, daß sich im Mittel daraus die normale Apexbewegung der Sonne ergibt. Geht man auch hier rein statistisch vor — wir unterdrücken die entsprechenden Tabellen und Figuren als unwesentlich —, so gelangt man zu folgenden Ergebnissen:

a) Die aus den Kalziumlinien abgeleiteten Geschwindigkeiten stimmen im Mittel überein mit den entsprechenden Komponenten der Sonnenbewegung und führen auf die gleichen durchschnittlichen Absolutgeschwindigkeiten, wie Campbell und Gyllenberg sie für die Gesamtheit der B-Sterne gefunden haben. Es ist nämlich:

$$K = -0,2 \text{ km/sec}, \qquad V = 6,5 \text{ km/sec bei Young}$$
$$+0,6 \quad \text{,,} \qquad\qquad 4,4 \quad \text{,,} \quad \text{,, Plaskett}$$

b) Die aus den übrigen Linien des Spektrums abgeleiteten Geschwindigkeiten geben einen positiven K-Effekt von beträchtlicher Größe und die durchschnittlichen Absolutgeschwindigkeiten nähern sich dem großen Wert für die Planetarischen Nebel. Es ist nämlich:

$$K = +6,6 \text{ km/sec}, \qquad V = 13,8 \text{ km/sec bei Young}$$
$$+4,5 \quad \text{,,} \qquad\qquad 24,0 \quad \text{,,} \quad \text{,, Plaskett}$$

Abschließend gewinnen wir also aus diesen orientierenden statistischen Betrachtungen den Eindruck, daß die „Kalziumwolken" relativ zum Sternsystem ruhen, daß bei den Sternen selbst aber besondere Verhältnisse vorliegen. Lassen wir die beobachteten Linienverschiebungen als reelle Radialgeschwindigkeiten gelten, dann fallen diese Geschwindigkeiten sowohl wegen ihrer Größe als auch wegen der einseitigen Bevorzugung positiver Werte aus dem allgemeinen Rahmen heraus.

Versucht man dem Kern des Problems durch genauere Rechnungen näher zu kommen, dann stößt man sofort auf erhebliche Schwierigkeiten. Das von Young gesammelte Material ist sehr heterogen und unvollständig. Schließt man nach den von Young selbst gegebenen Anmerkungen alle unsicheren Objekte aus, dann bleiben nur 15 Sterne übrig, die sich auf einen zu engen Spektralbereich verteilen (B 0 bis B 3), als daß sich bei der großen Ungenauigkeit der einzelnen Werte irgendein Gang zu erkennen geben könnte. Wir werden daher diese Zahlen hier nicht für sich diskutieren, sondern sie nur unten in anderem Zusammenhange mit verwerten.

Das umfangreichere und in sich homogenere Material Plasketts gestattet eine Unterteilung nach verschiedenen Gesichtspunkten. Wir behalten zunächst die Gruppierung nach Himmelsgegenden bei, in welcher Form Plaskett seine Resultate mitgeteilt hat. Der besseren Übersicht wegen stellen wir die Ergebnisse der angestellten Rechnungen in einer einzigen Tabelle zusammen. Unter n steht jeweils die Anzahl der zur Mittelbildung herangezogenen Sterne, in Klammern beigefügt die Anzahl der Sterne, bei denen die Geschwindigkeit V_* variabel ist, von denen aber keine Bahnbestimmung vorliegt. In diesen Fällen sind die Geschwindigkeiten von Plaskett nur auf ganze Kilometer angegeben, und in den folgenden Rechnungen wurde diesen Sternen das Gewicht 1, den anderen das Gewicht 2 erteilt. ε bedeutet den m. F.

einer Differenz $V_* - V_{Ca}$ von mittlerem Gewicht, so wie er formal bei der Ausgleichung hervorgeht; in Klammern beigefügt sind die Werte, welche nach den von PLASKETT angegebenen wahrscheinlichen Fehlern der einzelnen Geschwindigkeiten zu erwarten wären. Schließlich enthalten die beiden letzten Spalten der Tabelle die durchschnittlichen Absolutgeschwindigkeiten nach Befreiung von Apexbewegung und „K-Effekt". (Alles in km/sec.)

Tabelle 1

	n	$V_* - V_{Ca}$	$V_* - V_{\oplus}$	$V_{Ca} - V_{\odot}$	ε	$\bar{V}_*$	$\bar{V}_{Ca}$
Perseus....	6 (4)	$-29,6 \pm 4,9$	$-43,6 \pm 3,7$	$-12,6 \pm 3,1$	$\pm 12,1$ (2,5)	5,4	5,4
Monoceros..	6 (4)	$+20,3 \pm 2,2$	$+19,5 \pm 2,3$	$+ 0,0 \pm 0,4$	$\pm 6,6$ (2,1)	4,7	0,9
Cygnus I ..	5 (1)	$+ 5,2 \pm 4,4$	$+ 8,8 \pm 4,8$	$+ 4,2 \pm 0,7$	$\pm 10,7$ (3,3)	8,9	1,3
„ II..	5 (1)	$+ 8,8 \pm 2,2$	$+ 7,8 \pm 2,9$	$- 1,1 \pm 2,1$	$\pm 4,8$ (2,0)	5,3	4,1
O5e bis B2e	11 (5)	$- 7,7 \pm 10,0$	$-14,0 \pm 10,1$	$- 5,7 \pm 3,7$	$\pm 33,3$ (3,1)	31,1	10,5
O5 bis O7 .	7 (2)	$+18,2 \pm 3,2$	$+19,8 \pm 5,2$	$+ 2,5 \pm 2,1$	$\pm 8,6$ (2,8)	7,1	4,0
O8 und O9 .	11 (5)	$+15,6 \pm 5,9$	$+15,8 \pm 6,9$	$+ 0,2 \pm 2,0$	$\pm 23,8$ (2,1)	17,6	4,3
B1 und B2 .	7 (4)	$- 1,2 \pm 8,1$	$- 2,0 \pm 10,3$	$- 0,0 \pm 2,6$	$\pm 21,3$ (2,7)	22,2	4,2

In der Perseusgegend sind alle Differenzen negativ, und es erreicht vor allem $V_{Ca} - V_{\odot}$ wesentlich größere Werte als in irgendeiner der anderen Gruppen. PLASKETT hebt hervor, daß "the calcium lines in the stars in this region are not so strong and sharply defined ... than elsewhere"; vier von den Sternen haben variable, nur auf wenigen Beobachtungen beruhende Geschwindigkeiten. In der Monocerosgruppe fällt ein Stern stark heraus (B D $+ 11°$ 1204), bei dem die Differenz $V_* - V_{Ca} = - 42,7$ km/sec beträgt, während alle anderen Sterne positive Differenzen ergeben. Der Stern hat ein B-Spektrum mit Emissionslinien; er wurde aus dem Gruppenmittel, das er stark einseitig beeinflussen würde, ausgeschlossen. Das gleiche gilt von λ Cephei in der Gruppe Cygnus II. Er hat das Spektrum O6e und liefert eine Differenz $V_* - V_{Ca} = - 59,2$ km/sec, während alle anderen Differenzen dieser Gruppe zwischen $+ 9$ und $+ 17,5$ km/sec liegen.

Sieht man von diesen Ausnahmen, auf die wir sogleich zurückkommen werden, zunächst ab, dann lassen sich die folgenden Gemeinsamkeiten feststellen:

a) Die Differenzen $V_{Ca} - V_{\odot}$ sind durchweg kleiner als die Differenzen $V_* - V_{\odot}$.

b) Die Streuung der Differenzen $V_{Ca} - V_{\odot}$ ist wesentlich kleiner als die der $V_* - V_{\odot}$.

c) Die Sternlinien zeigen bei 3 Gruppen eine Rotverschiebung, bei der Perseusgruppe eine Violettverschiebung gegenüber den Kalziumlinien.

d) Die Kleinheit der Zahlen in·der letzten Spalte legt den Schluß nahe, daß die Geschwindigkeiten der „Kalziumwolken" der einzelnen Sterne einer Gruppe nicht unabhängig sind, sondern daß es sich um teilweise gleichgerichtete „Bewegungen" der „Kalziumwolken", vielleicht überhaupt jeweils nur um ein und dieselbe ausgedehnte Wolke handelt.

Gewisse andere Züge treten in Erscheinung, wenn wir zu einer Gruppierung nach Spektraltypen übergehen, wie dies in der 2. Abteilung der Tabelle geschehen ist. Man erkennt jetzt vor allem das besondere Verhalten der Sterne mit Emissionslinien: im Mittel eine Violettverschiebung der Sternlinien gegenüber den Kalziumlinien bei einer außerordentlich starken Streuung, die kaum erlaubt, dem Mittelwert eine einheitliche Deutung zu geben. Auch sonst ist die Streuung in den neuen Gruppen größer als in denen nach Himmelsgegenden, wodurch der oben gewonnene Eindruck verstärkt wird, daß es sich bei den auftretenden Differenzen in den gemessenen Radialgeschwindigkeiten um Gemeinsamkeiten in den Bewegungen räumlich benachbarter Sterne handle.

Im übrigen aber legt der Verlauf der Zahlen doch die Vermutung nahe, daß die systematische Verschiebung der Sternlinien gegen die Kalziumlinien bei den frühesten Typen größer ist und gegen die B-Sterne hin abnimmt, so daß es nicht ganz unberechtigt erscheint, hier nach tieferen Zusammenhängen zu suchen. Handelt es sich bei den auftretenden Differenzen zwischen V_* und V_{Ca} wirklich um eine vom Spektraltypus abhängige Erscheinung, dann muß sich diese deutlicher zu erkennen geben, wenn wir unser ganzes Material zusammenfassen, d. h. die Sterne aus YOUNGS Liste hinzunehmen und gleichzeitig innerhalb enger Bereiche des Spektraltypus mitteln. Die Sterne mit Emissionslinien scheiden dabei ganz aus; über ihr Verhalten wird noch zu sprechen sein. Die Tabelle 2 enthält alle nötigen Zahlenangaben.

Tabelle 2

Sp.	$V_* - V_{Ca}$ km/sec	Gew.	$B - R$ km/sec	$V_* - V_\odot$ km/sec	Gew.	$B - R$ km/sec	$V_{Ca} - V_\odot$ km/sec
O 5	$+ 25,4$	2	$+ 2,1$	$+ 25,2$	2	$+ 2,9$	$- 0,2$
6	$+ 13,1$	6	$- 6,9$	$+ 13,7$	6	$- 5,5$	$+ 0,6$
7	$+ 23,4$	3	$+ 6,6$	$+ 28,4$	3	$+ 12,4$	$+ 5,0$
8	$+ 18,0$	8	$+ 4,5$	$+ 17,3$	9	$+ 4,4$	$- 0,7$
9	$+ 12,8$	7	$+ 2,5$	$+ 9,4$	8	$- 0,3$	$- 3,4$
B 0	$+ 6,4$	3	$- 0,6$	$+ 3,5$	3	$- 3,1$	$- 2,9$
1	$+ 3,0$	14	$- 0,8$	$+ 4,1$	14	$+ 0,7$	$- 1,1$
2	$- 16,0$	7	$- 16,6$	$- 19,5$	7	$- 19,8$	$- 3,5$
3	$+ 3,7$	7	$+ 6,4$	$+ 5,1$	6	$+ 8,0$	$+ 1,4$
5	$+ 7,6$	1	$+ 16,7$	$+ 11,7$	2	$+ 10,9$	$+ 4,1$

Die Differenzen $V_* - V_{Ca}$ und $V_* - V_\odot$ lassen sich durch lineare Formeln, die Differenzen $V_{Ca} - V_\odot$ durch eine Konstante gut darstellen. Wählen wir als spektrale Einheit 0,1 Harvard-Klassen, so führt die Ausgleichung auf die folgenden Zusammenhänge:

$$V_* - V_{\mathrm{Ca}} = +7,0 \;\dotdiv\; 3,24 \;(\mathrm{Sp.} - \mathrm{B}\,0), \qquad \varepsilon_1 = \pm 18,9 \;\text{km/sec}$$
$$\pm 2,5 \quad \pm 1,02$$
$$V_* - V_\odot = +6,6 \;-\; 3,15 \;(\mathrm{Sp.} - \mathrm{B}\,0), \qquad \varepsilon_1 = \pm 23,5 \quad ,,$$
$$\pm 3,0 \quad \pm 1,24$$
$$V_{\mathrm{Ca}} - V_\odot = -0,44 \pm 0,80 \qquad\qquad \varepsilon_1 = \pm\; 6,2 \quad ,,$$

ε_1 ist der m. F. einer Differenz vom Gewicht 1. In der nebenstehenden Figur sind die Verhältnisse bildlich zum Ausdruck gebracht. Rechnung und Figur bestätigen vollkommen, daß die „Kalziumgeschwindigkeiten" unabhängig vom Typus stets identisch mit den Sonnengeschwindigkeiten sind, während die „Sterngeschwindigkeiten" beim Fortschreiten von B nach O positiv anwachsen. Können wir angesichts dieser merkwürdigen Gesetzmäßigkeit die von PLASKETT gegebene Deutung aufrecht erhalten, daß wir es bei den Differenzen $V_* - V_{\mathrm{Ca}}$ mit wirklichen relativen Radialgeschwindigkeiten zwischen einem Stern und einer „Kalziumwolke" zu tun haben?

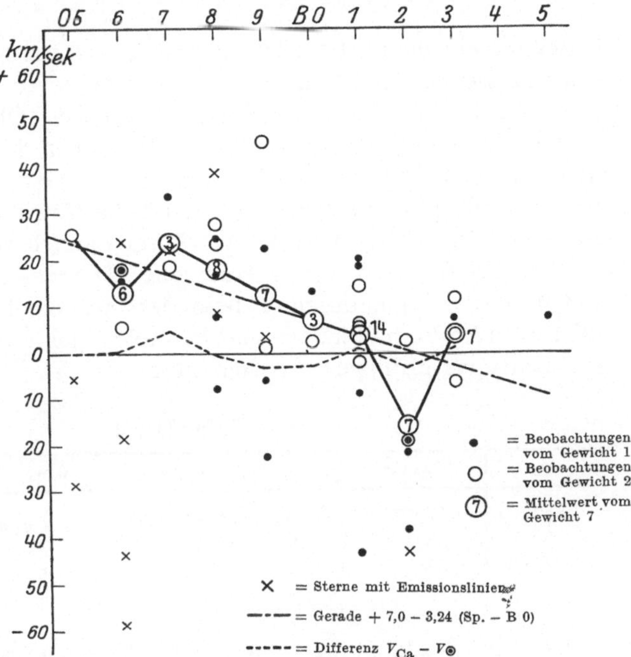

Abb. 3. Darstellung der Differenzen $V_* - V_{\mathrm{Ca}}$ (Ordinaten) in Abhängigkeit vom Spektraltypus (Abszissen).

Es ist allerdings zu bedenken, daß die zur Verfügung stehenden Objekte aus nur wenigen Gruppen am Himmel benachbarter Sterne stammen, daß also eine gemeinsame Bewegung nicht unbedingt von der Hand zu weisen ist. Aber gerade die wenigen negativen Differenzen, die vorkommen, sind über alle Gruppen verteilt und haben nur geringes Gewicht, während der lineare Abfall der Rotverschiebung mit dem Spektraltypus so auffallend ist, daß man darin doch etwas anderes als reelle Radialgeschwindigkeiten sehen möchte. Denkt man an einen Gravitationseffekt im Sinne EINSTEINS, wobei die

Kalziumwolke so weit ausgedehnt zu denken ist, daß für sie das Potential nur noch gering ist, dann erhält man für das Produkt $m^{\frac{2}{3}}\delta^{\frac{1}{3}}$, wo m die Masse, δ die mittlere Dichte der betreffenden Sterne ist, ziemlich hohe Werte, die entweder große Masse oder große Dichte oder beides erfordern, wie aus der folgenden kleinen Tabelle hervorgeht:

Tabelle 3

Sp.	$\sqrt[3]{m^2\delta}$	$\delta_{(m=100)}$	$m_{(\delta=1)}$
B 1	6,3	0,025	16
O 9	17,1	0,50	71
O 7	27,9	2,17	147
O 5	38,7	5,8	241

Unter δ steht die mittlere Dichte (Sonne = 1) unter der Voraussetzung einer Masse von 100 Sonnenmassen, unter m die Masse (Sonne = 1), wenn die Dichte gleich der Sonnendichte ist. Obwohl die Werte für m und δ keineswegs außerhalb des Bereiches der Möglichkeit liegen, ist dieser Deutung doch nicht allzu großes Gewicht beizulegen. Was aber auf alle Fälle bestehen bleibt, ist die Gewißheit, daß es sich hier um Sterne handelt, die von der Norm ganz beträchtlich abweichen.

Wir haben oben die Sterne mit Emissionslinien ausgeschieden und sie in der Figur durch $\times$ besonders hervorgehoben. Sie zeigen ein abweichendes Verhalten, wie man aus folgenden Mittelwerten, nach dem Spektraltypus gruppiert, ersehen mag.

Tabelle 4

Sp.	$V_* - V_{Ca}$	Gew.	$V_* - V_\odot$	Gew.	$V_{Ca} - V_\odot$	Gew.
O 5e	− 17 km/sec	2	− 25 km/sec	2	− 3 km/sec	3
6	− 33	6	− 43	6	− 7	7
7/8	+ 23	6	+ 16	6	− 7	6
O 9−B 2	− 12	3	− 9	3	+ 5	4

Über die Feststellung hinaus, daß diese Sterne, die sich über alle bei der Untersuchung der Sterne mit Absorptionslinien verwendeten Gruppen verteilen, sich dem allgemeinen Bilde nicht einordnen, läßt sich hier kaum etwas Weiteres aussagen; die Anzahl der Sterne ist zu klein, die Beobachtungsgenauigkeit im einzelnen viel zu gering, als daß die Unterteilung nach Spektralklassen irgendwelche Gesetzmäßigkeiten aufzudecken vermöchte.

Zum Schluß kommen wir auf die eingangs aufgeworfene Frage zurück: Lassen sich aus den diskutierten Linienverschiebungen Schlüsse ziehen auf die kosmogonische Stellung der O-Sterne und der Planetarischen Nebel? Wir stellen kurz alle Zahlen (in km/sec) zusammen, die uns in dieser Hinsicht wichtig erscheinen.

Linien / Sp.		P	O5 – O9	Bo – 3	B	A	F.	G	K	M	M1e – M8e
$\overline{V}$	Ca	—	6,8	4,2							
	Abs.	—	13,1	13,8	7,0	11,8	14,5	15,8	15,9	17,2	33,4
	Em.	30,3	31,1	—							30,9
K	Ca	—	− 0,4	− 0,3							
	Abs.	—	+ 16	0	+ 4,3	+ 0,1	+ 0,2	− 0,8	+ 3,6	+ 5,3	− 0,2
	Em.	+ 1,1	(− 12,7)	—							− 11,7

Der mittlere Teil der Tabelle enthält die oben bereits angeführten Zahlen GYLLENBERGS für $\overline{V}$ und K. An diese Zahlen haben wir links die für die B- und O-Sterne von uns gefundenen und die für die Planetarischen Nebel aus LICK XIII abgeleiteten angefügt, rechts die von MERRILL für die Me-Sterne angegebenen Werte, und zwar getrennt, je nachdem die Kalzium , die Absorptions- oder Emissionslinien benutzt wurden. Auf Grund dieser Zahlen gelangen wir zu folgenden Schlüssen:

a) Alle Objekte, in deren Spektren Emissionslinien auftreten (Planetarische Nebel, O-Sterne und langperiodische Veränderliche der Klasse M), zeigen dieselbe große Absolutgeschwindigkeit von 31 km/sec.

b) Soweit Absorptions- *und* Emissionslinien vorhanden sind, erscheinen die letzteren nach Violett verschoben (also qualitativ übereinstimmend mit den bei den Novae gemachten Erfahrungen).

c) Die Kalziumlinien ergeben keinen K-Effekt und setzen die Reihe der von M nach B abnehmenden Absolutgeschwindigkeiten bis O 5 fort.

d) Die Absorptionslinien führen auf einen von B 3 bis O 5 stark anwachsenden positiven K-Effekt und auf Absolutgeschwindigkeiten, die rund doppelt so groß sind wie für die B-Sterne im Mittel.

Wie immer wir auch die verschiedenen Zahlen deuten wollen, so stellen sie keine *normale* Fortsetzung der die Hauptmasse der Sterne umfassenden Spektralklassen dar. Hält man an PLASKETTS Deutung fest, dann handelt es sich um Sterne mit relativ großen Geschwindigkeiten, und ihr Vorkommen ist eng verknüpft mit kosmischen Staubwolken. Die Novae würden sich dann dieser Sterngruppe einordnen nach Maßgabe der Theorie v. SEELIGERS. Es ist aber auch sehr wohl möglich, daß wir es hier mit gewissen instabilen Zuständen zu tun haben, bei denen große Massen unter dem Einfluß relativ hohen Strahlungsdruckes eine teilweise Auflösung erleiden. Die Planetarischen Nebel wären in diesem Falle der Durchgangszustand, in dem die ausgestrahlte Materie ihre größte Ausdehnung erreicht hat, die Novae jene Objekte, bei denen das Instabilwerden sich „katastrophal" vollzieht. Eine Prüfung dieser verschiedenen Möglichkeiten wird erst möglich sein, wenn noch weiteres Material über die ruhenden Kalziumlinien vorliegen wird, vor allem für die Wolf-Rayet-Sterne und die Planetarischen Nebel. Gerade das Studium der an beide Enden der Spektralreihe vorläufig nur lose sich anknüpfenden Ausnahmeerscheinungen wird den Schlüssel liefern zu manchem Rätsel, das die Kosmogonie heute noch für uns birgt.

Die Bedeutung
von Farbenhelligkeitsdiagrammen
für das Studium der Sternhaufen.

Von **P. ten Bruggencate**, Göttingen.

Mit 8 Abbildungen.

1. Russell-Diagramme für Sternhaufen. Die Aufzeichnung von Russell-Diagrammen für kugelförmige und offene Sternhaufen hat sich als ein Mittel erwiesen, nicht nur Parallaxenabschätzungen für die Haufen zu erhalten, sondern auch tiefer in die kosmogonische Stellung der Sternhaufen einzudringen.

Es soll deshalb damit begonnen werden, die wichtigsten Punkte, die bei der Betrachtung der Diagramme auffallen, zu besprechen. Leider ist das Material, das mir zur Verfügung stand, sehr klein, so daß manche Schlüsse nur vorläufigen Charakter haben. Den Diagrammen liegen die folgenden Kataloge über Helligkeit und Farbe der einzelnen Sterne zugrunde:

Messier 13: SHAPLEY, Thirteen hundred stars in the Herculis cluster, Mt. Wilson Contr. 116.

Messier 3: SHAPLEY, Photometric catalogue of 848 stars in M. 3, Mt. Wilson Contr. 176.

Messier 11: SHAPLEY, The galactic cluster M. 11, Mt. Wilson Contr. 126.

Messier 37: v. ZEIPEL, Photometrische Untersuchungen der Sterngruppe M. 37. K. Sv. Vet. Ak. Hdl. Bd. 61, Nr. 15.

Messier 67: SHAPLEY, A catalogue of 311 stars in M. 67, Mt. Wilson Contr. 117.

Schon ein flüchtiger vergleichender Blick auf die fünf Diagramme zeigt, daß die fünf Haufen in drei Gruppen zerfallen: Messier 13 und Messier 3 bilden die erste Gruppe; M. 11 scheint mit M. 37 verwandt zu sein; M. 67 dagegen ist wohl eine Sterngruppe ganz anderer Art. M. 13 und M. 3 werden zu den kugelförmigen Haufen gezählt, M. 11 und M. 37 zu den offenen und M. 67 deutet SHAPLEY[1]) auf Grund von Vergleichen zwischen den Haufensternen und denen der Umgebung als „a condensation in an ordinary stellar field".

---

[1]) Mt. Wilson Contr. 126, p. 2.

Wir wollen nun die Diagramme für die Sternhaufen vergleichen mit dem Russell-Diagramm für das Sternsystem. Dieses zeigt, daß sich die Riesensterne unserer Umgebung im großen und ganzen in einem horizontalen Ast anordnen und daß sich der Ast der Zwerge zwischen den Spektralklassen A o und F o etwa bei der absoluten Helligkeit $+ 2^m.0$ nach rechts unten loslöst[1]). Hat man ein Doppelstern- oder mehrfaches Sternsystem, bei welchem die einzelnen Komponenten Riesen sind, aber verschiedene Masse besitzen, so ist es eine bekannte Erscheinung, daß die Verbindungslinien der Komponenten im Russell Diagramm eine schräge Stellung besitzen, die sich einfach dadurch erklärt, daß die kleinere Masse ihre Entwicklung rascher durchläuft als die große. Stellt man sich nun eine Gruppe von sehr vielen gleich alten Sternen vor, und zeichnet für eine solche Gruppe das Russell-Diagramm, so wird sich kein horizontal verlaufender, sondern ein geneigter Ast der Riesen ergeben. Wir wollen den geneigten Ast der Riesensterne oder die zunehmende Helligkeit bei röter werdender Farbe geradezu als Kriterium für eine Gruppe gleich alter Sterne ansehen. Bei verhältnismäßig jungen Sterngruppen ist es sehr wohl denkbar, daß ihr Russell-Diagramm nur den geneigten Ast der Riesen zeigt und ein Zwergast noch gar nicht vorhanden ist.

Die Russell-Diagramme für Sternhaufen werden natürlich mehr oder weniger beeinflußt durch fremde Sterne, die keine wirklichen Glieder der Gruppe sind. Die Störungen werden besonders groß bei den offenen Haufen.

Die Diagramme (I und II) der Kugelhaufen M. 13 und M. 3 zeigen ganz deutlich das Kriterium einer Gruppe gleich alter Sterne. Das gleiche zeigt auch der Haufen M. 11 (Diagramm III) und auch noch — aber weniger deutlich — der offene Sternhaufen M. 37 (Diagramm IV). Es ist wohl anzunehmen, daß dieser Haufen, der in der Aurigawolke eingebettet liegt, schon eine beträchtliche Anzahl fremder eingefangener Sterne verschiedenen Alters enthält, wodurch dann sofort eine viel gleichmäßigere Verteilung der Riesensterne nach Helligkeit und Farbe entsteht. Dies ist zweifellos der Fall bei M. 67 (Diagramm V), wo jedes Kriterium für den gemeinsamen Ursprung der Haufensterne fehlt. Das Russell-Diagramm für M. 67 bestätigt deshalb die SHAPLEYsche Vermutung, daß M. 67 etwas anderes ist als die gewöhnlichen offenen Haufen, nämlich nur eine Verdichtung in einem gewöhnlichen Sternfeld. Durch diese kurzen Andeutungen soll gezeigt werden, wie mit Hilfe der Russell-Diagramme vielleicht ein kontinuierlicher Übergang gewonnen werden kann von den stark konzentrierten Kugelhaufen bis zu den offensten Sterngruppen.

Der Hauptunterschied zwischen den Diagrammen der Kugelhaufen M. 13 und M. 3 und dem Russell-Diagramm für das Sternsystem besteht

---

[1]) Annual Report Mt. Wilson 1921, p. 270.

nicht in dem geneigten Riesenast — was wohl zuerst auffällt —, denn
er erklärt sich zwanglos aus der Annahme eines gleichen Alters aller

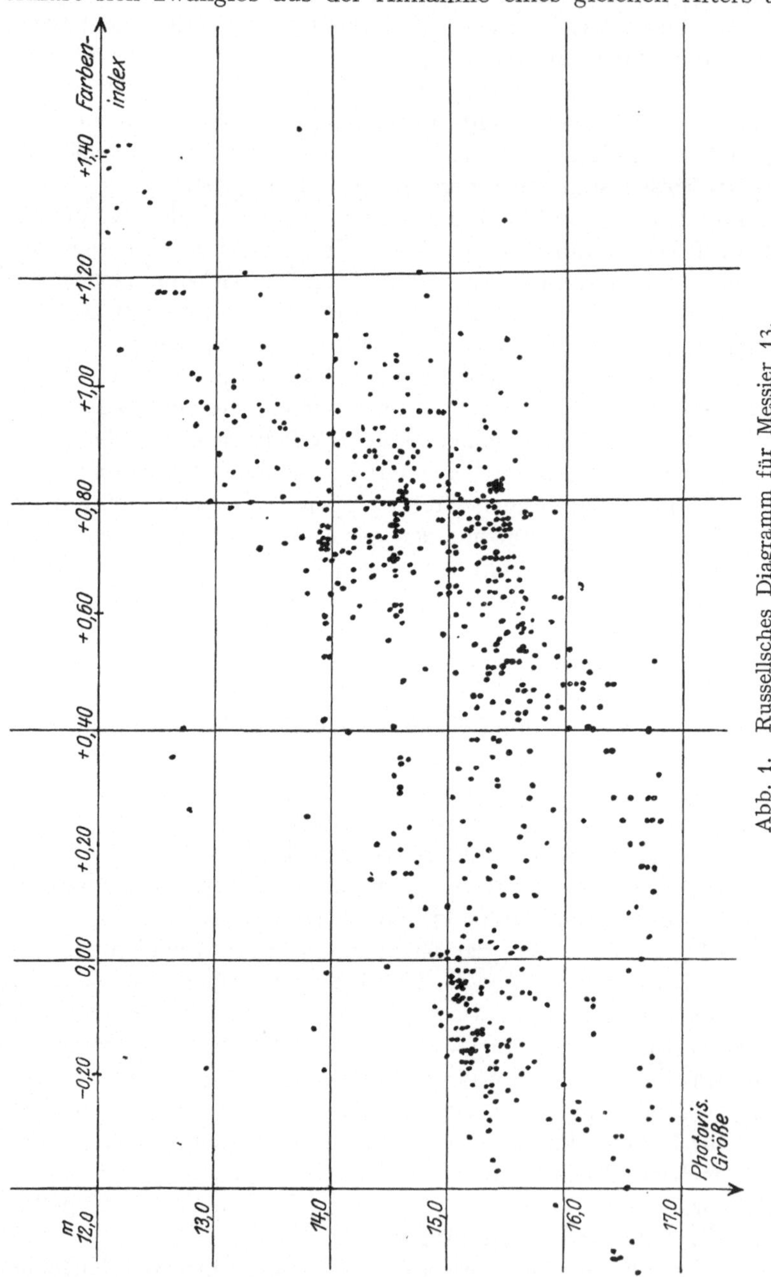

Abb. 1. Russellsches Diagramm für Messier 13.

Haufensterne. Er liegt vielmehr im Aufspalten des einen Astes von
Riesensternen bei beiden Haufen in zwei Äste bei einem Farben-

index von etwa $+ 0^{m}\!.70$, also bei den Spektraltypen F7 und F8. Dieser kritische Punkt ist bei M. 3 ganz deutlich ausgeprägt; bei M. 13 ist er

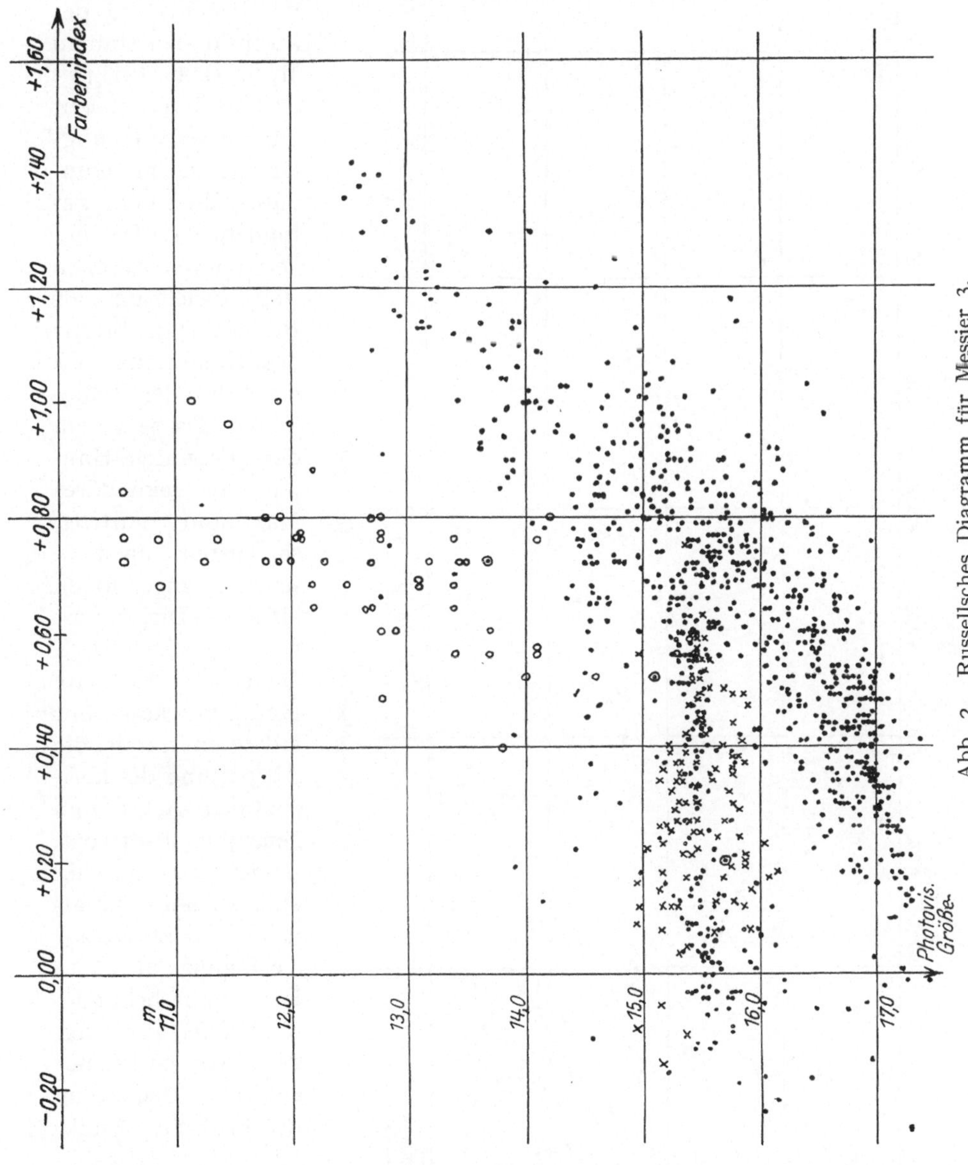

Abb. 2. Russellsches Diagramm für Messier 3.

weit weniger scharf zu erkennen wegen der geringen Anzahl vermessener, schwacher Sterne. Es ist jedoch höchst wahrscheinlich, daß seine Lage in beiden Haufen nach Farbe und absoluter Helligkeit die gleiche ist. Wir haben in den Diagrammen nichts anderes vor uns als eine zweidimensionale Darstellung des Shapleyschen Phänomens der kritischen Helligkeit in der Sternentwicklung[1]). Shapley hat die Häufigkeitskurven der Leuchtkräfte der Sterne in den Haufen M. 2, 3, 5, 13, 15[1]) und 68 [2]) untersucht und bei jeder Häufigkeitskurve ein sekundäres Maximum gefunden. Diese Häufigkeitskurven ergeben sich gleichsam als Projektion der Sterne des Diagramms auf die Achse der Helligkeiten. Da bei jedem der genannten Haufen ein sekundäres Maximum auftritt, so besteht auch in jedem . zugehörigen Russell-Diagramm der kritische Punkt. Denn das sekundäre Maximum kann nur auftreten durch die Abspaltung des horizontalen Astes vom geneigten Riesenast. Leider steht mir kein anderes Material als das hier benutzte zur Verfügung, um diese Frage auch bei anderen Haufen zu verfolgen. Auf die kosmogonische Bedeutung des kritischen Punktes in den Russell-Diagrammen der Kugelhaufen möchte ich im dritten Abschnitt

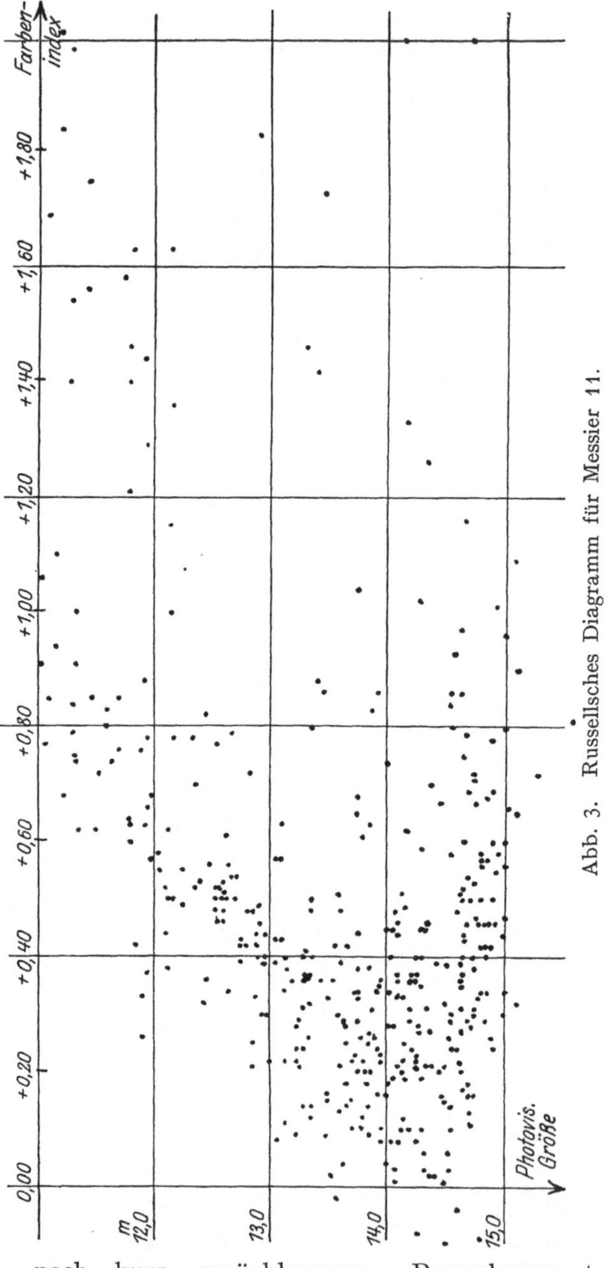

Abb. 3. Russellsches Diagramm für Messier 11.

noch kurz zurückkommen. Bemerkenswert ist es, festzustellen,

[1]) Shapley: Mt. Wilson Contr. 155. [2]) Shapley: Mt. Wilson Contr. 175.

daß er *nur* bei den Kugelhaufen auftritt und nicht bei den offenen
Haufen M. 11, 37 und 67.

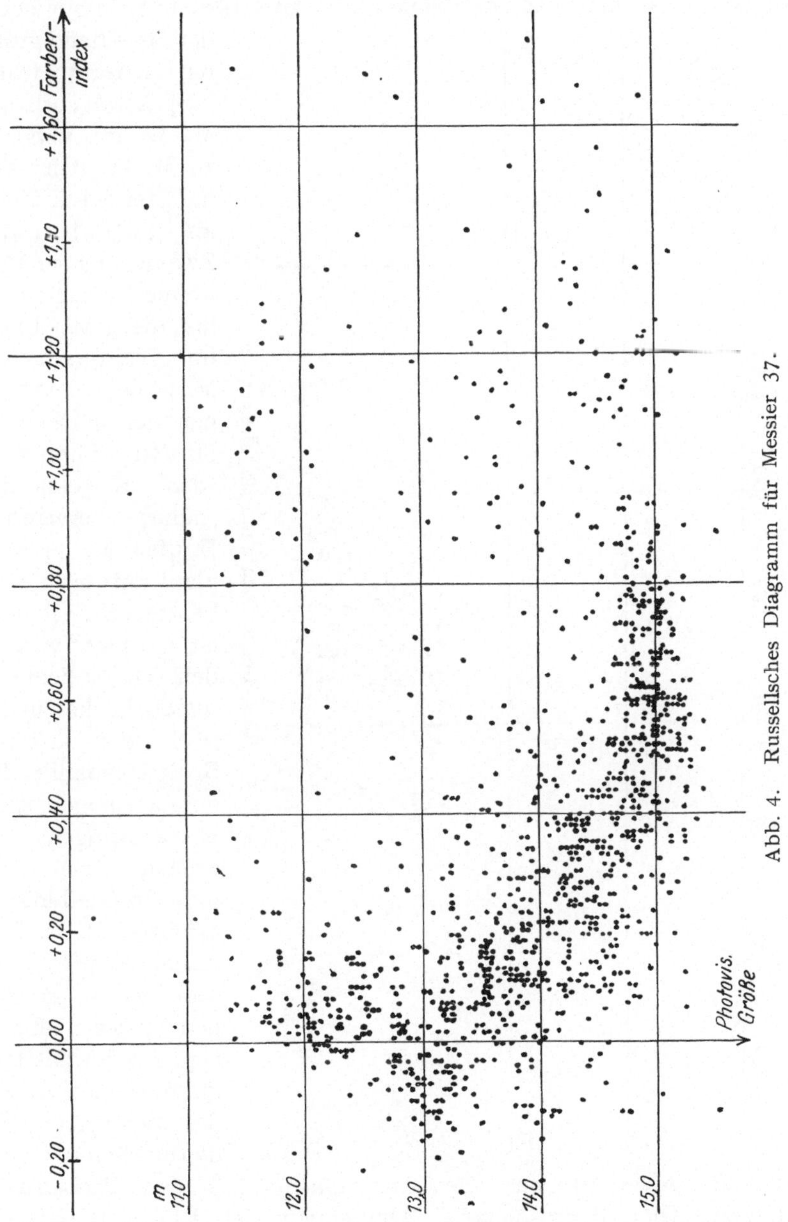

Abb. 4. Russellsches Diagramm für Messier 37.

Bei M. 11, der noch am meisten Verwandtschaft mit den typischen
Kugelhaufen hat, zeigt das Russell-Diagramm einen nach rechts unten
verlaufenden Ast von Sternen, der sich zwischen a5 und f5 vom geneigten

Riesenast loslöst. Dieser Ast stellt die Zwergsterne des Haufens dar. Bei M. 37 haben wir eine ganz ähnliche Erscheinung; der Ansatz der Zwergsterne ist mehr nach der blauen Seite zu verschoben, entsprechend der bedeutend größeren Anzahl früher Spektraltypen in M. 37 im Vergleich zu M. 11. Hier sind die Sterne schwächer als $13^{m}5$ bestimmt Zwerge. Es sei noch aufmerksam gemacht auf die große Lücke im Diagramm von M. 37 bei hellen Sternen der Farben a7 bis f10, während der den Kugelhaufen näher verwandte Haufen M. 11 diese nicht aufweist. Auch bei den Kugelhaufen ist sie nicht vorhanden. Dafür liegt bei diesen in diesem Bereich der kritische Spaltungspunkt. Bei den kosmogonischen Betrachtungen im dritten Abschnitt ist diese Feststellung von Wichtigkeit. Bei M.67, einem Haufen, der sehr arm ist an Sternen blauer Spektraltypen, zeigt das Diagramm einen Ansatz des Zwergastes bei den Farben f0 bis g5.

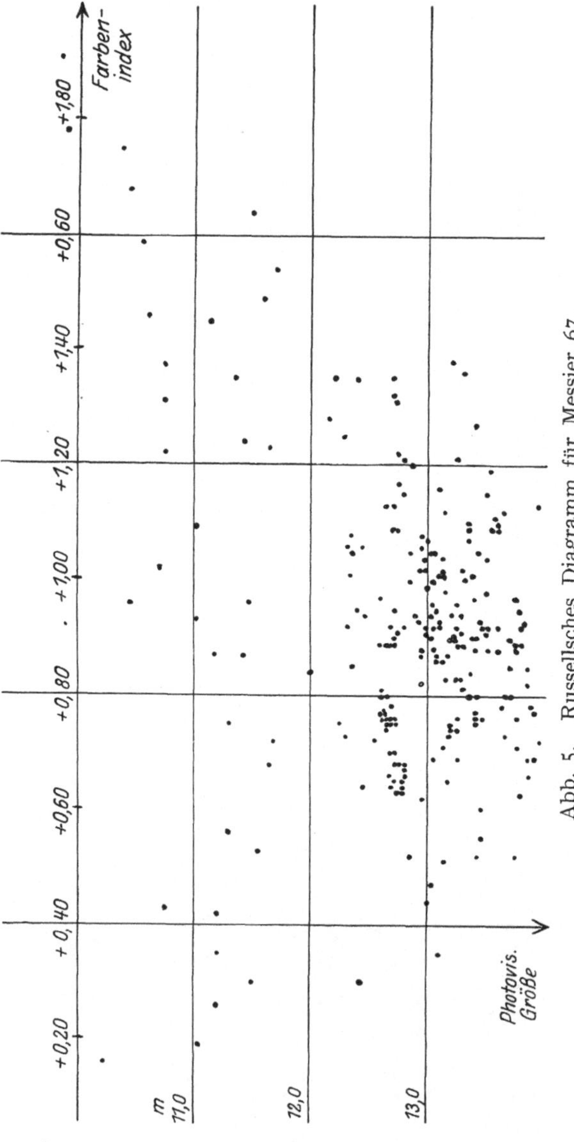

Abb. 5. Russellsches Diagramm für Messier 67.

2. Die Ableitung der Haufenparallaxen aus den Russell-Diagrammen. Die überraschende Ähnlichkeit des Russell-Diagramms von M. 37 mit dem Diagramm für die Sterne des Sternsystems, wie es z.B. im Annual Report of the Director of the Mt. Wilson Obs. für 1921 sich findet, hat mich zuerst auf die Möglichkeit hingewiesen, die

Russell-Diagramme von Sternhaufen zu Parallaxenabschätzungen zu benutzen.

Wenn auch hinsichtlich der Farbe der Ansatz des Zwergastes bei den drei Haufen M. 11, M. 37 und M. 67 etwas variiert — eben abhängig von der Reichhaltigkeit der Sterngruppe an blauen Spektraltypen —, so scheint mir doch die Hypothese, daß die Ablösung des Zwergastes vom Ast der Riesen bei allen Haufen bei der gleichen absoluten Helligkeit geschieht wie bei den Sternen des Sternsystems, große Wahrscheinlichkeit zu besitzen und jedenfalls brauchbar zu sein, um die Parallaxen der Größenordnung nach abzuschätzen.

Die Loslösung des Zwergastes vom horizontalen Riesenast findet bei den Sternen des Sternsystems bei einer absoluten Helligkeit von $+ 2^{m}0$ ($\pi - 0''1$) statt[1]). Bei den offenen Haufen M. 11, 37 und 67 ist es möglich, diejenige scheinbare Helligkeit ziemlich genau festzulegen, bei der sich der Ast der Zwergsterne vom Ast der Riesensterne loslöst. Indem man nun — nach der gemachten Hypothese — diesen abgelesenen scheinbaren Helligkeiten die absolute Helligkeit $+ 2^{m}0$ zuordnet, hat man eine Abschätzung der Parallaxe gewonnen. Bei den Kugelhaufen M. 13 und M. 3, für welche Diagramme gezeichnet werden konnten, liegt die Sache nicht so einfach. Hier kann man nur diejenige scheinbare Helligkeit angeben, oberhalb welcher sicher kein Abtrennen der Zwerge möglich ist. Dies liefert dann Maximalparallaxen. Besonders bei M. 13 sind viel zu wenig schwache Sterne vermessen, um sagen zu können, ob sich in dem beobachteten Helligkeitsintervall überhaupt schon ein Zwergast loslöst. Wieweit bei einer Gruppe gleich alter Sterne ein Ast der Zwerge ausgebildet ist, hängt wesentlich vom Alter der Sterngruppe ab. Ist das Alter der Gruppe kleiner als die Entwicklungsdauer eines Sternes normaler Masse vom roten Riesen bis zum f-Zwerg, so kann kein Zwergast auftreten. Die Schwierigkeit bei diesen Betrachtungen liegt darin, daß man keinerlei zuverlässige Abschätzung der Zeit bei den Vorgängen der Stellarastronomie machen kann. Es soll nur angedeutet werden, daß theoretisch sehr wohl eine Sterngruppe denkbar ist, deren Russell-Diagramm plötzlich abbricht. Bei den typischen Kugelhaufen ist es allerdings sehr unwahrscheinlich, daß ihre Russell-Diagramme keinen Zwergast enthalten wegen des stark konzentrierten Hintergrundes sehr schwacher Sterne, den die Kugelhaufen zeigen.

Im folgenden handelt es sich daher bei M. 13 und M. 3 um Maximalparallaxen, bei den offenen Haufen um eine Abschätzung der wirklichen Parallaxen. Aus den Diagrammen findet man, daß der absoluten Helligkeit $+ 2^{m}0$ in den Sternhaufen:

Messier	3	13	11	37	67
die scheinbaren Helligkeiten	16.0	15.5	14.0	13.0	13.0

---

[1]) Annual Report Mt. Wilson 1921, p. 270.

entsprechen. Aus der Gleichung

$$M = m + 5 + 5 \cdot \log \pi$$

erhält man dann sofort die Parallaxenwerte:

Messier	π		π (Shapley)	π (Schouten)
3	$\leqq$	0″00016	0″00007	0″00055
13	$\leqq$	00020	00009	00075
11	$=$	00025	00014	00055
37	$=$	00063	00040	00250
67	$=$	00063	00025	00100

Zum Vergleich sind in den letzten Spalten die Parallaxen von Shapley und Schouten angegeben.

Wir wollen nun umgekehrt vorgehen und bestimmen, bei welcher scheinbaren Helligkeit in den einzelnen Haufen die Trennung von Zwergen und Riesen stattfinden muß, unter Annahme der Shapleyschen und Schoutenschen Parallaxen.

Shapleys Parallaxen.

Messier	m (für $M = +2^{m}\!0$	Bemerkung
3	$+$ 17.75	Shapleys Parallaxen viel-
13	17.25	leicht zu klein
11	16.25	
37	14.00	Shapleys Parallaxen be-
67	15.00	stimmt zu klein

Schoutens Parallaxen.

Messier	m (für $M = + 2^{m}\!0$)	Bemerkung	
3	$+$ 13.30	fast lauter Zwerge	un-
13	12.60	lauter Zwerge	möglich
11	13.30	möglich	
37	10.00	lauter Zwerge, unmöglich	
67	12.00	möglich	

Die Bemerkungen in der letzten Spalte ergeben sich sofort aus den Deutungen, die wir den Diagrammen der Haufen im ersten Abschnitt gegeben haben.

Es ist interessant, noch die Parallaxe anzuführen, die Kopff für M. 13 gefunden hat[1]) bei Benutzung der Kapteyn-Schoutenschen Methode. In dem von ihm benutzten Helligkeitsintervall scheint die ausgeglichene Helligkeitskurve gut mit der Kapteynschen übereinzustimmen. Kopff hat nur die äußeren Teile des Haufens benutzt, um die unbestimmten Verhältnisse im Zentrum des Haufens zu ver-

---

1) Kopff: Astr. Nachr. 219, p. 311.

meiden, während in SCHOUTENS Arbeiten Zählungen über den ganzen Haufen vorgenommen wurden. Als Parallaxe von M. 13 fand er $0''\!.0003$, ein Wert, der der Größenordnung nach besser mit dem aus dem Farbenhelligkeitsdiagramm bestimmten Maximalwert $0''\!.0002$ übereinstimmt als die SHAPLEYSche oder SCHOUTENSche Parallaxe.

Das Studium der Farbenhelligkeitsdiagramme von Sternhaufen macht es wahrscheinlich, daß die SHAPLEYSchen Parallaxen mit 1.5 bis 2.0 multipliziert werden müssen. Dies deckt sich mit allen neueren Er-

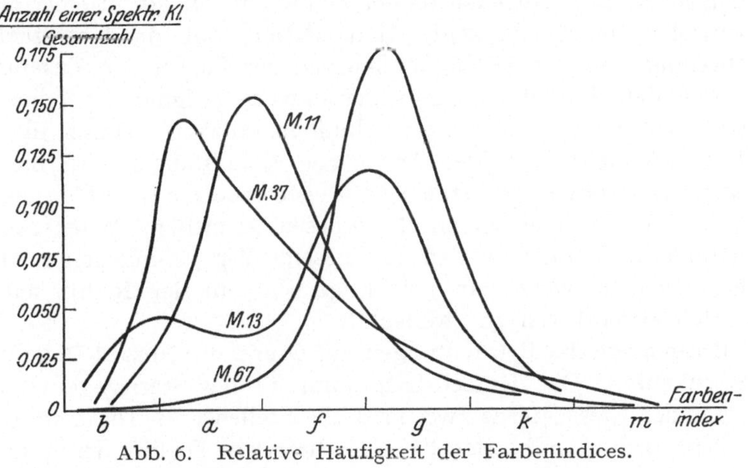

Abb. 6. Relative Häufigkeit der Farbenindices.

fahrungen. SHAPLEY selbst hält eine Division seiner Entfernungen mit 1.5 bis 2.0 für wahrscheinlich.

Anschließend an diese Betrachtungen möchte ich hier noch Häufigkeitskurven für Farbenindizes besprechen, wie ich sie mit Hilfe des zitierten Materials gezeichnet habe. Ähnliche Kurven finden sich auch in den Arbeiten SHAPLEYS und v. ZEIPELS. Die Kurven des Diagramms sind vergleichbar, da als Ordiniate gewählt wurde die Anzahl der Sterne einer Spektralklasse dividiert durch die Gesamtzahl der Haufensterne. Wir können an Hand des Diagramms die notwendigen Veränderungen der Häufigkeitskurve bei Sternhaufen studieren, die verschiedene Entfernung von uns besitzen und ein verschiedenes Alter haben. Beim typischen Kugelhaufen (M. 13), der auch die größte Entfernung besitzt unter den Haufen, die wir hier betrachten, haben wir ein sekundäres Maximum in der Häufigkeitskurve bei den frühen Spektraltypen b und a. Bei älteren Haufen werden mehr Zwerge vorhanden sein, die wir dann beobachten können, wenn uns die Haufen nahe genug sind. Solche Beispiele haben wir in den Haufen M. 11 und M. 37. Das sekundäre Maximum ist verschwunden, und aus diesem und dem Hauptmaximum von M. 13 hat sich ein einziges scharfes Maximum gebildet: bei M. 37 beim Farbenindex $+ 0^{m}\!.10$, bei M. 11 bei $+ 0^{m}\!.40$. Die Verschiebung des Maximums bei M. 11 nach röterer Farbe im Vergleich zu M. 37 rührt, wie das Russell-Diagramm für M. 11 und M. 37 zeigt, nicht her von einer Vermehrung der röteren Zwerge bei M. 11, aus der man auf eine kleinere Entfernung dieses Haufens schließen könnte, sondern nur daher, daß M. 11 überhaupt arm an Sternen der Farben b und a0—a5 ist. Wohl aber wird nun die weitere Verschiebung des Maximums bei M. 67 gegenüber M. 11 ausschließlich hervorgerufen durch das Überwiegen der Zwergsterne der Farben f und g, ein sicheres Kriterium dafür, daß M. 67 uns näher ist als der Haufen M. 11, der in einer Milchstraßenwolke eingebettet liegt. Man erkennt, daß die Untersuchung des Maximums in den Häufigkeitskurven der Farben, im Zusammenhang mit den Farbenhelligkeitsdiagrammen geeignet ist, eine Abschätzung zu liefern für relative Haufenparallaxen. Durch die notwendige Kombination beider Arten von Diagrammen wird die Abschätzung schwieriger. Trotzdem ist sie geeignet, die bei Ableitung von Haufenparallaxen eingehenden Hypothesen zu prüfen. Widersprechen die gefundenen Parallaxen den auf diesem Weg gefundenen relativen Entfernungen, so wird man berechtigt sein, an der Richtigkeit der gemachten Hypothesen zu zweifeln.

3. Kosmogonische Betrachtungen auf Grund der Russell-Diagramme für Sternhaufen. Das Russell-Diagramm für das Sternsystem[1]) weist beim Ast der Riesensterne zwei kritische Stellen auf. Die erste solche Stelle liegt im Spektralbereich F5—G8, die zweite liegt an der Spitze des Diagramms bei den B-Sternen. Im ersten Bereich finden sich fast gar keine normalen Riesensterne; es ist vielmehr der Bereich der langperiodischen δ-Cepheisterne. Der zweite Bereich reiht sich nach der blauen Seite an die B-Sterne an und enthält die O-Sterne und die planetarischen Nebel. Kosmogonisch kann man diese Tatsachen wohl in der folgenden Weise deuten[2]), was hier nur kurz skizziert werden soll.

Von Sternen normaler Masse, etwa zwischen 0.5 und 5 Sonnenmassen, weiß man, daß ihr Lebensweg durch den horizontalen Ast der Riesen führt, dort bei einem gewissen Spektraltypus in den geneigten Ast der Zwerge abbiegt. Bei welchem Typus dies Abbiegen eintritt, hängt von der Masse des Sternes ab. Diese Sterne bleiben stabil während ihrer ganzen Entwicklung oder bilden *normale* Doppelsternsysteme. Denken wir uns die Masse eines Sternes als wachsend, so wird die Möglichkeit des Instabilwerdens größer, erstens weil der Strahlungsdruck im Verhältnis zum Gasdruck wächst, und zweitens weil der Stern wegen der größeren Masse zu früheren Spektraltypen, also höheren Temperaturen fortschreiten kann. Es wird eine Masse geben, die ausreicht zur Erreichung der Typen B und O. Und von diesen Massen ist

---

[1]) Annual Report Mt. Wilson 1921, p. 270.
[2]) Kienle: Naturwissenschaften. 12. Jahrg., Heft 24.

anzunehmen, daß sie in diesem Stadium instabil werden und auf diese Weise zu den planetarischen Nebeln hinführen.

Sterne mit noch größerer Masse werden, da der Strahlungsdruck mit der Masse wächst, ehe sie überhaupt die B-Typen erreichen, schon instabil werden. Und als solche Massen haben wir wohl die δ-Cepheisterne anzusehen.

Kurz zusammengefaßt denken wir uns die Entwicklung der Sterne nach dem Russell-Diagramm für die Sterne des Sternsystems so, daß die größten Massen zu den δ-Cepheisternen führen; kleinere, aber doch noch übernormale Massen, die wir vielleicht Übergiganten nennen können, führen zu den O-Sternen und den planetarischen Nebeln. Die normalen Massen durchlaufen den horizontalen Riesenast, höchstens bis zu den Typen A5—A10 und biegen dann ab in den Ast der Zwerge. Es entstehen nun die folgenden Probleme: Gibt es zwischen den bei F und G liegenden instabilen δ-Cepheisternen und den an der Spitze stehenden O-Sternen noch andere instabile Zustände; was hat die große Lücke im Russell-Diagramm bei den A-Sternen zu bedeuten, und gibt es von den O-Sternen oder den Cepheiden irgendeinen Übergang in den Zwergast?

Die folgenden Ausführungen sollen zeigen, wie man vielleicht mit Vorteil zur Lösung der aufgeworfenen Fragen die Sternhaufen heranziehen kann. Da das Material, das mir zur Verfügung stand, ein sehr spärliches ist, so soll durch das Folgende nur ein Weg gewiesen werden, auf welchem man hoffen kann, weiter vorwärts zu kommen, wenn vollständigere Daten über Helligkeiten und Farben der Sterne in Sternhaufen zur Verfügung stehen.

Meine Untersuchung beschränkt sich im großen und ganzen auf den Haufen M. 3, von dem ein ausführlicher früher zitierter Katalog der Helligkeiten und Farben seiner Sterne existiert. Er enthält auch eine verhältnismäßig große Anzahl von Haufenveränderlichen, die, nach den Untersuchungen von SHAPLEY über die Periodenhelligkeitskurven der Cepheiden, eng zusammenhängen mit den kurzperiodischen Cepheiden (Periode kleiner als 1^d). In MT. WILSON Contr. 154 findet sich ein Farbenhelligkeitskatalog dieser Veränderlichen. Im Russell-Diagramm von M. 3 habe ich die Haufenveränderlichen als Kreuze eingezeichnet. Sie fallen genau zusammen mit dem sich vom geneigten Hauptast der gewöhnlichen Sterne loslösenden horizontalen Ast. Es ist bemerkenswert, daß nur ein einziger Veränderlicher auf der roten Seite des Spaltungspunktes, den wir den kritischen Punkt genannt haben, liegt. Und dessen Farbenindex ist $+ 0^m.75$, während der kritische Punkt beim Index $+ 0^m.70$ liegt. Andererseits erstrecken sich die Veränderlichen vor bis b0 ($- 0^m.40$), und zwar liegen alle, ebenso wie die große Mehrzahl der gewöhnlichen Sterne des horizontalen Astes, in einem Streifen von der Breite einer Größenklasse. Da diese Veränderlichen bis zu den b-Typen vorschreiten, müssen sie eine Masse besitzen, die größer ist als die normale Sternmasse. Wir nennen daher die gewöhnlichen Sterne des horizontalen Astes Übergiganten, die Sterne des geneigten Astes gewöhnliche Riesen. Ganz analog ist das Diagramm für den Kugelhaufen M. 68 (Diagramm VII) entworfen[1]). Leider enthält der benutzte Katalog nur sehr wenig Sterne. Aber immerhin ist eine ganz analoge Anordnung der gewöhnlichen Riesen und der Haufenveränderlichen, wie bei M. 3, sofort zu erkennen. Shapley hat erkannt, daß es die Haufenveränderlichen sind, die das sekundäre Maximum in den Häufigkeitskurven der Leuchtkräfte verstärken. Aber wie groß bei den einzelnen Haufen die Streuung der Veränderlichen nach Helligkeit und Farbe ist und ob sie vielleicht von Haufen zu Haufen variiert, ist aus der eindimensionalen Darstellung Shapleys nicht zu erkennen. Zu diesem Zweck habe ich das Diagramm VIII entworfen, in welchem die Veränderlichen von M. 3, M. 15[2]) und M. 68

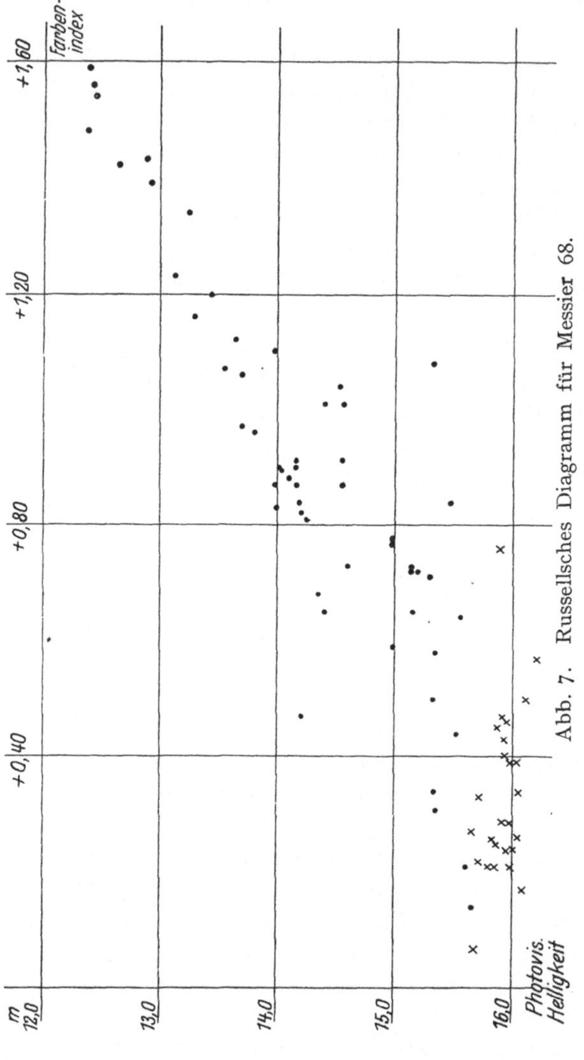

Abb. 7.   Russellsches Diagramm für Messier 68.

eingezeichnet sind nach Farbe und absoluter Helligkeit. Zur Umrechnung der beobachteten scheinbaren Helligkeiten der Veränderlichen in absolute wurden die mit 1,5 multiplizierten Parallaxen von Sharpley als die plausibelsten Parallaxenwerte benutzt. Das Diagramm beweist, daß sich die

[1]) Shapley: Mt. Wilson Contr. 175.
[2]) Shapley: Mt. Wilson Contr. 154, p. 15.

Veränderlichen dieser drei Haufen, und daher mit großer Wahrscheinlichkeit diejenigen aller Haufen, nach Farbe und Helligkeit völlig gleich verhalten. Durch einen größeren Kreis ist in dem Diagramm der kritische Punkt in M. 3 eingezeichnet. Man sieht, daß diesem Punkt auch in dem Haufen M. 15 und M. 68 die Rolle eines kritischen zukommt. So haben uns die Russell-Diagramme zu einem Punkt geführt, der in der Kosmogonie zweifellos eine große Bedeutung hat.

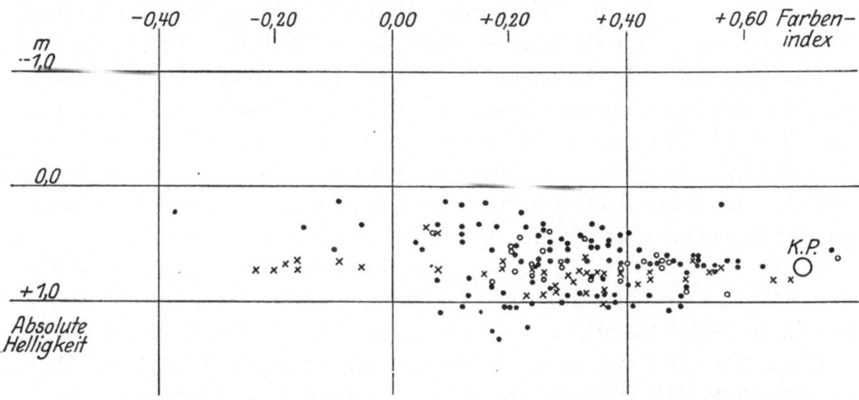

Abb. 8. Haufenveränderliche: • Messier 3, ✕ Messier 15, ○ Messier 68.

Die Folgerung, die wir aus den bisherigen Betrachtungen ziehen, ist die, daß die Übergiganten der kugelförmigen Sternhaufen nichts anderes sind als die Sterne des Sternsystems, die wir weiter oben als Übergiganten bezeichnet haben. Die Wahrscheinlichkeit ist groß, daß sich aus ihnen die O-Sterne und planetarischen Nebel entwickeln. Wir haben in den Sternen des horizontalen Astes und den Haufenveränderlichen ein Stadium im Entwicklungsprozeß eines Sternes großer Masse zum planetarischen Nebel. Bei solchen Sternen — sie können Haufenveränderliche sein — kann also schon beim kritischen Punkt eine Art Instabilität eintreten, entsprechend der zunehmenden Helligkeit der Riesensterne des Sternsystems vom Typus A5 bis vor zu den O-Sternen. Der kritische Punkt liegt aber gerade da, wo die blauesten δ-Cepheisterne im Russell-Diagramm liegen. D. h. wir haben in den Haufenveränderlichen instabile Zwischenzustände zwischen δ-Cepheisternen und planetarischen Nebeln gefunden; Zwischenzustände nur in dem Sinne gemeint, daß dort, wo das Russell-Diagramm des Sternsystems eine Lücke zeigt, bei den Farben A0—F5, in den Kugelhaufen instabile Sterne vorkommen. Die äußerst geringe Streuung der Haufenveränderlichen nach der Helligkeit läßt darauf schließen, daß diese Art von Instabilität nur bei einer ganz bestimmten Sternmasse auftreten kann. Die Zustände in den Kugelhaufen scheinen solche zu sein, daß solche Massen verhältnismäßig häufig — übrigens ganz verschieden bei einzelnen Haufen — auftreten können.

Wir stellen uns vor, daß beim Altern eines Haufens allmählich ein Auflösungsprozeß stattfindet, eine Ansicht, die stark gestützt wird durch die Untersuchungen von Jeans. Es ist deshalb sehr wohl möglich, daß die offeneren Haufen aus typischen Kugelhaufen durch Auflösung hervorgegangen sind. Nun enthalten aber die offenen Sterngruppen keine oder doch nur in seltenen Fällen Haufenveränderliche. Was ist der Grund dieser Erscheinung? Von den wenigen kurzperiodischen galaktischen Cepheiden weiß man, daß sie sehr große Geschwindigkeiten besitzen. Es ist also durchaus wahrscheinlich, daß beim Auflösungsprozeß ein verhältnismäßig großer Prozentsatz der forteilenden Sterne aus Haufenveränderlichen besteht. Andererseits ist es auch denkbar, daß die Haufenveränderlichen irgendwie im Laufe der Entwicklung in den Ast der Zwerge einbiegen. Ob und wie dies geschieht, können uns nur die Russell-Diagramme von Sternhaufen lehren, die Zwischenglieder zwischen Kugelhaufen und offenen Haufen bilden. Von hier aus ist auch, wenn überhaupt, ein Übergang zu finden zwischen den O-Sternen und dem Ast der Zwerge. Es ist doch eigentlich ein mehr oder weniger kontinuierlicher Übergang von den Russell-Diagrammen der typischen Kugelhaufen zu denen der offenen Haufen, wie z. B. M. 11 oder M. 37, zu erwarten. Die dazwischenliegenden Diagramme müssen zeigen, wie der Ast der Übergiganten und damit der kritische Punkt verschwindet, wenn man zu immer offeneren, d. h. im allgemeinen älteren Gruppen übergeht; denn der kritische Punkt fehlt völlig in den Diagrammen der offenen Haufen.

Viel schwieriger scheint ein Verfolgen der langperiodischen δ-Cepheisterne zu sein und an Hand von Diagrammen von Sternhaufen kaum durchzuführen. Man kennt zur Zeit, soviel ich weiß, nur in ω-Centauri, 47 Tucanae und M. 15 einzelne langperiodische Cepheiden. Alle drei sind sehr reichhaltige Sterngruppen.

Aus dem Katalog der spektroskopischen Parallaxen in Mt. Wilson Contr. 199 habe ich in das Diagramm von M. 3 noch die langperiodischen δ-Cepheisterne als Ringe und vier kurzperiodische als Punkte in Ringen eingezeichnet. Als Parallaxe von M. 3 wurde wieder das 1,5fache der Shapleyschen gewählt, also 0$\overset{''}{.}$00011. Man sieht, daß sie fast ausnahmslos außerhalb des Gebietes der Sterne des Haufens liegen. Von den vier kurzperiodischen Cepheiden fallen drei in das Gebiet der Haufenveränderlichen. Aus dem Diagramm müssen wir schließen, daß die Cepheiden, wenn überhaupt, nur aus den hellsten roten Riesen des Haufens hervorgehen können, was nicht unmöglich erscheint; es ist jedoch unwahrscheinlich, da die langperiodischen Cepheiden im Durchschnitt absolut heller sind, also auch massiger als die hellsten Sterne von M. 3. Nehmen wir die Möglichkeit an, so müßten sich in den Haufen, die zwischen den typischen Kugelhaufen und den offenen Haufen stehen, langperiodische Cepheiden vorfinden. Die Annahme jedoch, daß Sterne mit so großer Masse, daß sie gleich am Anfang ihrer Entwicklung instabil werden und zu den δ-Cepheisternen übergehen, kaum in normalen Sternhaufen vorkommen können, scheint mir nicht unwahrscheinlich zu sein.

Die künftigen Untersuchungen werden sich deshalb vor allem in der Richtung zu erstrecken haben, die Lücke im Beobachtungsmaterial zwischen den typischen Kugelhaufen und den typischen offenen Haufen zu überbrücken. Vom Studium der Zwischenglieder kann eine Lösung der Hauptprobleme der Kosmogonie erwartet werden, nämlich, wie die Weiterentwicklung der Haufenveränderlichen, kurzperiodischen und langperiodischen Cepheiden vor sich geht.

Kugelnebel, Spiralnebel und Flächenhelligkeit.

Eine statistische Studie.

Von **Carl Wirtz**, Kiel.

Mit 1 Abbildung.

1. Aus Jeans' Kosmogonie. J. H. JEANS hat in seiner Kosmogonie[1]) ein Weltbild entwickelt, nicht nur geschlossen und logisch in sich aufgebaut, sondern in dessen dynamischen Folgerungen auch Wahrnehmungen beschrieben, die zu dem Bau der JEANSschen Entwicklungstheorie nicht unmittelbar den Anlaß gegeben haben. JEANS umfaßt in gleich packender Form die kleine und die große astronomische Welt, die Entstehung des nahen Planetensystems und die des fernen Weltalls. Die Natur des Gegenstandes bringt es mit sich, daß eine direkte Bestätigung aus den Beobachtungen für kosmogonische Theorien nicht zu erbringen ist. Nicht das zeitliche Nacheinander der Theorie schaut das menschliche Geschlecht, nur das gleichzeitige Nebeneinander der Erscheinungen am Himmel kann für die Kosmogonie herangezogen werden. Indes auch hier für verschiedene Arten von Objekten in verschiedenem Range.

Unter den vielerlei Möglichkeiten, die Ergebnisse von JEANS' Stellardynamik mit der Erfahrung zu vergleichen, sei eine Klasse von Körpern herausgegriffen, der alle Kosmogonien eine wichtige Rolle zuschreiben, die Klasse der Nebel. Auch JEANS bedient sich ihrer immer wieder in fesselnder Form, gleichsam zur Kontrolle der ganzen Rechnungen. Bei JEANS geschieht das z. B., indem die feineren Züge der Umrißlinien einzelner Nebelindividuen studiert werden, und es zeigt sich dann eindrucksvoll die Übereinstimmung mit der rechnerisch abgeleiteten Figur.

In anderer Weise wird hier ein Versuch gemacht, einiges zu der Frage beizutragen, inwieweit die Kosmogonie von JEANS ein Abbild des zeitlichen Verlaufs der Entwicklungsvorgänge der Materie sei. Nicht ein einzelnes Individuum ist Gegenstand der Betrachtung, sondern Gesamtheiten verschiedener Kategorien von flächenhaften Nebel-

---

[1]) JEANS, J. H.: Problems of cosmogony and stellar dynamics. Cambridge 1919.

objekten. Es kann dann in dem statistischen Verhalten verschiedener direkt beobachtbarer Charakteristiken nachgesehen werden, ob die Aufeinanderfolge und die Übergänge der Arten diejenigen sind, die die Theorie erfordert. Kein Zweifel, die Schlüsse aus der Statistik zwingen und überzeugen nicht. Und würde auf die Weise vielleicht ein Einwand gegen die Kosmogonie von JEANS gefunden, so braucht man ihn nicht schwer wiegen zu lassen. Aber solche Erwägungen gelten für jegliche kosmogonische Forschungsart, gelten für die gesamte Wissenschaft überhaupt. Es existieren weder zwingende Einwände noch überzeugende Beweise.

Die Theorie von JEANS kann hier auch nicht im Grundzug skizziert werden. Nur einige Bemerkungen mögen die statistischen Rechnungen des Zusammenhanges halber begleiten.

Als erste sichtbare Großformen der Materie erscheinen die Kugelnebel; zu ihnen zählen auch die schwach linsenförmigen strukturlosen Nebel mit jenem scharfen Rand, dessen feinerer Linienzug mit der JEANSschen Theorie aufs schönste harmoniert. Auf die Kugelnebel folgt das Stadium der Spiralnebel in den verschiedenen Windungsstufen. Mit dem Ablauf der Entwicklung geht eine erhebliche Ausbreitung der ursprünglichen Gasmasse einher, derart, daß das schließliche Sternsystem einen sehr viel größeren Raum einnimmt als der Kugelnebel, das Anfangsglied in der Kette der Erscheinungen.

2. Kugelnebel. Unter den Charakteristiken für flächenhafte Objekte steht die Flächenhelligkeit nicht zuletzt an Wert. In Lund Meddelanden, Ser. II, Nr. 29 (1923) (zitiert als LM) liegen die photometrischen Messungen der Flächenhelligkeiten von 566 Objekten vor; diese photometrischen Werte sollen zu einigen anderen Charakteristiken in statistische Beziehung gebracht werden.

Zuerst die Kugelnebel. Es wurden aus LM diejenigen Nebel ausgewählt, die nach den Beschreibungen von H. D. CURTIS[1] sicher Kugelnebel und zum Teil schon im Katalog in LM als solche bezeichnet sind, dazu noch 3 Objekte aus anderen Quellen. Zwei Kugelnebel fortgelassen, für die in LM die Totalhelligkeit gemessen wurde. Für alle Flächenobjekte ist die statistische Beziehung zwischen Flächenhelligkeit und scheinbarem Durchmesser von unzweifelhaftem Interesse, schon wegen des Gesetzes der Unabhängigkeit der Flächenhelligkeit von der Entfernung.

Ingesamt waren es 35 Kugelnebel, für die in LM photometrische Flächenhelligkeiten vorkamen. Als Durchmesser (Dm) liegt die scheinbare große Achse zugrunde, wie sie sich im Mittel aus dem visuellen und dem photographischen Wert ergibt. Der visuelle Dm steht im Katalog in LM, der photographische wurde den Angaben von CURTIS[2]

---

[1] Publ. Lick Obs. Bd. 13, Teil I. 1918.
[2] Publ. Lick Obs. Bd. 13, Teil I. 1918.

oder F. G. Pease[1]) entnommen. Die weitaus überwiegende Mehrzahl der 35 Kugelnebel zeigte natürlich keine ovale Gestalt, nur 4 Nebel haben verschiedene Achsen. Dm in Bogenminuten ausgedrückt.

Flächenhelligkeit Mg und log Dm.

Argum. Mg			Argum. log Dm		
Mg	log Dm	n	log Dm	Mg	n
$10^{m}32$	0.36	5	9.54	$11^{m}96$	4
10.79	0.10	8	9.70	11.57	8
11.23	9.88	5	9.93	11.50	8
11.61	9.83	7	0.10	11.05	7
11.87	9.83	4	0.27	10.69	3
12.13	9.66	5	0.45	10.39	4

Der hellste Nebel NGC 221 ($Mg = 8^{m}90$, $Dm = 1'7$) ist weggelassen; es handelt sich um den kleinen Begleiter des Andromedanebels (Abstand $= 0°4$), mit dem er die gleiche Flächenhelligkeit aufweist. Rechnet man aus den 35 Wertepaaren den Korrelationskoeffizienten (Kk) r aus, so kommt:

$$r = -0.720 \pm 0.081 \text{ (m. F.)}$$

mit den Regressionslinien:

$$Mg = 11^{m}14 - 1^{m}842 \cdot \log Dm$$

$$\log Dm = \quad 3.11 - 0.281 \cdot Mg$$

Mittlere ausgleichende Linie:

$$Mg = 11^{m}10 - 2^{m}70 \cdot \log Dm$$

Bleibt in diesen Rechnungen NGC 221 fort, so ändert sich r nur innerhalb des m. F.:

$$r = -0.763 \pm 0.072 \; (n = 34)$$

mittlere ausgleichende Linie:

$$Mg = 11^{m}17 - 2^{m}26 \cdot \log Dm$$

Man sieht, zu $^3/_4$ des Betrages wird die Flächenhelligkeit durch den lg Dm bestimmt; die mittlere ausgleichende Linie stellt am besten den Streuweg der einzelnen Punkte dar. Und die Abhängigkeit läuft in dem Sinne, daß die Flächenhelligkeit mit dem scheinbaren Durchmesser ganz erheblich wächst: nimmt der Durchmesser um das 10 fache zu, so wird Mg um mehr als 2^{m} heller. Jedenfalls zeigt die absolute Größe von r, daß eine starke statistische Bindung beider Merkmale, Mg und Dm, aneinander existiert. Siehe die Abbildung.

---

[1]) Contrib. Mt. Wilson Obs. Nr. 132, 186. 1917, 1920.

Ein weiterer Gang tritt in den Mg der Kugelnebel nach galaktischer Breite B zutage. Die dritte Spalte enthält den mittleren scheinbaren Dm, dessen Verlauf mit B die negative Korrelation zwischen Flächengrößenklasse und Durchmesser wieder bestätigt.

Mg und scheinbarer Durchmesser der Kugelnebel und galaktische Breite.

B	Mg	Dm	n
76°	10^{m}86	1'.65	8
67	11.06	1.22	5
56	11.39	1.02	9
44	11.44	0.79	8
25	11.81	0.75	4

D. h. mit Annäherung an die Milchstraße werden die Kugelnebel rasch schwächer (und kleiner), in dem vorliegenden Intervall fast linear mit der Breite B. In der Milchstraße selbst fehlen die Kugelnebel, gleich den Spiralnebeln und Kugelsternhaufen.

3. Spiralnebel. Nun läßt sich gegen die Ergebnisse einwenden, daß die Auswahl der Kugelnebel nicht nach festen Linien habe erfolgen können, trotzdem sie, wie es eine statistische Behandlung voraussetzt, dem Zufall unterworfen war. Um diese Möglichkeit einer einseitigen Verschiebung zu prüfen, wurde die gleiche Diskussion auf eine Klasse von Körpern angewandt, deren statistisches Verhalten als Kollektivgegenstand durch frühere Untersuchungen[1] aus umfangreichem Material gut bekannt war, auf die Spiralnebel. Aus derselben Quelle wie vorhin bei den Kugelnebeln (CURTIS l. c.) wurden die Nebel festgestellt, die nach dem Anblick der photographischen Platte unzweifelhafte Spiralnebel sind und für die zugleich photometrische Flächenhelligkeiten in LM vorliegen. Es sind auf die Weise Spiral- und

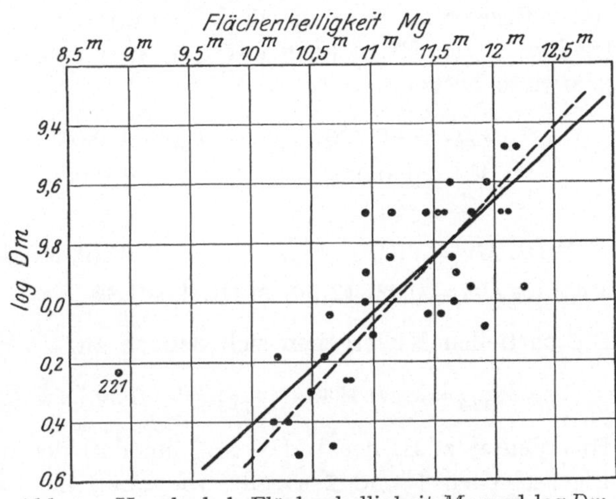

Abb. 1. Kugelnebel, Flächenhelligkeit Mg und log Dm.
Ausgleichende Linien { ———— mit NGC 221. — — — — ohne NGC 221. }

---

[1] LM, S. 25 ff.; „Aus der Statistik der Spiralnebel", Astr. Nachr. Bd. 221. 1924.

Kugelnebel aus den gleichen Beobachtungsreihen unter den gleichen Bedingungen herausgesucht. Das rechtfertigt die Erwartung, daß in beiden Fällen eine statistisch gleichartige Gesamtheit gewonnen worden ist.

Bei den Spiralnebeln kommt außer 1. Flächenhelligkeit Mg und 2. scheinbarem Durchmesser Dm als Charakteristik der Gestalt das 3. Achsenverhältnis A der großen zur kleinen Achse in Betracht, und diese drei Eigenschaften greifen in ihrer statistischen Abhängigkeit so ineinander, daß man auf die partiellen Kk übergehen muß, um die reinen Beziehungen je zweier Eigenschaften zueinander zu erhalten. Zur Herleitung der Kk wurden hier, der größeren Zahl der Spiralnebel halber, Verteilungstafeln angelegt, die in $\log Dm$ mit 0.2, in Mg mit $0^{m}\!.6$ und in A mit der Einheit fortschritten. Als Durchmesser wieder das Mittel der scheinb. gr. Achse aus visuellen und photographischen Beobachtungen, deren Ergebnisse schon in den Listen in LM enthalten sind. Bezeichnet man die drei Charakteristiken der Spiralnebel Mg, $\log Dm$, A der Reihe nach mit 1, 2, 3, so gehen die folgenden drei Kk paarweise hervor:

$$r_{12} = -0.120 \qquad r_{13} = +0.019 \qquad r_{23} = +0.529$$
$$\text{m. F.} \quad \pm 0.085 \qquad \pm 0.086 \qquad \pm 0.062$$
$$n = 134 \qquad\qquad 135 \qquad\qquad 133$$

Mittl. $Mg = 11^{m}\!.70$ $\qquad$ $11^{m}\!.68$ Mittl. $\log Dm = 0.432\,(2^{.}\!70)$

Mittl. $\log Dm = 0.430\,(2^{.}\!69)$ Mittl. $A = 3.39$ $\qquad$ 3.39

Die partiellen Kk ergeben sich danach zu:

$$r_{12.3} = -0.153 \qquad r_{13.2} = +0.097 \qquad r_{23.1} = +0.534$$

Hier deutet z. B. der Index 12.3 an, daß der betr. Kk zwischen den Eigenschaften 1 und 2 so gilt, wie wenn die Eigenschaft 3 konstant geblieben wäre; die Korrelation der beiden Größen 1 und 2 erscheint also gereinigt vom Einfluß der in die Beobachtungen hineinspielenden Größe 3. Analog verstehen sich 13.2 und 23.1.

Die starke Korrelation 2/3 (A/Gr. Achse) verdient nicht viel Aufmerksamkeit. Es handelt sich dabei in Wahrheit um eine V-förmige oder dreieckige Verteilung derart, daß Nebel mit kleinen Durchmessern (2) nie ein großes Achsenverhältnis A (3) besitzen. Eine Notwendigkeit, die aus der Wahrnehmungsmöglichkeit fließt; denn wenn kleine Nebel zu fadenförmigen Gebilden zusammenschrumpfen, werden sie nicht mehr entdeckt. Der gleiche Kk 2/3 hatte sich übrigens auch aus einem umfangreichen Spiralnebelmaterial in mehrfacher Diskussion ergeben[1].

---

[1] Astr. Nachr. Bd. 221. 1924, l. c.

Wichtiger sind die beiden ersten Kk 1/2 und 1/3, die ihren Ursprung in der räumlichen Anordnung und in der physikalischen Beschaffenheit des den Rechnungen zugrunde liegenden Kollektivgegenstandes suchen. Hier zeigt sich, die Korrelation Mg/A (1/3) ist verschwindend gering, sie ergibt den gleichen kleinen Betrag und auffallenderweise das gleiche Vorzeichen, wie sehr viel größere und ganz anders ausgeschnittene Abzählungen und Beobachtungen[1]). Das Vorzeichen $+$ sagt aus, daß die Flächenhelligkeit eines Spiralnebels abnimmt mit der Annäherung der Ebene der Spirale an die Gesichtslinie. Wären die Spiralnebel aber der Milchstraße koordinierte Sternsysteme von ähnlichem Aufbau, so müßte sich ein erheblicher Kk r_{13} mit negativem Vorzeichen ergeben. Vgl. dazu LM, S. 30.

Sehr klein, wenig den mittleren Fehler übersteigend, ist auch r_{12}. Wenn seine Realität nicht von vornherein verneint wird, so deshalb, weil dieser Kk immer wieder aus Reihen weit größeren Umfanges und verschiedener Art mit gleich kleinem Wert und mit demselben negativen Vorzeichen herauskam[2]). D. h. im statistischen Durchschnitt zeigen Flächenhelligkeit und Dm eine lockere Bindung in dem Sinne, daß der scheinbar größere Dm die Tendenz zu größerer Flächenhelligkeit hat. —

Ordnung der hellen Spiralnebel nach galaktischer Breite.

B	Mg	n	Dm	n
78°	$11^{m}.51$	30	4.28	31
65	11.73	32	3.27	32
55	11.81	29	3.29	30
42	11.70	25	2.91	24
22	11.72	15	3.16	15

Kein sonderlich ausgesprochener Verlauf, weder in Mg noch in Dm, der nicht dem für die Spiralnebel in früheren Untersuchungen gefundenen widerspricht. Dort fand sich: Maxima der Flächenhelligkeit an den Polen und bei der Milchstraße, Minima in mittleren Breiten[2]). Das hat man auch hier, wo die Helligkeitsunterschiede recht gering bleiben.

Die Prüfung der jetzt durch den Zufall des Vorkommens in zwei Verzeichnissen ausgewählten hellen Spiralnebel hat demnach gezeigt, daß sie in jeder statistisch erfaßbaren Hinsicht übereinstimmen mit der großen Masse der „kleinen Nebel", deren Verhalten als Kollektivgegenstand früher untersucht ist. Es besteht mithin die Annahme zu Recht, daß auch die durch genau denselben Zufall aus der Welt der

---

[1]) LM, S. 30; A. N. Bd. 221. 1924, l. c.
[2]) LM, S. 26; A. N. Bd. 221. 1924, l. c.

Kugelnebel ausgelesenen Objekte die statistischen Merkmale dieser besonderen Körpergruppe repräsentieren.

4. Statistischer Vergleich; kugelförmige Sternhaufen. Damit ist ein Vergleich zwischen Kugelnebeln und Spiralnebeln nach ihrem allgemeinen statistischen Verhalten möglich. Es zeigt sich, daß die Kugelnebel durchgreifend verschieden sind von den Spiralnebeln: bei den Kugelnebeln ein starker Gang der Flächenhelligkeit gleichsinnig mit dem scheinbaren Durchmesser, bei den Spiralnebeln nichts davon oder nur in sehr geringem Maße. Bei den Kugelnebeln ferner eine ausgesprochene Abnahme der Flächenhelligkeit mit der Annäherung an die Milchstraße, bei den Spiralnebeln nur eine lose und ganz andersartige Bindung zwischen Helligkeit und galaktischer Breite.

Dieses grundsätzlich verschiedene Gesamtverhalten der beiden Körperklassen spricht nicht dafür, daß die Spiralnebel die Fortsetzung einer mit den Kugelnebeln beginnenden Entwicklungsreihe sind.

Nimmt man an, daß die Kugelnebel sich im Verlauf der Entwicklung ausdehnen (JEANS) und daß die Entwicklungsstadien im Raum durchschnittlich gleichförmig gemischt sind, so erwartet man vielmehr, daß mit wachsendem scheinbaren Durchmesser eine Abnahme der Flächenhelligkeit eintritt. Eine Erscheinung, die man in unzweideutiger Klarheit bei den planetarischen Nebeln bemerkt[1]). Dort paßt dieses Phänomen durchaus zu den Vorgängen, die nach JEANS sich in einer rotierenden Gasmasse abspielen müssen.

Gibt es denn eine Körperklasse, der die Kugelnebel in ihren statistischen Eigenschaften als Kollektivgegenstand nahe stehen? Das sind merkwürdigerweise die kugelförmigen Sternhaufen. Bei ihnen begegnet man einem gleich hohen und gleichsinnigen Kk zwischen Flächenhelligkeit und $\log Dm$ wie bei den Kugelnebeln: $r = -0.639 \pm 0.070$ $(n = 71)$ [2]), und die Regressionslinien, l. c. nicht mitgeteilt, ergeben sich zu:

$$Mg = 10^{m}84 - 1^{m}304 \cdot \log Dm$$

$$\log Dm = \quad 3.76 - 0.313 \cdot Mg$$

Mittlere ausgleichende Linie:

$$Mg = 11^{m}42 - 2^{m}247 \cdot \log Dm$$

D. h. nahe Gleichheit, sogar in der Lage der mittleren ausgleichenden Linie, besonders wenn man bei den Kugelnebeln den erheblich aus dem Streuweg herausfallenden NGC 221 fortläßt. Auch in dem Zuge herrscht Übereinstimmung zwischen Kugelnebeln und Kugelhaufen, daß beide Gebilde mit der Annäherung an die Milchstraße an Helligkeit abnehmen, beide in ähnlich beträchtlichem Grade.

---

[1]) LM, S. 34; A. N. Bd. 219, S. 256. 1923.
[2]) A. N. Bd. 220, S. 294. 1924; ferner LM, S. 37.

Die Kugelnebel würden mithin nach ihren statistischen Merkmalen den kugelförmigen Sternhaufen zuzuweisen sein, einer Körperklasse, die in der kosmogonischen Theorie einem anderen Zweige, ja einem anderen Ende der Entwicklung angehört. Andererseits verlangt die Kosmogonie den Übergang der Kugelnebel in Spiralnebel, mit denen indes die Kugelnebel durch keinerlei statistische Eigenschaften verbunden sind.

In einem Punkte läßt sich Übereinstimmung zwischen statistischer Beobachtung und JEANS' Kosmogonie feststellen. Es darf angenommen werden, daß Kugelnebel und Spiralnebel gleichförmig gemischt im Weltraum angeordnet sind. Wird nun durch einen identischen zufälligen Prozeß eine Gesamtheit aus beiden Körperklassen ausgewählt, wie es in den vorliegenden Untersuchungen geschah, so werden die mittleren Abstände für Kugelnebel und Spiralnebel gleich sein. Eine Verschiedenheit der scheinb. Dm geht dann auf die linearen Dm zurück. Nun findet sich

für die 35 hellen Kugelnebel: geometr. Mittel, scheinb. $Dm = 0\overset{\prime}{.}91$

für die 134 hellen Spiralnebel: geometr. Mittel, scheinb. $Dm = 2\overset{\prime}{.}70$

D. h. die spätere Entwicklungsphase, der Spiralnebel, weist eine durchschnittlich dreimal größere Ausdehnung auf als das Urstadium, der Kugelnebel. Und JEANS' Theorie sagt in der Tat voraus, daß das Ausstoßen von Materie aus dem Äquator sich fast unbegrenzt fortsetzt, so daß die Dimensionen in der Äquatorebene mit der fortschreitenden Entwicklung des Nebels ständig wachsen müssen.

Wenn dann der durchschnittliche scheinb. Dm der kugelförmigen Sternhaufen sich wieder größer ergibt, als der der Spiralnebel, $4\overset{\prime}{.}11$ geometrisches Mittel (A. N. Bd. 220, S. 294. 1924), so hat das nichts mehr mit der hier verfolgten Linie der Entwicklung zu tun.

5. Eigenbewegungen von Spiralnebeln. Immer wieder — und ganz gewiß in jeder Kosmogonie — trifft man mit der Frage zusammen: Wo stehen im Weltraum die Spiralnebel? Ist es möglich, nach ihrer Anordnung und Ausdehnung, daß die Spiralen ferne Sternsysteme sind, dem Milchstraßensystem koordiniert? Da Parallaxen von Spiralnebeln direkt nicht bestimmt werden können, bieten den sichersten Anhalt die Eigenbewegungen, wenn deren Ermittelung einwandfrei gelänge. Mit hoher innerer Genauigkeit ist das — vorerst allerdings nur für 7 Objekte — durch die Arbeiten von A. VAN MAANEN[1]) am Mt. Wilson Observatory erreicht, und da nach einer Bemerkung des Autors „mit diesen Messungen von M 33 gegenwärtig die möglichen Untersuchungen über die inneren Bewegungen in Spiralnebeln zu einem Abschluß ge-

---

1) Contrib. Mt. Wilson Obs. Nr. 118, 213, 214, 242, 243, 255, 260. 1916, 1921, 1922, 1923.

langt sind", so kann man versuchen, das hochwertige Mt. Wilson-Material daraufhin nachzusehen, ob sich schon die Spur eines systematischen Anteils der Nebel-EB darin verrät. Der Grund dafür, daß bis auf weiteres neue Nebel-EB nicht abgeleitet werden können, liegt darin, daß frühere photographische Aufnahmen an den großen Reflektoren des Mt. Wilson- und Lick-Obervatory entweder nicht vorhanden sind, oder daß die älteren Bilder nicht befriedigen.

Die folgende Tabelle stellt aus den 7 Aufsätzen VAN MAANENS die jährlichen EB als Grundlage der Rechnung und einige weitere Daten für die 7 Spiralnebel zusammen.

EB und innere Bewegungen von 7 Spiralnebeln.

| NGC | Eigenbewegung | | Nebel-punkte | Vgl. ** | Zwischen-zeit | Innere Bewegung | | Scheinb. Dm |
	$\mu_\alpha \cos \delta$	μ_δ				μ_{Strom}	μ_{trans}	
598	$+ 0''\!.003$	$- 0''\!.004$	400	24	$12^{\text{a}}\!.1$	$+ 0''\!.020$	$- 0''\!.003$	$55' \times 40$
2403	$+ 0.002$	0.000	76	24	11^{a} u. 8^{a}	$+ 0.019$	$+ 0.007$	16×10
3031	$+ 0.014$	$- 0.005$	110	14	11.2	$+ 0.039$	$+ 0.007$	16×10
4736	$- 0.014$	0.000	32	15	9	$+ 0.021$	$+ 0.009$	5×3.5
5055	$+ 0.005$	$- 0.015$	98	21	12	$+ 0.019$	$+ 0.004$	8×3
5194	$+ 0.006$	$+ 0.001$	80	20	11.2	$+ 0.021$	$+ 0.003$	12×6
5457	$+ 0.005$	$- 0.013$	87	32	5^{a} u. 9^{a}	$+ 0.021$	0.000	16
						$+ 0.023$	$+ 0.004$	

Die scheinbaren Dm sind nach CURTIS (l. c.) angesetzt. Unter μ_{Strom} und μ_{trans} stehen die Komponenten der inneren Bewegung der Nebelknoten längs der Spiralarme und senkrecht dazu; das $+$-Zeichen bedeutet die Richtung vom Kern auswärts. Die Helligkeit der Vergleichsterne, im einzelnen nicht angegeben, bewegt sich zwischen $13^{\text{m}}\!.5$ und $17^{\text{m}}\!.7$, im Durchschnitt kann 16^{m} angenommen werden. Im ganzen sind wohl die Vergleichsterne etwas schwächer als die vermessenen Nebelpunkte. Die EB der Nebel verstehen sich relativ zu den Vergleichsternen. Nun sind deren EB zwar nicht bekannt; sie lassen sich aber auf Grund der Untersuchungen von P. J. VAN RHIJN[1]) abschätzen, und man findet dann mit kleiner Extrapolation um 2^{m}, daß für Sterne 16^{m} die zu erwartende EB schwerlich an $0''\!.002$ heranreicht, wahrscheinlich nicht über $0''\!.001$ hinausgeht.

Da stets viele Sterne benutzt sind, besteht nicht die Gefahr, daß nahe schwache Sterne unter die Vergleichsterne gerieten, die durch starke EB erkannt worden wären. Die gefundene EB der Nebel ist nun erheblich größer als die wahrscheinliche EB der Vergleichsterne.

6. Zielpunkt der Sonnenbewegung und Abstand für Spiralnebel, ein Versuch. Demnach hat auch die Frage einen Sinn: Auf welchen Zielpunkt der Sonnenbewegung führen diese durch hohe innere Ge-

---

[1]) Publ. astr. labor. Groningen, Nr. 34. 1923.

nauigkeit ausgezeichneten Nebel-EB? A, D seien die äquatorealen Koordinaten des Zielpunktes, q die Winkelbewegung des Sonnensystems (Milchstraßensystems), gesehen aus der mittleren Entfernung der 7 Spiralnebel. Trennt man zunächst die EB nach AR und Dekl, so folgt

$$\text{aus den } \mu_\alpha: \quad A = 28° \qquad\qquad q \cos D = 0''0064,$$
$$\text{aus den } \mu_\delta: \quad A = 36 \qquad D = +49° \qquad q = 0.0069$$

und aus der Vereinigung der EB in beiden Koordinaten:

$$A = 31° \qquad D = +38° \qquad q = 0.0071$$
$$\text{galaktisch} \quad L = 107 \qquad B = -21$$

Die Übereinstimmung der Lösungen aus den μ_α und den μ_δ in A und q — die Einsetzung von D liefert für die μ_α-Lösung $q = 0''0081$ — spricht dafür, daß diese Werte, trotz geringer Anzahl und ungünstiger Verteilung der Nebel, eine nicht ganz negative Beziehung zu den wirklichen Bewegungsvorgängen haben. Dafür kann man auch die Tatsache heranziehen, daß frühere Versuche mit sehr viel mehr (98) Nebel-EB, die aber als Einzelwerte nicht in die Realität sich erhoben, auf einen Punkt $A = 70°$, $D = +34°$ führten[1]), der etwa ebensoweit von dem jetzt erhaltenen abweicht, wie das bei den ersten Ableitungen des Apex der Sonnenbewegung aus den Bewegungen der Fixsterne der Fall war.

Noch mehr. Die Bearbeitung der sicher bestimmten und sehr großen Radialbewegungen der Spiralnebel[2]) liefert einen Zielpunkt der Sonnenbewegung ($A = 54°$, $D = +83°$), der immerhin noch in dasselbe sphärische Oktantendreieck fällt, wie die beiden andern aus den EB abgeleiteten Punkte. Auf der Erdkugel würden drei entsprechende Punkte z. B. etwa liegen in Spitzbergen, Portugal, Kleinasien. Während die Genauigkeit der ersten Bestimmungen des Vertex der beiden Sterntriften im irdischen Analogon ungefähr dem Areal der Sahara entsprach, lassen die ungenaueren und an Zahl weit geringeren Bewegungen der Spiralnebel noch die Streufläche innerhalb der Grenzen des europäischen Kontinentes zu. Nimmt man die Koordinaten aus Radialbewegungen und Transversal-EB als verschiedene Bestimmungen desselben Punktes der Sphäre, so kann man die mittlere Entfernung der Nebel abschätzen. In einer früheren Untersuchung der Radialbewegungen der Spiralnebel (l. c. A. N. Bd. 215) fand sich für die lineare Geschwindigkeit der Sonne (der Milchstraße) relativ zum Zentroid der Spiralnebel $V_o = 712 \text{ km}/1^s = 150.3$ astr. Einh./1^a. Vereinigt man diesen Wert mit dem aus den Mt. Wilson-Messungen berechneten q, so folgt als mittlere Parallaxe der 7 VAN MAANENSCHEN Nebel $\bar\pi = 0''000\,047$, entsprechend einem Abstand von 21 200 Par-

---

[1]) A. N. Bd. 204, S. 29. 1917.   [2]) A. N. Bd. 215, S. 351. 1922.

sec = 69 000 Lichtjahre. Das gilt für die hellen Nebel mit großen scheinbaren Durchmessern, für die nun lineare Durchmesser von der Größenordnung 200 Parsec = 600 Lichtjahre hervorgehen.

Ein Bedenken erhebt sich, das ebenfalls aus den Beobachtungen VAN MAANENS fließt. Hauptgegenstand der Mt. Wilson-Messungen war das Studium der inneren Bewegungen der Nebel, neben denen die *EB* der Nebel als eine allerdings überaus wertvolle Zugabe erscheinen. In die Tabelle oben sind auch diese scheinbaren inneren Bewegungen aufgenommen, und ein Blick darauf zeigt, daß sie für alle 7 Nebel vom gleichen Range sind, im Mittel $+ 0\rlap{.}''023/1^a$ längs der Spiralarme vom Nebelkern weggerichtet. Verglichen mit der eben abgeleiteten Entfernung ergibt sich daraus eine Geschwindigkeit von 2300 km/1^s, i. e. molekulare Geschwindigkeit. Geschwindigkeiten von dieser Ordnung in den Spektralarmen verlangt auch JEANS' kosmogonische Theorie. Aber es fällt auf, daß die scheinbare Stromgeschwindigkeit für Nebel der verschiedenen scheinbaren Durchmesser dieselbe ist. Man erwartet, daß die scheinbaren Geschwindigkeiten direkt ungefähr mit dem scheinbaren Durchmesser laufen. Nichts davon, und der eine nach oben herausfallende Wert, $+ 0\rlap{.}''039$ für NGC 3031, gehört zu einem Durchmesser mittlerer Größe. Die innere Genauigkeit der VAN MAANENSchen Messungen ist so hoch und durch die große Zahl der eingestellten Nebelknoten und Vergleichsterne so verstärkt, daß an der Tatsächlichkeit dieser Vorgänge im photographischen Bilde ein Zweifel nicht möglich ist.

Den kleinen EB VAN MAANENS gegenüber mag ein Hinweis auf die sehr viel größeren EB am Platze sein, die H. D. CURTIS[1]) aus zahlreicherem, der Entstehung nach mit VAN MAANEN gleichartigem Material ableitet. Für 66 große Spiralnebel ergab sich aus Platten des Crossleyreflektors (Licksternwarte) bei durchschnittlich 14 Jahren Zwischenzeit eine mittlere relative EB von $0\rlap{.}''033/1^a$; Helligkeit der je 5 bis 6 symmetrisch gelegenen Vergleichsterne 12^m bis 15^m. Besonders aufgeführt wird nur der Nebel NGC 253, der, gemessen durch 9 Nebelknoten auf Platten aus 1902, 1907 und 1915, die größte und ziemlich gut verbürgte EB zeigt: $\mu_\alpha \cos \delta = - 0\rlap{.}''072$, $\mu_\delta = + 0\rlap{.}''028$, $\mu = 0\rlap{.}''077$. Die übrigen 65 EB sind nicht explizite mitgeteilt. VAN MAANENS 7 Nebel haben dagegen eine mittlere EB von nur $0\rlap{.}''010$. Auffällig ist es wieder, daß für die verschiedenen Nebelkategorien die EB nach CURTIS von derselben Größenordnung sich ergeben:

> 10 große diffuse Nebel, mittlere jährl. EB . . . $0\rlap{.}''036$,
> 17 planetarische Nebel, mittlere jährl. EB . . . 0.028,
> 47 sehr kleine (Spiral-)Nebel, mittlere jährl. EB . 0.040,
> 66 große Spiralen, mittlere jährl. EB. 0.033,

---

[1]) Publ. astr. soc. Pac. Bd. 27, S. 214. Dez. 1915.

also Werte, wie sie auch für helle Sterne gelten. Man erwartet, daß zwischen der zweiten und dritten Nebelklasse eine völlige Änderung im Abstand der Objekte eintritt. Für die Platten vom 60 inch.-Reflektor des Mt. Wilson scheint diese Gleichförmigkeit der mittleren EB aller Nebelklassen nicht zu bestehen; denn, gegenüber der mittleren EB $0''{.}010$ für Spiralnebel, folgt für vier planetarische Nebel[1]) die mittlere EB zu $0''{.}037$.

Auf einem ganz anderen Wege gewinnt man eine merkwürdige Bestätigung für die Größenordnung des gefundenen Abstandes der Spiralnebel. Eine kurze Erwähnung des Ergebnisses mag genügen; denn vorläufig wird man darin noch nicht mehr sehen dürfen als eine beachtenswerte Zufälligkeit. Es handelt sich um die wichtige, der sicheren Beobachtung zugängliche Folgerung aus der Kosmologie DE SITTERS, daß ein Objekt einen Dopplereffekt zeigen muß, der von der Entfernung abhängt. Dadurch kann rückwärts die Entfernung bestimmt werden, wenn der Krümmungsradius R der DE SITTERschen Raum-Zeit-Welt bekannt ist. L. SILBERSTEIN[2]) hat eine Ableitung des Wertes R mit Hilfe von 7 Kugelsternhaufen versucht, für die der Abstand (nach SHAPLEY) und die Radialbewegung vorlagen; es sind Objekte mit großen Radialbewegungen ($\geqq 100$ km) ausgewählt, um den Einfluß der zufälligen Fehler (m. F. etwa $\pm$ 50 km) auf R zu verringern. SILBERSTEIN hatte schon früher gezeigt, daß sowohl negative als positive Bewegungen mit DE SITTERS Kosmologie verträglich sind; er erhält jetzt als Krümmungsradius der Welt $R = 6.0 \times 10^{12}$ astr. Einh. $= 95 \times 10^6$ Lichtjahre $= 29 \times 10^6$ Parsec.

Für 5 der 7 Spiralnebel VAN MAANENS hat weiter V. M. SLIPHER[3]) die Radialbewegungen gemessen, deren Mittel sich zu $|260|$ km ergibt. Damit findet man durch R als mittlere Entfernung dieser Spiralnebel 25 000 Parsec $= 82 000$ Lichtjahre (Parallaxe $= 0''{.}000\,040$). Das stimmt, so gut als man überhaupt erwarten kann, zu den aus den EB erschlossenen 21 200 Parsec. Dieser neue Abstand der Spiralnebel beruht also auf dem Beobachtungsdatum der Dopplerverschiebungen und auf dem durch die Abstände der Kugelsternhaufen und deren Radialbewegungen festgelegten Krümmungsradius der DE SITTERschen Welt. Es liegt nahe, aus der guten Übereinstimmung der beiden Entfernungswerte völlig verschiedener Herkunft für die Spiralnebel zu schließen, daß die SHAPLEYschen Distanzen der Kugelsternhaufen auch der Größenordnung, nicht nur der Reihung nach richtig sind, und daß DE SITTERS Kosmologie eine hinreichende Beschreibung des Weltraumes bildet.

---

[1]) MARSH, HANNAH M.: Pop. Astron. Bd. 32, S. 34. 1924.
[2]) Nature Bd. 113, S. 350. 1924.
[3]) EDDINGTON, A. S.: Math. theory of relativity, S. 162. Cambridge 1923.

Allerdings steckt in den beiden Ableitungen der Entfernungen der Spiralnebel ein gemeinsamer Teil, der im einen Falle explizite benutzt wird, im anderen nicht in die Erscheinung tritt. Es ist die Komponente in den Radialbewegungen, die sich als Geschwindigkeit der Sonne ($712\,\mathrm{km}/1^{\mathrm{s}}$) relativ zum System der Spiralnebel darstellte. Dennoch sind beide Abstandsschätzungen wesentlich unabhängig voneinander.

7. Vergleichung photometrischer Flächenhelligkeiten. Für stellarstatistische und astrophysikalische Untersuchungen macht die photometrische Flächenhelligkeit ausgedehnter Gebilde ein wichtiges Merkmal aus. Auch in den Abschnitten 2, 3, 4 dieses Aufsatzes wurde weitreichender Gebrauch davon gemacht. Lagen bisher nur wenige vereinzelte Messungen vor, so ist jetzt durch die Veröffentlichung in Lund Meddelanden, Serie II, Nr. 29 (1923) ein Katalog dieser Charakteristik für 566 Nebelflecken und Sternhaufen aufgestellt. Nun werden zwar auch schon im Text jener Publikation Vergleichungen mit früheren Messungen ausgeführt, doch vermögen die deshalb nicht viel auszusagen, weil die älteren Beobachtungen nicht die heute mögliche und wünschenswerte innere und systematische Genauigkeit besitzen. Von Bedeutung wäre es offenbar, wenn sich die umfangreiche Reihe LM durch eine moderne visuelle Beobachtungsreihe näher untersuchen ließe.

Eine wertvolle, wenn auch kleine, Reihe derartiger Messungen wird nun mitgeteilt von H. VOGT[1]), der am achtzölligen Refraktor in Heidelberg (Königstuhl) die Flächenhelligkeiten von 41 Nebeln und Sternhaufen bestimmte. Bei VOGT und LM kommen 26 vergleichbare gemeinsame Objekte vor, verteilt über 5^{m} Größenklassenintervall. Ein Gang der beiden Reihen VOGT (V) und LM mit dem (visuellen) scheinbaren Durchmesser braucht nicht angenommen zu werden, weil die Gruppenmittel gemäß der Streuung der Einzelwerte zufälligen Charakter tragen:

D i f f e r e n z (LM—V) u n d s c h e i n b a r e r Dm.

Dm	LM—V	n
0ʹ7	$+0^{\mathrm{m}}\!.14$	3
1.3	-0.03	5
2.2	-0.12	5
3.4	-0.15	5
5.4	$+0.10$	6
40	$+0.18$	2

Dagegen zeigen die beiderseitigen Größenklassen nicht ganz die gleiche Weite. Am bequemsten läßt sich die Beziehung mit Hilfe des Korrelationskoeffizienten ableiten, durch den zugleich auf einen Blick gezeigt

---

[1]) A. N. Bd. 221, S. 11. 1924.

wird, wie sich die innere Übereinstimmung beider Reihen stellt. Aus den 26 Wertepaaren erhält man für den Kk zwischen den Flächenhelligkeiten LM und V den Wert:

$$r = + 0.9862 \pm 0.0131$$

und weiter die Regressionslinien:

$$LM = \quad 1.48 + 0.8628 \cdot V,$$
$$V = - 1.37 + 1.1275 \cdot LM$$

Als beste Darstellung von LM durch V und zur Reduktion von V auf das System LM hat man die mittlere ausgleichende Linie:

$$LM = 10^m.09 + 0^m.875 \, (V - 10),$$

eine Linie, die sich am engsten dem schmalen Streuweg der nach den Koordinaten LM und V aufgetragenen Punkte anschließt. Die Skala V ist also enger als die Skala LM; denn auf dasselbe Helligkeitsintervall entfallen mehr Größenklassen nach V als nach LM; z. B. $5^m.00$ V $= 4^m.38$ LM.

Daß die Dehnung der beiden Größenklassensysteme gegeneinander in der Tat linear verläuft, lehrt schon die gruppenweise

Ordnung der Differenzen (LM—V)

nach Arg. LM			nach Arg. V		
LM	LM—V	n	V	LM—V	n
$9^m.10$	$+0^m.17$	5	$8^m.87$	$+0^m.29$	5
10.37	+0.06	5	10.37	−0.06	5
10.79	+0.01	5	10.78	+0.01	5
11.26	−0.03	5	11.35	−0.02	6
12.05	−0.19	6	12.35	−0.23	5

Überträgt man die Helligkeiten V durch die Formel auf das System LM und bildet den Unterschied der so reduzierten V gegen LM, so erhält man eine Streuung von $\pm 0^m.179$ und eine mittlere absolute Differenz von $0^m.140$ (Größenklassen System LM).

Wenn es in LM, S. 25, noch heißt: „Alle verglichenen Reihen bergen erhebliche Widersprüche in sich, die wohl nur zum Teil aus der natürlichen Schwierigkeit und der Unsicherheit der Auffassung von Nebelhelligkeiten entspringen", so trifft diese Bemerkung nicht mehr die Reihen V und LM, deren zufällige Abweichungen innerhalb „der natürlichen Schwierigkeit und der Unsicherheit der Auffassung von Nebelhelligkeiten" liegen.

Über die Beziehungen zwischen den verschiedenen Klassen der veränderlichen Sterne.

Von H. Ludendorff, Potsdam.

Die Erklärung des Lichtwechsels der veränderlichen Sterne ist eine Aufgabe, von deren Lösung wir noch sehr weit entfernt sind. Für keine einzige der verschiedenen Klassen dieser Himmelskörper ist es bisher möglich gewesen, die Ursache der Lichtschwankungen in befriedigender Weise anzugeben, außer für die Verfinsterungsveränderlichen, deren Helligkeit aber nur scheinbar veränderlich ist, und die daher gar nicht zu den veränderlichen Sternen im eigentlichen Sinne des Wortes zu rechnen sind; von ihnen wird im folgenden ganz abgesehen.

Die veränderlichen Sterne lassen sich in verschiedene Klassen einteilen, die sich für den weniger mit dem Material Vertrauten ziemlich scharf voneinander zu unterscheiden scheinen. Es sind die Klassen der langperiodischen Veränderlichen (Mira Ceti-Sterne), der kurzperiodischen (δ Cephei-Sterne) und der unregelmäßigen Veränderlichen. Dazu kommt noch die Klasse der neuen Sterne, die man mit zu den Veränderlichen rechnen muß.

Dem heutigen Stande der Wissenschaft genügt diese rohe Einteilung nicht mehr. Schon seit geraumer Zeit ist man zu der Erkenntnis gekommen, daß die Klasse der unregelmäßigen Veränderlichen Objekte so grundverschiedener Art umfaßt, daß man diese unmöglich zu derselben Klasse rechnen darf, oder daß dies wenigstens nur aus rein formalen Gründen statthaft ist. Es sind daher verschiedene Versuche gemacht worden, die Klasse der unregelmäßigen Veränderlichen in verschiedene neue Klassen zu zerlegen. Den gegenwärtigen Kenntnissen entspricht die Einteilung, die ich in der 6. und 7. Auflage von NEWCOMB-ENGELMANNS Populärer Astronomie[1]) gegeben habe. Sobald man eine Klasseneinteilung der Veränderlichen vorgenommen hat, wird es die nächste Aufgabe sein, zu untersuchen, ob und was für Beziehungen zwischen den verschiedenen Klassen bestehen. Das ist nicht ganz einfach, da wir, wie gesagt, die Ursache des Lichtwechsels in keinem Falle wirklich kennen. Die folgenden Zeilen haben den Zweck, eine

---

[1]) Leipzig: W. Engelmann, 1921 und 1922.

solche Untersuchung zu liefern, wobei alle Klassen der Veränderlichen in den Kreis der Betrachtung gezogen werden sollen. Eine derartige Gesamtdiskussion liegt meines Wissens bisher noch nicht vor. Daß ich sie schon jetzt unternehme, obwohl meine seit 1919 im Gange befindlichen allgemeinen Untersuchungen[1] über die veränderlichen Sterne noch keineswegs abgeschlossen sind, mag die festliche Gelegenheit entschuldigen. Ich habe zudem den Eindruck, daß die Fortführung dieser Untersuchungen wesentlich neue Gesichtspunkte für das hier behandelte Thema nicht ergeben wird.

Wir gehen bei unseren Betrachtungen aus von der Klasse der Mira Ceti-Sterne, also den Sternen mit einem meist mehrere Größenklassen umfassenden, einigermaßen regelmäßigen Lichtwechsel, der in Perioden von mehreren Monaten bis zu etwa 600^d vor sich geht. Bei den Lichtkurven dieser Sterne kann man mehrere Typen unterscheiden, die aber nicht scharf voneinander verschieden, sondern durch Übergangsformen miteinander verbunden sind. Ich definiere diese Typen wie folgt (siehe III, IV und V):

    a: eine unsymmetrische Lichtkurve, Anstieg deutlich steiler als Abstieg, keine konstante Phase im Minimum;

(a): ebenso, aber mit nahezu oder völlig konstanter Phase von beträchtlicher Dauer im Minimum und meist sehr steilem Helligkeitsanstieg;

    b: eine Lichtkurve, die einer Sinuskurve oder einer Zykloide ähnlich ist (keine wesentliche Unsymmetrie);

    c: eine Lichtkurve, die auch symmetrisch ist, bei der der Stern aber im Maximum längere Zeit völlig oder mit großer Annäherung konstant bleibt, während die Minima besser definiert sind;

    d: eine Lichtkurve, bei der das breite Maximum durch ein sekundäres Minimum geteilt wird, so daß sie Ähnlichkeit mit der von β Lyrae hat;

(d): eine Lichtkurve mit deutlichem Stillstand oder Buckel im Anstieg.

Wie schon erwähnt, läßt sich eine scharfe Grenze zwischen den verschiedenen Formen von Lichtkurven nicht ziehen; Übergangsformen werden mit a—b usw. bezeichnet. Nur sehr wenige Lichtkurven von Mira-Sternen passen nicht in das obige Schema.

Natürlich würde es exakter sein, die Formen der Lichtkurven irgendwie zahlenmäßig darzustellen; bei der oben geschilderten, roheren

---

[1] Bisher sind veröffentlicht: Über die mit R Coronae borealis verwandten veränderlichen Sterne (A. N. Bd. 209, S. 273. 1919); Untersuchungen über veränderliche Sterne I (A. N. Bd. 214, S. 77. 1921); II (A. N. Bd. 214, S. 217. 1921); III (A. N. Bd. 217, S. 161. 1922); IV (A. N. Bd. 219, S. 1. 1923); V (A. N. Bd. 220, S. 145. 1924); VI (A. N. Bd. 220, S. 241. 1924); diese 7 Arbeiten werden im folgenden mit A und I bis VI zitiert.

Klassifizierung hat man aber den Vorteil, daß man die Lichtkurven vieler Sterne bereits einordnen kann, für die eine zahlenmäßige Darstellung noch nicht gut möglich wäre, da die vorhandenen Beobachtungen dazu nicht ausreichen.

Die Spektra der Mira Ceti-Sterne gehören den Klassen Md (Absorptionsspektrum M mit darüber gelagerten Emissionslinien, hauptsächlich des Wasserstoffs, die gegen die Absorptionslinien etwas nach Violett verschoben sind), M und N, sowie in einigen Fällen den Klassen S, K und R an. Man hat merkwürdigerweise bis vor kurzem bei den übrigens nur sehr spärlich vorliegenden allgemeinen Untersuchungen über die Mira-Sterne auf die Verschiedenheiten der Spektralklassen gar keine Rücksicht genommen, also z. B. die Mira-Sterne der Spektralklasse N ruhig zusammen mit denen der Spektralklasse Md behandelt. Ich habe aber in III und IV nachgewiesen, daß die Eigenschaften des Lichtwechsels je nach der Spektralklasse beträchtliche Unterschiede aufweisen; insbesondere sind die Mira-Sterne der Klasse N stark von denen der Klasse Md verschieden. Die wichtigsten dieser Unterschiede sind folgende:

Die N-Sterne haben durchschnittlich längere Perioden als die Md-Sterne und systematisch kleinere Helligkeitsschwankungen als die Md-Sterne gleicher Periode. Unter den Lichtkurven der N-Sterne sind die symmetrisch gestalteten weit häufiger als bei den Md-Sternen gleicher Periode. (Bei den Md-Sternen hat sich, worauf ich hier nicht weiter eingehe, eine starke Korrelation zwischen Form der Lichtkurve und Periodenlänge gezeigt. Man darf also bei Vergleichungen der Lichtkurven von Sternen verschiedener Spektralklassen immer nur Sterne gleicher Perioden vergleichen.)

Nach dem Gesagten muß man die Mira-Sterne der Spektralklasse N, obwohl sie ohne Zweifel mit denen der Spektralklasse Md aufs engste verwandt sind, als besondere Unterklasse betrachten.

Die Mira-Sterne der Spektralklassen K, Ma und Mb (ohne helle Linien) haben durchschnittlich viel kürzere Perioden als die Md-Sterne, wenn man von W Cassiopeiae (mit abnormem Spektrum) und denjenigen unter den ersteren Sternen absieht, welche Lichtkurven der Form d besitzen. Die Mc-Sterne, wieder mit Ausschluß derjenigen mit d-Lichtkurven, stehen in ihrer Periodenlänge weit weniger hinter den Md-Sternen zurück. Die Helligkeitsamplituden der langperiodischen K-, Ma-, Mb- und Mc-Sterne sind durchschnittlich kleiner als die der Md-Sterne gleicher Periode. Was die Form der Lichtkurven angeht, so scheinen bei den K- und M-Sternen von weniger als 200^{d} Periode die Helligkeitsanstiege durchschnittlich etwas steiler zu sein als bei den Md-Sternen gleicher Periode, ferner kommen unter den M-Sternen ohne helle Linien Lichtkurven mit breitem Maximum (c, b—c, c—d) und solche der Form d relativ weit häufiger vor als bei den Md-Sternen.

Im ganzen sind aber die langperiodischen K- und M-Sterne von den Md-Sternen jedenfalls weniger verschieden als die N-Sterne, und man wird sie daher kaum als besondere Unterklasse, sondern nur als Grenzfälle zu betrachten haben, um so mehr, als bei einigen jener K- und M-Sterne zuweilen Emissionslinien auftreten (z. B. bei SX Herculis; RT Hydrae ist im Draper-Katalog als Mc bezeichnet, während MERRILL[1]) diesen Stern unter den von ihm beobachteten Md-Sternen hat). Auf die wenigen Mira-Sterne der Klassen S und R will ich hier nicht weiter eingehen.

Besonderes Interesse verdienen die Mira-Sterne mit Lichtkurven der Formen d und (a), da sie den Übergang zu gewissen Klassen der unregelmäßigen Veränderlichen zu bilden scheinen.

An einer schon zitierten Stelle habe ich die folgende Einteilung der unregelmäßigen Veränderlichen vorgeschlagen, die alle diese Sterne mit ganz wenigen Ausnahmen umfaßt:

1. Gewöhnliche, rote unregelmäßige Veränderliche (μ Cephei-Sterne).
2. RV Tauri-Sterne.
3. U Geminorum-Sterne.
4. R Coronae-Sterne.
5. Nova-ähnliche Veränderliche.

Daß die Mira-Sterne mit den μ Cephei-Sternen eng verwandt sind, wird wohl nicht bezweifelt werden. Unter den Md-Sternen gibt es zwar, abgesehen von einigen RV Tauri-Sternen (siehe unten), nur einen, nämlich S Persei, von dem mit Sicherheit behauptet werden kann, daß er unregelmäßig veränderlich ist. Aber die Veränderlichen der Spektralklasse M (ohne helle Linien) gehören z. T. zu den Mira-, z. T. zu den μ Cephei-Sternen, und der Unterschied scheint nur graduell zu sein. Z. B. bildet W Persei (Mc) einen Übergang. Die Periode ist bei diesem Stern[2]) im Mittel 496^{d}, die Einzelwerte schwanken aber zwischen 400^{d} und 600^{d}, die Form der Lichtkurve ist stark veränderlich. Im allgemeinen scheint die Sache so zu liegen, daß M-Sterne mit größeren Amplituden eine regelmäßige Periode haben, solche mit kleinen Amplituden dagegen meist oder mindestens sehr häufig unregelmäßig sind; ganz ähnlich ist es mit den veränderlichen N-Sternen. Leider sind die roten unregelmäßigen Veränderlichen meist nur wenig beobachtet worden, und die Beobachtungen sind im Verhältnis zu der kleinen Helligkeitsamplitude zu ungenau, so daß wir erst sehr wenig über den Lichtwechsel dieser Sterne wissen. Es ist keineswegs ausgeschlossen, daß wir bei ihnen noch Unterklassen unterscheiden müssen.

Die RV Tauri-Sterne, die zweite Klasse der unregelmäßigen Veränderlichen, sind dadurch gekennzeichnet, daß zwischen zwei Hauptminima in der Regel ein sekundäres Minimum liegt. Die Lichtkurve

---

1) Astrophysical Journal Bd. 58, S. 215. 1923.
2) Harvard Obs. Bull. Nr. 784.

ist aber sehr veränderlich, bisweilen bleibt ein Minimum aus, bisweilen sind die Nebenminima ebenso tief wie die Hauptminima, bisweilen vertauschen sich Haupt- und Nebenminima. Die Lichtkurve gleicht zeitweise dem β Lyrae-, zeitweise dem δ Cephei- oder ζ Geminorum-Typus, ja in einigen Fällen kommen auch Algol-ähnliche Lichtkurven vor. Es ist gebräuchlich, die übrigens stark veränderliche Periode von Hauptminimum zu Hauptminimum zu rechnen.

In II habe ich die RV Tauri-Sterne einer zusammenfassenden Diskussion unterworfen. Unter Fortlassung einiger zweifelhafter Fälle kann man folgende Sterne in diese Klasse rechnen (es bedeutet P die Periode, A die Gesamthelligkeitsamplitude, β die galaktische Breite):

Stern	P	A	β	Spektrum
R Sagittae	70^{d}	$1\overset{\mathrm{m}}{.}8$	$11°$	c G 0
V Vulpeculae . .	75	0.7	9	c G 5 p
RV Tauri	78	2.5	12	?
U Monocerotis . .	92	1.5	5	c G 0 p
TV Androm. . . .	127	1.7	16	?
R Scuti	140	4.5	3	Pec
RV Androm. . .	171	2.7	11	Md ?
RS Camelop. . .	(190)	0.6	34	Mb
Z Ursae maj. . .	198	1.9	58	Md
W Cygni	259	1.1	6	Mc
BM Scorpii . . .	?	0.9	2	K

Die Zugehörigkeit von RV Andromedae und W Cygni zur RV Tauri-Klasse ist noch nicht sichergestellt. In II ist Z Ursae majoris noch nicht zu dieser Klasse gerechnet, nach einer schönen Beobachtungsreihe von Brun[1] muß dies aber ohne Zweifel geschehen (Brun nimmt $P = 99^{\mathrm{d}}$ an, aber bisweilen bleibt ein Minimum aus oder ist nur schwach angedeutet; wenn man den Stern als RV Tauri-Stern betrachtet, muß man daher für P den doppelten Betrag ansetzen).

Über die Spektra ist noch folgendes zu bemerken: Die von R Sagittae, V Vulpeculae und U Monocerotis sind auf dem Mt Wilson-Observatorium photographiert worden[2]; die beiden letztgenannten Sterne haben auffallend schwache Wasserstoffabsorptionslinien, „a condition almost certainly due to the partial balancing of absorption and emission in these lines". Die Radialgeschwindigkeiten aller drei Sterne sind veränderlich. Das Spektrum von R Scuti variiert nach. Joy[3] zwischen G 5 e im Maximum der Helligkeit und K 2 e p (K 2 mit hellen Linien des Wasserstoffs und Absorptionsbanden des Titan-Oxyds, also eng mit Md verwandt) im Minimum. Die Radialgeschwindigkeit ist variabel und

---

[1] Bull. de l'Obs. de Lyon Bd. 6, S. 24. 1924.
[2] Annual Report of the Director of the Mount Wilson Obs. 1922, S. 234.
[3] Publ. of the Astr. Soc. of the Pacific 1922, S. 349.

befolgt nach R. H. Curtiss[1]) eine veränderliche Periode von 65^d bis 75^d (also offenbar gleich der Hälfte der in der Tabelle angegebenen mittleren Periode des Lichtwechsels, die, wie erwähnt, von Hauptminimum zu Hauptminimum gerechnet ist). Die Wasserstoff-Emissionslinien sind gegen die Absorptionslinien nach Violett verschoben, so wie dies bekanntlich auch bei den Md-Sternen der Fall ist.

Die obige Tabelle zeigt nun, daß die Spektra der RV Tauri-Sterne eine deutliche Abhängigkeit von der Periodenlänge besitzen, und zwar haben die Sterne mit langen Perioden Spektra, die mit denen der Mira-Sterne eng verwandt oder identisch sind. Die mit kurzen Perioden liegen nahe der Milchstraße, für die mit längeren trifft dies, wie bei den Mira-Sternen, nicht mehr durchweg zu.

Die RV Tauri-Sterne sind hiernach unzweifelhaft mit den Mira-Sternen verwandt, und zwar, der Form der Lichtkurve nach zu urteilen, wohl gerade mit denjenigen unter diesen, die Lichtkurven der Form d oder diesen verwandte besitzen. Als Beispiele für solche führe ich an R Normae ($P = 484^d$, Spektrum Mb, Kurve d), R Doradus (345^d, Mc, d), R Ursae min. (332^d, Mc, c—d), RT Hydrae (Mc, 255^d, c—d), R Centauri (568^d, Md, d). Wie ähnlich die Lichtkurven dieser Mira-Sterne und die der RV Tauri-Sterne zeitweise sind, geht z. B. daraus hervor, daß ich in II Z Ursae majoris auf Grund einer noch unvollständigen Kenntnis des Lichtwechsels zu den Mira-Sternen gerechnet habe. Der Hauptunterschied ist der, daß die Mira-Sterne der genannten Art einen weit regelmäßigeren Lichtwechsel als die RV Tauri-Sterne und auch längere Perioden haben. In der Periodenlänge scheinen sie sich an die RV Tauri-Sterne im Sinne wachsender Perioden anzuschließen.

Wir wenden uns nun zu der nächsten Klasse der unregelmäßigen Veränderlichen, den U Geminorum-Sternen. Bei diesen wächst die Helligkeit aus einem Minimum in unregelmäßigen Intervallen meist sehr rasch um mehrere Größenklassen an, um dann langsamer wieder herabzusinken. Das Verhalten dieser Sterne im Minimum ist nicht in allen Fällen bekannt, doch ist bei SS Cygni, U Geminorum und RU Pegasi die Helligkeit im Minimum nur wenig veränderlich. Die Sterne, die zu dieser Klasse gehören, sind wenig zahlreich. Es sind folgende:

SS Aurigae	BI Orionis
SS Cygni	RU Pegasi
U Geminorum	UV Persei
X Leonis	TW Virginis

Nach den Angaben in der „Geschichte und Literatur der veränderlichen Sterne" ist das Intervall zwischen aufeinanderfolgenden Aufhellungen:

---

[1]) Popular Astronomy 1924, S. 220.

bei SS Aurigae . . . 40^d bis 103^d,
„ SS Cygni . . . . 24^d „ 80^d (durchschnittlich 51^d),
„ U Geminorum . . 62^d „ 152^d,
„ BI Orionis . . . . 19 „ 26^d,
„ RU Pegasi . . . 60 „ 90^d,
„ TW Virginis . . . 23^d und größer,

und ferner nach anderweitigen Angaben:

bei X Leonis . . . . durchschnittlich 16^d,
„ UV Persei . . . . etwa 200^d und größer.

Eine Bevorzugung geringer galaktischer Breiten ist bei diesen Sternen kaum festzustellen.

Über die Spektra der U Geminorum-Sterne ist infolge ihrer Lichtschwäche wenig bekannt. SS Cygni und SS Aurigae haben im Maximum nahezu kontinuierliche Spektra mit schwachen, breiten Absorptionslinien des H und He. Im Minimum zeigt SS Cygni starke, sehr breite Emissionslinien dieser beiden Elemente[1]). Es liegt offenbar eine Ähnlichkeit mit dem Spektrum der neuen Sterne vor.

Aus dem Gesagten geht hervor, daß die Lichtkurven der U Geminorum-Sterne den Lichtkurven der Form (a) bei den Mira-Sternen ähnlich sind, nur treten bei letzteren die Aufhellungen regelmäßig ein. Es ist aber zu erwähnen, daß auch bei den Md-Sternen mit (a)-Kurven gelegentlich starke Unregelmäßigkeiten auftreten. So scheinen bei Z Tauri die Maxima manchmal auszubleiben oder nur schwach angedeutet zu sein. Bei RW Lyrae erheben sich die Maxima manchmal bis fast zur 9. Größe, manchmal aber auch nur bis zur 13., d. h. nur etwa $1-1^1/_2$ Größenklassen über die Minimalhelligkeit. Besonders die Ähnlichkeit der Lichtkurven einiger dieser Md-Sterne mit der von SS Cygni ist verblüffend, nur der zeitliche Maßstab der Kurven ist bei SS Cygni ein anderer. In der Regel ist allerdings die Dauer des konstanten Minimums relativ zur Dauer der Erhebung über dieses bei SS Cygni größer als bei den Md-Sternen; immerhin dauert bei V Camelopardalis ($P = 511^d$) der konstante Teil des Minimums mitunter 300^d und mehr, bei Z Tauri ($P = 500^d$) etwa 320^d, und es kommen andererseits bei SS Cygni Fälle vor, wo die Dauer der Minima ziemlich kurz ist.

Der Md-Stern[2]) kürzester Periode, der eine Lichtkurve der Form (a) hat, ist wohl Z Cygni mit $P = 263^d$. Der Periodenlänge nach schließen sich also die Md-Sterne mit (a)-Kurven an die U Geminorum-Sterne schön an, da bei UV Persei das kürzeste beobachtete Intervall zwischen

---

[1]) Annual Report of the Director of the Mount Wilson Obs. 1922, S. 234, und Popular Astronomy 1922, S. 103. Vgl. auch Harvard Annals Bd. 56, S. 210 bis 211.

[2]) Bei Mira-Sternen der übrigen Spektralklassen kommen Lichtkurven der Form (a) augenscheinlich nicht vor.

zwei Aufhellungen etwa 200^d beträgt (die Maxima bei UV Persei sind allerdings viel spitzer als bei den Md-Sternen mit (a)-Kurven). In ganz ähnlicher Weise schließen sich, wie wir sahen, die Mira-Sterne mit d-Kurven an die RV Tauri-Sterne an. Nebenbei sei hier erwähnt, daß auch der merkwürdige, von einem Gasnebel umgebene Md-Stern R Aquarii eine Lichtkurve der Form (a) hat.

Nach allem kann man sich des Eindrucks nicht erwehren, daß vielleicht eine Verwandtschaft zwischen den Md-Sternen mit Lichtkurven der Form (a) und den U-Geminorum-Sternen vorliegt. Dagegen spricht freilich der Umstand, daß SS Cygni und U Geminorum Spektra haben, die von denen der Mira-Sterne grundverschieden sind; aber vielleicht ist dies nicht ausschlaggebend, da bei den RV Tauri-Sternen die einzelnen Objekte sehr verschiedene Spektra haben, deren Beschaffenheit von der Periodenlänge abhängt. Ähnliches könnte auch bei den U Geminorum-Sternen der Fall sein.

Zwischen den Mira-Sternen und den beiden letzten Klassen der unregelmäßigen Veränderlichen, nämlich den R Coronae-Sternen und den Nova-ähnlichen Veränderlichen, scheint eine direkte Verwandtschaft nicht zu bestehen.

Sehr schwierig ist nun die Frage einer etwaigen Verwandtschaft zwischen den Mira- und den δ Cephei-Sternen. Unter den letzteren können wir deutlich zwei Gruppen unterscheiden, je nachdem die Periode mehr oder weniger als etwa 1^d beträgt. Wir wollen diese beiden Gruppen als lang- und als kurzperiodische δ Cephei-Sterne bezeichnen. Daß diese beiden Gruppen eng miteinander verwandt sind, ist nicht zu bezweifeln und wird allgemein angenommen. In der Tat sind die Beziehungen zwischen den Änderungen der Radialgeschwindigkeit und denen der Helligkeit bei beiden Gruppen die gleichen, ebenso ähneln sich die Lichtkurven, wenngleich Lichtkurven mit lang ausgedehntem, konstantem Minimum [Form (a)] nur bei den kurzperiodischen δ Cephei-Sternen vorkommen. Grundverschieden sind andererseits die galaktische Verteilung (sehr starke Bevorzugung der Milchstraße bei den langperiodischen, Verteilung über den ganzen Himmel bei den kurzperiodischen), die Radialgeschwindigkeiten (kleine bei ersteren, sehr große bei letzteren), in geringerem Grade auch die Spektren (die langperiodischen weisen durchschnittlich „spätere" Spektraltypen auf als die kurzperiodischen). Trotzdem muß an der Annahme der Verwandtschaft festgehalten werden, für die auch der Verlauf von SHAPLEYs bekannter „Period-Luminosity-Curve" spricht.

Die Mira-Sterne können nun natürlich nur mit den langperiodischen δ Cephei-Sternen direkt verwandt sein. *Gegen* eine solche Verwandtschaft sprechen folgende Gründe:

1. Die verschiedene galaktische Verteilung (die Mira-Sterne sind, abgesehen von den der Spektralklasse N angehörigen, über den ganzen

Himmel verteilt, die langperiodischen δ Cephei-Sterne zeigen eine sehr starke galaktische Konzentration).

2. Die von der Sonnenbewegung befreiten Radialgeschwindigkeiten der Mira-Sterne sind durchschnittlich ziemlich groß[1]), die der langperiodischen δ Cephei-Sterne dagegen nur klein.

3. Die Mira-Sterne gehören vorwiegend den Spektralklassen M und N, die langperiodischen δ Cephei-Sterne den Klassen F, G und K an.

4. Die Radialgeschwindigkeiten verhalten sich bei beiden Klassen von Sternen zum Lichtwechsel ganz verschieden. Die Radialgeschwindigkeiten der δ Cephei-Sterne sind stark veränderlich, das Maximum bzw. Minimum der Helligkeit fällt ungefähr mit dem negativen bzw. positiven Maximum der Radialgeschwindigkeit zusammen. Bei den Mira-Sternen ist die aus den Verschiebungen der Absorptionslinien folgende Radialgeschwindigkeit gar nicht oder nur gering variabel, und bei Mira Ceti selbst ist das Verhalten dieser Radialgeschwindigkeiten zur Lichtkurve gerade umgekehrt wie bei den δ Cephei-Sternen[2]). Die Emissionslinien der Mira-Sterne scheinen allerdings nach MERRILL bei einigen Sternen, ganz wie die Absorptionslinien der δ Cephei-Sterne, sehr bald nach dem Maximum der Helligkeit ein Maximum der Annäherungsgeschwindigkeit zu ergeben. Bei Mira Ceti selbst ist die Sachlage nach JOY komplizierter, doch ergeben bei diesem Stern die hellen Linien im Minimum ein positives Maximum der Radialgeschwindigkeit.

Von diesen Gründen, die eine Verwandtschaft zwischen den Mira- und δ Cephei-Sternen zunächst unwahrscheinlich machen, sind aber 1. bis 3. wenig stichhaltig, da ganz ähnliche Unterschiede auch zwischen den lang- und den kurzperiodischen δ Cephei-Sternen bestehen.

Die Verschiedenheit der Spektra könnte sogar vielleicht als Grund *für* die Verwandtschaft angeführt werden. Ordnet man nämlich die kurzperiodischen und die langperiodischen δ Cephei- sowie die Mira-Sterne nach ihrer Periodenlänge in eine fortlaufende Reihe und bildet (unter Auslassung der N-Sterne) die mittleren Spektren für gewisse Periodenintervalle, so erhält man für die Spektra eine stetig fortlaufende Reihe von etwa A2 bis M, und auch innerhalb der Klasse der Mira-Sterne werden die Sterne noch immer röter, je länger die Periode ist.

Stärker als 1. bis 3. spricht 4. gegen die Verwandtschaft; aber auch dieser Gegengrund wird wenigstens etwas abgeschwächt durch das oben geschilderte Verhalten der Emissionslinien bei einigen der Mira-Sterne.

Für die Verwandtschaft der δ Cephei- und der Mira-Sterne sprechen folgende Umstände:

1. Mira-Sterne und δ Cephei-Sterne sind Riesensterne; im Minimum der Helligkeit ist ihre Temperatur niedriger als im Maximum, wie aus Änderungen im Spektrum hervorgeht.

---

[1]) Vgl. MERRILL in Astrophysical Journal Bd. 58, S. 215. 1923.
[2]) Vgl. JOY in Popular Astronomy 1923, S. 645.

2. Die Lichtkurven beider Klassen haben ihrer Gestalt nach in vielen Fällen die größte Ähnlichkeit.

3. Bei den Mira-Sternen mit verhältnismäßig kurzen Perioden ist die Helligkeitsamplitude durchschnittlich kleiner als bei denen mit längeren Perioden; die ersteren nähern sich dadurch den δ Cephei-Sternen, die kleine Helligkeitsamplituden besitzen.

4. Wenn auch (vgl. I) Sterne mit Perioden zwischen 45^d (der ungefähren oberen Periodengrenze für die unzweifelhaft der δ Cephei-Klasse angehörigen Sterne) und 90^d (der unteren Periodengrenze der unzweifelhaften Mira-Sterne) selten sind, so gibt es doch einige Sterne in dieser Lücke, die dem Spektraltypus M angehören (soweit die Spektra bekannt sind), und die einen Übergang zwischen den beiden Klassen zu bilden scheinen. Ganz ähnlich liegen die Dinge bei dem Übergang von den kurz- zu den langperiodischen δ Cephei-Sternen.

5. Die Mira-Sterne besitzen eine deutliche Verwandtschaft mit den RV Tauri-Sternen. Andererseits läßt sich, wie wir sehen werden, auch eine solche zwischen den RV Tauri-Sternen und den langperiodischen δ Cephei-Sternen feststellen.

6. In der kleinen Magellanschen Wolke kommt neben δ Cephei-Sternen mit normalen Perioden auch ein Veränderlicher mit einer Periode von 127^d vor. Dieser paßt seiner Helligkeit nach zwanglos in die Kurve, welche die Beziehung zwischen Helligkeit und Periodenlänge für die übrigen δ Cephei-Sterne der kleinen Magellanschen Wolke darstellt. Man hat also hier einen δ Cephei-Stern von 127^d Periode vor sich, den man sonst auf Grund der langen Periode als Mira-Stern betrachten würde. Die Lichtkurve ist die typische für Mira-Sterne [1]); die photographische Amplitude ist aber nur 0.9 Größenklassen.

7. In einigen Fällen scheinen Mira-Sterne in Beziehungen zu kugelförmigen Sternhaufen zu stehen, so wie dies für δ Cephei-Sterne häufig zutrifft. Die wichtigsten Fälle dieser Art sind folgende: Der kugelförmige Sternhaufen 47 Tucanae enthält 7 langperiodische Veränderliche [2]), von denen die drei hellsten Perioden von 211^d, 203^d und 192^d haben, während die Perioden der vier anderen noch unbekannt sind. Nahe bei NGC 6584 liegen zwischen AR $18^h 12^m.1$ und $18^h 13^m.9$, Dekl $-52°51'$ und $-52°32'$ sechs Veränderliche dicht zusammen [3]), von denen drei wahrscheinlich Perioden von 361^d, 238^d, 112^d haben, während die anderen kurze, noch unbekannte Perioden besitzen. Innerhalb $1\,^1/_2°$ Entfernung von NGC 6451 liegen zwölf Veränderliche [4]), von denen drei kurzperiodische δ Cephei-Sterne sind, während die neun anderen Perioden zwischen 353^d und 186^d aufweisen. Ebenso liegen [5])

---

[1]) Harvard Obs. Circular Nr. 237.   [2]) Harvard Obs. Bull. Nr. 783.
[3]) Harvard Obs. Bull. Nr. 801.   [4]) Harvard Obs. Bull. Nr. 799.
[5]) Harvard Obs. Bull. Nr. 798.

nahe bei NGC 6093 fünf Veränderliche, von denen vier Perioden von 582^d bis 151^d haben, der fünfte dagegen eine solche von weniger als 1^d hat.

8. Der δ Cephei-Stern RU Camelopardalis $(P = 22^d)$ hat[1]) im Maximum ein Spektrum der Klasse K, im Minimum merkwürdigerweise ein solches der Klasse R mit Emissionslinien des Wasserstoffs, die, wie bei den Md-Mira-Sternen, gegen die Absorptionslinien nach Violett verschoben sind. Daß RU Camelopardalis ein typischer δ Cephei-Stern ist, geht aus dem Verhalten der Veränderungen der Radialgeschwindigkeit zu denen der Helligkeit hervor.

Trotz der Verschiedenheiten, von denen nur die unter 4. genannten wirklich von Belang sind, scheint es also, als ob zwischen den Mira- und den langperiodischen δ Cephei-Sternen eine Verwandtschaft bestände. Aber die Frage ist noch nicht entschieden, und es ist wohl möglich, daß die eben ausgeprochene Ansicht irrig ist.

Ob etwa eine direkte Verwandtschaft zwischen den δ Cephei- und den unregelmäßigen μ Cephei-Sternen besteht, ist eine Frage, die wir noch ganz unentschieden lassen müssen; wir kennen den Lichtwechsel der letzteren noch zu wenig. Dagegen bestehen sicher enge Beziehungen zwischen den langperiodischen δ Cephei-Sternen und den RV Tauri-Sternen mit Perioden von weniger als etwa 100^d. Letztere sind (vgl. die Tabelle auf S. 84) galaktisch wie erstere. Die Spektren der langperiodischen δ Cephei-Sterne gehören vorwiegend den Klassen acF und acG an, die RV Tauri-Sterne R Sagittae, V Vulpeculae und U Monocerotis haben cG-Spektren.

Die Lichtkurve von RV Tauri gleicht zeitweise sehr stark der eines δ Cephei-Sternes, wenn man für die Periode nicht 78^d (von Hauptminimum zu Hauptminimum), sondern 39^d (von Hauptminimum zu Nebenminimum) annimmt, nur ist die Lichtkurve sehr veränderlich; ähnlich ist die Sachlage bei V Vulpeculae.

Eine Verwandtschaft der langperiodischen δ Cephei-Sterne zu denjenigen U Geminorum-Sternen, bei denen die Aufhellungen rasch aufeinander folgen, namentlich X Leonis, ist ebenfalls keineswegs ausgeschlossen. Allerdings scheint es unter den langperiodischen δ Cephei-Sternen solche mit lang andauernder konstanter Phase im Minimum nicht zu geben, aber auch bei X Leonis dürfte diese Phase manchmal nur ziemlich kurz (wenn überhaupt vorhanden) sein. Eine Verwandtschaft der δ Cephei-Sterne mit den R Coronae-Sternen und den Nova-ähnlichen Veränderlichen dürfte dagegen nicht in Betracht kommen.

Wenden wir uns nun den beiden zuletzt genannten Klassen von Veränderlichen zu. Es scheint, als ob zwischen den U Geminorum-Sternen und den Nova-ähnlichen Veränderlichen Beziehungen bestehen. SS Cygni hat, wie wir sahen, ein Spektrum, das an das der neuen Sterne

---

[1]) Publ. Astr. Soc. of the Pacific 1919, S. 180.

in gewissen Phasen ihrer Erscheinung erinnert. Das Aufleuchten geschieht bei manchen U Geminorum-Sternen, z. B. bei UV Persei, außerordentlich plötzlich; das Maximum dauert bei diesem Sterne nur wenige Tage. Für die Nova-ähnlichen Veränderlichen ist T Pyxidis ein typischer Fall. Dieser Stern erhob sich aus einem nahezu konstanten Minimum plötzlich in den Jahren 1890, 1902 und 1920 zu größerer Helligkeit und zeigte dabei 1920 ein typisches Nova-Spektrum. Man könnte ihn einen U Geminorum-Stern nennen, der nur in sehr langen Intervallen aufleuchtet, ebensogut aber auch eine Nova, die mehrmals erschienen ist[1]). Zu den Nova-ähnlichen Veränderlichen ist wohl auch RS Ophiuchi[2]) zu rechnen, sowie auch η Carinae[3]), wenngleich hier die Verhältnisse anders liegen. In beiden Fällen weist das Spektrum auf nahe Beziehungen zu den neuen Sternen hin. Jedenfalls umfaßt die Klasse der Nova-ähnlichen Veränderlichen bisher nur wenige Objekte.

Sehr eigenartig sind die unregelmäßigen Veränderlichen der R Coronae-Klasse. Die R Coronae-Sterne im engeren Sinne sind dadurch charakterisiert, daß sie lange bei einer konstanten Helligkeit im Maximum verharren. Diese Zeiten normaler Helligkeit werden unterbrochen durch meist scharf einsetzende Minima, in denen die Sterne oft ebenfalls lange Zeit unter bedeutenden Lichtschwankungen verbleiben. Ich habe diese Sterne in der schon auf S. 81 zitierten Arbeit A eingehend diskutiert und dabei noch einige zu ihnen gerechnet, welche noch nicht längere Zeit hindurch in einem konstanten Maximum beobachtet worden sind, aber sonst doch große Ähnlichkeit mit den R Coronae-Sternen haben. Im ganzen erhielt ich in A eine Liste von 13 R Coronae-Sternen. Zu diesen kommen noch folgende Sterne hinzu, deren Lichtwechsel damals noch gar nicht oder erst mangelhaft bekannt war:

S Apodis. $\beta = 12°$. Spektrum R (vgl. Harvard Bull. Nr. 761).

AB Aurigae. $\beta = 8°$. Spektrum A0. „Situated on the edge of an obscured region. Probably, therefore, its abnormal variability is to be attributed to occultation by cosmic dust clouds“ (vgl. Harvard Bull. Nr. 798).

VY Pegasi. $\beta = 41°$. Spektrum unbekannt (vgl. Harvard Bull. Nr. 792)

RT Serpentis. $\beta = 9°$. Spektrum A8 mit hellem Hβ. (Helligkeitsbeobachtungen an zahlreichen Stellen veröffentlicht.)

HD 81137. $\beta = 1°$. Spektrum Ma mit einigen hellen Linien, die mit solchen im Spektrum von η Carinae identisch zu sein scheinen (vgl. Harvard Bull. Nr. 783).

Z Canis majoris. $\beta = 1°$. Spektrum Bp (Hβ, Hγ hell) (vgl. Harvard Circ. Nr. 225).

---

[1]) Lichtkurve s. Harvard Annals Bd. 84, Nr. 5, und Publ. of the Astr. Soc. of the Pacific 1921, S. 191.

[2]) Lichtkurve s. Harvard Annals Bd. 84, Nr. 5, und Publ. of the Astr. Soc. of the Pacific 1921, S. 191.

[3]) Lichtkurve s. Annals of the Cape Observatory Bd. 9, S. 78 B.

Die Zugehörigkeit von VY Pegasi und HD 81137 zur Klasse der R Coronae-Sterne ist noch zweifelhaft.

Alle die hier und in A angeführten R Coronae-Sterne liegen in geringen galaktischen Breiten außer R Coronae selbst und VY Pegasi; sie besitzen sehr verschiedene Spektren, aber auffallend zahlreiche unter ihnen haben zeitweise oder dauernd helle Linien, und einige haben im Spektrum Eigentümlichkeiten, die an die neuen Sterne erinnern. Ich habe in A versuchsweise die Hypothese aufgestellt, daß der Lichtwechsel durch Staub- und Nebelmassen zustande kommt, die sich an dem Stern vorbeibewegen oder durch die der Stern sich hindurchbewegt. Um einfache Verfinsterungsvorgänge kann es sich aber wegen der Eigentümlichkeiten der Spektren vieler dieser Sterne sicherlich nicht in allen Fällen handeln. Diese hier angedeutete Hypothese, die an Seeligers Theorie der neuen Sterne anklingt, gewinnt noch dadurch an Wahrscheinlichkeit, daß manche der R Coronae-Sterne in Nebeln liegen[1]). So scheint alles in allem eine Verwandtschaft mit den Nova-ähnlichen Veränderlichen und den neuen Sternen selbst vorhanden zu sein, wie ich schon in A näher ausgeführt habe.

Von wenigen Ausnahmen abgesehen, auf die ich noch etwas näher eingehen werde, lassen sich alle veränderlichen Sterne, deren Lichtwechsel hinreichend bekannt ist, in die besprochenen Klassen einordnen. Diese Klassen und ihre verwandtschaftlichen Beziehungen zueinander sind nun in nachstehendem Schema übersichtlich dargestellt.

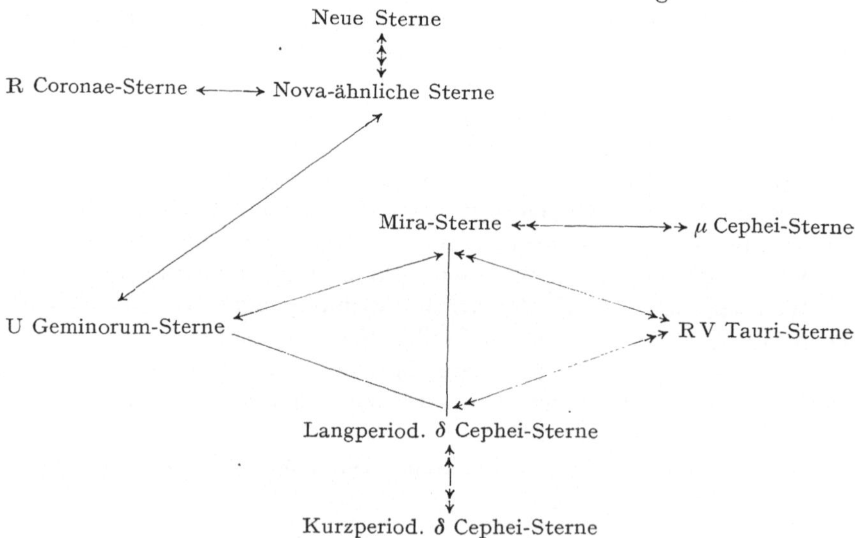

Die Stärke der Verwandtschaft zwischen den einzelnen Klassen wird durch die Art der Verbindungslinien zwischen den Bezeichnungen

[1]) Vielleicht sind die zahlreichen Veränderlichen im Orionnebel zu den R Coronae-Sternen zu rechnen; doch ist der Beweis hierfür noch nicht erbracht.

der Klassen gekennzeichnet. Linien mit zwei Spitzen an jedem Ende bedeuten eine unzweifelhafte Verwandtschaft, solche mit einer Spitze eine stark zu vermutende, und solche ohne Spitze eine zweifelhafte Verwandtschaft. Wo noch keine Verwandtschaft erkennbar ist, z. B. zwischen μ Cephei- und RV Tauri-Sternen, sind keine Verbindungslinien gezogen.

Es scheint nach dem Schema, als ob die veränderlichen Sterne (von den Verfinsterungsveränderlichen abgesehen) in gewissem Sinne eine Gesamtheit bilden, deren Glieder durch allmähliche Übergänge verbunden sind, so daß die Sachlage ähnlich, wenn auch weniger einfach, wie bei den Spektren der Sterne ist. (Vor allem sind bei den veränderlichen Sternen die Übergangsformen seltener, als dies bei den Spektren der Fall ist.) Freilich dürfen wir nicht vergessen, daß wir uns bei den veränderlichen Sternen noch auf sehr unsicherem Boden bewegen, und es liegt mir vollkommen fern, die obigen Ausführungen als endgültig zu betrachten. Es handelt sich hier vielmehr lediglich um einen ersten Versuch, die Verwandtschaftsverhältnisse zwischen den einzelnen Klassen der Veränderlichen einheitlich zu diskutieren.

Wie schon hervorgehoben, passen nur sehr wenige Veränderliche nicht in die hier angewandte Klasseneinteilung. Diese Ausnahmen betreffen meist Sterne mit kleinen Helligkeitsamplituden, und es ist wohl kein Zweifel, daß sich mancher von ihnen noch als zu einer der bekannten Klassen gehörig erweisen wird, wenn reicheres Beobachtungsmaterial vorliegt. Als gut verbürgte Ausnahmen nenne ich nur Z Andromedae und RW Aurigae. Ersterer Stern[1]) ist wohl mit den Nova-ähnlichen Veränderlichen verwandt. RW Aurigae ist von ENEBO viel beobachtet worden und erleidet ganz unregelmäßige Schwankungen zwischen den Größen 8.8 und < 12.2. Die Schätzungen verschiedener Beobachter sind oft sehr widerspruchsvoll. Vielleicht liegt Verwandtschaft mit der R Coronae-Klasse vor.

Auch die von GUTHNICK photoelektrisch gemessenen kurzperiodischen Veränderlichen mit sehr kleinen Helligkeitsamplituden, z. B. β Cephei, bieten hinsichtlich ihrer Klassifizierung noch Schwierigkeiten. Beziehungen zur δ Cephei-Klasse sind hier wohl vorhanden, es scheint mir aber nach den Ausführungen von EDITH E. CUMMINGS[2]) über β Cephei nicht angängig, diese Sterne einfach, wie es häufig geschieht, zur Klasse der kurzperiodischen δ Cephei-Sterne zu rechnen.

---

[1]) Lichtkurve s. Harvard Bull. Nr. 797.
[2]) Lick Observatory Bull. Nr. 349.

Stationäre Geschwindigkeitsverteilung im Sternsystem.

Fragment von **K. Schwarzschild** (†).

Das Planetensystem wird durch die Gravitation für außerordentlich lange Zeiten stabil in seinem jetzigen Zustand erhalten. Es ist möglich, daß auch im Fixsternsystem der Gravitation dieselbe Rolle zufällt, daß sich durch ihre Wirkung die verhältnismäßig reguläre linsenförmige Gestalt des Sternsystems dauernd erhält. Wie im Planetensystem müssen dann auch im Fixsternsystem die Anfangsgeschwindigkeiten der einzelnen Körper in besonderer Weise geordnet sein. Im Hinblick auf diese Möglichkeit soll nun folgende Aufgabe behandelt werden: Es sei ein Sternhaufen gegeben von ellipsoidischer Begrenzung und gleichförmiger Dichte (gleicher Anzahl und Masse der Sterne in der hinreichend groß genommenen Volumeinheit). Wie müssen die Anfangsgeschwindigkeiten der Sterne in einem solchen Haufen beschaffen sein, damit diese räumliche Verteilung unter der gegenseitigen Attraktion der Sterne dauernd erhalten bleibt?

Zur Begründung und Präzisierung der Aufgabe sei folgendes bemerkt: Der Sternhaufen soll seinen Dimensionen und seiner Massendichte nach natürlich etwa unserem Milchstraßensystem entsprechen. Ohne Gravitation würde die Anordnung des Milchstraßensystems bei den tatsächlichen Geschwindigkeiten der Sterne verhältnismäßig sehr vergänglich sein. Wenn z. B. die Sonne mit ihrer Geschwindigkeit von 20 km/sec geradeaus liefe, würde sie innerhalb 10^8 Jahren eine Strecke durchmessen, die einer Parallaxe von $0\overset{\text{..}}{''}0005$ entspricht und damit etwa dem Halbmesser des Sternsystems nach SEELIGER gleichkommt. Das Alter von Erde und Sonne ist zweifellos höher als 10^8 Jahre anzusetzen. Daß andererseits die Gravitation des Sternsystems erheblich sein muß, lehrt eine einfache Rechnung wie diese: Man nehme eine Kugel, deren Radius der Parallaxe $0\overset{\text{..}}{''}003$ entspricht. Die Massendichte innerhalb dieser Kugel sei so bemessen, daß je eine Sonnenmasse auf eine Kugel kommt, deren Radius der Parallaxe $1''$ entspricht. Eine Kreisbewegung an der Oberfläche der großen Kugel unter der Gravitation derselben erfolgt mit 20 km/sec Geschwindigkeit und innerhalb 10^8 Jahren.

Bei den Gravitationswirkungen selbst ist zu unterscheiden zwischen den *regulären Kräften*, die der Gesamtattraktion des Systems entspringen, wenn man die Sternmassen durch eine kontinuierliche Dichteverteilung ersetzt, und den *irregulären Kräften*, die von den sich zufällig nähernden Sternen der Umgebung ausgeübt werden. Letztere Kräfte wirken wie die Zusammenstöße der Moleküle in der Gatheorie, wenn auch bei der Seltenheit erheblicher Annäherungen zwischen zwei Fixsternen außerordentlich viel langsamer. (Direkte Zusammenstöße zwischen zwei Fixsternen sind so selten, daß sie nicht berücksichtigt zu werden brauchen.) Nach der unten folgenden Abschätzung würden die irregulären Kräfte erst nach Zeiten von der Größenordnung 10^{12} Jahren einen merklichen Energieaustausch und erst nach noch längeren Zeiten eine gleichmäßige Verteilung der Energie auf alle Freiheitsgrade zuwege bringen. Da nach dem oben Gesagten die regulären Kräfte viel erheblicher sind, so darf man in erster Annäherung die irregulären Kräfte neben den regulären vernachlässigen und jedem Stern eine bestimmte, nur wenig gestörte — damit ist nicht gemeint geschlossene — Bahn durch das Sternsystem zuschreiben.

Herr EDDINGTON[1]) hat ausgeführt, daß ein unmittelbarer Beweis für das Überwiegen der regulären über die irregulären Kräfte durch die Existenz von Sternströmen in der Art des Hyadenstroms gegeben ist. Wie man aus der Übertragung bekannter Betrachtungen über die Auflösung von Kometen[2]) erkennt, vermögen die regulären Kräfte einen solchen Sternstrom nicht zu zerstören, solange die Dichte des Sternstroms einige Male größer ist als die durchschnittliche Dichte der Sternverteilung, welche er durchsetzt. Die irregulären Kräfte hingegen wirken zerstörend auf ihn. Wenn also solche Sternströme noch jetzt existieren, so beweist dies, daß sich die Wirkung der irregulären Kräfte bisher noch nicht zu einem erheblichen Betrag hat summieren können.

Mit Hilfe unserer Abschätzung der irregulären Kräfte kann man über diesen Schluß Herrn EDDINGTONS noch hinausgehen, wenn man außerdem annimmt, daß die Sternströme ebenso alt sind wie das ganze Sternsystem, und nach dem Spektraltypus der Sterne des Hyadenstromes liegt kein Grund vor, ihn für besonders jung zu halten. Es folgt dann, daß das Alter des ganzen Sternsystems die Größenordnung von 10^{12} Jahren nicht überschreitet.

Es soll jetzt die eben benutzte Abschätzung der irregulären Kräfte ausgeführt werden. Dieselbe ist namentlich in prinzipieller Hinsicht lückenhaft, dürfte aber doch zu einer richtigen Größenordnung führen.

Man betrachte die Begegnung zweier Sterne der Massen m_1 und m_2. Die Geschwindigkeit ihres Schwerpunktes sei s; die ursprüngliche Relativgeschwindigkeit von m_1 gegen m_2 sei v. Der Winkel zwischen

---

1) Brit. Assoc. 1911. Stellar Distribution and Movements.
2) TISSERAND: Méc. celeste IV. Chap. XVI.

s und v sei φ. Dann gilt für die Geschwindigkeiten v_1 und v_2 beider Massen vor der Begegnung:

$$v_1^2 = s^2 + 2\,s\,v\,\frac{m_2}{m_1 + m_2}\cos\varphi + v^2\left(\frac{m_2}{m_1 + m_2}\right)^2$$

$$v_2^2 = s^2 - 2\,s\,v\,\frac{m_1}{m_1 + m_2}\cos\varphi + v^2\left(\frac{m_1}{m_1 + m_2}\right)^2$$

Die beiden Sterne beschreiben unter ihrer gegenseitigen Anziehung eine Hyperbel umeinander. Dabei bleibt die Schwerpunktsgeschwindigkeit s unverändert. Auch die Relativgeschwindigkeit v kehrt zu ihrem Ausgangswert zurück, nur ist sie aus der Richtung der einen Asymptote in die der anderen getreten. Ist $2\,w$ der (den Brennpunkt enthaltende) Winkel zwischen den beiden Asymptoten, so wird der Winkel φ' zwischen Relativ- und Schwerpunktsgeschwindigkeit nach der Begegnung: $\varphi' = \varphi + 180° - 2\,w$. Die Geschwindigkeiten v_1' und v_2' der Sterne nach der Begegnung werden damit:

$$v_1' = s^2 + 2\,s\,v\,\frac{m_2}{m_1 + m_2}\cos\varphi' + v^2\left(\frac{m_2}{m_1 + m_2}\right)^2$$

$$v_2' = s^2 - 2\,s\,v\,\frac{m_1}{m_1 + m_2}\cos\varphi' + v^2\left(\frac{m_1}{m_1 + m_2}\right)^2$$

Der Energieaustausch bei der Begegnung ist daher:

$$\Delta\varepsilon = m_1\,\frac{v_1^2 - v_1'^2}{2} = m_2\,\frac{v_2'^2 - v_2^2}{2} = s\,v\,\frac{m_1\,m_2}{m_1 + m_2}\,(\cos\varphi - \cos\varphi')$$

$$= 2\,s\,v\,\frac{m_1\,m_2}{m_1 + m_2}\sin\frac{\varphi' + \varphi}{2}\cdot\sin\frac{\varphi' - \varphi}{2} = 2\,s\,v\,\frac{m_1\,m_2}{m_1 + m_2}\sin\frac{\varphi + \varphi'}{2}\cdot\cos w$$

Der Asymptotenwinkel w läßt sich nun noch ausdrücken mittels der Relativgeschwindigkeit v und des Abstandes D der Asymptote vom Brennpunkt, wobei zu beachten ist, daß dieser Abstand zugleich die kleinste Entfernung ist, die die Sterne ohne Ablenkung durch die gegenseitige Gravitation erreicht hätten.

Wir wollen Sonnenmasse und Erdbahnhalbmesser als Einheit wählen und die Zeiteinheit gleich 58 Tagen setzen, so daß die Gausssche Konstante gleich 1 und die Einheit der Geschwindigkeit gleich der Erdgeschwindigkeit (30 km/sec) wird.

Bezeichnet man mit a und e die Halbachse und Exzentrizität der Hyperbel, so gilt nunmehr:

$$v^2 = \frac{m_1 + m_2}{a}, \qquad \cos w = \frac{1}{e}, \qquad D = a\,e\sin w = a\,\mathrm{tg}\,w = \frac{m_1 + m_2}{v^2}\,\mathrm{tg}\,w$$

Daraus folgt:

$$\mathrm{tg}\,w = \frac{D\,v^2}{m_1 + m_2}, \qquad \cos w = \frac{m_1 + m_2}{\sqrt{(m_1 + m_2)^2 + D^2\,v^4}}$$

Setzt man dies in den obigen Ausdruck für den Energieaustausch ein, so nimmt derselbe die für die weitere Rechnung geeignete Gestalt an:

$$\varDelta\varepsilon = 2\,m_1\,m_2\,\sin\left(\frac{\varphi+\varphi'}{2}\right)\frac{s\,v}{\sqrt{(m_1+m_2)^2+D^2\,v^4}}$$

Damit ist die einzelne Begegnung, so weit als hier erforderlich, erledigt, und es ist die Summenwirkung einer großen Reihe von Begegnungen zu bilden. Da für den einzelnen Stern von Begegnung zu Begegnung das Vorzeichen von $\varDelta\varepsilon$ wechselt und — wenigstens für Sterne mittlerer Geschwindigkeit — auf beide Vorzeichen gleichmäßig verteilt ist, so hat man die Quadratsumme $\sum(\varDelta\varepsilon)^2$ zu bilden und aus dieser die Wurzel zu ziehen.

Bei der Summation der Begegnungen ist ferner noch zu überlegen, bis zu welcher Größe D_0 der Minimaldistanz D man die Begegnungen als solche rechnen will. Es ist dabei zu bedenken, daß von der Anziehung jedes Sternes ein Teil in Abrechnung zu bringen ist, der durch die regulären Kräfte des kontinuierlichen Systems, durch welches wir das wirkliche ersetzen, wiedergegeben wird. Insofern ist also der obige Ausdruck für den Energieaustausch zu groß. Auf der anderen Seite werden auch die Begegnungen jenseits der Distanz D_0 noch kleine Beiträge zum Energieaustausch liefern. Man hat daher für D_0 einen Wert zu nehmen, bei dem der eine Fehler den anderen gerade aufhebt. Der richtige Wert von D_0 wird ein kleines Vielfaches von der Größenordnung der mittleren Sterndistanz sein. Es soll im folgenden $D_0 = 10^6$ (entsprechend $0\overset{''}{.}2$ Parallaxe) gesetzt werden. Es wird sich zeigen, daß es auf eine Zehnerpotenz in der Wahl von D_0 für unseren Zweck nicht ankommt.

Wir wollen nun die Wirkung der Begegnungen für die Zeiteinheit summieren. Zur Vereinfachung sollen alle Sternmassen gleich 1 gesetzt werden. Die mittlere Geschwindigkeit der Sterne sei V. Dann ist die mittlere Schwerpunktsgeschwindigkeit zweier Sterne $V/\sqrt{2}$, die mittlere Relativgeschwindigkeit $\sqrt{2}\cdot V$. Ist N die Anzahl der Sterne pro Volumeinheit, so wird bei dieser mittleren Relativgeschwindigkeit die Zahl der Begegnungen pro Zeiteinheit, bei welchen die Distanz D zwischen den Grenzen D und $D+dD$ liegt, gleich:

$$2\,\pi\,\sqrt{2}\,V\,N\,D\,d\,D$$

Die Häufigkeit jeder Relativgeschwindigkeit in den Grenzen v bis $v+dv$ werde nach dem MAXWELLschen Gesetz angesetzt zu:

$$\frac{3}{2}\sqrt{\frac{3}{\pi}}\cdot\frac{1}{V^3}\,e^{-\frac{3}{4}\frac{v^2}{V^2}}\cdot v^2\,d\,v$$

Dabei sind die Konstanten so gewählt, daß das mittlere Geschwindigkeitsquadrat gleich $2\,V^2$ wird, und daß die Integration über alle Werte von v die Summe 1 ergibt.

Als Mittelwert von $\sin^2 \dfrac{\varphi + \varphi'}{2}$ werde der Wert $\frac{1}{2}$, als Mittelwert von s^2 der Wert $\frac{1}{2} V^2$ eingeführt. Mit alledem erhält man als Summation für die Zeiteinheit:

$$\sum (\varDelta \varepsilon)^2 = 3 \sqrt{6\pi}\, N \iint \frac{e^{-\frac{3}{4}\frac{v^2}{V^2}} \cdot v^4\, d v\, D\, d D}{4 + v^4 D^2}$$

wobei das Integral über v von 0 bis ∞ und über D von 0 bis D_0 zu erstrecken ist. Die Integration nach D läßt sich ohne weiteres ausführen und ergibt:

$$\sum (\varDelta \varepsilon)^2 = \tfrac{3}{2} \sqrt{6\pi}\, N \int_0^\infty e^{-\frac{3}{4}\frac{v^2}{V^2}} \cdot \log \left(1 + \frac{v^4 D_0^2}{4}\right) d v$$

Setzt man $z^2 = \tfrac{3}{4} v^2 / V^2$ und $Y = \tfrac{2}{3} \cdot V^2 \cdot D_0$, so schreibt sich dies:

$$\sum (\varDelta \varepsilon)^2 = 3 \sqrt{2\pi}\, N V \int_0^\infty e^{-z^2} \log (1 + Y^2 z^4)\, d z$$

Die Ausführung letzteren Integrals wird einfach in Rücksicht auf den Umstand, daß Y eine große Zahl von der Ordnung 10^6 ist. Schreibt man nämlich:

$$\int_0^\infty e^{-z^2} \log (1 + Y^2 z^4)\, d z = 2 \log Y \int_0^\infty e^{-z^2}\, d z + \int_0^\infty e^{-z^2} \log \left(\frac{1}{Y^2} + z^4\right) d z$$

so nähert sich mit wachsendem Y das zweite Integral dem Wert:

$$C = 4 \int_0^\infty e^{-z^2} \log z\, d z$$

wofür ich durch eine rohe mechanische Quadratur finde: $C = -3{,}5$. Es gilt somit nahe:

$$\int_0^\infty e^{-z^2} \log (1 + Y^2 z^4)\, d z = \sqrt{\pi}\, \log Y - 3{,}5$$

und:

$$\sum (\varDelta \varepsilon)^2 = 3 \sqrt{2\pi}\, N V (\sqrt{\pi}\, \log Y - 3{,}5)$$

Man sieht, daß Y und damit die Distanz D_0 nur logarithmisch eingeht, so daß es, wie erwähnt, auf ihren genauen Wert nicht ankommt. Die Anzahl N der Sterne in der Volumeinheit werde bestimmt durch den Radius R der Kugel, auf welche durchschnittlich ein Stern kommt. Dann ist: $\dfrac{1}{N} = \dfrac{4\pi}{3} R^3$ und die vorstehende Formel geht über in:

$$\sum (\varDelta \varepsilon)^2 = \frac{9}{4} \sqrt{\frac{2}{\pi}}\, \frac{V}{R^3} (\sqrt{\pi}\, \log Y - 3{,}5)$$

Das ist das Ergebnis für die Summation über die Zeiteinheit (58 Tage).

Die Summation über t Tage ergibt, wenn man die numerischen Koeffizienten ausrechnet und Dezimallogarithmen einführt:

$$\sum (\varDelta \varepsilon)^2 = 46 \frac{V}{R^3} \left[\log (V^2 D_0) - 1{,}03\right] \cdot t$$

Es sind schließlich noch numerische Werte für V, D_0 und R einzuführen. Die mittlere Sterngeschwindigkeit V wird gleich der Erdgeschwindigkeit 1 gesetzt, D_0 sei gleich 10^6 (entsprechend $0{,}''2$ Parallaxe) und R gleich $2 \cdot 10^5$ (entsprechend $1''$ Parallaxe). Dann folgt:

$$\sum (\varDelta \varepsilon)^2 = 2{,}8 \cdot 10^{-14} t$$

Nach 10^{12} Jahren wird $\sum(\varDelta \varepsilon)^2 = 0{,}028$, die Wurzel daraus $= 0{,}17$ und mithin vergleichbar mit der durchschnittlichen Energie $\frac{1}{2} m V^2$ des einzelnen Sternes, welche nach den obigen Annahmen $- 0{,}5$ ist. Das ist das oben benutzte Resultat.

Wir kommen nun zu unserem eigentlichen Problem des Aufsuchens einer stationären Geschwindigkeitsverteilung in einem homogenen ellipsoidischen Sternhaufen.

Die Begrenzung des Sternhaufens habe in rechtwinkligen Koordinaten x, y, z die Gleichung:

$$\frac{x^2}{a^2} + \frac{y^2}{b^2} + \frac{z^2}{c^2} = 1 \tag{1}$$

so daß a, b, c die Achsen des begrenzenden Ellipsoids sind. Indem die Dichte innerhalb dieses Ellipsoids als konstant vorausgesetzt wird, folgt bekanntlich für das Gravitationspotential:

$$V = -\left(\frac{A^2}{2} x^2 + \frac{B^2}{2} y^2 + \frac{C^2}{2} z^2\right) + D \tag{2}$$

wo A, B, C, D Konstanten sind. Die Bewegungsgleichungen eines Punktes, der sich im Innern des Ellipsoids bewegt, sind daher:

$$\frac{d^2 x}{d t^2} = \frac{\partial V}{\partial x} = -A^2 x, \quad \frac{d^2 y}{d t^2} = \frac{\partial V}{\partial y} = -B^2 y, \quad \frac{d^2 z}{d t^2} = \frac{\partial V}{\partial z} = -C^2 z \tag{3}$$

Die Integration derselben ergibt:

$$x = \alpha \sin \vartheta_1, \qquad y = \beta \sin \vartheta_2, \qquad z = \gamma \sin \vartheta_3, \tag{4}$$

$$\vartheta_1 = A t + k_1, \qquad \vartheta_2 = B t + k_2, \qquad \vartheta_3 = C t + k_3, \tag{5}$$

wobei α, β, γ, k_1, k_2, k_3 Integrationskonstanten sind. Wir wählen α, β, γ positiv. Jede Koordinate führt also eine Sinusschwingung aus, deren Periode gleich $\frac{2\pi}{A}$ resp. $\frac{2\pi}{B}$ und $\frac{2\pi}{C}$ ist, deren Amplitude α resp. β, γ ist.

Es werde nun ausdrücklich vorausgesetzt, daß diese drei Perioden inkommensurabel untereinander seien. Man denke sich die drei Variablen ϑ für einen Augenblick als rechtwinklige Koordinaten aufgetragen und ziehe von ihnen stets solche Vielfache von 2π ab, daß sie immer zwischen 0 und 2π bleiben. Der Punkt der Koordinaten ϑ_1, ϑ_2, ϑ_3 wird dann innerhalb eines Würfels der Seitenlänge 2π bleiben und wird nach hinreichend langer Zeit diesen Würfel in gleichmäßiger Dichte ausfüllen. Der Stern, dessen Koordinaten x, y, z sind, wird demgemäß das Parallelepiped $|x| < |\alpha|$, $|y| < |\beta|$, $|z| < |\gamma|$ ausfüllen, dieses aber mit ungleichförmiger Dichte. Man hat:

$$d x \, d y \, d z = \alpha \beta \gamma \cos\vartheta_1 \cos\vartheta_2 \cos\vartheta_3 \, d\vartheta_1 \, d\vartheta_2 \, d\vartheta_3 \tag{6}$$

oder

$$d\vartheta_1 \, d\vartheta_2 \, d\vartheta_3 = \Delta \, d x \, d y \, d z$$

wobei

$$\Delta = \frac{1}{\sqrt{(\alpha^2 - x^2)(\beta^2 - y^2)(\gamma^2 - z^2)}} \tag{7}$$

ist. Der hier auftretende Faktor Δ von $d x \, d y \, d z$ gibt die — im Mittel über eine hinreichend lange Zeit genommene — Dichte an, mit welcher der Stern das von ihm durchlaufene Parallelepiped erfüllen würde. Da bei stationärer Dichteverteilung die Dichte in jedem Augenblick mit dem Mittelwert der Dichte über beliebig lange Zeiten identisch ist, so ist das zugleich der Beitrag, welchen der betrachtete Stern zur Massendichte in jedem Punkte des Systems liefert.

Wir betrachten jetzt die Gesamtheit aller Sterne des Systems. Die Anzahl derjenigen Sterne, deren Amplituden zwischen α und $\alpha + d\alpha$ resp. β und $\beta + d\beta$, γ und $\gamma + d\gamma$ liegen, sei:

$$\varphi(\alpha, \beta, \gamma) \, d\alpha \, d\beta \, d\gamma \tag{8}$$

Dann ist die gesamte Dichte an dem Punkte der Koordinaten x, y, z

$$\varrho(x, y, z) = \int \frac{\varphi(\alpha, \beta, \gamma) \, d\alpha \, d\beta \, d\gamma}{\sqrt{(\alpha^2 - x^2)(\beta^2 - y^2)(\gamma^2 - z^2)}} \tag{9}$$

Das Integral ist zu erstrecken über alle Amplituden, die größer sind als die betreffende Koordinate, weil Sterne kleinerer Amplitude die betreffende Koordinate überhaupt nicht erreichen, mithin auch nicht zur Dichte beitragen würden. Die Integrationen sind also von den Grenzen $\alpha^2 = x^2$, $\beta^2 = y^2$, $\gamma^2 = z^2$ bis ins Unendliche zu erstrecken.

Damit nun unser Sternsystem wirklich stationär seine Dichteverteilung beibehält, muß diese Dichte ϱ mit der zu Anfang angenommenen übereinstimmen. Es muß das vorstehende Integral für Punkte im Innern des Ellipsoids (1) gleich 1, außerhalb gleich 0 werden. Die Gleichung (9) ist demnach eine Integralgleichung, aus der die Funktion $\varphi(\alpha, \beta, \gamma)$ gemäß der vorgegebenen Funktion ϱ zu bestimmen ist.

Betrachtet man zunächst den eindimensionalen Fall

$$\varrho\,(x) = \int\limits_{\alpha=x}^{\infty} \frac{\varphi\,(\alpha)\,d\,\alpha}{\sqrt{\alpha^2 - x^2}}$$

so ist das eine bereits von ABEL behandelte Integralgleichung, deren Lösung folgendermaßen geschieht: Man multipliziere mit $\dfrac{x\,d\,x}{\sqrt{x^2 - \xi^2}}$ (ξ ein Parameter) und integriere von $x = \xi$ bis $x = \infty$. Das gibt

$$\int\limits_{x=\xi}^{\infty} \frac{\varrho\,(x)\,x\,d\,x}{\sqrt{x^2 - \xi^2}} = \int\limits_{x=\xi}^{\infty} \int\limits_{\alpha=x}^{\infty} \frac{\varphi\,(\alpha)\,x\,d\,x\,d\,\alpha}{\sqrt{(\alpha^2 - x^2)\,(x^2 - \xi^2)}}$$

Vertauscht man rechts die Integrationsordnung, so hat man bei festgehaltenem α zunächst über x von $x = \xi$ bis $x = \alpha$ zu integrieren, und dann über α von $\alpha = \xi$ bis $\alpha = \infty$. Da aber

$$\int\limits_{x=\xi}^{x=\alpha} \frac{x\,d\,x}{\sqrt{(\alpha^2 - x^2)\,(x^2 - \xi^2)}} = \frac{\pi}{2}$$

ist, so folgt:

$$\int\limits_{x=\xi}^{\infty} \frac{\varrho\,(x)\,x\,d\,x}{\sqrt{x^2 - \xi^2}} = \frac{\pi}{2} \int\limits_{\alpha=\xi}^{\infty} \varphi\,(\alpha)\,d\,\alpha$$

oder:

$$\varphi\,(\xi) = -\frac{2}{\pi} \frac{d}{d\,\xi} \int\limits_{x=\xi}^{\infty} \frac{\varrho\,(x)\,x\,d\,x}{\sqrt{x^2 - \xi^2}}$$

womit die Lösung gegeben ist.

Die dreidimensionale Integralgleichung läßt sich ganz analog behandeln und hat die Lösung:

$$\varphi\,(\xi,\eta,\zeta) = -\frac{8}{\pi^3} \frac{d^3}{d\,\xi\,d\,\eta\,d\,\zeta} \iiint\limits_{|x|\geqq\xi,\,|y|\geqq\eta,\,|z|\geqq\zeta} \varrho\,(x,y,z)\, \frac{x\,d\,x\cdot y\,d\,y\cdot z\,d\,z}{\sqrt{(x^2 - \xi^2)\,(y^2 - \eta^2)\,(z^2 - \zeta^2)}} \cdot \quad (10)$$

In unserem speziellen Falle hat man $\varrho = 1$ zu setzen und bis zu den durch das Ellipsoid

$$\frac{x^2}{a^2} + \frac{y^2}{b^2} + \frac{z^2}{c^2} = 1$$

bestimmten Grenzen zu integrieren.

Diese Integration ist leicht. Setzt man nämlich:

$$x^2 - \xi^2 = X^2\,, \qquad y^2 - \eta^2 = Y^2\,, \qquad z^2 - \zeta^2 = Z^2\,,$$

so geht das auszuführende Integral über in:

$$\iiint dX\, dY\, dZ$$

zu nehmen zwischen den Grenzen $X = Y = Z = 0$ und:

$$\frac{X^2}{a^2} + \frac{Y^2}{b^2} + \frac{Z^2}{c^2} = 1 - \frac{\xi^2}{a^2} - \frac{\eta^2}{b^2} - \frac{\zeta^2}{c^2}$$

Das ist aber nichts anderes als das Volumen eines Ellipsoids, dessen Achsen gleich resp. a, b, c multipliziert mit $\sqrt{1 - \dfrac{\xi^2}{a^2} - \dfrac{\eta^2}{b^2} - \dfrac{\zeta^2}{c^2}}$ sind. Das Volumen dieses Ellipsoids hat den Wert:

$$\frac{4}{3}\,\pi\,a\,b\,c\,\sqrt{1 - \frac{\xi^2}{a^2} - \frac{\eta^2}{b^2} - \frac{\zeta^2}{c^2}}^{\,3}$$

Setzt man diesen Wert für das Integral in (10) ein, so erhält man:

$$\varphi(\xi\,\eta\,\zeta) = \frac{32}{\pi^2\,a\,b\,c} \cdot \frac{\xi\,\eta\,\zeta}{\sqrt{1 - \dfrac{\xi^2}{a^2} - \dfrac{\eta^2}{b^2} - \dfrac{\zeta^2}{c^2}}^{\,3}}$$

oder:

$$\varphi(\alpha, \beta, \gamma) = \frac{32}{\pi^2\,a\,b\,c} \cdot \frac{\alpha\,\beta\,\gamma}{\sqrt{1 - \dfrac{\alpha^2}{a^2} - \dfrac{\beta^2}{b^2} - \dfrac{\gamma^2}{c^2}}^{\,3}} \tag{11}$$

Damit ist diejenige Verteilung der Anfangsbedingungen gefunden, bei welcher das homogene ellipsoidische Sternsystem stationär ist.

Es interessiert besonders, wie dabei die Geschwindigkeitsverteilung in jedem Punkte des Systems ist. Für die Geschwindigkeitskomponenten hat man:

$$\frac{dx}{dt} = u = A\sqrt{\alpha^2 - x^2}, \quad \frac{dy}{dt} = v = B\sqrt{\beta^2 - y^2}, \quad \frac{dz}{dt} = w = C\sqrt{\gamma^2 - z^2},$$

und damit, wenn man x, y, z festhält, u, v, w als unabhängige Variable statt α, β, γ einführt:

$$\frac{\varphi(\alpha, \beta, \gamma)\, d\alpha\, d\beta\, d\gamma}{\sqrt{(\alpha^2 - x^2)(\beta^2 - y^2)(\gamma^2 - z^2)}}$$

$$= \frac{32}{\pi^2\,a\,b\,c} \cdot \frac{du\, dv\, dw}{\sqrt{1 - \dfrac{x^2}{a^2} - \dfrac{y^2}{b^2} - \dfrac{z^2}{c^2} - \dfrac{u^2}{a^2\,A^2} - \dfrac{v^2}{b^2\,B^2} - \dfrac{w^2}{c^2\,C^2}}}$$

Gleich häufig sind hiernach Geschwindigkeiten, für welche der Ausdruck:

$$W^2 = \frac{u^2}{a^2\,A^2} + \frac{v^2}{b^2\,B^2} + \frac{w^2}{c^2\,C^2}$$

denselben Wert hat. Die Geschwindigkeitsverteilung ist demnach eine ellipsoidische. Die Achsen des Geschwindigkeitsellipsoids sind in jedem Raumpunkte den Achsen des Sternhaufens parallel. Die Häufigkeit nimmt mit wachsendem W zu und wird unendlich für $W^2 = 1 - \dfrac{x^2}{a^2} - \dfrac{y^2}{b^2} - \dfrac{z^2}{c^2}$. Letzterer Fall entspricht den Sternen, die bis an die Grenze des Systems gelangen. Über diesen Maximalwert von W hinausgehende Geschwindigkeiten kommen nicht vor. Der Maximalwert von W und damit die obere Grenze der vorkommenden Geschwindigkeiten wird um so kleiner, je mehr man sich der Grenze des Systems nähert. Auf der Grenze selbst werden alle Geschwindigkeiten Null.

Die Achsen des Geschwindigkeitsellipsoids findet man aus dem Ausdruck[1])

$$a^2 A^2 = 2\pi \int\limits_0^\infty \frac{1}{\left(1 + \dfrac{s}{a^2}\right)} \cdot \frac{ds}{\sqrt{\left(1 + \dfrac{s}{a^2}\right)\left(1 + \dfrac{s}{b^2}\right)\left(1 + \dfrac{s}{c^2}\right)}}$$

nebst den entsprechenden. Aus diesen Ausdrücken erkennt man leicht, daß das Geschwindigkeitsellipsoid im selben Sinne, aber weniger stark abgeplattet ist wie der Sternhaufen selbst.

Wenn man zum Fall des Rotationsellipsoids ($a = b$) übergehen will, so kann man voraussetzen, daß dabei immer noch ein kleiner Unterschied zwischen den Perioden $\dfrac{2\pi}{A}$ und $\dfrac{2\pi}{B}$ bleibt, welcher dieselben inkommensurabel macht. Unter dieser Voraussetzung geschieht der Übergang, indem man in der Tat einfach in den obigen Formeln $a = b$ setzt. Will man hingegen von vornweg strenge die Voraussetzung $a = b$, verbunden mit genauer Gleichheit der Perioden $\dfrac{2\pi}{A}$ und $\dfrac{2\pi}{B}$ machen, so ändert das Problem vollständig seinen Charakter. Es beschreibt dann jeder Stern in der Projektion auf die x, y-Ebene eine feste Ellipse und erfüllt im Laufe der Zeit nicht ein Parallelepiped, sondern den Mantel eines elliptischen Zylinders. Da in diesem Falle eine viel geringere Durchmischung eintritt, so wird die stationäre Geschwindigkeitsverteilung nicht mehr eindeutig bestimmt. Vielmehr zeigt sich, daß man die Abhängigkeit der Verteilungsfunktion $\varphi(\alpha, \beta, \gamma)$ von einer der beiden Variablen α, β bis auf gewisse Stetigkeitsforderungen beliebig ansetzen kann, worauf dann die von der anderen bestimmt ist. Man kann in diesem Falle eine stationäre Geschwindigkeitsverteilung angeben, bei der die Sterne — auf die x, y-Ebene projiziert — genau oder angenähert kreisförmige Bahnen beschreiben. Ebensowohl gibt es aber auch Geschwindigkeitsverteilungen, bei welchen die angenähert radialen Bewegungen vorwiegen.

---

[1]) Tisserand: Méc. céleste II, S. 45.

In Wirklichkeit wird nun die Forderung $a = b$ gewiß nicht strenge erfüllt sein. Wenn die Rotationssymmetrie aber sehr nahe erfüllt ist, so werden sich die Perioden $\dfrac{2\pi}{A}$ und $\dfrac{2\pi}{B}$ sehr wenig unterscheiden, und die oben vorausgesetzte Durchmischung der Sterne wird so lange Zeit in Anspruch nehmen, daß es eine Übertreibung wäre, das Prinzip des stationären Zustandes auch noch auf diese Zeit anzuwenden. Es ist also im Falle guter Rotationssymmetrie des Fixsternsystems sehr wohl möglich, daß die strenge Voraussetzung $a = b$ besser mit der Wirklichkeit stimmende Folgerungen gibt als der Grenzübergang aus dem dreiachsigen Ellipsoid.

Ganz abgesehen von dem hypothetischen Charakter unserer Forderung stationärer Geschwindigkeitsverteilung wäre eine Übertragung der obigen Ableitung auf das wirkliche Sternsystem schon deshalb bedenklich, weil die Dichte in demselben nicht gleichförmig ist. Es würde interessant sein, eine in dieser Beziehung sich der Wirklichkeit näher anschließende Verteilung zu bilden. Indessen glaube ich, daß die obige Ableitung zur Fixierung der Vorstellungen nützlich und in folgender Beziehung lehrreich ist.

Die bekannte Erscheinung einer der Milchstraße parallelen Vorzugsrichtung in den Eigenbewegungen der Fixsterne habe ich mir durch die Annahme zu erklären gesucht, daß die Sonne exzentrisch im Sternsystem steht und daß die Sterne angenähert kreisförmige Bahnen parallel der Milchstraßenebene beschreiben, wobei direkte und retrograde Bewegungen als gleichhäufig vorausgesetzt werden. Nach dieser Anschauung wäre die Richtung nach dem Zentrum des Systems senkrecht zur Richtung des bei $7^{\mathrm{h}} + 10°$ liegenden Vertex der Sternbewegungen zu suchen.

Herr Hertzsprung hat mir mündlich die Meinung ausgesprochen, daß die Sternbewegungen parabolischen Charakter haben und im wesentlichen radial gerichtet sein könnten, so daß wir das Zentrum in der Richtung der Vertexlinie selbst zu suchen hätten. Letztere Anschauung ist kürzlich von Herrn H. H. Turner publiziert und mit großem Schwung ausgeführt worden.

Die obige Rechnung zeigt nun, daß die stationäre Geschwindigkeitsverteilung in einem homogenen Ellipoisd die Richtung nach dem Zentrum nirgends verrät, sondern nur die Richtung der Hauptachsen des Ellipsoids wiedergibt. Wenn etwas Ähnliches für das wirkliche Sternsystem gilt, so muß die Richtung nach dem Zentrum des Systems weder mit der Vertexlinie noch mit einer Senkrechten dazu zusammenfallen.

Die geringe mittlere Geschwindigkeit der Heliumsterne führt Herr Eddington auf den Umstand zurück, daß sich diese im Durchschnitt in großen Entfernungen näher der äußeren Grenze des Sternsystems befinden. Herr Kapteyn hat die Schwierigkeiten dieser Anschauung hervorgehoben. Herr TURNER verweist die Heliumsterne umgekehrt mehr in das Zentrum des Systems. Halten wir uns an die oben gefundene Geschwindigkeitsverteilung, so ergibt sich folgendes: Kleine Geschwindigkeiten herrschen vor in den äußeren Teilen des Systems. Indessen kleine Energien (kleine Werte der Summen $\dfrac{\alpha^2}{a^2} + \dfrac{\beta^2}{b^2} + \dfrac{\gamma^2}{c^2}$, von kinetischer und potentieller Energie) finden sich nur im Innern des Systems. Wenn man also annimmt, daß die Heliumsterne von Anfang an kleine Gesamtenergie gehabt haben, so kommt man auf die TURNERsche Anschauung, gegen welche Widersprüche nicht vorliegen.

Beziehungen zwischen den unter sich getrennten Bewegungsformen im Gebiete der Himmelsmechanik.

Von **K. Bohlin**, Stockholm.

Die verschiedenen Entwicklungsarten, die in den Bewegungstheorien der Himmelskörper angewandt worden sind, unterscheiden sich wohl hauptsächlich durch die Wahl einer unabhängigen Veränderlichen. So rechnet ja die Theorie der Variation der Konstanten und im besonderen der Säkularstörungen mit der Zeit oder der mittleren Anomalie $g = nt + c$ als unabhängiger Veränderlichen. Aber schon im Zweikörperprobleme findet sich die Veranlassung, die Zeit durch die wahre Anomalie v oder durch die exzentrische u zu ersetzen, wobei je die Differentialrelationen:

$$d t = \frac{r^2}{\sqrt{c}} d v \quad \text{und} \quad n d t = \frac{r}{a} d u \tag{1}$$

als Substitutionen maßgebend sind. Hierzu ist als neuere Substitution zu erwähnen:

$$\text{const.} \, d t = r^{\frac{3}{2}} d u \tag{1a}$$

die gewissermaßen ein Mittel aus den Substitutionen (1) ist. Diese Substitution, die auf elliptische Funktionen führt, ist an und für sich durch die Analogie mit der mittleren Anomalie gerechtfertigt, welche letztere durch die Beziehung:

$$k \sqrt{1 + m} \, d t = a^{\frac{3}{2}} d g$$

gegeben ist.

Mit Anwendung der ersten der Substitutionen (1) ergibt sich unmittelbar die Integration der Differentialgleichungen des Zweikörperproblems in rechtwinkligen Koordinaten:

$$\frac{d^2 x}{d t^2} + k^2 (1 + m) \frac{x}{r^3} = 0 \, ; \quad \frac{d^2 y}{d t^2} + k^2 (1 + m) \cdot \frac{y}{r^3} = 0 \tag{2}$$

indem z. B. erhalten wird:

$$d \left(\frac{d x}{d t} \right) = - \frac{k^2 (1 + m)}{\sqrt{c}} \cdot \frac{x}{r} d v = - \frac{k^2 (1 + m)}{\sqrt{c}} \cos v \, d v = - \frac{k^2 (1 + m)}{\sqrt{c}} \cdot d \left(\frac{y}{r} \right)$$

Man bekommt so die „Geschwindigkeitsintegrale":

$$\frac{dx}{dt} = -\frac{k^2(1+m)}{\sqrt{c}}\cdot\frac{y}{r}+h$$

$$\frac{dy}{dt} = +\frac{k^2(1+m)}{\sqrt{c}}\cdot\frac{x}{r}+l$$

(3)

Benutzt man von jetzt an die zweite der Substitutionen (1), d. h. die exzentrische Anomalie, so verschwindet r im Nenner, und es wird z. B.:

$$dx = -\frac{k^2(1+m)}{an\sqrt{c}}\cdot y\,du+h\,dt$$

$$dy = +\frac{k^2(1+m)}{an\sqrt{c}}\cdot x\,du+l\,dt$$

woraus dann mit Leichtigkeit die Formeln:

$$x = a(\cos u - e);\quad y = u\sqrt{1-e^2}\sin u;\quad r = a(1-e\cos u)$$

herfließen.

Wie systematisch diese Vorgangsweise auch erscheinen mag, so läßt sich dieselbe jedoch nicht ohne weiteres für mehr als zwei Körper verallgemeinern. Daher die Auffassung, daß die Integrale der lebendigen Kraft und der Flächen, die solcher Verallgemeinerung fähig sind, vor jeder anderen Beziehung im Zweikörperprobleme den Vorzug haben.

I. Beziehungen zum Zweizentrenproblem.

1. Indessen wurde neuerdings auf gewisse Beziehungen im Zweikörperprobleme hingewiesen, die mit dem Zweizentrenproblem Verwandtschaft zeigen. Führt man in den ursprünglichen Differentialgleichungen (2) die neuen Bestimmungsstücke:

$$\xi^2 = \frac{x+iy}{a};\quad \eta^2 = \frac{x-iy}{a}$$

ein, so ergeben sich mit Anwendung des Vis viva-Integrales die einfachen Gleichungen:

$$\frac{d^2\xi}{du^2}+\tfrac{1}{2}\xi = 0;\quad \frac{d^2\eta}{du^2}+\tfrac{1}{2}\eta = 0$$

woraus:

$$\xi = \alpha\sin\frac{u-ia_0}{2};\quad \eta = -\alpha\sin\frac{u+ia_0}{2}$$

(4)

und damit:

$$\frac{r}{a} = \xi\eta = 1-\frac{\cos u}{\cos ia_0}\qquad\left(\alpha = \sqrt{\frac{-2}{\cos ia_0}}\right)$$

so daß die Exzentrizität als $e = 1 : \cos i\,a_0$ gegeben ist. Der Radius nimmt also den Wert „Null" an für die Argumente $u = \pm i\,a_0$.

2. Ohne auf Einzelheiten hier einzugehen, möge nun das Problem der zwei festen Zentra demgegenüber betrachtet werden. Hier erfolgt die bekannte Eulersche Auflösung durch Einführung der Summe und der Differenz der beiden Radien:

$$r + s = \mu; \qquad r - s = v \tag{5}$$

woraus auch:

$$x = \frac{\mu v + a^2}{2a}; \qquad x - a = \frac{\mu v - a^2}{2a}; \qquad \eta = \frac{\sqrt{(\mu^2 - a^2)(a^2 - v^2)}}{2a}$$

Diese Transformation hat neuerdings an aktueller Bedeutung gewonnen durch die Bemerkung[1]) von Prof. H. Samter über deren Zusammenhang mit der Thieleschen Transformation im Dreikörperproblem in enger Fassung, sowie durch die Untersuchungen, die Samter auf Veranlassung von Prof. E. Strömgren hierüber begonnen hat.

Setzt man mit Thiele:

$$v = a\cos u; \qquad \mu = a\cos i v \tag{6}$$

so wird nach einigen Reduktionen:

$$x + iy = a\cos^2\frac{u + iv}{2}; \qquad x - a + iy = -a\sin^2\frac{u + iv}{2}$$

$$x - iy = a\cos^2\frac{u - iv}{2}; \qquad x - a - iy = -a\sin^2\frac{u - iv}{2}$$

Hier ist a der Abstand der beiden *Focus*, und setzt man deshalb $a = 2a_1 e$ und außerdem:

$$\frac{x + iy}{a_1} = \xi^2; \qquad \frac{x - a + iy}{a_1} = \xi_1^2$$

$$\frac{x - iy}{a_2} = \eta^2; \qquad \frac{x - a - iy}{a_1} = \eta_1^2$$

so ist auch beispielsweise:

$$\xi_1 = \sqrt{-2e}\cdot\sin\frac{u + iv}{2}$$

$$\eta_1 = -\sqrt{-2e}\cdot\sin\frac{u - iv}{2} \tag{7}$$

Formeln, die den eben für das Zweikörperproblem betrachteten (4) analog sind. Von (4) zu (7) kommt man eben durch Variation der Konstante a_0, die in $-v$ übergeht.

3. Die Differentialgleichungen sind die folgenden:

$$\frac{d\mu}{\sqrt{(\mu^2 - a^2)\,M}} = \frac{2}{\mu^2 - v^2}\,dt; \qquad \frac{dv}{\sqrt{(a^2 - v^2)\,N}} = \frac{2}{\mu^2 - v^2}\,dt$$

---

[1]) Astr. Nachr. Nr. 5267.

wo M und N Polynome in μ und ν sind. Die Veränderlichen werden aber erst getrennt, wenn mit $+\,1$, $-\,1$ resp. μ^2, $-\,\nu^2$ multipliziert und addiert wird.

Anstatt t führte hier THIELE eine neue Veränderliche l durch die Substitution:

$$n\,d\,t = r\,s\,d\,l \tag{8}$$

in gewisser Analogie mit (1) ein.

Wenn aber von den beiden anziehenden Massen A und B die eine, z. B. A, verschwindet, so sollte die Substitution eigentlich in die im Zweikörperprobleme übliche, nämlich:

$$n\,d\,t = \frac{r}{a}\,d\,u$$

übergehen. Dies ist jedoch mit der Substitution (8) nicht der Fall und infolgedessen haben die Argumente mit der exzentrischen Anomalie keine Verwandtschaft. Die Aufgabe bleibt so von dem Zweikörperprobleme getrennt. Diese Trennung ist andererseits in dem Probleme dreier Körper in enger Fassung an und für sich vorhanden. Denn sobald eine Masse z. B. $A = 0$ wird, so gibt es keine Veranlassung, die Bewegung von C auf den bewegten Radius $A\,B$ zu beziehen; es entsteht eine feste elliptische Bahn von C um B. Den Übergang zwischen den beiden Fällen kann man sich kaum anders als durch Vermittlung einer Bewegung in bezug auf zwei feste Zentra vorstellen.

Im letzteren Falle ist nämlich der Übergang durch eine Modifikation von der Substitution (8) wohl denkbar. Setzt man nämlich anstatt (8):

$$a^2\,n\,d\,t = \frac{A\,a + B\,a}{A\,s + B\,r}\cdot r\,s\,d\,l \tag{9}$$

so geht dieselbe z. B. für $B = 0$ in die Substitution:

$$n\,d\,t = \frac{r}{a}\,d\,u$$

und das Argument l in die exzentrische Anomalie u über.

Auf Einzelheiten dieser Vorgangsweise soll hier nicht eingegangen werden. Nur ist es vielleicht von Interesse, anzugeben, wie sich hierdurch das Gleichungssystem verändert, nämlich in:

$$\frac{d\,\mu}{\sqrt{M_1}} - \frac{d\,\nu}{\sqrt{N_1}} = 0$$

$$\frac{\mu\,d\,\mu}{\sqrt{M_1}} - \frac{B - A}{B + A}\cdot\frac{\nu\,d\,\nu}{\sqrt{N_1}} = 4\,d\,l$$

Als besonderer Spezialfall tritt hier die Bedingung $A = B$ auf. Eine früher für die geradlinige Bewegung dreier Körper in Vorschlag gebrachte Substitution[1]:

$$c\, d\, t = \frac{x\, y}{x + y}\, d\, u$$

deckt sich mit der Substitution (9) für $A = B$.

2. Einige Beziehungen des allgemeinen Dreikörperproblems.

4. Auf den Radius $A\, B$ als x-Achse bezogen, haben die Bewegungsgleichungen des dritten Körpers C im Dreikörperprobleme enger Fassung die Form:

$$\frac{d^2 x}{d\, t^2} - 2\, n\, \frac{d\, y}{d\, t} - n^2 (x - \alpha) = - \frac{A\, x}{r^3} - \frac{B\, (x - a)}{s^3}$$

$$\frac{d^2 y}{d\, t^2} + 2\, n\, \frac{d\, x}{d\, t} - n^2\, y = - \frac{A\, y}{r^3} - \frac{B\, y}{s^3} \tag{10}$$

wobei $\alpha = \dfrac{B}{A + B} \cdot a$ und $\alpha + \beta = a$ gesetzt sein möge. Hierzu existiert das sog. Jacobische Integral, welches, obschon es mit dem Vis viva-Integrale ähnliche Ableitung und Gestalt gemeinsam hat, mit dem letzteren nicht zu verwechseln ist. Die Thielesche Transformation dieser Gleichungen wurde von ihm nur für den Fall $A = B$ angegeben. Der allgemeine Fall ist schon etwas mehr verwickelt und enthält eine Anzahl von $\dfrac{A - B}{A + B}$ abhängige Glieder. Die entsprechenden Gleichungen sind diesem Aufsatze, jedoch ohne Ableitung, als Zusatz zur Orientierung beigefügt.

5. Obschon es gewagt erscheinen mag, auf das allgemeine Dreikörperproblem im Rahmen dieses Aufsatzes einzugehen, möchte ich es doch nicht unterlassen, über einige neuerdings in bezug auf die Lagrangesche Reduktion dieses Problems zum Abschluß geführten Entwicklungen hier zu berichten. Bei dieser Reduktion ist man bestrebt, die Bestimmung der Radien r_1, r_2, r_3 getrennt zu führen, was indessen im allgemeinen nur durch Hinzuziehung einer vierten Größe s durch die Gleichung:

$$\frac{d\, s}{d\, t} = - [m_1\, p_1\, q_1 + m_2\, p_2\, q_2 + m_3\, p_3\, q_3] \tag{11}$$

geschehen kann. Die r_1, r_2, r_3 werden durch:

$$p_1 = \frac{r_2^2 + r_3^2 - r_1^2}{2}; \qquad p_2 = \frac{r_3^2 + r_1^2 - r_2^2}{2}; \qquad p_3 = \frac{r_1^2 + r_2^2 - r_3^2}{2}$$

---

[1] Analytische Merkmale des Dreikörperproblems. Astr. iaktt. o. unders. Observatorium, Stockholm. Bd. 9, Nr. 1, S. 56.

ersetzt und die Bezeichnungen:

$$q_1 = \frac{1}{r_2^3} - \frac{1}{r_3^3}; \quad q_2 = \frac{1}{r_3^3} - \frac{1}{r_1^3}; \quad q_3 = \frac{1}{r_1^3} - \frac{1}{r_2^3}$$

zur Vereinfachung gebraucht. Bei der Weiterführung der Rechnungen ergeben sich fast von selbst als Fundamentale die Größen:

$$R = \frac{1}{2}\left(\frac{r_1^2}{m_1} + \frac{r_2^2}{m_2} + \frac{r_3^2}{m_3}\right) = \frac{1}{2}\left(\mu_1 p_1 + \mu_2 p_2 + \mu_3 p_3\right)$$

$$\Lambda^2 = p_1 p_2 + p_2 p_3 + p_3 p_1 \qquad \text{(Doppelte Dreiecksfläche)}$$

$$Q = d p_1 d p_2 + d p_2 d p_3 + d p_3 d p_1$$

und es wurde gezeigt[1]), daß die oben erwähnte Größe s eine solche Bedeutung hat, daß:

$$R s + S = - k \Delta \cos i \qquad (11\mathrm{a})$$

wo i die Neigung der Dreiecksfläche gegen die unveränderliche Ebene ist. Für die Bewegung in einer Ebene ist $i = o$, und weil:

$$S = \frac{1}{2}\left[\frac{P_1}{m_1} + \frac{P_2}{m_2} + \frac{P_3}{m_3}\right]$$

und dabei:

$$P_1 = p_2 d p_3 - p_3 d p_2; \; P_2 = p_3 d p_1 - p_1 d p_3; \; P_3 = p_1 d p_2 - p_2 d p_1 \qquad (12)$$

wird in diesem Falle die Bestimmung der Radien tatsächlich getrennt geführt werden können.

Die Rechnung mit diesen auf das Dreieck bezüglichen Größen läßt sich systematisch durchführen. Einige Resultate ergeben sich fast unmittelbar, so z. B. die schon längst bekannte Gleichung:

$$\frac{d^2 R}{d t^2} = U + h \qquad (13)$$

mit dem Potential $U = \sum \frac{1}{m r}$. Verhältnismäßig einfach ist auch die Differentialgleichung für Δ:

$$\frac{d^2 \Delta}{d t^2} + \Delta\left[\frac{m_2 + m_3}{r_1^3} + \frac{m_3 + m_1}{r_2^3} + \frac{m_1 + m_2}{r_3^3}\right] = \frac{1}{2}\frac{s^2 + Q}{\Delta} \qquad (14)$$

Durch einige Reduktionen, die hier fortgelassen sein mögen, zeigt sich, daß Δ im Nenner auf der rechten Seite, wie naturgemäß ist, sich aufhebt, wobei ein aus dR, $d\Delta$ und Q zusammengesetzter Ausdruck sich entwickelt. Die Größe Δ erscheint infolge der Einfachheit dieser Gleichung als geeignetes Bestimmungsstück, das neben R in Betracht kommen kann. Im Spezialfalle eines gleichseitigen Dreiecks macht auch die Größe Δ keine Schwierigkeit. Neuerdings wurde noch die Auf-

---

[1]) Vetenskaps Akademiens Handlingar Bd. 42, Nr. 9. Stockholm 1907.

merksamkeit darauf gelenkt, daß im Probleme enger Fassung, wenn nämlich die eine Bewegung kreisförmig ist, z. B. wenn $r_3 = a$ und eine Masse verschwindet, infolge der dann geltenden Beziehung:

$$\varDelta = a\,y$$

das Dreieck einfach auf die y-Koordinate zurückgeführt werden kann.

Bei näherer Betrachtung geht dann die $\varDelta$-Gleichung (14) in die folgende:

$$\frac{d^2 y}{d t^2} + 2n \frac{d x}{d t} - n^2 y = - \frac{A y}{r^3} - \frac{B y}{s^3}$$

über. Dies ist die zweite der Gleichungen (10). Es befremdet dann zunächst, daß es im allgemeinen Dreikörperprobleme an einer mit $\varDelta$ verwandten Größe und damit an einer zu (14) entsprechenden Differentialgleichung fehlt. Der Zusammenhang in dieser Beziehung zeigt sich aber alsbald der zu sein, däß die zweite Differentialgleichung (10), nämlich für die x-Koordinate, unter den gemachten Voraussetzungen aus der Lagrangeschen Gleichung (11):

$$\frac{d s}{d t} = - \sum m\,p\,q$$

herfließt. Die allgemeine Behandlung des Problems schließt also einen Hinweis auf den reduzierten Spezialfall ein.

6. Die Gleichung der lebendigen Kraft erhält, wenn wir uns hier vorerst an die Bewegung in einer Ebene halten, die einfache Form:

$$\frac{1}{2R} \left[\left(\frac{d R}{d t} \right)^2 + \frac{M}{\mu} \left(\frac{d \varDelta}{d t} \right)^2 - \frac{M}{\mu} Q \right] = 2M \sum \frac{1}{m\,r} + h - \frac{k^2}{2R} \qquad (15)$$

$M = $ Summe, $\mu = $ Produkt der Massen. In dieser Gleichung ist das Flächenintegral durch die Größe $k^2 : 2\,R$ mit berücksichtigt.

7. Wollte man auf Grund dieser Vis-viva-Gleichung die Differentialgleichungen ableiten, so müßten die Operationen in bezug auf r_1, r_2, r_3 oder p_1, p_2, p_3 ausgeführt werden, weil zu R und $\varDelta$ kein drittes Bestimmungsstück wohldefinierter Art zugeordnet ist. Der Umstand, daß dies Verfahren die Differentialgleichungen richtig wiedergibt, jedoch nur bis auf gewisse Glieder, die besondererweise in Wegfall kommen, berührt gewissermaßen die allgemeinen mechanischen Prinzipien und eine kurze Erläuterung des Verhaltens dürfte daher nicht ohne Interesse sein.

Gibt man bei Einführung der üblichen Bezeichnungen:

$$2T = \frac{1}{2R} \left[\left(\frac{d R}{d t} \right)^2 + \frac{M}{\mu} \left(\frac{d \varDelta}{d t} \right)^2 - \frac{M}{\mu} Q \right]$$

$$2U_1 = 2M \sum \frac{1}{m\,r} - \frac{k^2}{2R}$$

der Vis viva-Gleichung (15) die Form $\left(H = \dfrac{h}{2}\right)$:

$$T - U_1 = H$$

und führt man die Rechnung in bezug auf die Gleichungen:

$$\frac{d}{dt}\left(\frac{\partial T}{\partial p_1^1}\right) - \frac{\partial T}{\partial p_1} = \frac{\partial U_1}{\partial p_1}$$

$$\frac{d}{dt}\left(\frac{\partial T}{\partial p_2^1}\right) - \frac{\partial T}{\partial p_2} = \frac{\partial U_1}{\partial p_2} \tag{16}$$

$$\frac{d}{dt}\left(\frac{\partial T}{\partial p_3^1}\right) - \frac{\partial T}{\partial p_3} = \frac{\partial U_1}{\partial p_3}$$

aus, so ergeben sich nachher leicht drei andere Gleichungen, deren erste dadurch erhalten wird, daß man in (16) sukzessive mit

$$p_1, \quad p_2, \quad p_3$$

multipliziert und addiert. Es ergibt sich in Übereinstimmung mit (13):

$$\frac{d^2 R}{dt^2} = U + h$$

Multipliziert man wieder in (16) der Reihe nach mit

$$\frac{1}{m_1}, \quad \frac{1}{m_2}, \quad \frac{1}{m_3}$$

so kommt die Gleichung:

$$\left.\begin{aligned}\frac{d^2 \Delta}{dt^2} - \frac{1}{R}\frac{dR}{dt}\frac{d\Delta}{dt} + \Delta\left[\sum \frac{m_2 + m_3}{r_1^3} + \frac{2U + h}{R} - \frac{k^2}{R^2}\right] &= 0 \\ &= \frac{kS}{R^2}\end{aligned}\right\}$$

die indessen fehlerhaft ist. Durch einige Reduktionen ergibt sich nämlich aus der Gleichung (14) auf der rechten Seite statt „Null" das oben hinzugesetzte Glied:

$$\frac{kS}{R^2}$$

Multipliziert man wiederum in (16) der Reihe nach mit

$$\mu_2 r_3^2 - \mu_3 r_2^2; \quad \mu_3 r_1^2 - \mu_1 r_3^2; \quad \mu_1 r_2^2 - \mu_2 r_1^2$$

so ergibt sich die Gleichung:

$$\left.\begin{aligned}\frac{dS}{dt} - \frac{S}{R}\frac{dR}{dt} - R\sum mpq &= 0 \\ &= -\frac{k}{R}\left(R\frac{d\Delta}{dt} - \Delta\frac{dR}{dt}\right)\end{aligned}\right\}$$

Gemäß (11) und mit Rücksicht auf (11 a) soll doch rechter Hand anstatt „Null" das nebenbei angesetzte Glied

$$-\frac{k}{R}\left(R\frac{d\Delta}{dt} - \Delta\frac{dR}{dt}\right)$$

stehen.

8. Das Auftreten dieser Anomalien liegt daran, daß solche Gleichungen wie (16) nicht immer unbedingt richtig sind. Denn man kann diesen Gleichungen die allgemeinere Form:

$$\frac{d}{dt}\left(\frac{\partial T}{\partial p_1^1}\right) - \frac{\partial T}{\partial p_1} = \frac{\partial U_1}{\partial p_1} + f P_1$$

$$\frac{d}{dt}\left(\frac{\partial T}{\partial p_2^1}\right) - \frac{\partial T}{\partial p_2} = \frac{\partial U_1}{\partial p_2} + f P_2 \qquad (16a)$$

$$\frac{d}{dt}\left(\frac{\partial T}{\partial p_3^1}\right) - \frac{\partial T}{\partial p_3} = \frac{\partial U_1}{\partial p_3} + f P_3$$

geben, ohne daß bei beliebiger Bestimmung des Faktors f die Vis viva-Gleichung verändert wird, weil die Größen P_1, P_2, P_3 gemäß (12) die Form von Flächengeschwindigkeiten haben, so daß die Beziehung:

$$P_1 dp_1 + P_2 dp_2 + P_3 dp_3 = 0$$

erfüllt ist.

Daraus, daß ebenso die Beziehung:

$$P_1 p_1 + P_2 p_2 + P_3 p_3 = 0$$

stattfindet, folgt weiter, daß auch die Gleichung:

$$\frac{d^2 R}{dt^2} = U + h$$

von den Zusatzgliedern in (16a) unberührt ist, während in den beiden anderen Gleichungen die Zusatzglieder in (16a), bei der Annahme

$$f = \frac{1}{2}\frac{M}{\mu}\frac{k}{R^2 \Delta} \qquad (17)$$

die oben erwähnten Zusatzbeträge auf der rechten Seite dieser Gleichungen erzeugen.

Zusatz.

Die vollständigen Differentialgleichungen in dem Dreikörperproblem in enger Fassung mit elliptischen Koordinaten $(dt = rs\,dl)$.

$$\frac{d^2\mu}{dl^2} - \frac{\mu}{\mu^2 - a^2}\left(\frac{d\mu}{dl}\right)^2 + 2n\frac{\mu^2 - \nu^2}{4}\cdot\frac{\sqrt{\mu^2 - a^2}}{\sqrt{a^2 - \nu^2}}\cdot\frac{d\nu}{dl}$$

$$- n^2\frac{\mu^2 - a^2}{4}\left\{\frac{\mu(\mu^2 - a^2)}{2} + \frac{1}{2}\frac{A - B}{A + B}\left[\begin{array}{l}a^2(\mu + \nu)\\ + \nu[\tfrac{3}{2}(\mu^2 - a^2) + \tfrac{1}{2}(a^2 - \nu^2)]\end{array}\right]\right\}$$

$$= [\tfrac{1}{2}(A + B) + C\mu](\mu^2 - a^2);$$

$$\frac{d^2\nu}{dl^2} + \frac{\nu}{a^2 - \nu^2}\left(\frac{d\nu}{dl}\right)^2 - 2n\frac{\mu^2 - \nu^2}{4}\frac{\sqrt{a^2 - \nu^2}}{\sqrt{\mu^2 - a^2}}\cdot\frac{d\mu}{dl}$$

$$- n^2\frac{a^2 - \nu^2}{4}\left\{\frac{\nu(a^2 - \nu^2)}{2} - \frac{1}{2}\frac{A - B}{A + B}\left[\begin{array}{l}a^2(\mu + \nu)\\ - \mu[\tfrac{3}{2}(a^2 - \nu^2) + \tfrac{1}{2}(\mu^2 - a^2)]\end{array}\right]\right\}$$

$$= [-\tfrac{1}{2}(A - B) - C\nu](a^2 - \nu^2)$$

(2 C ist die zum Potential hinzugefügte Konstante im Integral der lebendigen Kraft.)

Zur Bestimmung effektiver Wellenlängen der Sterne.

Von **G. Eberhard**, Potsdam.

Die Bestimmung effektiver Wellenlängen von Sternen wird neuerdings immer häufiger zur Anwendung gebracht, seitdem man fand, daß diese recht brauchbare Farbenäquivalente darstellen. Aber gewisse Schwierigkeiten sind trotz mancher eingehenden Untersuchungen immer noch geblieben. Es hat sich nämlich gezeigt, daß jeder Beobachter (und ebenso jedes Instrument) seine eigene Skala effektiver Wellenlängen hat, die in sich richtig sein kann, sich aber nicht ohne weiteres mit anderen Skalen vergleichen oder auf diese reduzieren läßt. Bei Benutzung von Refraktoren als Aufnahmsinstrumente pflegt man solche Unterschiede in der Hauptsache als eine Folge der verschiedenen Achromatisierung der Objektive oder, wenn identische Objektive benutzt werden, als Folge verschiedener Fokusierung anzusehen[1]. In der Tat können, namentlich bei älteren Objektivtypen, sehr erhebliche Differenzen in den effektiven Wellenlängen (und natürlich auch in den anderen Farbenäquivalenten) auf diese Weise entstehen, wie sich leicht einsehen läßt. Besondere Komplikationen treten auf, wenn das Objektiv außerdem noch eine beträchtliche Bildfeldwölbung besitzt. Photographiert man z. B. mit einem Himmelskartenrefraktor der üblichen Art eine Gruppe Sterne von gleichem Spektraltypus, nachdem man die Platte so gestellt hat, daß in ihren mittleren Teilen gute Schärfe der Sternbildchen erzielt wird, so werden sich infolge der starken Bildfeldwölbung dieser Objektive nicht die photographisch wirksamsten Strahlen der nach dem Rande der Platte zu stehenden Sterne auf der Platte vereinigen, sondern die Strahlen größerer Wellenlänge, d. h. die Sterne am Rande der Platte werden so wirken, als ob sie anders gefärbt wären als die Sterne in den mittleren Teilen. Dieser Färbungsunterschied erreicht bei einem Himmelskartenrefraktor einen solchen Betrag, daß ein A-Stern, wenn er nahe dem Rande der Platte steht, wie ein M- oder N-Stern auf die lichtempfindliche Schicht wirkt, wie Versuche von Prof. BIRCK mit dem Potsdamer Himmelskartenrefraktor ergaben. (Bei der Berechnung der Refraktion sollte man

---

[1] ROSENBERG: Über den Einfluß der Fokusierung auf die photographisch wirksamen Wellenlängen, Astr. Nachr. 5109, Bd. 213, S. 329.

8*

dies immer berücksichtigen.) Wesentlich günstiger in dieser Hinsicht wirken die modernen Objektive mit geebnetem Gesichtsfeld (Triplets).

Aber auch bei der Benutzung von Reflektoren erhielten, wider Erwarten, verschiedene Beobachter stark voneinander abweichende Skalen effektiver Wellenlängen, wie neuerdings einige größere Beobachtungsreihen, z. B. die von M. WOLF[1]), zeigen. Ja selbst mehrfache, mit einem Reflektor von demselben Beobachter unter gleichen Bedingungen, aber an verschiedenen Abenden ausgeführte Bestimmungen weisen Differenzen auf, die weit größer sind, als daß sie sich durch Ungenauigkeit der Messung selbst erklären ließen. Luftunruhe allein kann man dafür nicht verantwortlich machen. Zwar zeigen die schwachen Bilder bei unruhiger Luft etwas größere effektive Wellenlängen als bei ruhiger[2]), aber diese Unterschiede sind nicht so groß, zumal man bei besonders ungünstiger Luftbeschaffenheit derartige Aufnahmen nicht machen wird.

Störungen in den optischen Teilen des Aufnahmeapparates (Deformationen der Spiegel und Objektive) sowie Fehler in den Gittern dürften zwar auch Abweichungen hervorbringen, aber solche optischen Ursachen reichen nicht aus, um die großen Unterschiede zu erklären, die z. B. zwischen den Reflektoraufnahmen von BERGSTRAND und M. WOLF bestehen. Man hat überdies auch mit fehlerfreien Gittern solche großen Unstimmigkeiten gefunden.

Ohne Zweifel werden aber solche erzeugt durch systematisch verschiedene Auffassung der kleinen Beugungsspektren durch die Beobachter. So stellt z. B. M. WOLF auf „den Schwerpunkt der kurzen fischförmigen Figuren" ein[3]), während HERTZSPRUNG[4]) und LINDBLAD[5]) die „geschwärzte Gesamtfläche des Spektrums halbieren". Gerade die sorgfältige Untersuchung von WOLF zeigt, welche Schwierigkeiten dem Beobachter aus der Gestalt der kleinen Spektren und der Lichtverteilung in ihnen entstehen können. Aber auch diese Erklärung genügt nicht völlig, denn wie oben schon erwähnt, treten häufig für den mit demselben optischen Apparat arbeitenden Beobachter recht erhebliche Differenzen zwischen verschiedenen Aufnahmen auf, wo also eine verschiedene Auffassung des Bildchens nicht in Frage kommen kann.

Es bleibt schließlich nur noch klarzulegen, welche Rolle die photographische Platte bei der Bestimmung effektiver Wellenlängen spielt, und ich habe daher das Verhalten dieser einer eingehenden Untersuchung

---

[1]) Astr. Nachr. 5092, Bd. 213, S. 49 ff. Versuche mit dem Objektivgitter.

[2]) HERTZSPRUNG, E: Publ. d. Astrophysikal. Observatoriums zu Potsdam Nr. 63, S. 17.

[3]) l. c. S. 51, auch 54, 58.

[4]) l. c. S. 8.

[5]) LINDBLAD; Die photographisch effektive Wellenlänge als Farbenäquivalent der Sterne. Ark. f. Matem., Astron. o. Fysik Bd. 13, Nr. 26, S. 8.

unterzogen. Da Aufnahmen mit den gebräuchlichen Stabgittern viel zu klein für eine exakte Ausmessung mit dem Mikrophotometer sind, habe ich die Spektra mit Hilfe eines kleinen Sternspektrographen hergestellt. Als Lichtquelle diente eine auf konstanter Stromstärke gehaltene mattierte Metallfadenlampe. Vor der Öffnung des die Lampe einschließenden lichtdichten Kastens befanden sich noch einige matte Glasscheiben, die das Licht der Lampe gleichmäßiger verteilten. Vor der äußeren Mattscheibe und eng an ihr anliegend ließen sich zehn feine Drahtgitter bekannter Absorption zur meßbaren Änderung der Intensität einschieben, und zwar schwächte jedes Gitter rund 0^m5 mehr als das vorhergehende. Mittels dieser Vorrichtung ist auf jede Platte eine Reihe von Spektren mit wachsender Intensität, aber bei konstanter Belichtungszeit hergestellt worden.

Nachdem sich im Verlauf meiner photographisch-photometrischen Versuche ergeben hatte, daß die Dicke der Emulsionsschicht einen starken Einfluß auf das gesamte Verhalten der Platte dem Lichte gegenüber hat, benutzte ich zwei Sorten Schleußnerplatten *derselben* Emulsion, deren Schichtdicken 0,009 mm bzw. 0,050 mm betrug. Die Herstellung dieser Platten verdanke ich Dr. LÜPPO CRAMER. Die Aufnahmen sind mit VON HÜBLschem Glyzinentwickler völlig ausentwickelt (40 Minuten bei $+ 18°$) und nach Fertigstellung in einem Mikrophotometer vermessen worden.

Als erstes Ergebnis folgte, daß die Lage des Schwärzungsmaximums *ganz unabhängig* von der einwirkenden Lichtintensität ist. Es lag bei der Platte mit dünner Schicht zwischen $\lambda\,445$ und $\lambda\,448$, bei der mit dicker Schicht bei $\lambda\,462$. Aus dieser unveränderlichen Lage ist zu schließen, daß weder bei der HERTZSPRUNG-LINDBLADschen noch bei der WOLFschen Auffassung der Beugungsspektren das Schwärzungsmaximum selbst gemessen wird, sonst würde ja keine Reduktion auf normale Helligkeit anzubringen sein, die sich bei allen bisher vorliegenden Beobachtungsreihen als nötig erwiesen hat. Es zeigt sich aber ferner, daß die Farbenauffassung der Platte nicht unerheblich von der Dicke der Emulsionsschicht abhängig ist derart, daß die dicke Schicht eine stärkere Empfindlichkeit für Licht längerer Wellen besitzt als die dünne. Sternaufnahmen auf einer Platte mit dicker Schicht müssen also eine größere effektive Wellenlänge und natürlich auch einen größeren Farbindex aufweisen als auf einer dünnen Schicht derselben Emulsion. Da nun Platten auch aus ein und derselben Packung nur selten gleichmäßige Schichtdicken besitzen, so ist es begreiflich, daß man bei wiederholten Aufnahmen derselben Sterngruppe häufig voneinander verschiedene effektive Wellenlängen erhält. Da ferner fast jede Platte teils unregelmäßige, teils regelmäßig verlaufende Fehler (keilförmige Schicht) in der Dicke hat, so werden bei wiederholten Aufnahmen nicht selten größere Abweichungen einzelner Sterne vorkommen, oder die effektiven Wellenlängen der Sterne der einen Aufnahme werden vom Ort des Sterns auf der Platte abhängige Differenzen gegen die einer zweiten Platte besitzen. Beide Fälle, besonders ersterer, sind häufig beobachtet worden.

Werden die Platten nicht völlig ausentwickelt, was wohl gewöhnlich der Fall ist, so ändert sich nichts an dem vorstehenden Ergebnisse, als daß die Schwärzungsmaxima etwas nach den kleineren Wellenlängen hinrücken, wie folgende Tabelle zeigt:

Schwärzungsmaximum der dünnen dicken Schicht		Rodinal 1:20 5 Minuten
$\lambda\,443$	$\lambda\,459$	$+17,5°$
$\lambda\,444$	$\lambda\,458$	$+13,5°$

Es war nun von Interesse zu sehen, was sich ergibt, wenn man die vorliegenden Aufnahmen so behandelt, wie es meist bei der Bestimmung effektiver Wellenlängen geschieht, wo, wie bereits erwähnt, die geschwärzte Gesamtfläche halbiert wird. Zu diesem Zweck habe ich die Schwärzungen der in engen Intervallen mikrophotometrisch durchgemessenen Spektren als Ordinaten, die Wellenlängen als Abszissen in Millimeterpapier eingetragen und die resultierenden Punkte durch eine sich möglichst eng anschließende glatte Kurve verbunden. Man erhält auf diese Weise ebenso viele Kurven gleicher einwirkender Intensitäten, als man Spektralaufnahmen auf der Platte hat. Liest man nun aus der Zeichnung die Wellenlängen auf dem auf- wie auf dem absteigenden Ast der Kurve ab, die einer bestimmten Schwärzung entsprechen, und nimmt das Mittel aus diesen Ablesungen, so erhält man für jede einwirkende Intensität die Größe, welche von den meisten Beobachtern als effektive Wellenlänge bezeichnet wird. In der folgenden Tabelle sind die Werte für die Intensitäten $4^{m}\!.5$ bis $1^{m}\!.0$ und für die Schwärzungen 35, 39, 44, 47 des Meßkeiles (die Schwärzungen wachsen mit den Zahlen) enthalten.

Glycin (+18°) 40 Minuten.

m	λ effekt. Schleussner dick				λ effekt. Schleussner dünn		
	$S=35$	$S=39$	$S=44$	$S=47$	$S=35$	$S=39$	$S=44$
	$\mu\mu$	$\mu\mu$	$\mu\mu$	$\mu\mu$	$\mu\mu$	$\mu\mu$	$\mu\mu$
4,5	457,5	—	—	—	—	—	—
4,0	454	458	—	—	447,5	447,5	—
3,5	450	453	457	460,5	444,5	444	—
3,0	449	450,5	454	456	443	442,5	445,5
2,5	449,5	449,5	451,5	453,5	442	442	442
2,0	—	—	—	—	442,5	443	441
1,0	—	—	—	—	446	447	442,5
Schwärzungsmaximum:	$\lambda\,462$				$\lambda\,448$		

Die Tabelle zeigt wieder auf das deutlichste, daß die effektive Wellenlänge nicht mit dem Schwärzungsmaximum zusammenfällt, ferner daß sie stark mit der Intensität des einwirkenden Lichtes variiert. Diese Variation würde die Reduktion der effektiven Wellenlänge auf normale Helligkeit (Bildstärke) darstellen. Der Verlauf ist, wie aus der Tabelle folgt, gleichfalls stark von der Schichtdicke abhängig. Wenn man nun noch berücksichtigt, daß die verschiedenen Beobachter nicht nur verschiedene Bildstärken ihren Reduktionen als normale zugrunde legen, sondern auch die durch die Schwärzung bestimmten Grenzen der Sternscheibchen verschieden annehmen (z. B. wird sie der eine bei $S = 35$, der andere bei stärkerer Schwärzung, $S = 39$, sehen), so ergibt sich ohne weiteres aus der Tabelle, daß sehr große Unstimmigkeiten zwischen den einzelnen Beobachtern vorkommen können, auch wenn sie alle Reflektoren benutzt haben. Ganz analog gestalten sich übrigens die Verhältnisse, wenn man nicht ganz ausentwickelte Aufnahmen verwendet. Als Beispiel diene folgende Tabelle:

Rodinal 1 : 20 (+13,5°) 5 Minuten.

m	λ effekt. Schleussner dick			λ effekt. Schleussner dünn		
	$S=32,7$ $\mu\mu$	$S=35$ $\mu\mu$	$S=39$ $\mu\mu$	$S=32,7$ $\mu\mu$	$S=35$ $\mu\mu$	$S=39$ $\mu\mu$
4,0	458	—	—	—	—	—
3,5	452	—	—	446	—	—
3,0	448,5	452	—	442,5	—	—
2,5	448	449,5	—	441	442	—
2,0	447	447	—	440	440,5	—
1,5	448,5	447,5	455	440,5	440,5	440
1,0	449,5	448	451	442	441	439,5
0,5	452	449	449	444	441,5	438,5
0,0	454	451	449,5	446	442,5	440

Schwärzungsmaximum: $\lambda\,458$ $\lambda\,444$

In qualitativ ganz gleicher Weise wie die Schleussnerplatten verhalten sich die Platten anderer Fabriken, nur liegt das Schwärzungsmaximum zum Teil bei anderen Wellenlängen. Ich fand Werte für dasselbe zwischen $\lambda\,445$ und $\lambda\,485$.

Die photographische Platte übt somit einen großen, ungemein komplizierten Einfluß auf die Bestimmung effektiver Wellenlängen aus, und wenn man außerdem, wie oben besprochen, noch die verschiedene Auffassung der Spektren durch den sie ausmessenden Beobachter in Rechnung zieht, so dürften die starken Verschiedenheiten der Skalen mehrerer Beobachter ihre Erklärung in dem Zusammenwirken dieser beiden Ursachen finden. Bei dieser Sachlage kann natürlich nur eine rein *empirische* Reduktion der verschiedenen Skalen auf eine einheitliche Normalsequenz in Betracht kommen, daher ist der Vorschlag von Bergstrand und Rosenberg[1]) warm zu empfehlen.

---

[1]) Vorschlag zur Aufstellung einer Normalsequenz zur Bestimmung effektiver Wellenlängen. Astr. Nachr. 5159, Bd. 215, S. 447 ff.

Über die zwei Sternströme.

Von **Arnold Kohlschütter**, Potsdam.

Mit 4 Abbildungen.

1. Geschichtlicher Überblick bis zum gegenwärtigen Stand des Problems. Seit der Entdeckung der ersten Eigenbewegungen von Fixsternen ist der Frage nach Gesetzmäßigkeiten in den Bewegungen der Sterne dauernd das regste Interesse entgegengebracht worden. Schon HERSCHEL benutzte die geringe Zahl von nur 13 Sternen, deren Eigenbewegungen ihm zur Verfügung standen, um zu untersuchen, ob aus einer zufälligen Bewegungsverteilung ein systematisches Glied herausgefunden werden könnte. In der Tat leitete er schon aus diesen wenigen Sternen eine bevorzugte Bewegungsrichtung ab, die er naturgemäß als das Spiegelbild der Bewegung des Sonnensystems, von dem aus die Eigenbewegungen gemessen sind, auffaßte. Zahlreiche spätere Untersuchungen mit größerem Material bestätigten dieses Ergebnis, indem sich nur dies eine systematische Glied aus einer sonst zufälligen Bewegungsverteilung merklich hervorhob. So hat dieses Problem der Bestimmung des Zielpunktes der Sonnenbewegung, das „Apexproblem", mehr als ein Jahrhundert lang in dieser Fragestellung, die wir jetzt als zu eng gefaßt aufgeben müssen, bestanden, und viele eingehende Bearbeitungen erfahren.

Es war um die letzte Jahrhundertwende, als ein wesentlicher Schritt vorwärts getan wurde. Zunächst gelang es KOBOLD, festzustellen, daß man mit der Annahme, es lagere sich über die zufällige Verteilung der Sternbewegungen nur *ein* systematischer Effekt (die Sonnenbewegung), nicht auskommt, sondern daß auch nach Rücksichtnahme auf die Sonnenbewegung weitere systematische Abweichungen bestehen bleiben. In langjährigen Untersuchungen ist es dann KAPTEYN, durch Sammlung und kritische Bearbeitung sehr umfangreichen Beobachtungsmaterials, gelungen, die Existenz dieser Abweichungen über allen Zweifel sicherzustellen. In seiner „Zweistromhypothese" gab KAPTEYN dieser neuen Erscheinung eine Deutung. Er stellte sich vor, daß es im Sternsystem zwei Ströme gibt, die sich gegeneinander bewegen, von denen jeder in sich eine zufällige Geschwindigkeitsverteilung zeigt. Das Resultat der Überlagerung der Geschwindigkeiten ist eine bevorzugte Bewegungsrichtung im Raume, die sog. „Vertexrichtung", die mit der Richtung zusammenfällt, in welcher sich die beiden Ströme gegeneinander bewegen.

Diese Vorzugsrichtung liegt genau in der Milchstraße, nach den galaktischen Längen 164° bzw. 344° gerichtet. Beide Sterngruppen sind räumlich vollständig miteinander vermischt und zeigen außer der Bewegung durchaus keine unterschiedlichen Merkmale, indem spektrale Verteilung, absolute Größe, Entfernungen und alle anderen untersuchten Eigenschaften bei beiden merklich gleich sind. Es ist dies eine nicht zu erwartende auffallende Tatsache, welche zunächst Zweifel an der Richtigkeit der KAPTEYNschen Auffassung erzeugen konnte; aber die Wucht des aus den Bewegungen gelieferten Materials war so groß, daß diese Bedenken zunächst zurücktreten mußten.

Neben KAPTEYNS Zweistromtheorie wird meist die gleichzeitig entstandene SCHWARZSCHILDsche „Ellipsoidtheorie" gestellt. Insofern nicht ganz mit Recht, als KAPTEYN eine physikalische Deutung zu geben versuchte, während die Ellipsoidtheorie nur eine mathematische Darstellung der beobachteten Bewegungsanomalie ohne jeden Erklärungsversuch bedeutet. SCHWARZSCHILD nimmt nicht zwei Ströme, sondern nur einen einzigen Schwarm mit einer Abart MAXWELLscher Geschwindigkeitsverteilung an, von der Eigentümlichkeit, daß das Streuungsmaß der Geschwindigkeitsverteilung in den drei Koordinatenrichtungen verschieden ist. Die Bearbeitung des Beobachtungsmaterials nach der Ellipsoidtheorie liefert wieder eine bevorzugte Richtung, in welcher die Streuung der Geschwindigkeiten am größten ist, und diese Richtung fällt naturgemäß mit den KAPTEYNschen Vertices zusammen.

In neuester Zeit hat STRÖMBERG am Mount-Wilson-Observatorium mit dem jetzt zur Verfügung stehenden umfangreichen Material die räumlichen Bewegungen der Sterne untersucht. Ihm verdanken wir einen wesentlichen Fortschritt insofern, als er zeigte, daß man jetzt nicht mehr mit der für KAPTEYN und SCHWARZSCHILD ausreichenden Annahme einer größeren Beweglichkeit der Sterne in einer Richtung auskommt, sondern daß darüber hinaus noch eine Asymmetrie besteht. Eine Andeutung dieser Asymmetrie liegt schon in der seit vielen Jahren bekannten, aber in ihrer Bedeutung nicht voll erkannten und gewürdigten Tatsache, daß die größten Sterngeschwindigkeiten einseitig nur nach einer Seite des Himmels gerichtet sind, also nicht dieselbe Verteilung der Richtungen zeigen wie die kleineren Sterngeschwindigkeiten.

Man denke sich die Geschwindigkeit jedes einzelnen Sternes durch einen Vektor im Raume dargestellt. Als Ausgangspunkt dieser Vektoren wählt man zweckmäßig das Sonnensystem, weil wir von ihm aus die Sterngeschwindigkeiten messen; an und für sich ist jedoch die Wahl des Ausgangspunktes gleichgültig, und man könnte ebensogut jeden anderen Punkt wählen, der gegen die Sonne mit einer bestimmten Geschwindigkeit in einer bestimmten Richtung bewegt ist. Die Endpunkte der Vektoren erfüllen den Raum mit wechselnder Dichte, und den durch diese Dichteverteilung festgelegten Körper bezeichnen wir als „Geschwindigkeitskörper", die Flächen gleicher Dichte als „Geschwindigkeitsflächen". Die von STRÖMBERG geleistete große Arbeit bestand zunächst darin, diese Geschwindigkeitsflächen aus dem Beobachtungsmaterial ohne jede vorweggenommene Annahme über die Gestalt der Flächen abgeleitet zu haben. Es zeigte sich, daß die Geschwindigkeitsflächen im allgemeinen ellipsoid-ähnlich sind, wobei die größte Achse, wie man erwarten konnte, in die Richtung der KAPTEYNschen Vertices fällt; es erscheint demnach gerechtfertigt, von vornherein mit „Geschwindigkeitsellipsoiden" zu rechnen, was natürlich eine wesentliche Erleichterung bedeutet. Indem nun STRÖMBERG für verschiedene Gruppen von Sternen, die teilweise eine physikalische Zusammengehörigkeit, teilweise einen Bewegungszusammenhang bedeuteten, „Geschwindigkeitsellipsoide" ableitete, gab sich eine Asymmetrie kund, deren Wesen in den wichtigsten Zügen folgendermaßen geschildert werden kann: Wenn die Beobachtungen einen Unterschied in den Achsen der Ellipsoide verbürgen lassen, so liegt die größte Achse stets in Richtung der KAPTEYNschen Vertices. Aber die Mittelpunkte der Ellipsoide fallen nicht zusammen, sondern liegen auf einer Geraden, die in der galaktischen Ebene, und zwar senkrecht zur Vertexrichtung liegt. Die Mittelpunkte der Ellipsoide scheinen um so weiter abzurücken, je größer die Ausdehnung der Ellipsoide ist.

2. Die nächsten Sterne. Da die bisher bekannten systematischen Abweichungen von einer zufälligen Geschwindigkeitsverteilung parallel zur Milchstraße verlaufen — sowohl die KAPTEYNschen Ströme als auch die STRÖMBERGsche Asymmetrie liegen in dieser Ebene —, wollen wir uns auf die Betrachtung der in der galaktischen Ebene liegenden Bewegungskomponenten beschränken. Die von der Sonne aus gemessenen Geschwindigkeiten der einzelnen Sterne denken wir uns daher auf die galaktische Ebene projiziert, und diese Projektionen in ein Koordinatensystem eingetragen, dessen positive x-Richtung nach der galaktischen Länge 0°, und dessen positive y-Richtung nach der Länge $+ 90°$ zeigt. Die Untersuchung der Geschwindigkeitsverteilung besteht nun in der Diskussion einer Fläche $z\,(x, y)$, wobei z die Anzahl der in einem Flächenelement (x, y) endigenden Geschwindigkeitsvektoren bedeutet. Diese Fläche $z\,(x, y)$ kann direkt aus den beobachteten Geschwindigkeiten der einzelnen Sterne abgeleitet werden, was am zweckmäßigsten zeichnerisch durch Eintragung von „Höhenlinien" geschieht.

In der nebenstehenden Abb. 1 sind diese Höhenlinien für die uns nächsten Sterne wiedergegeben. Die Grundlagen hierfür habe ich einer kürzlich von J. HAAS veröffentlichten Zusammenstellung und Diskussion der uns näher als 20 Sternweiten stehenden Sterne entnommen, in welcher der Verfasser auch den allgemeinen Charakter der Geschwindigkeitsverteilung treffend schildert. Die Anzahl dieser Sterne — zur Zeit sind nur 186 bekannt — ist so gering, daß man ohne Ausgleichung keine

vernünftigen Höhenlinien bekommt; es wurden deshalb die durch Abzählung einzelner Felder erhaltenen Zahlen nicht direkt benutzt, sondern
die Mittel benachbarter Felder wiederholt gebildet, bis ein einigermaßen
glatter Flächenverlauf erhalten wurde. Als Mittelpunkt des Koordinatensystems wurde der Geschwindigkeitsschwerpunkt der betrachteten Sterne
gewählt. Die auf die galaktische Ebene projizierte Geschwindigkeit der
Sonne ist durch einen kleinen Kreis bezeichnet. Durch ein Kreuz ist
das Maximum der Geschwindigkeitsfläche dieser Sterne gekennzeichnet,

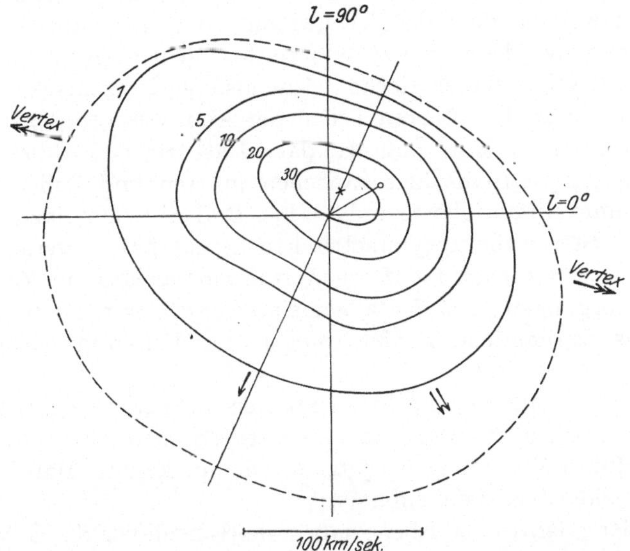

Abb. 1. Galaktische Geschwindigkeitsverteilung der nächsten Sterne.

dessen Koordinaten rechnerisch abgeleitet wurden und die Werte ergaben: galaktische Länge 66°, Abstand vom Nullpunkt 15 km/sec.
Die Richtung der KAPTEYNschen Vertices ist durch Doppelpfeile angegeben.

In bezug auf weiter nach außen liegende Höhenlinien ist die Abb.
unvollständig: Drei schnell bewegte Sterne, die außerhalb der Abb.
liegen würden, sind durch Pfeile angedeutet, so daß hiernach die weiteren
Höhenlinien eine sehr große Ausdehnung nach unten und links unten
bekommen müßten. Auf diese drei Sterne kann natürlich kein Gewicht
gelegt werden; daß sie aber den Charakter der äußeren Höhenlinien
richtig andeuten, weiß man aus anderem Material. So hat z. B. STRÖM-
BERG das entsprechende Bild für alle Sterne gegeben, deren Geschwindigkeit relativ zur Sonne größer als 100 km/sec ist. Es zeigt sich, daß alle
diese Sterne (etwa 85 an Zahl) ausnahmslos in der linken unteren Hälfte
der Abbildung liegen, begrenzt durch eine Gerade, die durch den Nullpunkt geht und ungefähr parallel zur Vertexrichtung verläuft. Ich habe

hiernach den Verlauf der äußersten Höhenlinien in der Abb. 1 durch eine gestrichelte Linie angedeutet.

Es sei ein Wort über die Vollständigkeit des Bildes eingeschaltet. Die Auswahl der benutzten Sterne ist im wesentlichen deswegen unvollständig, weil unsere Beobachtungen bei einer gewissen scheinbaren Helligkeit und daher auch bei einer gewissen absoluten Helligkeit abbrechen, also absolut schwache Sterne fehlen. Ist nun zu erwarten, daß durch Hinzufügen schwächerer Sterne der Prozentsatz der hinzukommenden Sterne überall im Bild derselbe ist? Keinesfalls. Denn es besteht eine starke Abhängigkeit der Geschwindigkeit von der absoluten Helligkeit, indem schwache Sterne systematisch schneller bewegt sind. Durch Hinzunahme schwächerer Sterne würde also der Prozentsatz hinzukommender Sterne in den inneren Teilen der Abb. wesentlich geringer sein als in den äußeren. Man muß also darauf achten, daß in bezug auf die Verteilung der Sternanzahlen zwischen inneren und äußeren Partien das Diagramm ein unsicheres und falsches Bild gibt. Ob hinzukommende schwächere Sterne die unsymmetrischen Eigenschaften verändern würden, läßt sich natürlich nicht sagen; wir müssen uns zur Zeit mit der Annahme begnügen, daß der allgemeine Charakter der Unsymmetrien durch das vorhandene Beobachtungsmaterial richtig wiedergegeben wird.

Obwohl nur aus den nächsten Sternen abgeleitet, zeigt die Abb. 1 doch im Prinzip alles, was man an systematischen Abweichungen von einer zufälligen Verteilung bis jetzt überhaupt kennt. Man kann es in folgenden Sätzen zusammenfassen:

I. In der galaktischen Ebene haben die Höhenlinien der Geschwindigkeitsverteilung ovale Gestalt, wobei die längere Ausdehnung in Richtung der Kapteynschen Vertices liegt.

II. Geschwindigkeitsschwerpunkt (Mittelwert) und Maximum der Geschwindigkeitsfläche („mode") fallen nicht zusammen, sondern liegen senkrecht zur Vertexrichtung voneinander getrennt (Strömbergs Asymmetrie). Dies kann auch durch die Tatsache ausgedrückt werden, daß große Sterngeschwindigkeiten nur *einseitig* in einer Richtung senkrecht zur Vertexrichtung vorkommen.

Nachdem man Punkt II jetzt als über allen Zweifel verbürgt ansehen kann, muß seine Wirkung auf die Kapteynsche Auffassung der zwei Sternströme beachtet werden. Und dieser Punkt ist in der Tat verhängnisvoll für die Zweistromtheorie in ihrer alten Form. Wenn jeder der beiden Sternströme irgendeine in sich symmetrische Geschwindigkeitsverteilung zeigt, wie es die Zweistromtheorie annimmt, so kann niemals durch die Übereinanderlagerung der beiden Geschwindigkeitsverteilungen eine Unsymmetrie entstehen, welche senkrecht zur Strombewegung orientiert ist. *Es muß daher notwendigerweise die Zweistromtheorie in ihrer alten Form aufgegeben werden.*

3. Theoretische Geschwindigkeitsverteilungen verschiedener Ordnung in genäherter Darstellung. Bei der Überlegung, ob es einen Ausweg zur Rettung der alten Zweistromtheorie geben könnte, zeigte sich eine merkwürdige Eigentümlichkeit, die mich zunächst sehr überraschte und die ich deswegen im folgenden wiedergeben möchte; nicht so sehr in der Meinung, daß dadurch eine Lösung des Problems gefunden werden könnte, als vielmehr nur des allgemeinen Interesses wegen und um zu zeigen, wie außerordentlich vorsichtig man in der Deutung von beobachteten Geschwindigkeitsverteilungen sein muß. Es ist nämlich das endgültige Bild der gesamten Geschwindigkeitsverteilung sehr stark davon abhängig, welche Geschwindigkeitsverteilung man einzeln in jedem der beiden Sternströme voraussetzt. Dies geht so weit, daß man unter gewissen, im Rahmen des vorhandenen Beobachtungsmaterials zulässigen Annahmen über die Einzelverteilung der Geschwindigkeiten ein Gesamtbild der Verteilung erhalten kann, dessen Höhenlinien in der Nähe des Maximums ausgesprochen länglich in einer Richtung verlaufen, die senkrecht zur Bewegungsrichtung der beiden Sternströme liegt. In solchem Fall würden also die beiden Sternströme sich nicht in der Richtung der Vertices, sondern genau senkrecht dazu bewegen! Es ist klar, daß dann die Schwierigkeit, die die alte Zweistromtheorie zu Fall bringt, nicht besteht, denn die STRÖMBERGsche Asymmetrie würde nun in Richtung der gegenseitigen Strombewegung liegen, wie es sein muß, während andererseits die beobachtete Dehnung der Höhenlinien senkrecht zu dieser Richtung ebenfalls von den zugrunde liegenden Annahmen gefordert wird.

Wir wollen uns jetzt die Aufgabe stellen, rechnerisch den Verlauf der Höhenlinien zu untersuchen, wenn wir zwei voneinander unabhängige Sternströme voraussetzen, von denen jeder in sich eine bestimmte Verteilung der Geschwindigkeiten zeigt. Wir untersuchen also die Fläche $z = z_1 + z_2 = f_1(x, y) + f_2(x, y)$, wobei f_1 und f_2 die Geschwindigkeitsverteilungen in den beiden Sternströmen bezeichnen. Bei MAXWELLscher Verteilung würden die Funktionen f_1 und f_2, abgesehen von Konstanten, die Form $e^{-h^2(x^2+y^2)}$ haben, in größerer Allgemeinheit wollen wir aber die Form $e^{-h^n (x^2+y^2)^{\frac{n}{2}}}$ ansetzen, die mit wachsendem n eine steiler abfallende Verteilung bedeutet. Bleibt man im Gebiet der Maxima dieser Funktionen, so kann man eine Näherung durch eine Potenzentwicklung einführen, und wir wollen zur ersten Orientierung uns mit den ersten Gliedern einer solchen Entwicklung begnügen. Wir setzen daher:

$$f_1(x, y) = a_1 - b_1\, r_1^n, \qquad\qquad f_2(x, y) = a_2 - b_2\, r_2^n$$

wobei:

$$r_1^2 = x^2 + y^2, \qquad\qquad r_2^2 = (x - d)^2 + y^2$$

sei, wir legen also das Zentrum der Geschwindigkeitsverteilung der
ersten Gruppe in den Koordinatenanfangspunkt, das der zweiten Gruppe
auf die positive x-Achse in den Abstand d vom Nullpunkt. Die resultierende Flächengleichung beider Gruppen lautet:

$$z = a_1 + a_2 - b_1\, r_1^n - b_2\, r_2^n$$

Für den Fall $n = 2$, d. h. die Näherung an eine MAXWELLsche Verteilung, ergibt sich ein einfaches Resultat: Das Maximum der Geschwindigkeitsfläche z liegt bei den Koordinaten $y_0 = 0$ und $x_0 = \dfrac{d\,b_2}{b_1 + b_2}$.

Die Differentialgleichung der Höhenlinien lautet $\dfrac{\partial z}{\partial x}\,dx + \dfrac{\partial z}{\partial y}\,dy = 0$,
woraus folgt, daß alle Höhenlinien Kreise mit dem Maximum als Zentrum sind.

Die Geschwindigkeitsfläche interessiert uns hauptsächlich in der
Nähe des Maximums, weil dort die Anzahl der beobachteten Sterne am
größten ist und ein Vergleich der theoretischen Kurve mit den beobachteten Höhenlinien am sichersten ist. Auch gilt ja unsere Näherung
nur für die Nähe des Maximums. Wir ersetzen in diesem Gebiet die
Höhenlinien durch Kegelschnitte, deren Hauptachsen wegen der Symmetrielage der beiden Geschwindigkeitsgruppen nach der x- und y-Richtung orientiert sein müssen. Bezeichnen wir diese Hauptachsen mit s_x
und s_y, so ist $\left(\dfrac{s_y}{s_x}\right)^2$ gleich dem Verhältnis der entsprechenden Krümmungsradien der Flächen. Unter Einführung von $\dfrac{\partial^2 z}{\partial x^2} = r$ und $\dfrac{\partial^2 z}{\partial y^2} = t$
haben wir also:

$$\frac{s_y}{s_x} = \sqrt{\frac{r}{t}} = \sqrt{n-1}$$

Das Verhältnis der Hauptachsen ist also in einfachster Weise nur von n abhängig. Man hat demnach für das Maximum der Fläche folgendes Schema
der Höhenlinien (Abb. 2).

Hierin liegt folgendes Resultat: Das Bild der gesamten Geschwindigkeitsverteilung von
zwei Sterngruppen, die gegeneinander bewegt sind, hängt
stark davon ab, welche Geschwindigkeitsverteilung innerhalb jeder einzelnen Sterngruppe besteht. Für n kleiner
als 2, d. h. für Geschwindigkeitsverteilungen, die flacher
als die MAXWELLsche verlaufen,

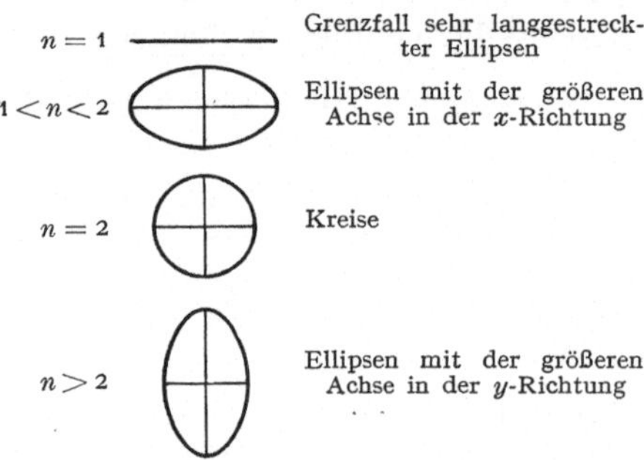

Abb. 2. Höhenlinien in der Nähe des Maximums für Geschwindigkeitsverteilungen verschiedener Ordnung.

zeigen die Höhenlinien der Gesamtverteilung im Maximum der Fläche eine Verlängerung in Richtung der Bewegung der beiden Sterngruppen. Dagegen für n größer als 2, also steile Einzelverteilung in jeder Gruppe, liegen die großen Achsen der Ellipsen senkrecht zur Bewegung der Sterngruppen, es kann also hier durch das Gesamtbild fälschlich eine Bewegung zweier Gruppen vorgetäuscht werden, die genau senkrecht zur tatsächlichen Bewegung liegt.

4. Geschwindigkeitsverteilungen nach Exponentialfunktionen verschiedener Ordnung. Gehen wir jetzt von der Näherung zu Exponentialfunktionen selbst über, so wollen wir zunächst $n = 2$ setzen, also in jeder der beiden Sterngruppen eine MAXWELLsche Verteilung annehmen. Die Gesamtverteilung sei durch die Gleichung gegeben:

$$z = e^{-(x^2 + y^2)} + a \cdot e^{-h^2[(x-d)^2 + y^2]}$$

Der Geschwindigkeitsschwerpunkt der gesamten Geschwindigkeitsverteilung, definiert durch die Gleichung $x_0 \int\limits_{-\infty}^{+\infty} z\,dx = \int\limits_{-\infty}^{+\infty} x z\,dx$, und, entsprechend für die y-Koordinate, hat die Koordinaten:

$$x_0 = \frac{ad}{a+h} \quad \text{und} \quad y_0 = 0$$

Sind x_1 und y_1 die Koordinaten des Maximums der Fläche, so ist $y_1 = 0$, während x_1 durch die Gleichung gegeben ist:

$$x_1 \cdot e^{-x_1^2} + ah^2(x_1 - d)e^{-h^2(x_1-d)^2} = 0$$

Schwerpunkt und Maximum fallen im allgemeinen nicht zusammen, sondern liegen in Richtung der Bewegung der beiden Sterngruppen voneinander entfernt. Nur für $a = 1$ und $h = 1$ fallen beide Punkte zusammen. Weiter interessiert uns das Verhältnis der beiden Hauptachsen der elliptischen Höhenkurven in der Nähe des Maximums. Wir erhalten hierfür:

$$\left(\frac{s_y}{s_x}\right)^2 = 1 - \frac{2}{d} x_1(d - x_1)[x_1 + h^2(d - x_1)]$$

Da für die allein in Frage kommenden „vernünftigen" Fälle x_1 kleiner als d ist, folgt hieraus, daß s_y kleiner als s_x ist; die Höhenkurven in der Nähe des Maximums sind also in Richtung der Bewegung der beiden Sterngruppen verlängert. Für kleine d werden die Kurven mehr und mehr Kreise, wodurch wir uns dem oben behandelten Näherungswert anschließen.

Nun wollen wir uns einer Exponentialfunktion vierter Ordnung zuwenden. Die Gesamtverteilung der Geschwindigkeiten sei durch die Fläche dargestellt:

$$z = e^{-(x^2 + y^2)^2} + a \cdot e^{-h^4[(x-d)^2 + y^2]^2}$$

Wiederum hat der Geschwindigkeitsschwerpunkt die Koordinaten:

$$x_0 = \frac{ad}{a + h} \quad \text{und} \quad y_0 = 0$$

Die Koordinaten x_1 und y_1 des Maximums der Fläche sind gegeben durch die Gleichungen:

$$y_1 = 0$$

und:

$$x_1^3 e^{-x_1^4} + ah^4(x_1 - d)^3 e^{-h^4(x_1 - d)^4} = 0$$

Maximum und Schwerpunkt fallen wieder für $a = 1$ und $h = 1$ zusammen, liegen aber im allgemeinen in Richtung der gegenseitigen Bewegung der beiden Sterngruppen voneinander getrennt. Das Verhältnis der Hauptachsen der elliptischen Höhenlinien in der Nähe des Maximums wird:

$$\left(\frac{s_y}{s_x}\right)^2 = 3 - \frac{4}{d} x_1(d - x_1) \cdot [x_1^3 + h^4(d - x_1)^3]$$

woraus zu ersehen ist, daß im allgemeinen, wenn d und damit auch x_1 nicht zu groß sind, s_y größer als s_x ist. Für $h = 1$ und $a = 1$, d. h. gleich steile und gleich hohe Verteilung in den beiden Einzelgruppen, wird $x_1 = \frac{1}{2}d$ und daher:

$$\frac{s_y}{s_x} = \sqrt{3 - \frac{d^4}{4}}$$

Für kleine Werte von d wird $\frac{s_y}{s_x} = 1.73$, d. h. wir erhalten eine starke Verlängerung der Höhenlinien in der Nähe des Maximums der Geschwindigkeitsverteilung, und zwar in der Richtung, die senkrecht zur Bewegungsrichtung der beiden Sterngruppen steht.

Um ein anschauliches Bild des Verlaufes der Höhenlinien für solche Geschwindigkeitsverteilungen zu geben, sind in Abb. 3 und 4 zwei einfache Fälle gezeichnet. Abb. 3 stellt den Fall einer Geschwindigkeitsverteilung vierter Ordnung dar mit den Konstanten: $h = 1$, $a = 0.5$, $d = 1$. Schwerpunkt und Maximum liegen hier bei den Werten $x_0 = 0.33$ bzw. $x_1 = 0.43$, während das Achsenverhältnis $\frac{s_y}{s_x}$ den Wert 1.66 hat. Abb. 4 stellt den Fall $h = 2$, $a = 1$, $d = 0.8$ dar, wofür $x_0 = 0.27$, $x_1 = 0.60$ und $\frac{s_y}{s_x} = 1.67$ werden. Diese als schematische Beispiele gegebenen theoretischen Verteilungen gleichen der beobachteten Verteilung insofern, als sie auch die zwei wesentlichen Eigenschaften aufweisen: 1. Trennung vom Schwerpunkt und Maximum, und 2. längliche Gestalt der Höhenlinien in der Nähe des Maximums in einer zur Richtung Schwerpunkt-Maximum senkrechten Richtung.

Ein Sinn kann einer solchen Hypothese über die Geschwindigkeitsverteilung nicht zugesprochen werden. Das kommt aber auch nicht in Frage, denn Sinn könnte man zur Zeit nur einer MAXWELLschen Verteilung zusprechen. Eine MAXWELLsche Verteilung könnte aber nur angenommen werden, wenn wir einen stationären Zustand der Bewegungen im Sternsystem hätten, und das ist zweifellos nicht der Fall. Über die tatsächliche Verteilung können wir also zur Zeit nur die Beobachtungen selbst befragen, und ein analytischer Ansatz ist nur daran gebunden, daß kein Widerspruch zu dem Beobachtungsmaterial auftreten darf.

5. Schlußwort. Zusammenfassend sagen wir: Die alte KAPTEYNsche Zweistromtheorie ist nicht mehr haltbar, seitdem eine Asymmetrie in der Geschwindigkeitsverteilung der

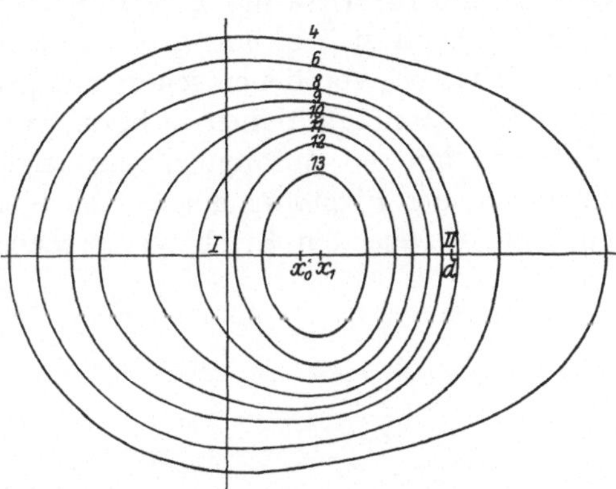

Abb. 3. Geschwindigkeitsverteilung vierter Ordnung nach den Konstanten $h = 1$, $a = 0.5$, $d = 1$

Sterne verbürgt ist, welche senkrecht zur Vertexrichtung, d. h. zur vermeintlichen Strombewegung, orientiert ist. Diese Asymmetrie ist das Wesentliche in der Geschwindigkeitsverteilung und darf bei einer Deutung nicht außer acht gelassen werden. Will man eine Zweistromauffassung beibehalten, so zwingt diese Asymmetrie dazu, eine gegenseitige Bewegung der zwei Ströme senkrecht zur Vertex-Richtung anzunehmen. Gegenüber dieser Asymmetrie kommt erst in zweiter Linie die Tatsache in Frage, daß die Geschwindigkeitskörper eine in der Vertex-Richtung längliche Gestalt haben, denn die Form der Geschwindigkeitskörper ist außerordentlich stark

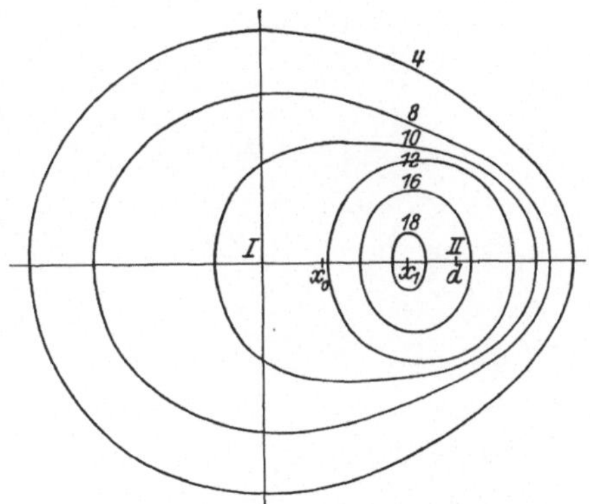

Abb. 4. Geschwindigkeitsverteilung vierter Ordnung nach den Konstanten $h = 2$, $a = 1$, $d = 0.8$

davon abhängig, welche Geschwindigkeitsverteilung einzeln in jedem der beiden Ströme vorausgesetzt wird. Man kann künstlich, wie gezeigt wurde, Verteilungen konstruieren, die in den inneren Gebieten sogar eine stark längliche Ausdehnung senkrecht zur Strombewegung zeigen.

In der Tat scheint jedoch eine Vorstellung, die nur zwei Ströme annimmt — etwa Feldsterne eines größeren galaktischen Systems, zwischen denen die Sterne unseres engeren Systems schwimmen —, kaum auszureichen. Schon in unserer Nähe ist offenbar das Bild der Bewegungen der Sterne wesentlich durch zwei spezielle Sterngruppen, die Taurus-Gruppe und die Ursa major-Gruppe, beeinflußt, und das ganze Sternsystem scheint sich in eine große Anzahl solch einzelner Gruppen auflösen zu lassen, von denen schon eine größere Zahl bekannt ist und die eine große Mannigfaltigkeit in bezug auf räumliche Ausdehnung, Sternfülle und inneren Zusammenhang zeigen. Für das Studium der Bewegungen der Fixsterne dürfte eine hierauf fußende Arbeitshypothese zur Zeit am meisten Erfolg versprechen.

Zur Statistik der Kometen und Planeten im Zusammenhang mit der Verteilung der Sterne.

Von S. Oppenheim, Wien.

Das Problem der Verteilung der Kometen, speziell ihrer Perihele, nach der Ellipsoid-Methode hat zuerst A. SVEDSTRUP[1]) in Angriff genommen. Die von ihm erhaltenen Resultate, die ich einer kurzen Diskussion[2]) unterzog, sind ganz merkwürdiger Art. Nicht, wie man auf Grund der Anschauung erwarten sollte, nach der die Kometen ständige Mitglieder des Sonnensystems sind, hat das Verteilungsellipsoid ihrer Perihele in seinen Hauptebenen eine Beziehung zur Ekliptik, vielmehr hängt es mit der Milchstraße und dem geheimnisvollen Vertex der Sternbewegungen zusammen. Dieses eigentümliche Resultat bewog mich, die Rechnungen A. SVEDSTRUPS an dem reichhaltigeren Beobachtungsmaterial, das zur Zeit gegeben und übersichtlich und wohlgeordnet in dem von Hofrat WEISS besorgten und im Astronomischen Kalender der Sternwarte Wien, Jahrgang 1909, veröffentlichten Verzeichnis der Bahnelemente aller bisher berechneten Kometen vorliegt, zu wiederholen.

Hierbei möchte ich in herzlicher Dankbarkeit der werktätigen Hilfe gedenken, die mir mein ehemaliger Schüler, H. ÉLBOGEN in Wien, gewährte, der den größten Teil der weitläufigen Rechnungen durchführte und sie mir zur Veröffentlichung übergab. Nur dadurch war es mir möglich, der Untersuchung eine breitere Grundlage zu geben, indem ich sie nicht bloß auf die der Verteilung der Kometenperihele beschränkte, sondern auch auf deren Bahnpole ausdehnte, ferner noch, um gleichsam einen empirischen Maßstab für die Genauigkeit der erhaltenen Ergebnisse zu gewinnen, die Perihele und Bahnpole der kleinen Planeten in die Rechnung einbezog, dabei von der Voraussetzung ausgehend, daß die Ellipsoide der letzteren genau nach der Ekliptik orientiert sein dürften.

Die Methode selbst besteht bekanntlich darin, zunächst die einzelnen Vektoren der Perihele nach den Formeln:

$$l = \cos \omega \cos \Omega - \sin \omega \sin \Omega \cos i$$
$$m = \cos \omega \sin \Omega + \sin \omega \cos \Omega \cos i$$
$$n = \sin \omega \sin i$$

---

[1]) Astr. Nachr. Bd. 107, S. 113. 1883. [2]) Astr. Nachr. Bd. 216, S. 47. 1922.

der Bahnpole dagegen nach:

$$l' = - \sin \Omega \sin i, \qquad m' = \cos \Omega \sin i, \qquad n' = \cos i$$

zu berechnen, aus ihnen die Summen:

$$A = \Sigma (l^2) \qquad B = \Sigma (m^2) \qquad C = \Sigma (n^2)$$
$$D = \Sigma (m n) \qquad E = \Sigma (n l) \qquad F = \Sigma (l m)$$

die Summierung erstreckt über alle in Frage kommenden Einzelkörper, zu bilden, wodurch die Gleichung des Ellipsoids in der Form:

$$A x^2 + B y^2 + C z^2 + 2 D z y + 2 E z x + 2 F x y = 1$$

erscheint, und dieses endlich auf die Hauptachsen zu reduzieren. Deren Richtung geben die Pole der gesuchten extremalen größten Kugelkreise, um die sich die Perihele oder Bahnpole am dichtesten gruppieren.

Bei der Aufstellung der Ellipsoidgleichung benützte ich die aus der Relation $l^2 + m^2 + n^2 = 1$ fließende Beziehung $A + B + C = $ der Anzahl der bei der Berechnung verwendeten Einzelkörper und dividierte alle Koeffizienten $AB \ldots F$ durch diese Zahl. Dadurch erzielte ich, daß stets $A + B + C = 1$ wurde und so die kubische Gleichung zur Bestimmung der Hauptachsen, die in Determinantenform:

$$\begin{vmatrix} A - k^2 & F & E \\ F & B - k^2 & D \\ E & D & C - k^2 \end{vmatrix} = 0$$

lautet, in der vereinfachten Form erscheint:

$$k^3 - k^2 + P k - Q = 0$$

1. Die Bahnpole der kleinen Planeten. Die Rechnung wurde hier auf die ersten 200 der kleinen Planeten beschränkt, deren Bahnelemente im Berliner Jahrbuch gegeben sind. Sie gab die Koeffizienten:

$$A = \quad 0.01174 \qquad B = \quad 0.09969 \qquad C = \quad 0.96857$$
$$D = - 0.000662 \qquad E = + 0.016483 \qquad F = + 0.001397$$

Aus ihnen die kubische Gleichung:

$$k^3 - k^2 + 0.030400\, k - 0.00021666 = 0$$

mit den Wurzeln und den ihnen entsprechenden Hauptachsen:

$$k_1 = 0.96885 \quad k_2 = 0.01992 \quad k_3 = 0.01123$$
$$a_1 = 1.002 \quad a_2 = 7.085 \quad a_3 = 9.438$$

während für ihre Richtungen, bezogen auf die Ekliptik, folgte:

$$\lambda_1 = \quad 357°8 \quad \lambda_2 = 170°5 \quad \lambda_3 = \quad 80°5$$
$$\beta_1 = + 89°0 \quad \beta_2 = + 1°0 \quad \beta_3 = - 0°1$$

Damit erweist sich das Verteilungsellipsoid der Bahnpole der kleinen Planeten als eine stark abgeplattete Scheibe, deren eine Hauptebene mit der Ekliptik zusammenfällt, und deren beiden anderen innerhalb dieser nahe nach den Längen 90° und 180° und den Breiten 0° gerichtet sind. Nimmt man als Bedingung an, daß dieses Zusammenfallen ein vollständiges sein solle, so folgt für die Genauigkeit der erhaltenen Zahlen in Länge und Breite etwa:

$$\varDelta \lambda = \pm 10° \qquad \varDelta \beta = \pm 1°$$

Aus der Kleinheit des von k freien Gliedes in der kubischen Gleichung läßt sich ferner schließen, daß die mit der Ekliptik zusammenfallende Hauptebene des Ellipsoids, d. i. die mittlere Bahnebene der kleinen Planeten, fast genau durch die Sonne als den Anfangspunkt des Koordinatensystems, das den Bahnelementen zugrunde liegt, hindurchgeht. Ihr sphärischer Abstand von der Sonne berechnet sich aus:

$$\sin^2 d = 0.00021666 \qquad \text{zu} \qquad d = 0°50'$$

2. Die Perihele der kleinen Planeten. Für diese begnügte ich mich bei der Rechnung bloß mit den ersten 100 unter ihnen und erhielt:

$$A = 0.50424 \qquad B = 0.47995 \qquad C = 0.01581$$
$$D = + 0.002946 \qquad E = - 0.002648 \qquad F = + 0.013323$$

mit der kubischen Gleichung:

$$k^3 - k^2 + 0.25379\,k - 0.003816 = 0$$

deren Wurzeln und daraus bestimmten Hauptachsen:

$$k_1 = 0.51000 \qquad k_2 = 0.47345 \qquad k_3 = 0.01655$$
$$a_1 = 1.400 \qquad a_2 = 1.453 \qquad a_3 = 7.953$$

sowie deren Hauptrichtungen:

$$\lambda_1 = 23°9 \qquad \lambda_2 = 113°4 \qquad \lambda_3 = 294°0$$
$$\beta_1 = + 0.4 \qquad \beta_2 = + 0.5 \qquad \beta_3 = + 89.6$$

sind. Daraus folgt, daß das Verteilungsellipsoid nahe die Form eines verlängerten Rotationsellipsoides hat, das in einer Hauptebene nach der Ekliptik orientiert ist, in den beiden anderen aber nicht mehr nach den Längen 0° und 90° gerichtet, sondern um einige Grade in positivem Sinne verschoben erscheint. Man wird wohl nicht fehlgehen, hierin den störenden Einfluß zu erkennen, den der Planet Jupiter, dessen Perihellänge 12°9 beträgt, auf die Gesamtmasse der kleinen Planeten ausübt. Setzt man als Bedingung fest, daß diese Orientierung nach $\lambda_1 = 12°9$, $\beta_1 = + 0°$, $\lambda_2 = 102°9$, $\beta_2 = + 0°$ genau zutreffen soll, so ergibt sich, wie oben, als Genauigkeitsabschätzung der erhaltenen Zahlen:

$$\varDelta \lambda = \pm 10° \qquad \varDelta \beta = \pm 1°$$

In gleicher Weise findet man als sphärischen Abstand der Hauptebene des Ellipsoids von der Sonne, berechnet aus $\sin^2 d = 0.003816$, den noch immer sehr kleinen Wert:

$$d = 3°4$$

3. Die Perihele der periodischen Kometen. Als periodische Kometen nimmt A. Svedstrup in sehr weitgehendem Sinne alle jene an, deren Exzentrizität $e < 0.995$ oder deren Umlaufszeit $U < 1000$ Jahre ist. An dieser Annahme wurde auch hier festgehalten. Von den im Verzeichnis des Hofrat Weiss enthaltenen 372 Kometen sind in diesem Sinne 65 als periodisch zu bezeichnen, und mit ihnen wurde die Berechnung des Ellipsoids durchgeführt. Sie gab:

$$A = \quad 0.44250 \qquad B = \quad 0.46113 \qquad C = \quad 0.09637$$
$$D = +0.02192 \qquad E = +0.03596 \qquad F = +0.06800$$

Die kubische Gleichung, deren Wurzeln, Hauptachsen und Hauptrichtungen fanden sich daraus zu:

$$k^3 - k^2 + 0.28474\,k - 0.01852 = 0$$

$$k_1 = \quad 0.5243 \qquad k_2 = \quad 0.3836 \qquad k_3 = \quad 0.0921$$
$$a_1 = \quad 1.381 \qquad a_2 = \quad 1.615 \qquad a_3 = \quad 3.295$$

$$\lambda_1 = \quad 48^\circ.4 \qquad \lambda_2 = 138^\circ.2 \qquad \lambda_3 = \quad 204^\circ.0$$
$$\beta_1 = +5.4 \qquad \beta_2 = -2.4 \qquad \beta_2 = +84.1$$

Das damit charakterisierte Ellipsoid zeigt Ähnlichkeit mit dem analogen der kleinen Planeten. Es ist wie dieses nach der Ekliptik orientiert, aber wohl schon etwas zu weit über das Perihel des Jupiter hinaus verschoben. Hält man auch hier an der Bedingung fest, daß das Ellipsoid genau nach dieser Richtung eingestellt sein soll, so folgen für die Genauigkeit der erhaltenen Zahlen die weitaus größeren Fehler:

$$\Delta\lambda = \pm 30^\circ \qquad \Delta\beta = \pm 5^\circ$$

4. Die Perihele der parabolischen Kometen. Von den nach Abzug der 65 periodischen Kometen übrigbleibenden 306 nichtperiodischen sind 144 in direkter und 162 in retrograder Bewegung. H. Elbogen berechnete deren Verteilungsellipsoide sowohl getrennt für die ersteren (in der Folge mit I bezeichnet), wie für die zweiten (mit II), als auch für alle 306 zusammengenommen (als III gekennzeichnet) Er fand:

	I	II	III
$A =$	0.28189	0.29015	0.28626
$B =$	0.42323	0.38161	0.40120
$C =$	0.29487	0.32824	0.31254
$D =$	$+0.00666$	-0.01682	-0.00577
$E =$	$+0.01210$	$+0.03222$	$+0.02275$
$F =$	-0.02771	$+0.03731$	$+0.00671$

Die aus diesen Werten abgeleiteten kubischen Gleichungen, deren Wurzeln, die ihnen entsprechenden Hauptachsen und Hauptrichtungen sind:

$$\text{I} \quad k^3 - k^2 + 0.32627\,k - 0.03488 = 0$$
$$\text{II} \quad k^3 - k^2 + 0.32851\,k - 0.03537 = 0$$
$$\text{III} \quad k^3 - k^2 + 0.32911\,k - 0.03566 = 0$$

	I	II	III
$k_1 =$	0.42884	0.39536	0.40160
$k_2 =$	0.30024	0.34634	0.32612
$k_3 =$	0.27092	0.25829	0.27228
$a_1 =$	1.527	1.590	1.578
$a_2 =$	1.825	1.699	1.751
$a_3 =$	1.921	1.908	1.916

	I	II	III
$\lambda_1 =$	$100°\!.5$	$72°\!.5$	87.3
$\beta_1 = +$	1.9	$-\;5.1$	$-\;3.0$
$\lambda_2 =$	187.0	171.7	182.4
$\beta_2 = -$	62.2	$-\;62.2$	$-\;59.8$
$\lambda_3 =$	$191°\!.5$	159.3	175.5
$\beta_3 = +$	27.7	$+\;27.3$	$+\;30.0$

oder nach Transformation in Rektaszension und Deklination:

	I	II	III
$\alpha_1 =$	$101°\!.6$	$71°\!.2$	87.1
$\delta_1 = +$	24.9	$+\;17.2$	$+\;20.4$
$\alpha_2 =$	147.0	138.2	147.2
$\delta_2 = -$	56.5	$-\;51.7$	$-\;53.2$
$\alpha_3 =$	201.8	172.8	189.0
$\delta_3 = +$	20.9	$+\;33.1$	$+\;29.1$

Der Anblick der Zahlen sagt vor allem, daß die zwei Verteilungsellipsoide der Perihele der direkten (I) und retrograden (II) Kometen weder
in der Größe ihrer Hauptachsen noch in ihren Hauptrichtungen nennenswerte Unterschiede zeigen. Die aus ihnen abgeleiteten Mittelwerte (III)
dürften daher die genaueren Angaben darstellen. Er sagt ferner, daß das
Verteilungsellipsoid fast sphärisch ist. Daraus ist zu schließen, daß die
Materie, aus denen sich die Kometen gebildet haben, nahezu zentrisch
symmetrisch die Sonne umgibt. Aber den Richtungen λ und β oder α und
δ kommt nunmehr keine Beziehung zur Ekliptik zu. Im Gegenteil, wie
man aus den unter III für α_3 und δ_3 angeführten Werten findet, zeigt
sich, daß eine Hauptebene des Ellipsoids mit der Ebene der Milchstraße
zusammenfällt, für deren Pol gegenwärtig als beste Werte etwa:

$$\alpha = 192° \qquad \delta = +\,28°$$

anzunehmen sind. Ebenso zeigt die Richtung der kleinsten Achse bzw.
größten Wurzel der kubischen Gleichung mit den Werten $\alpha_1 = 87°\!.1$,
$\delta_1 = +\,20°\!.4$ mit sehr großer Annäherung nach dem geheimnisvollen

Vertex der Sternbewegungen, für den nach verschiedenen Methoden und aus verschiedenen Beobachtungsdaten Zahlen zwischen den Grenzen

$$\alpha = 80° \text{ bis } 100°, \qquad \delta = + 3° \text{ bis } + 28°$$

berechnet wurden.

Es ist schwer, dieses eigentümliche Resultat richtig zu deuten, da es mit der Anschauung, daß alle Kometen, also auch die parabolischen unter ihnen, ständige Mitglieder unseres Sonnensystems sind, in Widerspruch steht. Denn diese würde eine Einstellung nach der Ekliptik, wie es tatsächlich bei den periodischen Kometen der Fall ist, auch bei den nichtperiodischen fordern. Vielleicht dürfte hier der Standpunkt als der annehmbarste erscheinen, der aussagt, daß die Kometen jetzt noch, da sie sich schon enger dem Sonnensystem angegliedert haben, ihre ursprüngliche der der Sterne analoge Verteilung beibehielten, indem sie die Ebene der Milchstraße und in ihr die Richtung nach dem Vertex bevorzugten, während die kurzperiodischen unter ihnen, durch Jupiter und die anderen großen Planeten eingefangen, sich bereits der Ekliptik anschlossen. Möglich, daß da ein Prozeß vorliegt, der langsam und stetig, jedenfalls aber in säkularen Perioden verläuft und damit enden wird, bis alle Kometenmaterie eine Anordnung nach der Ekliptik erlangen dürfte.

5. Die Bahnpole der parabolischen Kometen. Die Koeffizienten der Ellipsoidgleichungen für diese, in der gleichen Reihenfolge wie die der Perihele angesetzt, I (direkte), II (retrograde Kometen) und III (Mittelwerte), sind nach den Rechnungen von H. ELBOGEN:

	I	II	III
$A =$	0.35120	0.27804	0.31246
$B =$	0.26920	0.31397	0.29291
$C =$	0.37960	0.40799	0.39463
$D =$	$- 0.00071$	$+ 0.00832$	$+ 0.00408$
$E =$	$- 0.00617$	$- 0.02547$	$- 0.01638$
$F =$	$- 0.01457$	$- 0.01230$	$- 0.01337$

Die aus ihnen abgeleiteten kubischen Gleichungen, deren Wurzeln und die ihnen entsprechenden Hauptachsen und Hauptrichtungen lauten:

$$\text{I} \quad k^3 - k^2 + 0.32974\,k - 0.035778 = 0$$
$$\text{II} \quad k^3 - k^2 + 0.32795\,k - 0.035338 = 0$$
$$\text{III} \quad k^3 - k^2 + 0.32995\,k - 0.035965 = 0$$

	I	II	III
$k_1 =$	0.38200	0.41406	0.39704
$k_2 =$	0.35168	0.31502	0.31616
$k_3 =$	0.26632	0.27092	0.28590
$a_1 =$	1.618	1.554	1.585
$a_2 =$	1.686	1.782	1.778
$a_3 =$	1.941	1.922	1.870

$$
\begin{array}{ccccccc}
 & \text{I} & \text{II} & \text{III} & & \text{I} & \text{II} & \text{III} \\
\end{array}
$$

	I	II	III		I	II	III
$\lambda_1 =$	$174°2$	$151°3$	$162°3$	$\alpha_1 =$	$240°0$	$236°6$	$239°3$
$\beta_1 =$	$+\ 77.6$	$+\ 77.3$	$+\ 78.1$	$\delta_1 =$	$+\ 64.8$	$+\ 69.5$	$+\ 67.3$
$\lambda_2 =$	350.0	102.7	331.6	$\alpha_2 =$	345.9	103.3	329.4
$\beta_2 =$	$+\ 12.3$	$-\ 8.4$	$+\ 11.7$	$\delta_2 =$	$+\ 7.4$	$+\ 14.4$	$+\ 0.1$
$\lambda_3 =$	80.2	14.1	62.1	$\alpha_3 =$	79.2	9.2	59.5
$\beta_3 =$	$+\ 0.9$	$+\ 12.3$	$+\ 2.2$	$\delta_3 =$	$+\ 24.0$	$+\ 14.2$	$+\ 22.7$

Was an diesen Zahlen vor allem auffällt, ist die Vertauschung der zwei Hauptrichtungen, 2 und 3, in den beiden Gruppen I und II, der direkten und retrograden Kometen. Es ist näherungsweise:

$$\lambda_2\beta_2 \text{ der Gruppe } I = \lambda_3\beta_3 \text{ der Gruppe } II$$
$$\lambda_3\beta_3 \quad \text{,,} \quad \text{,,} \quad I = \lambda_2\beta_2 \quad \text{,,} \quad \text{,,} \quad II$$

so, als ob die beiden ihnen entsprechenden Hauptachsen der beiden Ellipsoide I und II ihre gegenseitige Lage miteinander vertauscht hätten und nur in der ersten Hauptrichtung $\lambda_1\beta_1$ identisch sind. Eine notwendige Folge dieses eigentümlichen Ergebnisses ist, daß das aus ihnen gezogene Mittel, angeführt unter III, da es die beiden Richtungen 2 und 3 in nicht zutreffender Weise miteinander verknüpft, nur in der ersten -$\lambda_1\beta_1$- nicht aber in den beiden anderen richtig ist. Um auch in ihnen den richtigen Mittelwert zu erhalten, muß erst das Ellipsoid II so transformiert werden, daß es in allen seinen Hauptrichtungen mit denen von I zur Deckung kommt.

Diese Tranformation ist in der folgenden Weise durchzuführen. Wird die reduzierte Ellipsoidgleichung in der Form:

$$k_1 \xi^2 + k_2 \eta^2 + k_3 \zeta^2 = 1$$

angenommen und werden die Richtungscosinus der drei Hauptachsen k_1, k_2 und k_3 mit a_{ik}, $(i, k = 1, 2, 3)$ bezeichnet, so erhält man die Koeffizienten der ursprünglichen Gleichung nach den Formeln:

$$
\begin{aligned}
A &= k_1 a_{11}^2 &&+ k_2 a_{21}^2 &&+ k_3 a_{31}^2 \\
B &= k_1 a_{12}^2 &&+ k_2 a_{22}^2 &&+ k_3 a_{32}^2 \\
C &= k_1 a_{13}^2 &&+ k_2 a_{23}^2 &&+ k_3 a_{33}^2 \\
D &= k_1 a_{12} a_{13} &&+ k_2 a_{22} a_{23} &&+ k_3 a_{32} a_{33} \\
E &= k_1 a_{13} a_{11} &&+ k_2 a_{23} a_{21} &&+ k_3 a_{33} a_{31} \\
F &= k_1 a_{11} a_{12} &&+ k_2 a_{21} a_{22} &&+ k_3 a_{31} a_{32}
\end{aligned}
$$

und daher für die transformierte Gleichung mit Vertauschung der beiden Achsen k_2 und k_3:

$$
\begin{aligned}
A' &= k_1 a_{11}^2 &&+ k_3 a_{21}^2 &&+ k_2 a_{31}^2 \\
B' &= k_1 a_{12}^2 &&+ k_3 a_{22}^2 &&+ k_2 a_{32}^2 \\
C' &= k_1 a_{13}^2 &&+ k_3 a_{23}^2 &&+ k_3 a_{33}^2 \\
D' &= k_1 a_{12} a_{13} &&+ k_3 a_{22} a_{23} &&+ k_2 a_{32} a_{33} \\
E' &= k_1 a_{13} a_{11} &&+ k_3 a_{23} a_{21} &&+ k_2 a_{33} a_{31} \\
F' &= k_1 a_{11} a_{12} &&+ k_3 a_{21} a_{22} &&+ k_3 a_{31} a_{32}
\end{aligned}
$$

Die Rechnung danach gab die Zahlenwerte, die nunmehr an Stelle von II zu treten haben:

$$A' = \quad 0.31715 \qquad B' = \quad 0.27457 \qquad C' = \quad 0.40828$$
$$D' = +\,0.016327 \qquad E' = -\,0.019732 \qquad F' = +\,0.007082$$

und ihnen entsprechen tatsächlich die gleichen Wurzeln wie ursprünglich, nur in der Anordnung k_1, k_3 und k_2 und daher auch als Hauptrichtungen:

$$\lambda_1 = \quad 156°3 \qquad \lambda_2 = \quad 14°1 \qquad \lambda_3 = \quad 102°7$$
$$\beta_1 = +\,77.3 \qquad \beta_2 = +\,12.3 \qquad \beta_3 = -\quad 8.4$$

Nun erst können die Koeffizienten der beiden Ellipsoide I und II gemittelt werden, und man erhält als neue Mittelwerte, in der Folge mit (III) bezeichnet:

$$\text{(III)} \quad A = \quad 0.33317 \qquad B = \quad 0.27205 \qquad C = \quad 0.39478$$
$$D = +\,0.083095 \qquad E = -\,0.013348 \qquad F = -\,0.003107$$

Sie führen zu der neuen kubischen Gleichung, ihren Wurzeln, Hauptachsen und Hauptrichtungen:

$$k^3 - k^2 + 0.32931\,k - 0.035707 = 0$$
$$k_1 = 0.39904 \qquad k_2 = 0.32884 \qquad k_3 = 0.27212$$
$$a_1 = 1.583 \qquad a_2 = 1.744 \qquad a_3 = 1.917$$
$$\text{(III)} \quad \lambda_1 = \quad 161°2 \qquad \lambda_2 = \quad 358°7 \qquad \lambda_3 = \quad 88°0$$
$$\beta_1 = +\quad 77.5 \qquad \beta_2 = +\,11.8 \qquad \beta_3 = -\quad 3.6$$

Oder in Reklaszension und Deklination umgewandelt:

$$\text{(III)} \quad \alpha_1 = \quad 237°9 \qquad \alpha_2 = \quad 354°0 \qquad \alpha_3 = \quad 87°8$$
$$\delta_1 = +\,67.4 \qquad \delta_2 = +\quad 10.2 \qquad \delta_3 = +\,19.8$$

Erst dies sind die richtigen Größenwerte, die an Stelle der oben unter III angeführten zu setzen sind und die Gesamtmittel aller 306 in Rechnung gebrachten Kometen darstellen. Speziell zeigt sich wieder in α_3 und δ_3 die Richtung nach dem Vertex fast vollständig identisch mit der gleichen Richtung in dem Verteilungsellipsoid der Kometenperihele; die beiden anderen dagegen α_1 und δ_1 sowie α_2 und δ_2 sind zunächst ganz unbekannt, d. h. mit keiner irgendwie anderwärts bekannten in Übereinstimmung zu bringen.

6. Die Eigenbewegung der Fixsterne. Nun soll die Nutzanwendung dieser Rechnungen auf einige Probleme der Stellarastronomie folgen. Es ist bekannt, daß viele unter ihnen auf die Bestimmung von Ellipsoiden aus gewissen Beobachtungsdaten führen und in deren Berechnung das Endergebnis ihrer Untersuchung finden. Beispiele hierzu sind die BESSEL-KOBOLDsche Methode zur Bestimmung des Apex der Sonnenbewegung aus den Eigenbewegungen der Fixsterne, die SCHWARZSCHILDsche Ellipsoidhypothese zur Ableitung des Vertex der Sternbewegungen, besonders in der ihr von C. CHARLIER gegebenen Form, nach der diese Ableitung aus den Momenten zweiter Ordnung der Eigenbewegungen der Sterne, d. h. ihrer Streuung, erfolgt, wie auch endlich Untersuchungen über die scheinbare Verteilung der Sterne am Himmel, wonach man die aus photographischen Aufnahmen gewonnenen Sternzahlen in FOURIERsche Reihen auflöst und sich bei der Darstellung mit den Gliedern zweiter Ordnung begnügt.

Merkwürdigerweise führen alle in dieser Richtung durchgeführten Rechnungen[1]), ebenso wie bei denen über die Verteilung der Bahnpole der nichtperiodischen Kometen, auf zwei Ellipsoide, mit einem Paar identischer und den beiden anderen gegeneinander um 90° gedrehten Hauptrichtungen, scheinbar so, als ob sich das ganze System der Fixsterne in zwei Gruppen teilen würde, denen eben zwei in der erwähnten rein geometrischen Beziehung zueinanderstehende Ellipsoide entsprechen. Diese beiden Gruppen von Sternen sind räumlich voneinander geschieden. Zur ersten gehören Sterne, die zumeist dem Zuge der Milchstraße am Himmel folgen bis zu einer Grenze von etwa $\pm 40°$ galaktischer Breite, zur zweiten wiederum mit wenigen Ausnahmen alle jene, welche nunmehr nördlich und südlich der Milchstraße innerhalb der höheren galaktischen Breiten von $\pm 40°$ bis zu den Polen $\pm 90°$ liegen. Eine Zweiteilung des Himmels, die keinesfalls mit der KAPTEYNschen Hypothese der zwei Schwärme zu identifizieren ist. Denn nach dieser sind die Sterne der beiden Schwärme nicht räumlich voneinander getrennt, sondern durchdringen sich wechselseitig und erfüllen in einem fast konstanten Zahlenverhältnis alle Teile des Himmels. Ob sie aber in Analogie mit der Zweiteilung der Kometen in direkte und retrograd bewegte auch auf direkte und retrograde Bewegungen der Sterne zurückzuführen ist, würde einer besonderen Untersuchung[2]) bedürfen, auf die jedoch hier nicht eingegangen werden soll.

Vielmehr sei das Ziel der folgenden Untersuchung, auf die Frage aufmerksam zu machen, ob es denn eigentlich gestattet ist, bei der durch diese Zweiteilung des Himmels bedingten Doppeldeutigkeit der berechneten Ellipsoide das einfache Mittel ihrer drei Hauptrichtungen zu nehmen und es als Mittelwert für die Gesamtheit der Sterne gelten zu lassen, oder ob nicht zunächst das eine Ellipsoid gegen das zweite so zu drehen ist, bis alle drei Hauptrichtungen zusammenfallen, worauf man erst berechtigt ist, deren Mittelwerte zu berechnen. Das Beispiel der Verteilung der Bahnpole der Kometen weist darauf hin, daß der letztere Vorgang der richtigere ist.

---

[1]) Vgl. Vierteljahrschr. d. Astr. Ges. Bd. 56, S. 191. 1921.

[2]) Es sei in dieser Richtung verwiesen auf: OPPENHEIM: Eine Erweiterung der AIRYschen Methode zur Bestimmung des Apex. A. N. Bd. 209, S. 149. 1919; sowie H. ELBOGEN: Über die Eigenbewegungen der Sterne von verschiedenem Spektraltypus und aus ihnen berechnete Parallaxen. A. N. Bd. 215, S. 393. 1921.

7. Das Bessel - Koboldsche Ellipsoid. Bei der Bessel-Koboldschen Methode der Apexbestimmung, d. h. dem nach ihr zu berechnenden Ellipsoid, am besten als Momentenellipsoid zu bezeichnen, kommt die Berücksichtigung dieser Doppeldeutigkeit weniger in Betracht. Man legte eben den Hauptwert der Untersuchung auf die Bestimmung des Apex als den in den beiden Ellipsoiden zusammenfallenden Hauptrichtungen, und beachtete die beiden anderen fast gar nicht, vielleicht wegen der Unmöglichkeit, ihnen nach der Zweischwarmhypothese eine bestimmte geometrische Deutung zu geben. Immerhin dürfte eine Angabe über die beiden Mittelwerte, die die Rechnung liefert, je nachdem ob eine Vertauschung der Achsen des zweiten Ellipsoids gegen das erste, um diese miteinander in volle Übereinstimmung zu bringen, vorgenommen wird oder nicht, nicht ohne Interesse sein, besonders, da die Grundlagen der Rechnung, die Koeffizienten $A, B \ldots F$, schon von mir in der Notiz: Über die Bahnebene der Sonne (A. N. Bd. 202, S. 89, 1916) veröffentlicht sind. Dort erhielt ich, wie in Kürze mitgeteilt werden möge, auf Grundlage der von Charlier in den Meddelanden der Sternwarte Lund (Serie II, Nr. 9. 1913) besorgten Zusammenstellung der Eigenbewegung der Sterne aus dem Bossschen Katalog, für die Ellipsoide der beiden Sterngruppen I und II, und ihren Mittelwert, III, ohne Vertauschung der Achsen, die Werte ihrer Hauptrichtungen zu:

	I	II	III
$\alpha_1 =$	$150°4$	$35°6$	$110°2$
$\delta_1 = +$	40.3	$+ 44.5$	$+ 66.5$
$\alpha_2 =$	28.7	159.4	4.6
$\delta_2 = +$	31.8	$+ 29.5$	$+ 6.2$
$\alpha_3 =$	274.5	269.4	270.9
$\delta_3 = +$	33.5	$+ 31.2$	$+ 31.0$

während nach Vertauschung der Achsen im Ellipsoid II die neuen Mittelwerte (III) folgen, die von den oberen, III, ziemlich stark abweichen, mir aber jedenfalls die richtigeren zu sein scheinen:

	I	II	(III)
$\alpha_1 =$	$150°4$	$159°4$	$155°1$
$\delta_1 = +$	40.3	$+ 29.5$	$+ 35.2$
$\alpha_2 =$	28.7	35.6	31.4
$\delta_2 = +$	31.8	$+ 44.5$	$+ 38.1$
$\alpha_3 =$	274.5	269.5	271.5
$\delta_3 = +$	33.5	$+ 31.2$	$+ 32.3$

Die mittlere Apexrichtung, $\alpha_3 = 271°5$, $\delta_3 = + 32°3$, steht mit der von Charlier nach der Airyschen Methode gefundenen, $\alpha_3 = 273°2$ $\alpha_3 = + 31°2$, im besten Einklang.

8. Das Schwarzschild-Charliersche Streuungsellipsoid. Von größerer Bedeutung wird erst die Frage nach der Berücksichtigung der

Doppeldeutigkeit der Ellipsoide beim CHARLIERschen Streuungsellipsoid. Bei diesem werden stets alle drei Hauptrichtungen ihrer Achsen berechnet und mitgeteilt. Als Beispiele seien unter anderen erwähnt die Rechnungen von WICKSELL[1]), die sich auf die Eigenbewegungen der Sterne auf Grundlage des Bossschen Kataloges beziehen, dann die von GYLLENBERG[2]), die die gleiche Aufgabe, aber die Radialgeschwindigkeiten der Sterne betreffend und gesondert nach den drei Spektraltypen B und A, F und G sowie K und M, behandeln, sowie die analoge Untersuchung von STRÖMBERG[3]): The distribution of the velocities of stars of spectral types F to M. In ihnen wird auf die Zweiteilung der Sterne am Himmel und die Doppeldeutigkeit der ihr entsprechenden Ellipsoide keine Rücksicht genommen. Es scheinen mir daher die von ihnen mitgeteilten Zahlen nur in der den beiden Ellipsoiden gemeinsamen Richtung, d. i. der nach dem Vertex, zutreffend zu sein, nicht mehr aber in den beiden anderen, von denen eine, wie bekannt, nach dem Pole der Milchstraße zeigt. Jedenfalls bedürfen alle diese Rechnungen einer Revision.

Ich habe nur die Rechnungen von WICKSELL einer solchen unterzogen und erhielt aus den von ihm mitgeteilten Daten für die Ellipsoide der beiden Sterngruppen, I und II, sowie nach Vertauschung der Achsen, (II), und das aus I und (II) gemittelte Ellipsoid, III, die Werte, denen ich die von WICKSELL gefundenen Zahlen, mit W bezeichnet, beischließe:

	I	II	(II)	III	W
$\alpha_1 =$	$279°.6$	$278°.2$	$278°.2$	$278°.8$	$274°.3$
$\delta_1 = -$	6.8	$- 19.2$	$- 19.2$	$- 12.9$	$- 12.4$
$\alpha_2 =$	177.9	212.2	174.6	172.4	159.1
$\delta_2 = -$	59.4	$+ 49.4$	$- 34.1$	$- 50.8$	$- 62.5$
$\alpha_3 =$	193.6	174.6	212.2	198.4	189.2
$\delta_3 = +$	29.6	$- 34.1$	$+ 49.4$	$+ 36.2$	$+ 24.1$

Die recht bedeutenden Differenzen in den Angaben unter III und W weisen auf die Notwendigkeit hin, bei der Berechnung des Mittels, das für die Gesamtheit der Sterne des ganzen Himmels gelten soll, auf den Unterschied in den beiden Ellipsoiden I und II Rücksicht zu nehmen. Die ohne Berücksichtigung dieser Tatsache durchgeführten Rechnungen sind als mit einem systematischen Fehler behaftet zu bezeichnen, und die von den Rechnern mitgeteilten mittleren Fehler scheinen mir daher ganz illusorisch zu sein.

Als gleich fehlerhaft wären auch die Dichtigkeitsrechnungen STRÖMBERGS in der schon oben erwähnten Abhandlung anzusehen, sowie die aus ihnen konstruierten Diagramme, die die Sternverteilung für die

[1]) Meddelanden der Sternwarte Lund, Serie II, Nr. 12. 1915.
[2]) Dieselben: Serie II, Nr. 13. 1915.
[3]) Astrophysical Journal Bd. 56, 283. 1922.

einzelnen Spektralklassen darstellen, soweit sie sich auf eine harmonische Analyse von Sternzahlen gründen. Denn auch bei dieser Analyse muß die Zweiteilung der Sterne beachtet werden. Jede der zwei Sterngruppen gibt eine Reihe, und beide unterscheiden sich voneinander durch das Vorzeichen in den Gliedern zweiter Ordnung. Leider teilt H. Strömberg gar keine der Beobachtungsdaten mit, auf die sich seine Berechnungen stützen. Eine Revision derselben konnte daher nicht durchgeführt werden.

Aus dem gleichen Grunde der Nichtbeachtung der zwei Gruppen, in die sich die Gesamtheit der Sterne am Himmel teilt, ist nach meinem Dafürhalten auch die Kritik, die Eddington[1]) an der Bessel-Kobold-schen Methode der Apexberechnung übt, nicht berechtigt, wenn er sagt: „this method makes no discrimination between the two ways, in which a star may move along its great circle", und fortsetzt: „Two stars moving in exactly opposite directions will have the same pole." Wohl etwa den gleichen Pol der Eigenbewegung, aber nicht notwendigerweise das gleiche Ellipsoid.

9. Der Vertex der Sternbewegungen. Für die Koordinaten des Vertex ergeben die verschiedenen Methoden seiner Berechnung Werte, die ungefähr zwischen den Grenzen:

$$\alpha = 260° \text{ bis } 280°, \qquad \delta = -10° \text{ bis } 20°$$

liegen und, in galaktischen Koordinaten reduziert, auf:

$$l = 330° - 340°, \qquad b = 0°$$

führen. Es liegt also der Vertex in der Ebene der Milchstraße. Andererseits ist man, wie bekannt, bei der Untersuchung über die Verteilung der Kugelsternhaufen auf die gleiche Länge $l = 325°$ als jene Richtung gekommen, um die sich diese häufen, während die entgegengesetzte $l = 145°$ von ihnen ganz gemieden wird. Wie weiter bekannt, sollen nach der Anschauung Shapleys die Kugelhaufen ein größeres galaktisches System bilden, in das jenes, von dem unsere Sonne einen Teil bildet, wie eine lokale Anhäufung eingebettet ist. Dann aber muß die Richtung, in der der Mittelpunkt des großen Systems von der Sonne aus gesehen erscheint, in die der größten Konzentration der Kugelhaufen, d. h. in die Länge $325°$, fallen, und der Vertex wäre sonach mit dieser Zentrumsrichtung identisch. — Auf die Frage jedoch, ob diese Übereinstimmung der beiden Richtungen nur ein Zufall ist, gibt die Stellarastronomie zur Zeit noch keine Antwort. Sie kann nur die Aussage machen, daß sie in der Definition der Vertex nach der Zwei-schwarmhypothese, wonach er die relative Bewegungsrichtung des einen gegen den anderen Schwarm darstellt, nicht begründet ist.

---

[1]) Eddington, A. S.: Stellar movements and the structure of the universe. S. 84. London 1914.

Aber noch mehr! Von den drei Richtungen, zu denen man bei der Berechnung des Momentenellipsoides kommt, fällt die beiden Ellipsoiden gemeinsame in die des Apex der Sonnenbewegung. Die zweite, durch $\alpha_2 = 155°1$, $\delta_2 = +35°2$ bestimmte, gibt nach Reduktion in galaktische Koordinaten $l = 157°5$, $b = +57°9$ und führt damit nach Projektion auf die Ebene der Milchstraße ($b = 0°$) wieder zu der geheimnisvollen Länge $l = 157°$ bzw. $337°$, sowie nach deren Rücktransformation in Äquatorkoordinaten zu $\alpha = 90°7$, $\delta = +20°2$, d. i. dem Vertex, der damit als Projektion der zweiten Hauptrichtung des Momentenellipsoids auf die Milchstraße erscheint.

Hierdurch ist ein Zusammenhang zwischen den beiden Ellipsoiden der Stellarastronomie, dem Momenten- und dem Streuungsellipsoid, festgelegt, der darin besteht, daß eine Hauptrichtung des ersteren, projiziert auf die Milchstraße, in eine Hauptrichtung des zweiten fällt, ohne daß aber damit das Geheimnis des Vertex als gelöst angesehen werden kann.

Zum Strahlungsgleichgewicht der Sterne.

Von **H. v. Zeipel**, Upsala.

Von einer rotierenden Gasmasse sagt man, daß sie sich im mechanischen Gleichgewicht befindet, wenn die Schwerkraft genau durch den Auftrieb des Totaldruckes aufgehoben wird. Die Schwerkraft ist die Resultante der Attraktions- und der Zentrifugalkraft. Der Totaldruck ist die Summe des Gas- und des Strahlungsdruckes. Ein gasförmiger Körper von gegebener Masse, mittlerer Dichte und Rotationsgeschwindigkeit kann sich auf unendlich viele Arten in mechanischem Gleichgewicht befinden. Jeder solche Gleichgewichtszustand ist durch einen gewissen Funktionszusammenhang zwischen der Dichte und dem Schwerepotential oder zwischen dem Gasdruck und der Temperatur charakterisiert. Aber dieser Zusammenhang kann willkürlich gewählt werden. Bei adiabatischem Gleichgewicht ist der Gasdruck einer Potenz der Temperatur proportional. Die adiabatischen Gleichgewichtszustände sind besonders eingehend von EMDEN in seiner Arbeit „Gaskugeln" studiert worden.

1. Die Strahlungsintensität. Die Einführung des Begriffes „Strahlungsgleichgewicht" durch SCHWARZSCHILD[1]) bedeutete einen wichtigen Schritt zur Lösung des Problems. Jedes Element des gasförmigen Körpers wird nach allen Richtungen von Energie durchstrahlt. Emission und Absorption innerhalb des Gases erfolgen im Einklang mit KIRCHHOFFS und STEFANS Gesetzen. Wenn der Strahlungszustand stationär, d. h. unabhängig von der Zeit ist, so sagt man, daß Strahlungsgleichgewicht stattfindet. Wegen der Energieausstrahlung des Sternes in den Raum muß hierbei ständig neue Energie im Innern des Sternes gebildet oder freigemacht werden.

SCHWARZSCHILD leitete die Differentialgleichung für die Intensität I eines Strahles von gegebener Richtung innerhalb der Gasmasse ab. Seine Differentialgleichung kann man schreiben:

$$\frac{dI}{ds} + I k \varrho = \mu T^4 k \varrho \tag{1}$$

Hier bedeutet $\dfrac{dI}{ds}$ den Differentialquotienten der Intensität in der Richtung des Strahles, ϱ die Dichte des Gases, T seine Temperatur,

---

[1]) Göttinger Nachrichten 1906, S. 41. — Sitzungsber. d. preuß. K. Akad. d. Wiss. 1914, S. 1183.

k seinen Absorptionskoeffizient und $\mu = 1{,}685 \cdot 10^{-5}$ eine physikalische Konstante. Wenn ϱ, T und k gegebene Funktionen der Lage und also auch der Wegstrecke s längs der Richtung des Strahles sind, so wird I vollständig durch die lineare Differentialgleichung (1) sowie durch die Anfangsbedingung $I = 0$ für $s = 0$ bestimmt, welche Bedingung bedeutet, daß die Einstrahlung aus dem Raum gleich Null ist. Die Lösung läßt sich in Integralform schreiben. Aber beim Studium des Innern der Sterne ist es zweckmäßiger, MILNE's[1]) Entwicklung:

$$I = \mu T^4 - \frac{1}{k\varrho}\frac{d\mu T^4}{ds} + \frac{1}{k\varrho}\frac{d}{ds}\left(\frac{1}{k\varrho}\frac{d\mu T^4}{ds}\right) - \cdots \tag{2}$$

anzuwenden, welche, wie unmittelbar ersichtlich ist, formell (1) genügt. Die ersten Glieder dieser Entwicklung nehmen innerhalb des Sternes nicht allzu nahe seiner Oberfläche überaus schnell ab. Nur die drei ersten Glieder kommen für verschiedene Zwecke zur Anwendung. Man hat:

$$\frac{d}{ds} = \alpha\frac{d}{dx} + \beta\frac{d}{dy} + \gamma\frac{d}{dz} \tag{3}$$

wo α, β, γ die Richtungscosinus des Strahles mit der x-, y- und z-Achse sind.

2. Die Energieerzeugung. SCHWARZSCHILD studierte nur das Strahlungsgleichgewicht in der Nähe der Oberfläche des Sternes. Hier konnte die Energieerzeugung als gleich Null angenommen werden. EDDINGTON[2]) ging weiter und suchte das Strahlungsgleichgewicht im Innern der Sterne zu erforschen. Hierbei stellte er die Differentialgleichung des Strahlungsgleichgewichtes auf. Dieselbe stellt einen Differentialausdruck für die Energieerzeugung dar. Mit $4\pi\varepsilon$ wird die Energieproduktion pro Sekunde und Gramm innerhalb des Sternes bezeichnet. Ist P die Emission und Q die Absorption der Masseneinheit pro Sekunde, so ist offenbar:

$$4\pi\varepsilon = P - Q \tag{4}$$

Nach KIRCHHOFFS und STEFANS Gesetzen ist:

$$P = 4\pi k\mu T^4$$

Weiter ist:

$$Q = k\int I\,d\Sigma \tag{5}$$

wo der Elementarkegel $d\Sigma$ die Fläche der ganzen Einheitssphäre durchläuft. Nach Einführung der Entwicklung (2), wo (3) gilt, in den Integral-

---

[1]) Monthly Notices of R. A. S. Bd. 83, S. 122.

[2]) Monthly Notices of R. A. S. Bd. 77, S. 16, 596; Zeitschr. f. Physik Bd. 7, S. 351.

ausdruck (5) für Q findet man unter Berücksichtigung der wohlbekannten Relationen:

$$\int d\Sigma = 4\pi \qquad\qquad \int \alpha\, d\Sigma = 0\,,\ldots$$

$$\int \beta\gamma\, d\Sigma = 0\,,\ldots \qquad\qquad \int \alpha^2\, d\Sigma = \frac{4\pi}{3}\,,\ldots$$

daß der Ausdruck (4) für ε geschrieben werden kann:

$$\varepsilon = -\frac{1}{3\varrho}\sum_{3}\frac{d}{dx}\frac{d\mu T^4}{k\varrho\, dx} + \ldots \tag{6}$$

Auf der rechten Seite dieser Formel enthält die Summe $\sum_{3}$ drei Glieder, entsprechend x, y und z.

Die Gleichung (6), welche mit großer Approximation im Innern des Sternes nicht allzu nahe seiner Oberfläche gilt, ist die Differentialgleichung des Strahlungsgleichgewichtes. Wenn die Energieerzeugung ε und der Absorptionskoeffizient k bekannte Funktionen von ϱ und T wären, so würde (6) in Verbindung mit der Gleichung des mechanischen Gleichgewichtes zur Bestimmung des Gleichgewichtszustandes führen. Aber ε ist völlig unbekannt. Man weiß nicht einmal, ob die Energieerzeugung durch die Abnahme der potentiellen Energie infolge der Transformation der Elemente verursacht wird, oder ob Materie in Energie übergeht. Auch über die Natur des Absorptionskoeffizienten herrscht Unsicherheit, insofern verschiedene Theorien etwas verschiedene Ausdrücke für k geben.

Gegenüber diesen Schwierigkeiten entschloß sich Eddington zu der Annahme, daß ε und k innerhalb des Sternes konstant seien.

Dagegen nahm A. Kohlschütter[1] an, daß die Energieerzeugung proportional der Temperatur, k aber konstant sei.

Eddingtons Theorie für das Gleichgewicht kugelförmiger Sterne wurde von Milne[2] auf rotierende Sterne mit geringer Abplattung ausgedehnt. Milne hält an Eddingtons Hypothese fest, daß ε konstant ist.

Im folgenden wird unter einer sehr allgemeinen und glaubhaften Voraussetzung der generelle Ausdruck für die Energieerzeugung innerhalb einer rotierenden Gasmasse in Strahlungsgleichgewicht abgeleitet. Als Grundhypothese wird angenommen, daß die Natur des Gases, d. h. seine Zusammensetzung, in jeder Niveaufläche konstant ist, aber von einer Niveaufläche zur anderen variieren kann. Dann sind auch Ionisationsgrad, mittleres Atomgewicht, Absorptionskoeffizient und Energieerzeugung in der Niveaufläche konstant. Diese Grundhypothese ist bedeutend allgemeiner als die Hypothesen von Eddington und Kohlschütter. Ferner wird der Gasmasse eine Eigenschaft zugeschrieben,

---

[1] Publ. d. Astrophys. Obs. z. Potsdam Nr. 78. 1922.
[2] Monthly Notices of R. A. S. Bd. 83, S. 118.

die während der Entwicklung des Sternes fortbesteht, wenn sie nur einmal vorhanden war. Mit dieser allgemeinen Grundhypothese als Ausgangspunkt wird unten gezeigt, daß die Energieerzeugung innerhalb eines rotierenden Sternes mit Notwendigkeit dem Gesetz:

$$4\pi\varepsilon = B\left(1 - \frac{\omega^2}{2\pi G\varrho}\right) \qquad (7)$$

folgen muß, wo ω die Rotationsgeschwindigkeit, G die Attraktionskonstante ($G = 6{,}6576 \cdot 10^{-8}$) und B eine Konstante innerhalb des Sternes ist.

3. Die Strahlungskraft und der Strahlungsdruck. Eine Masseneinheit, die von einem Strahlenbündel von der Intensität I und innerhalb eines Kegels $d\Sigma$ um eine Richtung α, β, γ passiert wird, absorbiert pro Sekunde die Energiemenge $\delta Q = I k d\Sigma$. Hierbei wird die Masseneinheit — im Einklang mit der elektromagnetischen Theorie und LEBEDEWS Experiment — durch eine Strahlungskraft von der Größe $c^{-1} \cdot \delta Q$ in der Richtung des Strahles beeinflußt. ($c = 3 \cdot 10^{10}$ bezeichnet die Geschwindigkeit des Lichtes.)

Die totale Strahlungskraft wird von der nach allen Richtungen gehenden Strahlung erzeugt. Die Projektion dieser Kraft auf einen Vektor s', dessen Richtungscosinus α', β', γ' sind, ist also:

$$S_{s'} = \frac{k}{c}\int I \sum_3 \alpha\,\alpha' \cdot d\Sigma$$

$$= -\frac{k}{c}\frac{4\pi}{3}\frac{1}{k\varrho}\sum_3 \alpha'\frac{d\mu T^4}{dx}$$

$$= -\frac{1}{\varrho}\frac{d\dfrac{a}{3}T^4}{ds'}$$

wo

$$a = \frac{4\pi\mu}{c} = 7{,}060 \cdot 10^{-15}$$

Definiert man den Strahlungsdruck $\tilde{\omega}$ durch die Formel:

$$\tilde{\omega} = \frac{a}{3}T^4 \qquad (8)$$

so ist also:

$$\varrho S_{s'} = -\frac{d\tilde{\omega}}{ds'}$$

4. Die Gleichung des mechanischen Gleichgewichtes. Bezeichnet p den Gasdruck, $\tilde{\omega}$ den Strahlungsdruck, ϱ die Dichte und Φ das Schwerepotential, so ist die Gleichung des mechanischen Gleichgewichtes:

$$d(p + \tilde{\omega}) = \varrho\,d\Phi \qquad (10)$$

Die Gleichung der Niveaufläche ist:

$$\Phi = \mathrm{konst}$$

Aus (10) folgt, daß auch der Totaldruck $p + \tilde{\omega}$ auf der Niveaufläche konstant ist. Demnach besteht ein Funktionszusammenhang:

$$p + \tilde{\omega} = P(\Phi) \tag{11}$$

Dann ergibt (10):

$$\varrho = \frac{d(p + \tilde{\omega})}{d\Phi} = P'(\Phi) \tag{12}$$

Also ist auch die Dichte ϱ in der Niveaufläche konstant.

In Übereinstimmung mit v. d. Waals Zustandsgleichung ist der Gasdruck p durch die Formel:

$$p = \frac{R}{m} \frac{\varrho}{1 - b\varrho} T \tag{13}$$

gegeben, wo $R = 8{,}2962 \cdot 10^7$ die Gaskonstante ist. v. d. Waals Konstante b ist der umgekehrte Wert der Gasdichte bei größtmöglicher Kompression. Bei idealen Gasen ist $b = 0$. m bezeichnet das mittlere Molekulargewicht des Gasgemisches. Bei Ionisation werden die freien Elektronen mit als Moleküle gerechnet. Bei starker Ionisation ist m bei fast allen Gasen nahe $= 2$.

Aus unserer Grundhypothese, daß die Natur des Gases in der ganzen Niveaufläche die gleiche ist, folgt, daß m und b in der letzteren konstant sind. Demnach ergibt sich auf Grund von (11), (13), (8), (12), daß die Temperatur T in der Niveaufläche konstant ist.

Das gleiche gilt zufolge unserer Grundhypothese auch für den Absorptionskoeffizienten k.

Innerhalb einer Gasmasse, die sich in Strahlungsgleichgewicht befindet und deren Natur in der Niveaufläche konstant ist, sind also m, k, b, ϱ, T, p, $\tilde{\omega}$ Funktionen von Φ.

5. Die Energieerzeugung innerhalb einer rotierenden Gasmasse, deren Natur in der Niveaufläche konstant ist. Ich greife auf Eddingtons Ausdruck (6) für die Energieerzeugung zurück.

Man hat:

$$\frac{1}{k\varrho} \frac{dT^4}{dx} = \frac{1}{k\varrho} \frac{dT^4}{d\Phi} \frac{d\Phi}{dx}$$

$$\frac{d}{dx}\left(\frac{1}{k\varrho} \frac{dT^4}{dx}\right) = \frac{1}{k\varrho} \frac{dT^4}{d\Phi} \frac{d^2\Phi}{dx^2} + \frac{d}{d\Phi}\left(\frac{1}{k\varrho} \frac{dT^4}{d\Phi}\right)\left(\frac{d\Phi}{dx}\right)^2$$

Der Ausdruck (6) wird also:

$$\varepsilon = -\frac{\mu}{3\varrho}\left\{\frac{1}{k\varrho} \frac{dT^4}{d\Phi} \Delta\Phi + \frac{d}{d\Phi}\left(\frac{1}{k\varrho} \frac{dT^4}{d\Phi}\right)\left(\frac{d\Phi}{dn}\right)^2\right\} \tag{14}$$

denn für das Quadrat des Differentialquotienten von Φ in der Richtung der Normale der Niveaufläche gilt:

$$\left(\frac{d\,\Phi}{d\,n}\right)^2 = \sum_3 \left(\frac{d\,\Phi}{d\,x}\right)^2$$

was nur bedeutet, daß die Schwerekomponenten $\dfrac{d\,\Phi}{d\,x}$, $\dfrac{d\,\Phi}{d\,y}$, $\dfrac{d\,\Phi}{d\,z}$ sind.

Aus Poissons Gleichung folgt:

$$\Delta\Phi = -4\pi G\varrho + 2\omega^2 \tag{15}$$

wo ω die Rotationsgeschwindigkeit der Gasmasse ist. Demnach ist $\Delta\Phi$ in der Niveaufläche konstant. Dasselbe gilt also von dem ganzen ersten Glied im Ausdruck (14).

Es läßt sich zeigen, daß die Schwere $\dfrac{d\,\Phi}{d\,n}$ bei einer *rotierenden* Flüssigkeits- oder Gasmasse in mechanischem Gleichgewicht notwendig in der Niveaufläche variabel ist. Den Beweis übergehen wir hier.

Aus unserer Grundhypothese folgt, daß ε in der Niveaufläche konstant ist. Also ist bei Strahlungsgleichgewicht für eine *rotierende* Gasmasse, deren Natur in der Niveaufläche konstant ist:

$$\frac{d}{d\,\Phi}\left(\frac{1}{k\,\varrho}\,\frac{d\,T^4}{d\,\Phi}\right) = 0$$

d. h.

$$\varrho\,d\,\Phi = C\cdot\frac{a}{3}\,\frac{d\,T^4}{k} \tag{16}$$

wo C innerhalb des Körpers konstant ist.

Auf Grund von (15) und (16) erhält nun der Ausdruck (14) für die Energieerzeugung die Form:

$$\varepsilon = \frac{c\,G}{C}\left(1 - \frac{\omega^2}{2\pi G\varrho}\right) \tag{17}$$

Dieses Gesetz für die Energieerzeugung ist vom Strahlungsgesetz Kirchhoffs und Stefans sowie von der Zustandsgleichung des Gases ganz unabhängig. Nur die Schwere ist dafür maßgebend. Bei Ableitung von Formel (17) wurde auch keine spezielle Annahme bezüglich der Quantitäten m, b und k gemacht. Dieselben können völlig arbiträre Funktionen der Natur und des Zustandes des Gases sein. Ebensowenig wurde etwas über die Rotationsgeschwindigkeit oder allgemeine Form des Systems angenommen. Letzteres kann aus einem oder mehreren Teilkörpern mit willkürlichen Massen und mittleren Dichten bestehen. Unsere Resultate haben also eine sehr große Tragweite. Sie können auf gewöhnliche Riesen- und Zwergsterne, aber auch auf Übergangsformen zwischen einfachen und Doppelsternen, sowie auf Doppel- und

Multipelsterne mit gebundener Rotation angewandt werden und gelten auch für schalenartige oder ringförmige Gleichgewichtsfiguren.

Bei verschwindender Rotation reduziert sich das Gesetz (17) auf EDDINGTONS Hypothese, daß ε innerhalb des Sternes konstant ist. Indem wir von dem allgemeinen Falle ($\omega^2 \neq 0$) zu dem besonderen ($\omega^2 = 0$) gingen, ist es uns also gelungen, eine festere Grundlage für EDDINGTONS Hypothese zu finden, als es bisher möglich war. Es ist bemerkenswert, daß EDDINGTONS Hypothese keine Folge unserer fundamentalen Annahme ist, wenn wir nur Sterne ohne Rotation betrachten; denn für solche ist die Schwere in der Niveaufläche konstant. Wir können dann nicht zu der Relation (16) gelangen.

Aus dem Obigen folgt, daß ε, falls m, b und k in der Niveaufläche konstant sind, nur unter derjenigen Bedingung innerhalb des ganzen Sternes konstant sein kann, daß entweder $\omega^2 = 0$ oder $\varrho = $ konstant in der ganzen Gasmasse ist.

Die Energieproduktion ist nach dem Zentrum der Gasmasse zu positiv, solange $\varrho > \dfrac{\omega^2}{2\pi G}$. Es ist ungewiß, ob die Formel (17) auch in den äußersten Schichten des Sternes Geltung behält. In einem Zwischengebiet, wo $\varrho < \dfrac{\omega^2}{2\pi G}$, ist indessen die Energieerzeugung negativ. Dies bedeutet, daß hier Energie aufgespeichert oder in Materie verwandelt wird. Das fragliche Gebiet erstreckt sich tatsächlich ziemlich weit nach innen, auch bei nahezu sphärischen Sternen. Für wenig abgeplattete Idealgassterne gilt die Formel:

$$\frac{\bar{r} - r_0}{\bar{r}} = 0{,}9031 \left(\frac{\omega^2}{2\pi G \varrho_m}\right)^{\frac{1}{3}} + \ldots = 0{,}9846 \cdot (\text{Abplattung})^{\frac{1}{3}} + \ldots$$

Hier bezeichnet $\bar{r}$ den mittleren Radius des Sternes und r_0 den der Niveaufläche, für die $\varrho = \dfrac{\omega^2}{2\pi G}$.

Für die Sonne ist:

$$\frac{\omega^2}{2\pi G \varrho_m} = 10^{-5}$$

und also:

$$\frac{\bar{r} - r_0}{\bar{r}} = 0{,}019$$

Für einen Idealgasstern mit der Abplattung 10^{-3} ist:

$$\frac{\bar{r} - r_0}{\bar{r}} = 0{,}098\,46$$

In so großen Abständen von der Oberfläche gelten die Theorie und die Formel (17) mit Sicherheit.

Die Energieproduktion der Volumeneinheit $4\pi\varrho\varepsilon$ ist aus zwei Gliedern zusammengesetzt. Das erste positive Glied ist proportional der

Dichte ϱ, das zweite negative Glied ist innerhalb des Sternes konstant und proportional des Quadrates der Rotationsgeschwindigkeit. Ob diese mathematische Zerlegung auch einen physikalischen Grund hat, das mag dahingestellt bleiben.

6. Der Zusammenhang zwischen dem Gasdruck und der Temperatur. Auf Grund von (16) und (8) erhält die Gleichung des mechanischen Gleichgewichts (10) die Form:

$$d\,p = \frac{a}{3}\left(\frac{C}{k} - 1\right)d\,T^4 = \left(\frac{C}{k} - 1\right)d\,\tilde\omega \tag{18}$$

Wenn die Abhängigkeit des Absorptionskoeffizienten vom Zustand des Gases bekannt wäre, so würde (18) den Zusammenhang zwischen p und T ergeben.

EDDINGTON nahm anfangs k als konstant an. Später setzte er[1]), mehr in Übereinstimmung mit modernen Emissionstheorien, voraus, daß $k \sim \frac{\varrho}{m} T^{-3}$. Dieses Gesetz läßt sich auch schreiben:

$$k = K\,p\,T^{-4} = \frac{a}{3}\,K \cdot \frac{p}{\tilde\omega} \tag{19}$$

wo K eine Konstante ist.

Wenn der Absorptionskoeffizient dem Gesetz (19) folgt, so erhält der Zusammenhang zwischen dem Gasdruck und der Temperatur die Form:

$$p = \frac{a}{3}\,\frac{\beta}{1-\beta}\,T^4 \tag{20}$$

wo β eine Konstante ist, welche C ersetzt zufolge der Substitution:

$$C = \frac{a}{3}\,K\,\frac{\beta}{(1-\beta)^2}$$

v. d. WAALS Gleichung (13) verglichen mit (20) gibt die Relation:

$$\frac{R}{m}\,\frac{\varrho}{1-b\varrho} = \frac{a}{3}\,\frac{\beta}{1-\beta}\,T^3 \tag{21}$$

Nach Integration von (16) erhält man den Zusammenhang:

$$\Phi - \tilde\Phi = \frac{a}{3}\,\frac{b}{1-\beta}\,T^4 + \frac{R}{m}\,\frac{4}{\beta}\,T \tag{22}$$

Die Integrationskonstante $\tilde\Phi$ bezeichnet den Wert des Schwerepotentials auf der Oberfläche des Körpers.

Wird T aus (21) und (22) eliminiert, so ergibt sich die Gleichung:

$$\frac{b\varrho\,(1-\frac{3}{4}\,b\varrho)^3}{(1-b\varrho)^4} = \frac{a\,b}{3\cdot 4^3}\left(\frac{m}{R}\right)^4\frac{\beta^4}{1-\beta}(\Phi - \tilde\Phi)^3 \tag{23}$$

---

[1]) Zeitschr. f. Physik Bd. 7, S. 351.

Bei Idealgasen vereinfachen sich die Formeln (21), (22) und (23) zu:

$$\varrho = \frac{m}{R}\frac{a}{3}\frac{\beta}{1-\beta}\,T^3 \tag{24}$$

$$\Phi - \tilde{\Phi} = \frac{R}{m}\frac{4}{\beta}\,T \tag{25}$$

$$\varrho = \frac{a}{3\cdot 4^3}\left(\frac{m}{R}\right)^4\frac{\beta^4}{1-\beta}\,(\Phi - \tilde{\Phi})^3 \tag{26}$$

7. Die Bestimmung des Schwerepotentials. Zufolge der Gleichungen von Poisson und Laplace genügt das Schwerepotential Φ der partiellen Differentialgleichung:

$$\varDelta\,\Phi + 4\,\pi\,G\,\varrho - 2\,\omega^2 = 0 \tag{27}$$

wo $\varDelta$ Laplaces Symbol ist. Außerhalb der Teilkörper ist $\varrho = 0$. Innerhalb derselben gilt (23) oder (26).

Wenn die Anzahl der Teilkörper n ist, so kommen in der Differentialgleichung (27) $2\,n + 1$ Parameter vor, nämlich in erster Linie die für alle Körper gemeinsame Rotationsgeschwindigkeit ω und außerdem für jeden Teilkörper ein Parameterpaar β und $\tilde{\Phi}$.

Die gesuchte Lösung muß derart sein, daß die Differenz zwischen Φ und dem Zentrifugalpotential nebst ihren Differentialquotienten erster Ordnung im ganzen Raum endlich und kontinuierlich ist.

Tatsächlich bestimmen diese Bedingungen die Lösung vollständig. Als die $2\,n + 1$ Fundamentalparameter können die Rotationsgeschwindigkeit, die Massen und die mittleren Dichten gewählt werden. Auf ihnen beruhen alle Eigenschaften des Systems in bezug auf Massenverteilung, Form, Strahlung usw.

Über die Grenzkurven und ihre Einhüllende im asteroidischen Dreikörperproblem bei elliptischer Bahn des störenden Körpers.

Von **A. Wilkens**, Breslau.

Wenn im Problem der drei Körper eine der drei Massen als verschwindend, die beiden anderen aber als beliebig betrachtet werden, so gestattet das Integral der lebendigen Kraft, unter der Voraussetzung einer kreisförmigen Bewegung der beiden nicht verschwindenden Massen umeinander, in der HILLschen Form des JACOBISchen Integrals die Festlegung einer für alle Zeiten unüberschreitbaren Grenze für den Asteroiden. Die Grenze ist definiert, sobald zu irgendeiner Zeit die relative Lage und Geschwindigkeit des Asteroiden in bezug auf die beiden von 0 verschiedenen Massen vorgelegt sind. Die Grenzkurve hat dann ihre einfachste Darstellung in der DARWINSchen Form:

$$\frac{2}{r_0} + r_0^2 + m'\left(\frac{2}{\Delta} + \Delta^2\right) = J \tag{1}$$

wo r_0 und Δ die Bipolarkoordinaten des Asteroiden in bezug auf die Sonne und den Jupiter als Pole sind, wobei m' unter der Annahme der Sonnenmasse als Einheit die Jupitermasse und J die JACOBISche Konstante fixiert. Ist aber die Bahn des Jupiter kein Kreis, sondern wie von nun ab angenommen werden soll, eine Ellipse, so wird die der obigen Grenzkurvengleichung entsprechende Gleichung eine Funktion nicht nur der Koordinaten, sondern auch der Geschwindigkeit der Koordinaten und explizite auftretender Zeit werden; zu untersuchen ist dann die Frage, ob es möglich ist, die neue und komplizierte Gleichung so zu transformieren, daß sie nur noch von den Koordinaten und der expliziten Zeit abhängt, so daß alsdann jedem Zeitpunkt eine bestimmte Grenzkurve entsprechen würde. Dann aber wäre die Möglichkeit gegeben, die Einhüllende aller zeitlichen Grenzkurven zu ermitteln, als Erweiterung des Grenzkurvenproblems für den Fall einer elliptischen Bahn des störenden Körpers. Die Durchführbarkeit dieser Skizze ist das Ziel der vorliegenden Arbeit.

§ 1. Die Aufstellung der Bedingungsgleichungen. Ist die Sonne der Anfangspunkt eines festen rechtwinkligen Koordinatensystems, in dessen

xy-Ebene die Bewegung des Jupiter wie die des Asteroiden vor sich gehen soll, so lauten die Differentialgleichungen der Koordinaten x und y des Asteroiden, wenn x' und y' die Koordinaten des Jupiter sind und die Gausssche Konstante gleich 1 gesetzt wird:

$$\left.\begin{aligned} \frac{d^2 x}{d t^2} &= -\frac{x}{r^3} + \frac{\partial \Omega}{\partial x} \\[2mm] \frac{d^2 y}{d t^2} &= -\frac{y}{r^3} + \frac{\partial \Omega}{\partial y} \end{aligned}\right\} \tag{2}$$

wo:

$$\Omega = \text{Störungsfunktion} = m'\left(\frac{1}{\Delta} - \frac{x\,x' + y\,y'}{r'^3}\right)$$

und:

$$\Delta^2 = (x'-x)^2 + (y'-y)^2, \qquad r^2 = x^2 + y^2, \qquad r'^2 = x'^2 + y'^2$$

Folglich ergibt sich nach Anbringung der Multiplikatoren $2\dfrac{dx}{dt}$ und $2\dfrac{dy}{dt}$ und darauffolgender Addition und Integration der Gleichungen (2) für das Geschwindigkeitsquadrat:

$$V^2 = \left(\frac{dx}{dt}\right)^2 + \left(\frac{dy}{dt}\right)^2 = \frac{2}{r} + 2\int\left(\frac{\partial \Omega}{\partial x}\cdot\frac{dx}{dt} + \frac{\partial \Omega}{\partial y}\cdot\frac{dy}{dt}\right)dt + C \tag{3}$$

wo $C = $ Integrationskonstante. An Stelle des festen Koordinatensystems trete jetzt ein bewegliches System ξ, η derart, daß die ξ-Achse ständig mit dem Radiusvektor des Jupiter zusammenfällt; wegen der vorausgesetzten elliptischen Bewegung des Jupiter ist die Bewegung der ξ-Achse dann eine ungleichmäßige. Ist v' die von der Richtung der x-Achse aus gezählte Länge des Jupiter, so ist die Beziehung der beiden Koordinatensysteme fixiert durch:

$$\left.\begin{aligned} x &= \xi\cos v' - \eta\sin v' \\ y &= \xi\sin v' + \eta\cos v' \end{aligned}\right\} \tag{4}$$

so daß, ausgedrückt durch ξ, η und deren Geschwindigkeiten:

$$V^2 = \left(\frac{dx}{dt}\right)^2 + \left(\frac{dy}{dt}\right)^2 = \left(\frac{d\xi}{dt}\right)^2 + \left(\frac{d\eta}{dt}\right)^2 + r^2\left(\frac{dv'}{dt}\right)^2 + 2\left(\xi\frac{d\eta}{dt} - \eta\frac{d\xi}{dt}\right)\frac{dv'}{dt} \tag{5}$$

Folglich ist die Geschwindigkeit im bewegten System, nach (3) und (5):

$$\left.\begin{aligned} \left(\frac{d\xi}{dt}\right)^2 + \left(\frac{d\eta}{dt}\right)^2 &= C + \frac{2}{r} + 2\int\left(\frac{\partial \Omega}{\partial x}\cdot\frac{dx}{dt} + \frac{\partial \Omega}{\partial y}\cdot\frac{dy}{dt}\right)dt - r^2\left(\frac{dv'}{dt}\right)^2 \\[2mm] &\quad - 2\left(\xi\frac{d\eta}{dt} - \eta\frac{d\xi}{dt}\right)\frac{dv'}{dt} \end{aligned}\right\} \tag{6}$$

Diese Bedingungsgleichung, deren rechte Seite für die Grenzkurve später gleich 0 zu setzen ist, stellt nun eine Beziehung zwischen den Koordinaten

ξ, η, deren Geschwindigkeiten, der Zeit t, die in r' und v' enthalten ist, und der Konstanten C dar, so daß sie die Form hat:

$$f\left(\xi,\eta,\frac{d\xi}{dt},\frac{d\eta}{dt},t,C\right)=0 \tag{7a}$$

Dabei treten die Differentialquotienten $\dfrac{d\xi}{dt}$ und $\dfrac{d\eta}{dt}$ gemäß (4) auch implizite in $\dfrac{dx}{dt}$ und $\dfrac{dy}{dt}$ auf. Ließen sich die Geschwindigkeiten $\dfrac{d\xi}{dt}$ und $\dfrac{d\eta}{dt}$ eliminieren, so daß die Gleichung (7a) in die Form:

$$g(\xi,\eta,t)=0 \tag{7b}$$

überginge, so würde sie sich von der DARWINschen Form im Spezialfalle einer Kreisbahn des Jupiter nur durch das Auftreten von t unterscheiden. Indem t als Parameter betrachtet wird, gehört zu jedem t eine bestimmte Grenzkurve; die Vermittlung der Einhüllenden aller Grenzkurven ist dann an die Bedingung:

$$\frac{\partial g}{\partial t}=0 \tag{7c}$$

gebunden, die dann in Verbindung mit (7b) die Koordinaten ξ und η als Funktion von t resp. nach Elimination von t die Gleichung der Einhüllenden selbst als Funktion von ξ und η allein ergibt.

Nach der Beziehung (4) folgt zunächst für die Flächengeschwindigkeit

$$x\frac{dy}{dt}-y\frac{dx}{dt}=\xi\frac{d\eta}{dt}-\eta\frac{d\xi}{dt}+r^2\frac{dv'}{dt} \tag{8}$$

andererseits ist nach (2):

$$x\frac{d^2y}{dt^2}-y\frac{d^2x}{dt^2}=\frac{d}{dt}\left(x\frac{dy}{dt}-y\frac{dx}{dt}\right)=x\frac{\partial\Omega}{\partial y}-y\frac{\partial\Omega}{\partial x} \tag{9}$$

so daß folglich:

$$x\frac{dy}{dt}-y\frac{dx}{dt}=c+\int\left(x\frac{\partial\Omega}{\partial y}-y\frac{\partial\Omega}{\partial x}\right)dt \tag{9a}$$

wo $c=$ Integrationskonstante. Folglich ergibt sich nach (8) und (9a):

$$\xi\frac{d\eta}{dt}-\eta\frac{d\xi}{dt}=c-r^2\frac{dv'}{dt}+\int\left(x\frac{\partial\Omega}{\partial y}-y\frac{\partial\Omega}{\partial x}\right)dt \tag{10}$$

womit zunächst die die Ableitungen $\dfrac{d\xi}{dt}$ und $\dfrac{d\eta}{dt}$ explizite enthaltenden Teile der rechten Seite von (6) eliminiert werden.

Die unter dem Integral von (6) in $\dfrac{dx}{dt}$ und $\dfrac{dy}{dt}$ implizite enthaltenen Ableitungen $\dfrac{d\xi}{dt}$ und $\dfrac{d\eta}{dt}$ sind folgendermaßen eliminierbar: Da Ω nach der Definition (2) als eine Funktion von x, y, x' und y' allein aufgefaßt werden kann, so ist die totale Ableitung:

$$\frac{d\Omega}{dt} = \frac{\partial\Omega}{\partial x}\frac{dx}{dt} + \frac{\partial\Omega}{\partial y}\frac{dy}{dt} + \frac{\partial\Omega}{\partial x'}\cdot\frac{dx'}{dt} + \frac{\partial\Omega}{\partial y'}\frac{dy'}{dt} \tag{11}$$

während die partielle Ableitung:

$$\frac{\partial\Omega}{\partial t} = \frac{\partial\Omega}{\partial x'}\frac{dx'}{dt} + \frac{\partial\Omega}{\partial y'}\frac{dy'}{dt} \tag{11a}$$

weil die Zeit explizite nur in den Jupiterkoordinaten enthalten ist.

Folglich wird das zu eliminierende Integral:

$$\int\!\!\left(\frac{\partial\Omega}{\partial x}\cdot\frac{dx}{dt} + \frac{\partial\Omega}{\partial y}\cdot\frac{dy}{dt}\right)dt = \Omega - \int\!\!\left(\frac{\partial\Omega}{\partial x'}\frac{dx'}{dt} + \frac{\partial\Omega}{\partial y'}\frac{dy'}{dt}\right)dt + C' \tag{12}$$

wo $C' = \text{const.}$ d. h. eine von $\dfrac{d\xi}{dt}$ und $\dfrac{d\eta}{dt}$ jetzt unabhängige, allein noch von ξ, η und t abhängende Funktion, indem Ω, $\dfrac{\partial\Omega}{\partial x'}$, $\dfrac{\partial\Omega}{\partial y'}$ nur Funktionen von ξ, η und t, und ferner $\dfrac{dx'}{dt}$ und $\dfrac{dy'}{dt}$ Funktionen von t allein sind. Folglich nimmt die Bedingungsgleichung (6) die folgende Form an, wenn $C + 2C'$ in die Bezeichnung C zusammengefaßt werden:

$$\left.\begin{aligned}
\left(\frac{d\xi}{dt}\right)^2 + \left(\frac{d\eta}{dt}\right)^2 &= C - 2c\frac{dv'}{dt} + \frac{2}{r} + 2\Omega + r^2\left(\frac{dv'}{dt}\right)^2\\
&\quad - 2\int\!\!\left(\frac{\partial\Omega}{\partial x'}\cdot\frac{dx'}{dt} + \frac{\partial\Omega}{\partial y'}\frac{dy'}{dt}\right)dt - 2\frac{dv'}{dt}\int\!\!\left(x\frac{\partial\Omega}{\partial y} - y\frac{\partial\Omega}{\partial x}\right)dt
\end{aligned}\right\} \tag{13}$$

Da die Koordinaten ξ, η auch in den Funktionen unter den Integralen auftreten, verbunden mit Funktionen von t, so stellt die gleich 0 gesetzte rechte Seite von (13) eine Integralgleichung dar, während die dem Spezialfalle $e' = 0$ entsprechende Grenzkurvengleichung (1) nach DARWIN eine gewöhnliche Funktionalgleichung ist. Ferner ist für den allgemeinen Fall, wo $e' \neq 0$, das Auftreten der neuen unabhängigen Konstanten c der Flächen bemerkenswert, die sich, wenn $e' = 0$ und deshalb $\dfrac{dv'}{dt} = \text{const}$ ist, mit C zu einer einzigen willkürlichen Konstanten verbindet. Das Nichtvorkommen der Integrale in (13) im Falle einer Kreisbahn des Jupiter legt den Gedanken nahe, daß die betreffenden Integrale sich für $e' = 0$ fortheben müssen, was sogleich bewiesen werden soll; dabei soll e' als gemeinsamer Faktor der Integrale in Evidenz gesetzt werden.

Ist $e' = 0$, so ist $r' = a' = \text{const}$, ferner $\dfrac{dv'}{dt} = n' = \text{const.}$ und

$$\left.\begin{aligned}
x' &= a'\cos l'\\
y' &= a'\sin l'
\end{aligned}\right\} \tag{14}$$

wo l' die mittlere Länge, so daß:

$$\left.\begin{aligned} \frac{dx'}{dt} &= -y'n' \\[2mm] \frac{dy'}{dt} &= x'n' \end{aligned}\right\} \tag{14a}$$

Ferner ist nach der Definition (2) für Ω:

$$\left.\begin{aligned} \frac{\partial\Omega}{\partial x'} &= m'\left[\frac{x-x'}{\Delta^3} - \frac{x}{r'^3} + 3(xx'+yy')\frac{x'}{r'^5}\right] \\[2mm] \frac{\partial\Omega}{\partial y'} &= m'\left[\frac{y-y'}{\Delta^3} - \frac{y}{r'^3} + 3(xx'+yy')\frac{y'}{r'^5}\right] \end{aligned}\right\} \tag{15}$$

so daß, wenn $e' = 0$, unter Verwendung von (14 a):

$$\frac{\partial\Omega}{\partial x'}\frac{dx'}{dt} + \frac{\partial\Omega}{\partial y'}\frac{dy'}{dt} = -m'n'(xy'-yx')\left(\frac{1}{\Delta^3} - \frac{1}{a'^3}\right) \tag{16}$$

Da ferner:

$$\left.\begin{aligned} \frac{\partial\Omega}{\partial x} &= m'\left(\frac{x'-x}{\Delta^3} - \frac{x'}{r'^3}\right) \\[2mm] \frac{\partial\Omega}{\partial y} &= m'\left(\frac{y'-y}{\Delta^3} - \frac{y'}{r'^3}\right) \end{aligned}\right\} \tag{17}$$

so folgt:

$$x\frac{\partial\Omega}{\partial y} - y\frac{\partial\Omega}{\partial x} = m'(xy'-yx')\left(\frac{1}{\Delta^3} - \frac{1}{r'^3}\right) \tag{18}$$

Folglich ist im Falle $e' = 0$:

$$\frac{\partial\Omega}{\partial x'}\frac{dx'}{dt} + \frac{\partial\Omega}{\partial y'}\cdot\frac{dy'}{dt} + n'\left(x\frac{\partial\Omega}{\partial y} - y\frac{\partial\Omega}{\partial x}\right) = 0 \tag{19}$$

was zunächst zu beweisen war.

Folglich reduziert sich die Bedingungsgleichung (13):

$$\left(\frac{d\xi}{dt}\right)^2 + \left(\frac{d\eta}{dt}\right)^2 = 0$$

auf die folgende Form, falls $c' = 0$:

$$C - 2cn' + \frac{2}{r} + 2\Omega + r^2n'^2 = 0 \tag{13a}$$

Setzen wir noch der Einfachheit halber $a' = 1$, so ist der Faktor n'^2 von r^2 in (13 a): $n'^2 = 1 + m'$. Da ferner Ω wegen der Beziehung $xx' + yy' = rr'\cos(r,r')$ auf die Form:

$$\Omega = m'\left(\frac{1}{\Delta} - \frac{r}{r'^2}\cos(r,r')\right) \tag{2a}$$

gebracht werden kann, so läßt sich die in Ω in (13 a) noch enthaltene Zeit t eliminieren, weil

$$\Delta^2 = r^2 + r'^2 - rr'\cos(r.r') \tag{2b}$$

so daß:

$$\cos(r, r') = \frac{r^2 + r'^2 - \Delta^2}{2\, r\, r'}$$

so daß folglich:

$$\Omega = m'\left(\frac{1}{\Delta} - \frac{1}{2\, r'^3}(r^2 + r'^2 - \Delta^2)\right) \tag{2c}$$

Die Substitution dieses Ausdruckes in (13 a) führt dann aber direkt auf die Darwinsche Form der Grenzkurvengleichung:

$$\frac{2}{r} + r^2 + m'\left(\frac{2}{\Delta} + \Delta^2\right) = J \tag{13 b}$$

wo $J = -C + 2\,c\,n' + m'$ die Jacobische Konstante fixiert.

Nach dieser Reduktion im Spezialfälle $e' = 0$ ist es nun leicht, im allgemeinen Falle $e' \neq 0$ den gemeinsamen Faktor e' der Integrale in (13) evident zu machen und die Gleichung (13) dementsprechend zu vereinfachen. Da:

$$\left.\begin{aligned} x' &= r'\cos v' \\ y' &= r'\sin v' \end{aligned}\right\} \tag{20}$$

so werden die Geschwindigkeiten von x' und y', wenn $\dfrac{d\,v'}{d\,t} = n' + \Delta\,n'$ gesetzt wird, wo n' die konstante mittlere Bewegung des Jupiter: $n' = \sqrt{1 + m'}$ und $\Delta\,n'$ der mit $e' = 0$ verschwindende Teil von $\dfrac{d\,v'}{d\,t}$ ist, definiert durch $\dfrac{d\,\Delta\,n'}{d\,t} = \dfrac{d^2 v'}{d\,t^2}$:

$$\left.\begin{aligned} \frac{d\,x'}{d\,t} &= -y'\,n' - y'\,\Delta\,n' + \frac{x'}{r'}\cdot\frac{d\,r'}{d\,t} \\[2mm] \frac{d\,y'}{d\,t} &= +x'\,n' + x'\,\Delta\,n' + \frac{y'}{r'}\cdot\frac{d\,r'}{d\,t} \end{aligned}\right\} \tag{21}$$

wo $\dfrac{d\,r'}{d\,t}$ mit $e' = 0$ verschwindet.

Folglich wird auf Grund der Beziehungen (15) und (21):

$$\left.\begin{aligned} \frac{\partial\Omega}{\partial x'}\cdot\frac{d\,x'}{d\,t} + \frac{\partial\Omega}{\partial y'}\cdot\frac{d\,y'}{d\,t} = m'\Bigg[&-n'(x\,y' - y\,x')\left(\frac{1}{\Delta^3} - \frac{1}{r'^3}\right) \\ &-\Delta\,n'(x\,y' - y\,x')\left(\frac{1}{\Delta^3} - \frac{1}{r'^3}\right) + (x\,x' + y\,y')\left(\frac{1}{\Delta^3} - \frac{1}{r'^3}\right)\frac{1}{r'}\cdot\frac{d\,r'}{d\,t} \\ &-\frac{r'}{\Delta^3}\cdot\frac{d\,r'}{d\,t} + 3\,(x\,x' + y\,y')\frac{1}{r'^4}\cdot\frac{d\,r'}{d\,t}\Bigg] \end{aligned}\right\} \tag{22}$$

Mithin ergibt sich noch mit Rücksicht auf (18) für die Summe der in (13) auftretenden Integrale:

$$\left.\begin{aligned}
&\int\left(\frac{\partial\,\Omega}{\partial\,x'}\frac{d\,x'}{d\,t}+\frac{\partial\,\Omega}{\partial\,y'}\frac{d\,y'}{d\,t}\right)d\,t+\frac{d\,v'}{d\,t}\int\left(x\frac{\partial\,\Omega}{\partial\,y}-y\frac{\partial\,\Omega}{\partial\,x}\right)d\,t\\[2mm]
&=-m'\int(x\,y'-y\,x')\left(\frac{1}{\varDelta^3}-\frac{1}{r'^3}\right)\varDelta\,n'\,d\,t\\[2mm]
&\quad+m'\varDelta\,n'\int(x\,y'-y\,x')\left(\frac{1}{\varDelta^3}-\frac{1}{r'^3}\right)d\,t\\[2mm]
&\quad+m'\int\Bigl[(x\,x'+y\,y')\left(\frac{1}{\varDelta^3}-\frac{1}{r'^3}\right)\frac{1}{r'}-\frac{r'}{\varDelta^3}\\[2mm]
&\quad+3\,(x\,x'+y\,y')\frac{1}{r'^4}\Bigr]\frac{d\,r'}{d\,t}\,d\,t
\end{aligned}\right\} \qquad (23)$$

Die beiden ersten Glieder rechter Hand von (23) sind in ein einziges Doppelintegral zusammenfaßbar, denn sie haben zusammen die Form:

$$D=-\int u\,v\,d\,t+u\int v\,d\,t \qquad (24)$$

wenn:

$$u=\varDelta\,n' \qquad \text{und} \qquad v=m'\,(x\,y'-y\,x')\left(\frac{1}{\varDelta^3}-\frac{1}{r'^3}\right)$$

so daß nach der Regel der partiellen Integration:

$$D=\int(\textstyle\int v\,d\,t)\,d\,u=m'\int\Bigl[\int(x\,y'-y\,x')\left(\frac{1}{\varDelta^3}-\frac{1}{r'^3}\right)d\,t\Bigr]\cdot\frac{d\,\varDelta\,n'}{d\,t}\,d\,t \qquad (24\,\text{a})$$

Folglich wird die in (13) auftretende Summe:

$$\left.\begin{aligned}
&-2\int\left(\frac{\partial\,\Omega}{\partial\,x'}\frac{d\,x'}{d\,t}+\frac{\partial\,\Omega}{\partial\,y'}\cdot\frac{d\,y'}{d\,t}\right)d\,t-2\frac{d\,v'}{d\,t}\int\left(x\frac{\partial\,\Omega}{\partial\,y}-y\frac{\partial\,\Omega}{\partial\,x}\right)d\,t\\[2mm]
&=-2\,m'\int\Bigl[\int(x\,y'-y\,x')\left(\frac{1}{\varDelta^3}-\frac{1}{r'^3}\right)d\,t\Bigr]\frac{d\,\varDelta\,n'}{d\,t}\,d\,t\\[2mm]
&\quad-2\,m'\int\Bigl[\frac{x\,x'+y\,y'}{r'}\left(\frac{1}{\varDelta^3}-\frac{1}{r'^3}\right)-\frac{r'}{\varDelta^3}+3\,\frac{x\,x'+y\,y'}{r'^4}\Bigr]\frac{d\,r'}{d\,t}\,d\,t
\end{aligned}\right\} (23\,\text{a})$$

Hiermit ist eine einfache Form für die beiden in (13) auftretenden Integrale erreicht, und ferner ist, weil die Faktoren rechter Hand von (23 a), nämlich $\dfrac{d\,\varDelta\,n'}{d\,t}=\dfrac{d^2\,v'}{d\,t^2}$ im 1. und $\dfrac{d\,r'}{d\,t}$ im 2. Integral mit $e'=0$ verschwinden, zugleich gezeigt, wie e' als Faktor aus beiden Integralen in (13) abspaltbar ist; die explizite Darstellung von $\dfrac{d\,\varDelta\,n'}{d\,t}$ und $\dfrac{d\,r'}{d\,t}$ folgt später im Zusammenhange mit der unmittelbaren Anwendung.

Setzen wir schließlich mit Rücksicht auf die Beziehungen (4) und (20):

$$\left.\begin{aligned}
x\,x'+y\,y'&=\xi\,r'\\
x\,y'-y\,x'&=-\eta\,r'
\end{aligned}\right\} \qquad (25\,\text{a})$$

und zur Abkürzung:

$$-\frac{d^2 v'}{d t^2}\int \eta\, r'\left(\frac{1}{\varDelta^2}-\frac{1}{r'^3}\right) d t = l \left.\vphantom{\begin{array}{c}1\\1\\1\end{array}}\right\}$$

$$\left[\xi\left(\frac{1}{\varDelta^3}-\frac{1}{r'^3}\right)-\frac{r'}{\varDelta^3}+3\,\frac{\xi}{r'^3}\right]\frac{d r'}{d t} = m \left.\vphantom{\begin{array}{c}1\\1\end{array}}\right\} \qquad (25\,\mathrm{b})$$

so nimmt unsere Bedingungsgleichung (13) die folgende Endform an:

$$C - 2c\,\frac{d v'}{d t}+\frac{2}{r}+2\,\Omega+r^2\left(\frac{d v'}{d t}\right)^2-2\,m'\!\int l\,d t - 2\,m'\!\int m\,d t = 0 \qquad (13\,\mathrm{c})$$

Sollen in (13 c) die Größen r, $\varDelta$ und r' nebst t allein als Variable auftreten, so sind außer der Funktion Ω, die schon in (2c), S. 158, durch r, $\varDelta$ und r' dargestellt ist, noch die in l und m auftretenden ξ und η als Funktionen von r, $\varDelta$ und r' darzustellen. Da ξ die Projektion von r auf r' und $\eta^2 = r^2 - \xi^2$, so folgt:

$$\xi\,r' = r\,r'\cos(r,r') = \tfrac{1}{2}\,(r^2 + r'^2 - \varDelta^2) \left.\vphantom{\begin{array}{c}1\\1\end{array}}\right\}$$

$$\eta\,r' = \tfrac{1}{2}\,\sqrt{[(r+r')^2-\varDelta^2]\,[\varDelta^2-(r-r')^2]} \qquad (26)$$

Hiermit ist dann das Ziel erreicht, die Gleichung (13 c) als eine Beziehung zwischen r, $\varDelta$ und t aufzufassen, mit t als Parameter, in bezug auf den die einhüllende Grenzkurve abzuleiten ist. Differentiation von (13 c) nach dem Parameter t, soweit t explizite auftritt oder implizite in r' und v' enthalten ist, ergibt als Bedingungsgleichung der Einhüllenden:

$$-2c\,\frac{d^2 v'}{d t^2}+2\,\frac{\partial\,\Omega}{\partial t}+2\,r^2\,\frac{d v'}{d t}\,\frac{d^2 v'}{d t^2}-2\,m'\,l-2\,m'\,m = 0 \qquad (13\,\mathrm{d})$$

Diese Gleichung muß für $e' = 0$ identisch verschwinden, was auch zutrifft, indem $\dfrac{d^2 v'}{d t^2}$, l und m bereits nach obigem mit $e' = 0$ verschwinden und weil $\dfrac{\partial\,\Omega}{\partial t}$, da gemäß (2 c), S. 158, $\dfrac{\partial\,\Omega}{\partial t}=\dfrac{\partial\,\Omega}{\partial r'}\cdot\dfrac{d r'}{d t'}$ wegen des Faktors $\dfrac{d r'}{d t}$ mit $e' = 0$ ebenfalls verschwindet.

Das Problem ist damit zunächst prinzipiell gelöst, indem die beiden Grundgleichungen (13 c) und (13 d) bei Elimination des Parameters t die gesuchte Gleichung der Einhüllenden aller zeitlichen Grenzkurven in r und $\varDelta$ als Bipolarkoordinaten ergeben, wie für die DARWINsche Grenzkurve im Spezialfalle $e' = 0$. Die Elimination von t stößt aber nun praktisch bei beliebig vorgelegtem e', zumal die beiden Gleichungen (13 c) und (13 d) Integralgleichungen sind, allgemein auf ganz erhebliche Schwierigkeiten, so daß zwecks Lösung des Problems durch Reihenentwicklungen eine notwendige Beschränkung auf kleine, planetare Exzentrizitäten des störenden Körpers stattfinden muß.

§ 2. Entwicklung der Grundgleichungen nach kleinen Parametern.
Die Bedingungsgleichung (13 c) enthält an kleinen resp. klein zu haltenden Parametern zunächst nur die im folgenden stets als klein vorausgesetzte Exzentrizität e' des Jupiter. Setzen wir $r = r_0 + \varDelta r$, so soll $\varDelta r$ in jedem Falle als klein und mit e' verschwindend vorausgesetzt werden, so daß r_0 dem Betrage von r bei kreisförmiger Jupiterbahn entspricht. Dann soll die Gleichung (13 c) zunächst unter Voraussetzung beliebiger, nicht kleiner Masse m' nach Potenzen von e' und $\varDelta r$ entwickelt werden, und zwar bis zum 2. Grade einschließlich, wobei e' und $\varDelta r$ als von der gleichen Größenrechnung betrachtet werden.

Zu entwickeln ist dann: $\dfrac{d\,v'}{d\,t}$, $\dfrac{1}{r}$, $\varOmega$, $r^2\left(\dfrac{d\,v'}{d\,t}\right)^2$, l und m.

Man findet sukzessive:

$$\frac{d\,v'}{d\,t} = n'\left(1 + 2\,e'\cos M' + \frac{5}{2}\,e'^2\cos 2\,M'\right) \tag{27a}$$

wo M' die mittlere Anomalie des Jupiter darstellt.

$$\frac{1}{r} = \frac{1}{r_0} - \frac{1}{r_0^2}\,\varDelta r + \frac{1}{r_0^3}\,(\varDelta r)^2 \tag{27b}$$

$$2\,\varOmega = m'\left(\frac{2}{\varDelta} + \frac{\varDelta^2}{r'^3} - \frac{r^2}{r'^3} - \frac{1}{r'}\right) \tag{27c}$$

gemäß (2 c) S. 158, erhält, weil $a' = 1$ und:

$$r' = 1 - e'\cos M' + \tfrac{1}{2}\,e'^2 - \tfrac{1}{2}\,e'^2\cos 2\,M'$$

geordnet nach Potenzen von e' und $\varDelta r$, die Form:

$$2\,\varOmega = m'\left(\frac{2}{\varDelta} + \varDelta^2\right) - m'\,r_0^2 - m' + 3\,m'\,\varDelta^2\,e'\cos M' - 3\,m'\,r_0^2\,c'\cos M'$$
$$- 2\,r_0\,m'\,\varDelta r - m'\,e'\cos M' + \tfrac{3}{2}\,m'\,e'^2\,\varDelta^2 + \tfrac{9}{2}\,m'\,\varDelta^2\,e'^2\cos 2\,M'$$
$$- \tfrac{3}{2}\,m'\,r_0^2\,e'^2 - \tfrac{9}{2}\,m'\,r_0^2\,e'^2\cos 2\,M' - 6\,m'\,r_0\,e'\,\varDelta r\cos M' - m'\,(\varDelta r)^2$$
$$- m'\,e'^2\cos 2\,M'$$

$$r^2\left(\frac{d\,v'}{d\,t}\right)^2 = n'^2\,[r_0^2 + 4\,r_0^2\,e'\cos M' + 2\,r_0\,\varDelta r + 2\,e'^2\,r_0^2 \left.\vphantom{\frac{a}{b}}\right\}$$
$$+ 7\,e'^2\,r_0^2\cos 2\,M' + (\varDelta r)^2] + 8\,r_0\,e'\,\varDelta r\cos M' \left.\vphantom{\frac{a}{b}}\right\} \tag{27d}$$

$$- 2\,m'\,l = 2\,m'\,\frac{d^2\,v'}{d\,t^2}\int\left(\frac{1}{\varDelta^3} - \frac{1}{r'^3}\right)\eta\,r'\,d\,t \tag{27e}$$

Da $\dfrac{d^2\,v'}{d\,t^2}$ bereits vom 1. Grade in e' ist, sind die Faktoren $\dfrac{1}{\varDelta^3} - \dfrac{1}{r'^3}$ und $\eta\,r'$ nur bis zum 1. Grade in e' und $\varDelta r$ zu entwickeln. Bei Entwicklung der 2. Gleichung (26) ergibt sich:

$$\eta\,r' = \eta_0 - \frac{1}{4}\frac{r_0 - 1}{\eta_0}\,[(1 + r_0)^2 - \varDelta^2]\,[\varDelta r + e'\cos M'] \left.\vphantom{\frac{a}{b}}\right\}$$
$$+ \frac{1}{4}\frac{r_0 + 1}{\eta_0}\,[\varDelta^2 - (1 - r_0)^2]\,[\varDelta r - e'\cos M'] \left.\vphantom{\frac{a}{b}}\right\} \tag{28}$$

wo:

$$\eta_0 = (\eta\, r')_{e'=0} = \tfrac{1}{2}\sqrt{[(1+r_0)^2 - \varDelta^2]\,[\varDelta^2 - (1-r_0)^2]} \qquad (28\,\mathrm{a})$$

In dem Integral in (27e) ist dann nur insoweit nach t zu integrieren, als t explizite in M' auftritt. Es wird also zunächst:

$$\left.\begin{aligned}
-2m'l &= 2m'e'n'^2\eta_0\left(\frac{1}{\varDelta^3}-1\right)[-2t\sin M' - 5e't\sin M'] \\
&\quad -4m'n'^2 e'f_2\,\varDelta\,r\,t\sin M' - 4m'n'e'^2 g_1\sin^2 M' \\
&\quad +12m'n'e'^2\eta_0\sin^2 M'
\end{aligned}\right\} \qquad (27\,\mathrm{e})$$

wo die Koeffizienten f_2 und g_1 in dem späteren Formelsystem (32) definiert sind. Mit Rücksicht auf die Beziehungen:

$$\int n't\sin M'\,dt = -t\cos M' + \frac{1}{n'}\sin M'$$

$$\int n't\sin 2M'\,dt = -\frac{1}{2}t\cos 2M' + \frac{1}{4n'}\sin 2M'$$

wird dann:

$$\begin{aligned}
-2m'\int l\,dt &= -4m'e'\eta_0\left(\frac{1}{\varDelta^3}-1\right)(-n't\cos M' + \sin M') \\
&\quad -10m'e'^2\eta_0\left(\frac{1}{\varDelta^3}-1\right)\left(-\frac{1}{2}n't\cos 2M' + \frac{1}{4}\sin 2M'\right) \\
&\quad -2m'e'^2(g_1-3\eta_0)\left(n't - \frac{1}{2}\sin 2M'\right) \\
&\quad -4m'e'f_2\,\varDelta\,r\left(-\frac{t}{n'}\cos M' + \frac{1}{n'^2}\sin M'\right)
\end{aligned}$$

Wird schließlich im letzten Gliede von (13c):

$$-2m'm = -2m'\frac{dr'}{dt}\left[\xi\left(\frac{1}{\varDelta^3}-\frac{1}{r'^3}\right) - \frac{r'}{\varDelta^3} + \frac{3\,\xi}{r'^3}\right] \qquad (27\,\mathrm{f})$$

die Größe ξ mittels (26) eliminiert und die Entwicklung der Funktionen von r und r' nach $\varDelta\,r$ und e' vorgenommen, so erhält man:

$$\begin{aligned}
-2m'\int m\,dt &= 2m'e'\left[\frac{1}{2\varDelta^3}(r_0^2-1-\varDelta^2) + r_0^2+1-\varDelta^2\right]\cos M' \\
&\quad +2m'e'\left(\frac{r_0}{\varDelta^3}+2r_0\right)\varDelta\,r\cos M' \\
&\quad +m'e'^2\left[\frac{1}{\varDelta^3}\left(\frac{3}{4}r_0^2 - \frac{3}{4}\varDelta^2 - \frac{1}{4}\right) + 3r_0^2 - 3\varDelta^2 + 2\right]\cos 2M'
\end{aligned}$$

Die in $\int l\,dt$ und $\int m\,dt$ auftretenden Terme in $\varDelta\,r$ haben in (13c), das Produkt $m'e'$ als Faktor, so daß, wenn m' klein von der 1. Ordnung die betreffenden Glieder von der 3. Ordnung klein und deshalb bei Festhaltung der Genauigkeitsgrenze bis zur 2. Ordnung zu vernachlässigen wären. Ist m' nicht klein von der 1. Ordnung, so genügt es aber bei Beschränkung auf die 2. Ordnung, in die betreffenden Terme in Δr einen Näherungswert 1. Ordnung zu substituieren. Bei Substitution der Ausdrücke (27 a) bis (27 f) in (13 c) erhält man schließlich die in der folgenden Tabelle (31) zusammengestellten Glieder der einzelnen Ordnungen:

0. Ordnung:

$$C - 2 c n' - m' + \frac{2}{r_0} + r_0^2 + m' \left(\frac{2}{\Delta} + \Delta^2 \right)$$

1. Ordnung:

$$- 4 c n' e' \cos M' - 2 \frac{\Delta r}{r_0^2} + 3 m' \Delta^2 e' \cos M' - 3 m' r_0^2 e' \cos M' - m' e' \cos M'$$

$$+ 4 r_0^2 e' (1 + m') \cos M' + 2 r_0 \Delta r + 4 m' e' \eta_0 \left(\frac{1}{\Delta^3} - 1 \right) n' t \cos M'$$

$$- 4 m' e' \eta_0 \left(\frac{1}{\Delta^3} - 1 \right) \sin M' + 2 m' e' g_2 \cos M'$$

2. Ordnung:

$$- 5 c n' e'^2 \cos 2 M' + 2 \frac{(\Delta r)^2}{r_0^3} + \frac{3}{2} m' \Delta^2 e'^2 + \frac{9}{2} m' \Delta^2 e'^2 \cos 2 M'$$

$$- \frac{3}{2} m' r_0^2 e'^2 - \frac{9}{2} m' r_0^2 e'^2 \cos 2 M' - 6 m' r_0 e' \Delta r \cos M' - m' e'^2 \cos 2 M'$$

$$+ 2 e'^2 r_0^2 (1 + m') + 7 e'^2 r_0^2 (1 + m') \cos 2 M' + (\Delta r)^2$$

$$+ 8 r_0 (1 + m') e' \Delta r \cos M' + 5 m' e'^2 \eta_0 \left(\frac{1}{\Delta^3} - 1 \right) n' t \cos 2 M'$$

$$- \frac{5}{2} m' e'^2 \eta_0 \left(\frac{1}{\Delta^3} - 1 \right) \sin 2 M' - 2 m' e'^2 (g_1 - 3 \eta_0) n' t$$

$$+ m' e'^2 (g_1 - 3 \eta_0) \sin 2 M' - 4 m' f_2 n' e' \left[- t \cos M' + \frac{1}{n'} \sin M' \right] \Delta r$$

$$+ m' e'^2 g_3 \cos 2 M' + \frac{2 m'}{n'} e' f_1 \Delta r \cos M'$$

$$\left. \right\} \quad (31)$$

unter Benutzung der folgenden Abkürzungen:

$$f_1 = n' \left(\frac{r_0}{\Delta^3} + 2 r_0 \right)$$

$$f_2 = n'^2 \left(\frac{1}{\Delta^3} - 1 \right) \left[- \frac{1}{4} \frac{r_0 - 1}{\eta_0} \left((1 + r_0)^2 - \Delta^2 \right) + \frac{1}{4} \frac{r_0 + 1}{\eta_0} \left(\Delta^2 - (1 - r_0)^2 \right) \right]$$

$$g_1 = \left(\frac{1}{\Delta^3} - 1 \right) \left[- \frac{1}{4} \frac{r_0 - 1}{\eta_0} \left((1 + r_0)^2 - \Delta^2 \right) - \frac{1}{1} \cdot \frac{r_0 + 1}{\eta_0} \left(\Delta^2 - (1 - r_0)^2 \right) \right]$$

$$g_2 = \frac{1}{2 \Delta^3} (r_0^2 - 1 - \Delta^2) + r_0^2 1 + - \Delta^2$$

$$g_3 = \frac{3}{4} (r_0^2 - \Delta^2) \frac{1}{\Delta^3} - \frac{1}{4} \cdot \frac{1}{\Delta^3} + 3 (r_0^2 - \Delta^2) + 2$$

$$\left. \right\} \quad (32)$$

Zu den obigen Gliedern in (31) müssen nun aus folgendem Grunde noch weitere von e' abhängende Glieder hinzugefügt werden. Die Grenzkurve wird konstruiert, indem über r' als Basis ein Dreieck mit r und $\varDelta$ als den beiden übrigen Seiten errichtet wird. Bei $e' \neq 0$ ist nun aber die Basis r' variabel, und außerdem setzt die Fixierung von r' als Basis die Kenntnis von M' resp. die Annahme von M' als unabhängige Variable voraus; diese letztere Annahme ist aber, wie aus dem Folgenden hervorgehen wird, unbrauchbar. Man kann der genannten Voraussetzung aber leicht aus dem Wege gehen und die Konstruktion der Kurve so ausführen, daß die Basis, wie im Spezialfall $e' = 0$, konstante Länge hat und die Pole der Bipolarkoordinaten unabhängig von der Zeit festliegen. Man führt zu diesem Zweck statt der Entfernung $\varDelta$ des Kurvenpunktes P vom Jupiter J die neue Entfernung D desselben Punktes P von dem auf der Verbindungslinie Sonne-Jupiter gelegenen Punkt K ein, der von der Sonne S den konstanten Abstand 1 hat. Da der Abstand JK von der Ordnung der Exzentrizität e' ist, sind $PJ = \varDelta$ und $PK = D$ nur um Größen der Ordnung e' voneinander verschieden. Wird der Winkel $PKJ = \psi$ gesetzt, so wird:

$$r^2 = D^2 + 1 - 2D \cos\psi \tag{33}$$

ferner ist im Dreieck PKJ:

$$\varDelta^2 = D^2 + (1 - r')^2 - 2D(1 - r')\cos\psi \tag{34}$$

so daß die Substitution von ψ aus (33) in (34) die gesuchte Beziehung ergibt:

$$\varDelta^2 = D^2 + (1 - r')^2 - (1 - r')(1 + D^2 - r^2) \tag{35}$$

worin der 2. und 3. Term der rechten Seite mit $e' = 0$ verschwinden. Die Substitution des unter (27c) gegebenen Ausdruckes:

$$1 - r' = e' \cos M' + \tfrac{1}{2} e'^2 \cos 2M' - \tfrac{1}{2} e'^2$$

und (b) von $r = r_0 + \varDelta r$ in (35) ergibt bei Entwicklung wieder bis zum 2. Grade in e' und $\varDelta r$ den endgültigen Ausdruck:

$$\left.\begin{aligned}\varDelta^2 = D^2 &+ (r_0^2 - 1 - D^2)\, e' \cos M' + \tfrac{1}{4} e'^2 (2 + D^2 - r_0^2) \\ &+ \tfrac{1}{4} e'^2 \cos 2M' (r_0^2 - D^2) + 2r_0 \varDelta r\, e' \cos M'\end{aligned}\right\} \tag{36a}$$

Für die außer $\varDelta^2$ noch benötigten Funktionen $\dfrac{1}{\varDelta}$ und $\dfrac{1}{\varDelta^3}$ erhält man weiter die Ausdrücke:

$$\left.\begin{aligned}\frac{1}{\varDelta} = \frac{1}{D} &- \frac{1}{2}\frac{1}{D^3}(r_0^2 - 1 - D^2)\, e' \cos M' \\ &- \frac{1}{4}\frac{1}{D^3} e'^2 (2 + D^2 - r_0^2) - \frac{1}{4}\frac{1}{D^3} e'^2 (r_0^2 - D^2) \cos 2M' \\ &- \frac{r_0}{D^3} \varDelta r\, e' \cos M' + \frac{3}{16}\frac{1}{D^5}(r_0^2 - 1 - D^2)^2\, e'^2 \\ &+ \frac{3}{16}\frac{(r_0^2 - 1 - D^2)^2}{D^5}\, e'^2 \cos 2M'\end{aligned}\right\} \tag{36b}$$

$$\frac{1}{\varDelta^3} = \frac{1}{D^3} - \frac{3}{2}\frac{1}{D^5}(r_0^2 - 1 - D^2)\, e' \cos M' \tag{36c}$$

wobei $\dfrac{1}{\varDelta^3}$ nur bis zum 1. Grade entwickelt zu werden braucht, weil $\dfrac{1}{\varDelta^3}$ erst in denjenigen Termen in (31), die bereits von der 1. Ordnung sind, vorhanden ist. Schließlich bleibt noch die Entwicklung der von $\varDelta$ abhängigen Funktion η_0 als Funktion von D bis zum 1. Grade in e'; man findet:

$$
\left.
\begin{aligned}
\eta_0 &= \eta_0^0 + \eta_0^1 e' \cos M' \\[2mm]
\text{wo:}\qquad \eta_0^0 &= \tfrac{1}{2}\sqrt{[(1+r_0)^2 - D^2][D^2 - (1-r^0)^2]} \\[2mm]
\eta_0^1 &= \frac{1}{2}\,\frac{(r_0^2 - D^2)^2 - 1}{\sqrt{[(1+r_0)^2 - D^2][D^2 - (1-r_0)^2]}}
\end{aligned}
\right\} \tag{37}
$$

Ersetzen wir jetzt alle Funktionen von $\varDelta$ in (31) durch die soeben entwickelten Funktionen von D, so treten zu den Termen (31) die folgenden Zusatzglieder hinzu, nachdem in (31) nach Substitution der betreffenden Funktionen von D die Bezeichnung $\varDelta$ überall durch D ersetzt zu denken ist:

$$
\left.
\begin{aligned}
&\textit{1. Ordnung:} \\[2mm]
&-m'\frac{1}{D^3}(r_0^2 - 1 - D^2)\,e'\cos M' + m'(r_0^2 - 1 - D^2)\,e'\cos M' \\[4mm]
&\textit{2. Ordnung:} \\[2mm]
&-\frac{1}{2}m'\frac{1}{D^3}(2 + D^2 - r_0^2)\,e'^2 - \frac{1}{2}m'e'^2\frac{1}{D^3}(r_0^2 - D^2)\cos 2M' \\[2mm]
&-2m'\frac{r_0}{D^3}\varDelta r\,e'\cos M' + \frac{3}{8}m'\frac{(r_0^2 - 1 - D^2)^2}{D^5}e'^2 \\[2mm]
&+\frac{3}{8}m'\frac{(r_0^2 - 1 - D^2)^2}{D^5}e'^2\cos 2M' + \frac{1}{2}m'e'^2(2 + D^2 - r_0^2) \\[2mm]
&+\tfrac{1}{2}m'e'^2(r_0^2 - D^2)\cos 2M' + 2m'r_0\varDelta r\,e'\cos M' \\[2mm]
&+3m'e'^2(r_0^2 - 1 - D^2)\cos^2 M' \\[2mm]
&+3m'\frac{e'^2}{D^5}(r_0^2 - 1 - D^2)\eta_0^0\sin 2M' - \frac{6m'e'^2}{D^5}(r_0^2 - 1 - D^2)\eta_0^0 M'\cos^2 M' \\[2mm]
&-2m'e'^2\frac{\eta_0^1}{D^3}\sin 2M', + 4m'e'^2\frac{\eta_0^1}{D^3}M'\cos^2 M' + 2m'e'^2\eta_0^1\sin 2M' \\[2mm]
&-4m'e'^2\eta_0^1 M'\cos^2 M' - \frac{3}{2}m'\frac{e'^2}{D^5}(r_0^2 - 1)(r_0^2 - 1 - D^2)\cos^2 M' \\[2mm]
&+\frac{1}{2}m'\frac{e'^2}{D^3}(r_0^2 - 1 - D^2)\cos^2 M' - 2m'e'^2(r_0^2 - 1 - D^2)\cos^2 M'
\end{aligned}
\right\} \tag{38}
$$

Da $r = r_0 + \varDelta r$ und $\varDelta = D + D_1$ gesetzt war, wo $\varDelta r$ und D_1 mit $e' = 0$ verschwinden, so sind r_0 und D durch die DARWINsche Gleichung:

$$
\frac{2}{r_0} + r_2^0 + m'\left(\frac{2}{D} + D^2\right) = J \tag{38a}
$$

aneinander gebunden, wo $J = -C + 2cn' + m'$.

§ 3. Die einhüllende Grenzkurve. Die Gesamtheit der Glieder 1. und 2. Ordnung führt nach (31) und (38) allgemein zu der folgenden Gleichung, nachdem die Glieder 0. Ordnung für sich verschwinden, was in (38 a) zum Ausdruck gebracht ist:

$$\alpha\, \Delta r + \beta(\Delta r)^2 = \gamma \sin M' + \delta \cos M' + \varepsilon\, M' \cos M' + \zeta \sin 2 M' \\ + K \cos 2 M' + \lambda + \mu\, M' + \nu\, M' \cos 2 M' \quad\Big\} \quad (39)$$

wo die Koeffizienten die folgende Bedeutung haben:

$$\alpha = 2\left[r_0 - \frac{1}{r_0^2} + m' e' r_0 \left(2 - \frac{1}{D^3} \right) \cos M' + 4 r_0 e' \cos M' \right]$$

$$\beta = 1 + \frac{2}{r_0^3}, \quad \gamma = 4 m' e' \left(\frac{1}{D^3} - 1 \right) \eta_0^0 + \frac{4 m' e'\, \Delta r}{n'^2} f_2$$

$$\delta = 4 e' n' (c - n' r_0^2) - \frac{2 e'\, \Delta r\, m'\, f_1}{n'}$$

$$\varepsilon = - 4 m' e' \eta_0^0 \left(\frac{1}{D^3} - 1 \right) - \frac{4 m' e'\, \Delta r}{n'^2} f_2$$

$$\zeta = e'^2 \left[- 3 m' \frac{\eta_0^0}{D^5} (r_0^2 - 1 - D^2) + 2 m' \frac{\eta_0^1}{D^3} \right.$$

$$\left. + \frac{5}{2} m' \eta_0^0 \left(\frac{1}{D^3} - 1 \right) - m' (g_1 - 3 \eta_0^0) - 2 m' \eta_0^1 \right]$$

$$\varkappa = e'^2 \left[5 c n' - \frac{9}{2} m' D^2 + \frac{9}{2} m' r_0^2 + m' - 7 r_0^2 (1 + m') - m' g_3 \right.$$

$$+ \frac{1}{2} m' \frac{1}{D_3} (r_0^2 - D^2) - \frac{3}{8} \frac{1}{D^5} (r_0^2 - 1 - D^2)^2 - \frac{1}{2} m' (r_0^2 - D^2)$$

$$+ \frac{3}{4} m' \frac{1}{D^5} (r_0^2 - 1)(r_0^2 - 1 - D^2) - \frac{1}{4} m' \frac{1}{D^3} (r_0^2 - 1 - D^2)$$

$$\left. - \frac{1}{2} m' (r_0^2 - 1 - D^2) \right] \qquad\qquad (40)$$

$$\lambda = e' \left[- \frac{3}{2} m' e' D^2 + \frac{3}{2} m' r_0^2 e' - 2 r_0^2 (1 + m') e' \right.$$

$$+ \frac{1}{2} m' \frac{1}{D^3} (2 + D^2 - r_0^2) e' - \frac{3}{8} m' e' \frac{1}{D^5} (r_0^2 - 1 - D^2)^2$$

$$- \frac{1}{2} m' e' (2 + D - r_0^2) + \frac{3}{4} m' e' \frac{1}{D^5} (r_0^2 - 1)(r_0^2 - 1 - D^2)$$

$$\left. - \frac{1}{4} m' e' \frac{1}{D^3} (r_0^2 - 1 - D^2) - \frac{1}{2} m' e' (r_0^2 - 1 - D^2) \right]$$

$$\mu = e'^2 \left[2 m' (g_1 - 3 \eta_0^0) + 3 m' \frac{1}{D^5} (r_0^2 - 1 - D^2) \eta_0^0 - 2 m' \frac{\eta_0^1}{D^3} + 2 m' \eta_0^1 \right]$$

$$\nu = e'^2 \left[- 5 m' \eta_0^0 \left(\frac{1}{D^3} - 1 \right) + 3 m' \frac{1}{D^5} (r_0^2 - 1 - D^2) \eta_0^0 - 2 m' \frac{\eta_0^1}{D^3} + 2 m' \eta_0^1 \right]$$

In (39) ist $n' t = M'$ gesetzt worden, indem die Zeit vom Momente des Periheldurchganges des Jupiter gezählt werden soll. Aus der Zusammenstellung (40) der Koeffizienten folgt, daß α und β allgemein von der 0. Ordnung, außer wenn r_0 nahe 1, wobei α von der 1. Ordnung wird; ferner sind γ, δ und ε allgemein von der mindestens 1., λ, $\varkappa$, μ,

ν und λ von der 2. Ordnung. Auf Abänderungen der Größenordnungen bei speziellen Kurvenpunkten und bei kleiner Masse wäre spezielle Rücksicht zu nehmen.

Die 2. Bedingungsgleichung für die einhüllende Grenzkurve folgt aus (39) mittels partieller Differentiation nach M'.

Mit Rücksicht auf die Abhängigkeit des Koeffizienten α von M' erhält man die Bedingungsgleichung:

$$-2\left[m'r_0\left(2 - \frac{1}{D^3}\right) + 4 r_0\right]\Delta r \, e' \sin M' = \gamma \cos M' - \delta \sin M' \left.\begin{array}{l} \\ + \, \varepsilon \cos M' - \varepsilon M' \sin M' + 2\zeta \cos M' - 2\varkappa \sin 2 M' \\ | \, \mu \, | \, \nu \cos 2 M' - 2\nu M' \sin 2 M' \end{array}\right\} \tag{41}$$

Aus den Gleichungen (39) und (41) folgt nun zunächst allgemein wegen des Auftretens der Glieder in M', $M' \cos M'$, $M' \sin M'$, $M' \cos 2M'$ und $M' \sin 2 M'$, daß die Grenzkurven und ihre Einhüllende allgemein säkularen Veränderungen unterliegen, worauf am Schluß nochmals zurückzukommen ist. Ist m' nicht klein, so sind die säkularen Veränderungen wegen des Koeffizienten ε bereits von der 1. Ordnung in e'; ist m' klein von der 1. Ordnung, so werden die säkularen Veränderungen wegen des Koeffizienten ε von der 2. Ordnung: $m'e'$. Wenn die Kurve ferner bei kleinem m' speziell in der Nähe der LAGRANGEschen Dreieckspunkte gelegen sein sollte, welcher Fall mit Rücksicht auf die Planeten der Jupitergruppe des Sonnensystems spezielles praktisches Interesse hat, werden die säkularen Veränderungen von der 3. Ordnung: $m'e'(1 - D)$, weil D in der Nähe der Librationspunkte nahe gleich 1 ist.

Wird m' nicht als klein angenommen und berücksichtigt man in 1. Näherung nur die Glieder 1. Ordnung, so ist nach (39):

$$\Delta r = \frac{1}{\alpha}\left[\gamma \sin M' + \delta \cos M' + \varepsilon M' \cos M'\right] \tag{42}$$

und entsprechend nach (41):

$$0 = (\gamma + \varepsilon)\cos M' - \delta \sin M' - \varepsilon M' \sin M' \tag{43}$$

Da nach (40) die Summe $\gamma + \varepsilon = 0$ ist, so folgt aus (43), falls erstens $\sin M' \neq 0$, also $M' \neq 0$ und $180°$:

$$M' = -\frac{\delta}{\varepsilon} = \frac{\delta}{\gamma} \tag{44}$$

substituieren wir $\delta = -\varepsilon M'$ in (42), so wird:

$$\Delta r = \frac{\gamma}{\alpha}\sin M' = -\frac{\gamma}{\alpha}\sin\left(\frac{\delta}{\varepsilon}\right) \tag{45}$$

Der Algorithmus zur Ableitung der Koordinaten der einhüllenden Grenzkurve ist also der, daß man nach Vorlegung der JACOBIschen Konstanten J, indem die Energiekonstante C und bei $e' \neq 0$ auch die Flächenkonstante c mittels der für $t = 0$ geltenden Koordinaten x, y und Geschwindigkeiten $\frac{dx}{dt}$, $\frac{dy}{dt}$ nach (3) und (9a) als gegeben zu betrachten sind, zunächst D als unabhängige Variable wählt. Dann ergibt sich zuerst r_0 aus der Gleichung $r_0^2 - \dfrac{2}{r_0} + m'\left(D^2 + \dfrac{2}{D}\right) = J$. Hat man dann weiter die Koeffizienten α, $\gamma = -\varepsilon$ und δ mittels der bisher bekannten Größen nach (40) ermittelt, so folgt schließlich M' nach (44) und $\varDelta r$ nach (45), woraus dann die zu D gehörige andere Bipolarkoordinate $r = r_0 + \varDelta r$ folgt. Da m' nicht als klein galt, muß e' klein sein, wenn nicht die Säkularglieder bereits in kurzem Zeitraum die Mitnahme von Gliedern höherer Ordnung erforderlich machen sollen. Bemerkenswert ist, daß M' nach (44), weil δ wie γ proportional e' sind, von e' ganz unabhängig ist, solange e' klein angenommen wird. Dagegen ist $\varDelta r$ wegen γ proportional e' und verschwindet, wie es sein muß, mit e'; wenn aber der Nenner α in (45) selbst von der Ordnung e' wird, indem $1 - r_0$ von der Ordnung e' wird, so daß infolgedessen $\dfrac{\gamma}{\alpha}$ in 1. Näherung von e' unabhängig würde, ist das Glied $\alpha\,\varDelta r$ in (39) von vorneweg von der 2. Ordnung, und man muß deshalb in diesem Falle von vorneweg überhaupt die Glieder 2. Ordnung in Rücksicht ziehen.

Ist zweitens (43) dadurch erfüllt, daß $\sin M' = 0$, d. h. $M' = 0$ resp. $180°$, so wird die einhüllende Grenzkurve dementsprechend entweder durch eine spezielle zeitliche Grenzkurve oder auch durch zwei spezielle zeitliche Grenzkurven gebildet, wobei im letzteren Falle die eine Grenzkurve die innere, die andere die äußere Einhüllende fixiert, wie bei einer um den Mittelpunkt rotierenden Ellipse der Kreis um den Mittelpunkt mit der kleinen Halbachse als Radius die innere Einhüllende und der Kreis mit der großen Halbachse als Radius die äußere Einhüllende bildet. Dann ist nach (42) in 1. Näherung:

$$\varDelta r = \frac{\delta}{\alpha} \quad \text{resp.} \quad -\frac{\delta}{\alpha} \tag{46}$$

Zusammenfassend ergibt sich als Resultat: Bei nichtverschwindender Exzentrizität e' der Jupiterbahn gehört zu jedem Zeitpunkt eine Grenzkurve, deren Abweichungen von der für $e' = 0$ geltenden Grenzkurve für die Bewegung des Asteroiden nach (39) zu ermitteln sind. Die Einhüllende der zeitlichen Grenzkurven ergibt sich mittels der Bedingungsgleichung (41) unter Zuhilfenahme von (39) nach dem dort beschriebenen Algorithmus. Die Lösung ist aber, da Koordinaten und Zeit in der elliptischen Jupiterbewegung nur durch Reihenentwicklungen nach e' aneinander gebunden sind, auf kleine Exzentrizitäten beschränkt, zumal die Lösung auch in e' multiplizierte Säkularglieder aufweist, so daß auch eine zeitliche Beschränkung der Lösung besteht. Diese Beschränkung wird aber um so geringer, je kleiner e', um mit $e' = 0$ ganz zu verschwinden, in welchem Falle die HILLsche Grenzkurve die für unbegrenzte Zeiten gültige Lösung bildet.

Sur une propriété
géométrique des trajectoires des bolides
dans l'atmosphère terrestre.

Par **Kyrille Popoff**, Sofia-Bulgarie.

Avec 3 figures.

Nous voulons attirer l'attention des astronomes sur une propriété géométrique des trajectoires des bolides dans l'atmosphère[1]) qui peut fournir facilement les éléments nécessaires pour l'étude de leurs orbites au dehors de l'atmosphère.

Nous conservons à la loi de résistance de l'air au mouvement sa forme la plus générale et nous prenons comme point de départ le théorème de H. Poincaré que dans des conditions bien déterminées les intégrales sont des fonctions holomorphes des paramètres qui figurent dans les équations différentielles et par conséquent peuvent être développées en séries de puissances entières des paramètres.

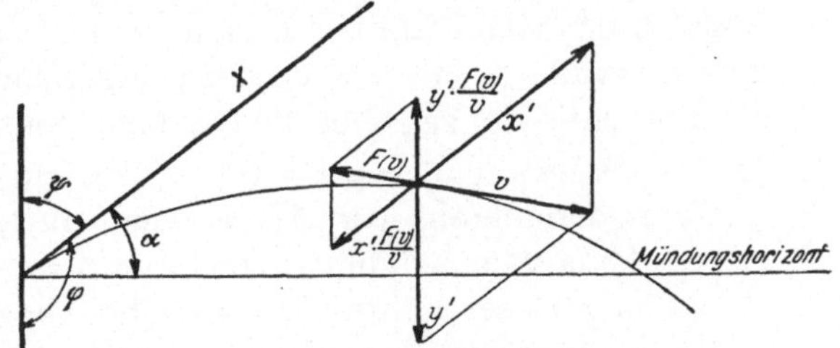

Fig. 1. Les composantes de la force de la résistance d'après les axes de coordonnées introduits dans ce travail.

Considérons le mouvement d'un point dans un milieu résistant et soit $\Phi(v)$ la force de résistance pour l'unité de masse. En choisissant un point quelconque de la trajectoire pour origine, l'axe des x étant dirigé suivant la vitesse de ce point et l'axe des y suivant la verticale descendante, les côtés du triangle des forces composantes de $\Phi(v)$ seront dans le rapport $v : x' : y'$ et par conséquent les équations du mouvement seront:

$$\frac{d^2 x}{d t^2} = -\frac{d x}{d t}\frac{\Phi(v)}{v} \qquad \frac{d^2 y}{d t^2} = g - \frac{d y}{d t}\frac{\Phi(v)}{v} \tag{1}$$

---

[1]) Popoff, Kyrille: Über eine Eigenschaft der ballistischen Kurve und ihre Anwendung auf die Integration der Bewegungsgleichungen. Zeitschr. f. angew. Mathem. u. Mechanik Bd. 1, S. 96—106. 1921. — Popoff, Kyrille: Sur l'intégration des équations de la Balistique dans des conditions générales de la résistance. Cpt. rend. hebdom. des séances de l'Acad. des sciences. p. 337—340. 16 août 1922. — Popoff, Kyrille: Sur une propriété des orbites. Astr. Nachr. Bd. 211, Nr. 5059. August 1920. — Popoff, Kyrille u. G. Stainoff: Über eine Methode zur Höhenbestimmung. Meteorolog. Zeitschr. 1920, H. 9, S. 263.

avec les conditions initiales: $t = 0$, $x = 0$, $y = 0$, $x' = v_0$, $y' = 0$. Ici la vitesse v est donnée par:

$$v^2 = x'^2 + y'^2 - 2\,x'y'\sin\alpha$$

où $x' = \dfrac{dx}{dt}$, $y' = \dfrac{dy}{dt}$ et α est l'angle que fait la vitesse à l'origine avec l'horizon.

Sur $\Phi(v)$ nous ne faisons que les hypothèses suivantes: 1. Cette fonction est positive pour des valeurs positives de v et croît en même temps que v; 2. $\dfrac{\Phi(v)}{v} = f(v)$ peut être développée suivant les puissances positives et entières de v dans tout le plan de v.

Une étude préliminaire[1]) de ces équations montre que $(v_0 , 0)$ étant la position initiale du point (x', y') ce point reste pendant tout le temps (même pour $t = \infty$) dans le premier quadrant du plan à une distance finie; sa distance minima à l'origine pour les différentes valeurs de $\sin\alpha$ de l'intervalle $(-1 , 1 - \varepsilon)$ reste supérieure à une quantité positive r^1, différente de zéro, et l'on a pour toute la durée du mouvement $\sqrt{x'^2 + y'^2} > r^1$ et $v^2 = x'^2 + y'^2 - 2\,x'y'\sin\alpha > A^2$.

Nous désignerons par A_1^2 la valeur minimale de v^2 pour $\sin\alpha = 0$ et par A_2^2 sa valeur minimale pour $\sin\alpha = -1$. Dans ces conditions v et par là $f(v)$ seront des fonctions holomorphes de x', y' et $\sin\alpha$ au voisinage de chaque point des courbes intégrales qui correspondent aux valeurs réelles de $\sin\alpha$ de l'intervalle $(-1 , +1 - \varepsilon)$ et d'après le théorème de Poincaré il existe un t_1 tel que pour toute valeur de $t \leq t_1$ les intégrales x' et y' seront des fonctions holomorphes de $\sin\alpha$ tant que $|\sin\alpha| < 1 - \varepsilon$.

On aura par conséquent

$$x' = \varphi_{01}(t) + \varphi_{11}(t)\sin\alpha + \varphi_{21}(t)\sin^2\alpha + \ldots + \varphi_{n1}(t)\sin^n\alpha + \ldots ,$$

$$y' = \psi_{01}(t) + \psi_{11}(t)\sin\alpha + \psi_{21}(t)\sin^2\alpha + \ldots + \psi_{n1}(t)\sin^n\alpha + \ldots$$

pour $t \leq t_1$, les φ_{n1} et ψ_{n1} étant des fonctions de t seulement.

Nous désignerons par x_1' et y_1' les valeurs de x' et y' pour $t = t_1$ et nous ferons dans les équations du mouvement (1) le changement de variables:

$$x' = x_1' + \xi', \quad y' = y_1' + \eta', \quad t = t_1 + \tau.$$

On obient ainsi:

$$\frac{d\xi'}{d\tau} = -(x_1' + \xi')\,f(v), \qquad \frac{d\eta'}{d\tau} = g - (y_1' + \eta')\,f(v)$$

avec les conditions initiales $\tau = 0$, $\xi' = 0$, $\eta' = 0$.

---

[1]) Popoff, Kyrille: Sur l'intégration des équations de la balistique extérieure. Journ. des mathémat. pures et appliquées (sous presse).

Pour des valeurs réelles de $\sin \alpha$ le point $({}^1_i x + \xi', y'_1 + \eta')$ reste à une distance supérieure à r^1 de l'origine; d'autre part x'_1 et y'_1 sont des fonctions holomorphes de $\sin \alpha$ tant que $|\sin \alpha| < 1 - \varepsilon$. Par conséquent v^2, qui est une fonction holomorphe de x', y' et $\sin \alpha$, sera aussi une fonction holomorphe de ξ', η' et $\sin \alpha$. Soit $\sin \alpha_1$ une racine de l'équation en $\sin \alpha$:

$$v^2 = (x'_1 + \xi')^2 + (y'_1 + \eta')^2 - 2 (x'_1 + \xi') (y'_1 + \eta') \sin \alpha = 0$$

Cette racine ne peut pas être réelle puisque dans ce cas le premier membre qui représente la valeur de v^2 dans le mouvement réel pour $\alpha = \alpha_1$ aura une valeur plus grande que A^2. Le module m_1 de $\sin \alpha_1$ sera donc différent de zéro et par conséquent $f(v)$, qui est une fonction holomorphe de v, sera aussi une fonction holomorphe de ξ', η' et de $\sin \alpha$ tant que $|\sin \alpha| < m_1$ et d'après le théorème de POINCARÉ les intégrales ξ', η' seront des fonctions holomorphes de $\sin \alpha$ tant que $|\sin \alpha| < m_1$ et τ ne dépasse pas une certaine valeur τ_1 assignée par ce théorème.

On aura ainsi:

$$\xi' = \varphi_{02}(\tau) + \varphi_{12}(\tau) \sin \alpha + \ldots + \varphi_{n2}(\tau) \sin^n \alpha + \ldots$$
$$\eta' = \psi_{02}(\tau) + \psi_{12}(\tau) \sin \alpha + \ldots + \psi_{n2}(\tau) \sin^n \alpha + \ldots$$

pour $\tau \leq \tau_1$.

Puisque $\xi'(0) = 0$ et $\eta'(0) = 0$ pour toute valeur de $\sin \alpha$, on doit avoir $\varphi_{02}(0) = 0$ et $\psi_{02}(0) = 0$.

On prendra le point $x'_2 = x'_1 + \xi'(\tau_1)$, $y'_2 = y'_1 + \eta'(\tau_1)$ comme nouveau point de repère et l'on continuera ainsi de proche en proche à l'infini. Dans les étapes successives on peut bien être obligé de rétrécir le rayon de convergence de ces séries d'après les puissances entières de $\sin \alpha$, mais ce rayon restera toujours différent de zéro puisque la fonction:

$$(x'_i + \xi')^2 + (y'_i + \eta')^2 - 2 (x'_i + \xi') (y'_i + \eta') \sin \alpha$$

où x'_i et y'_i sont les valeurs de x', y' obtenues dans la $i^{\text{-me}}$ étape comme fonctions holomorphes de $\sin \alpha$ dans un domaine $|\sin \alpha| < m_i$, ne pourra pas s'annuler pour $\sin \alpha_i = 0$, puisque dans ce cas elle représente le carré de la vitesse dans le mouvement réel qui correspond à l'angle de projection $\alpha = 0$ et ce carré de vitesse est supérieur à v_0^2 puisque dans le cas $\alpha = 0$ la vitesse ne fait qu'augmenter. Cette racine ne peut non plus tendre vers zéro quand i tend vers l'infini, puisque l'expression:

$$(x'_i + \eta')^2 + (y'_i + \eta')^2 - 2 (x'_i + \xi') (y'_i + \eta') \sin \alpha$$

est une fonction continue de $\sin \alpha$ qui pour $\sin \alpha = 0$ prend une valeur finie plus grande que v_0^2.

Dans le cas où $f(v)$ serait une fonction paire de v elle sera holomorphe comme v pour toutes les valeurs de $\sin \alpha$ et par conséquent les développements ci-dessus sont valables pour toutes les valeurs de $\sin \alpha$.

On obtient ainsi:

$$\frac{dx}{dt} = x_0'(t) + x_1'(t)\sin\alpha + \ldots + x_n'(t)\sin^n\alpha + \ldots \left.\vphantom{\frac{dx}{dt}}\right\}$$

$$\frac{dy}{dt} = y_0'(t) + y_1'(t)\sin\alpha + \ldots + y_n'(t)\sin^n\alpha + \ldots \qquad (A^1)$$

pour toute valeur de t de zéro à l'infini, où

$$x_n'(t) = \varphi_{n1}(t) \qquad \text{pour des valeurs de } t \text{ de } 0 \text{ à } t_1$$

$$= \varphi_{n1}(t_1) + \varphi_{n2}(t - t_1) \qquad \text{pour des valeurs de } t \text{ de } t_1 \text{ à } t_2 = t_1 + \tau_1$$

$$= \varphi_{n1}(t_1) + \varphi_{n2}(t_2 - t_1) + \varphi_{n3}(t - t_2) \text{ pour des valeurs de } t \text{ de } t_2 \text{ à } t_3 \text{ etc.}$$

On a de même pour $y_n'(t)$.

Le développement (A^1) étant valable pour toute valeur de t entre 0 et T, où T peut être pris arbitrairement grand et pour $|\sin\alpha| \leq R < 1$, on doit avoir une quantité positive M telle que pour chaque valeur de t de l'intervalle $(0, T)$ on ait:

$$|x_n' R^n| < M, \qquad |y_n' R^n| < M$$

Pour $|\sin\alpha| < R$ les termes de (A^1) étant plus petits que les termes correspondants de la série:

$$\sum M \left(\frac{\sin\alpha}{R}\right)^n$$

on peut intégrer les séries (A') terme à terme entre les limites 0 et t et l'on obtient ainsi:

$$x = x_0(t) + x_1(t)\sin\alpha + x_2(t)\sin^2\alpha + \ldots \left.\vphantom{\big|}\right\}$$
$$y = y_0(t) + y_1(t)\sin\alpha + y_2(t)\sin^2\alpha + \ldots \qquad (A)$$

Pour obtenir un autre développement des intégrales nous mettrons v sous la forme:

$$v = \sqrt{x'^2 + y'^2 - 2x'y'\sin\alpha} = \sqrt{(x' + y')^2 - 4x'y'\sin^2\frac{\varphi}{2}}$$

où $\varphi = \dfrac{\pi}{2} + \alpha$. Ici encore v reste une fonction holomorphe de x', y' est $\sin^2\dfrac{\varphi}{2}$ tant que $-1 \leq \sin\alpha < 1 - \varepsilon$.

Des considérations tout à fait semblables à celles que nous venons de faire permettent d'établir que dans tout intervalle $(0, T)$ de la variable t les quantités x, y, x', y' sont des fonctions holomorphes de $\sin^2\dfrac{\varphi}{2}$ tant que $\left|\sin^2\dfrac{\varphi}{2}\right|$ reste plus petit qu'une quantité déterminée R différente de zéro et que par conséquent on aura aussi les développements:

$$x = x_0(t) + x_1(t)\sin^2\frac{\varphi}{2} + x_2(t)\sin^4\frac{\varphi}{2} + \cdots + x_n(t)\sin^{2n}\frac{\varphi}{2} + \cdots \left.\vphantom{\big|}\right\}$$

$$y = y_0(t) + y_1(t)\sin^2\frac{\varphi}{2} + y_2(t)\sin^4\frac{\varphi}{2} + \cdots + y_n(t)\sin^{2n}\frac{\varphi}{2} + \cdots \qquad (B)$$

pour toute valeur finie de t.

On aura un troisième développement en mettant v sous la forme:

$$v = \sqrt{(x' - y')^2 + 4x'y' \sin^2 \frac{\psi}{2}}$$

où $\psi = \dfrac{\pi}{2} - \alpha$.

Dans ce cas aussi tant que x' n'est pas devenu égal à y', v sera une fonction holomorphe de x', y' et $\sin^2 \dfrac{\psi}{2}$ et par conséquent on aura:

$$\left.\begin{aligned}
x &= x_0(t) + x_1(t)\sin^2 \frac{\psi}{2} + x_2(t)\sin^4 \frac{\psi}{2} + \cdots + x_n(t)\sin^{2n} \frac{\psi}{2} + \cdots \\
y &= y_0(t) + y_1(t)\sin^2 \frac{\psi}{2} + y_2(t)\sin^4 \frac{\psi}{2} + \cdots + y_n(t)\sin^{2n} \frac{\psi}{2} + \cdots
\end{aligned}\right\} \quad (C)$$

ce développement étant valable de $t = 0$ jusqu'au moment où x' devient égal à y'.

Dans le cas où $f(v)$ sera une fonction paire de v, ces derniers développements aussi sont valables pour tout intervalle de t.

Il est facile de se rendre compte que les développements (A) et (B) sont d'une convergence très rapide. En effet pour le calcul effectif des coefficients de ces développements il faut développer $f(v)$ et pour cela il faut commencer par développer v. Or nous avons:

$$v = \sqrt{x'^2 + y'^2}\sqrt{1 - \frac{2x'y'}{x'^2 + y'^2}\sin\alpha} = (x' + y')\sqrt{1 - \frac{4x'y'}{(x' + y')^2}\sin^2 \frac{\varphi}{2}}$$

D'autre part on a:

$$(x' - y')^2 \geqq 0$$

d'où:

$$x'^2 + y'^2 \geqq 2x'y' \quad \text{ou bien} \quad (x' + y')^2 \geqq 4x'y'$$

et puisque x' et y' sont des quantités positives et finies qui ne s'annulent pas à la fois on aura:

$$1 \geqq \frac{2x'y'}{x'^2 + y'^2} \geqq 0 \quad \text{ou bien} \quad 1 \geqq \frac{4x'y'}{(x' + y')^2} \geqq 0$$

Par conséquent, tant que $|\sin\alpha| < 1$ ou bien $\left|\sin^2 \dfrac{\varphi}{2}\right| < 1$, on peut développer v et de même $f(v)$, qui est une fonction holomorphe de v, en séries suivant les puissances entières et positives de:

$$\frac{2x'y'}{x'^2 + y'^2}\sin\alpha \quad \text{ou bien de} \quad \frac{4x'y'}{(x' + y')^2}\sin^2 \frac{\varphi}{2}$$

Ces développements convergent très rapidement puisque pour $t = 0$ et $t = \infty$ les facteurs:

$$\frac{2x'y'}{x'^2 + y'^2} \quad \text{et} \quad \frac{4x'y'}{(x' + y')^2}$$

sont nul (pour $t = 0$ on a $y' = 0$ et pour $t = \infty$ on a $x' = 0$ et y' très grand). Pour des petites valeurs de t ces facteurs sont petits de l'ordre de $\dfrac{y'}{x'}$ et pour des grandes valeurs de t ils sont petits de l'ordre de $\dfrac{x'}{y'}$.

Cette convergence rapide des séries de v conduit à la convergence rapide des séries x et y. On aura une approximation suffisante en ne retenant que les premiers termes de ces développements et en posant simplement:

$$x = x_0(t), \qquad y = y_0(t)$$

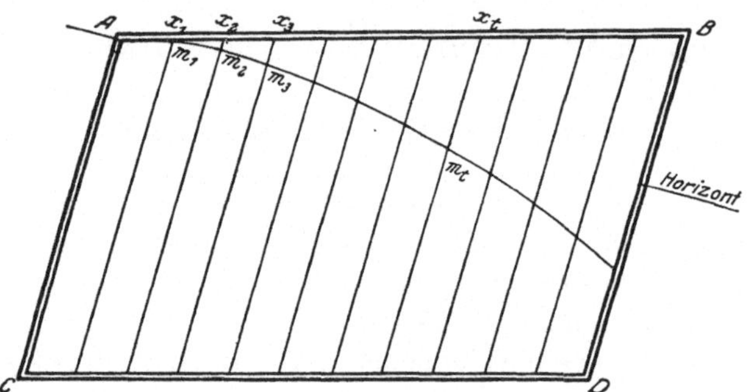

Fig. 2. Appareil servant à obtenir les trajectoires qui correspondent à la même vitesse initiale et aux différentes valeurs de α par des transformations simples de l'une d'elles.

Dans ce cas x et y ne dépendent que du temps et de v_0. Prenons un cadre parallélogramme $ABCD$, dont les côtés peuvent tourner autour des points A, B, C, D et portons sur ses côtés opposés AB et CD les points x_1, x_2, $x_3 \ldots x_t$ à des distances $x_0(1)$, $x_0(2)$, $x_0(3) \ldots x_0(t)$ des points A et C. Tendons entre les points correspondants des fils sur lesquels on porte les points m_1, m_2, $m_3 \ldots m_t$, à des distances $y_0(1)$, $y_0(2)$, $y_0(3) \ldots y_0(t)$ des points correspondants x_1, x_2, $x_3 \ldots x_t$ de AB. Les points m_1, m_2, $m_3 \ldots m_t \ldots$ ainsi obtenus seront alignés sur une trajectoire qui correspond à l'angle $\alpha = \sphericalangle BAC - \dfrac{\pi}{2}$.

En déformant le cadre on obtient toutes les trajectoires qui correspondent à la même vitesse initiale v_0 et à des valeurs de α entre $-\dfrac{\pi}{2}$ et $+\dfrac{\pi}{2}$.

Le temps que met le point pour arriver à un point m_t sur n'importe laquelle des trajectoires ne change pas avec la déformation du cadre et une lecture simple sur AB ferra connaître ce temps.

Avec un tel appareil simple nous sommes arrivés à transformer l'une dans l'autre d'une manière très satisfaisante les trajectoires pour la sixième charge des »*Schaubilder für 10,5 cm Gebirgshaubitze L/12*«

de *Krupp*, ce qui montre que notre première approximation donne des trajectoires qui s'approchent beaucoup des trajectoires réelles.

Dans les développements ci-dessus nous avons négligé dans la première approximation les variations de la densité de l'air dans les

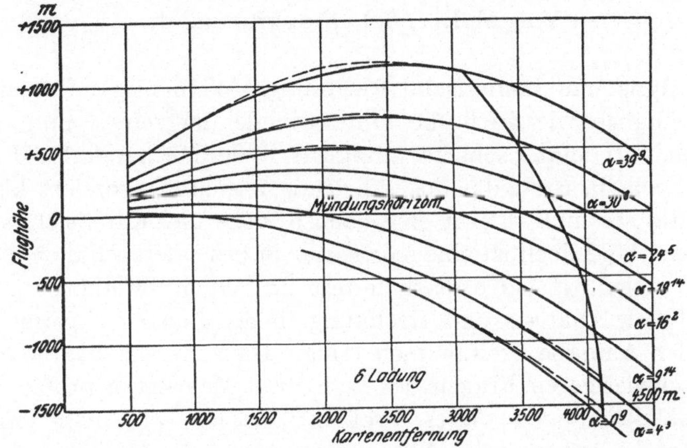

Fig. 3. Cette figure reproduit les trajectoires pour la sixième charge des ,,Schaubilder für 10,5 cm Gebirgshaubitze *L*/12'' de Krupp. Les lignes pointillées représentent les trajectoires obtenues en déformant la trajectoire $\alpha = 16^2$ de Krupp par l'appareil reproduit dans la figure 2.

différentes couches de l'atmosphère. La considération de ces variations se fera d'après les mêmes principes. Pour le calcul effectif des coefficients de nos développements nous renvoyons à notre mémoire dans la Zeitschrift für angewandte Mathematik und Mechanik.

Probleme der rechnenden Himmelsmechanik.

Von **M. Brendel**, Frankfurt a. M.

1. Analysis und numerische Rechnung. Wohl nirgends treten die Zusammenhänge wie auch die Unterschiede der reinen und der angewandten Mathematik schärfer zutage als in der Mechanik des Himmels. Sind doch die Fragen, die uns die letztere stellt, besonders klar und durchsichtig; es muß darauf eine ebenso klare und durchsichtige Antwort geben. Und doch ist eine solche nur in besonderen Fällen möglich; vollkommen gelingt sie eigentlich nur im reinen Zweikörperproblem. Hier sind sich Analysis und Rechnung in der Behandlung und in der Lösung der Aufgabe vollkommen einig. Ganz anders liegt die Sache bei den schwierigeren Fragen des Drei- und Mehrkörperproblems. Die Differentialgleichungen dieses Problems haben, wie man weiß, der Analysis nicht nur manche Anregung geboten, sondern dadurch auch manchen schönen Erfolg veranlaßt. Beiden Problemgruppen, dem Zwei- und dem n-Körperproblem, gemeinsam sind die seit lange bekannten Integrale der Bewegung des Schwerpunkts, der Energie und der Flächengeschwindigkeiten. Während diese Integrale aber beim Zweikörperproblem von großem Nutzen sind und den Weg für die vollständige Lösung vorbereiten, erscheint dagegen jeder Weg, von ihnen aus der Lösung des Mehrkörperproblems näherzukommen, verschlossen, wenn es gilt, die Bewegung *aller Körper* des Systems zu finden. Es zeigt sich nämlich, daß in den Ausdrücken für alle diese Integrale die Koordinaten eines jeden Körpers mit dessen Masse multipliziert vorkommen. Je kleiner die Masse eines Körpers ist, je weniger er also zur Festlegung des Schwerpunkts, zum Wert der Energie und zu den Flächengeschwindigkeiten beiträgt, desto weniger genau ist seine Bewegung durch diese Integrale bestimmt, und wir erfahren aus ihnen gar nichts über die Bewegung eines Körpers, dessen Masse verschwindet. Und doch ist gerade das Problem der Bewegung eines Körpers mit verschwindender Masse nicht nur theoretisch das am nächsten liegende, sondern auch das in der Natur sich am häufigsten darbietende. Man mag dies Mißverhältnis etwa vergleichen mit einem System gewöhnlicher linearer Gleichungen mit sehr kleiner oder verschwindender Determinante; ist diese sehr klein, so ist zwar die analytische Lösung möglich, aber für die Rechnung versagt sie; ist die Determinante Null, so versagen beide.

Eine Ausnahme macht das im Falle des „eingeschränkten" Problems geltende JACOBIsche Energieintegral, das wenigstens zu wichtigen Untersuchungen über gewisse Stabilitätsfragen (Grenzkurven) dient[1]).

Einen größeren Nutzen kann die rechnende Astronomie schon aus anderen neueren analytischen Errungenschaften ziehen; hier bieten die Ergebnisse gewisser Existenzuntersuchungen, wie die über periodische Lösungen, dem Studium wirklicher Bewegungen wenigstens brauchbare Ausgangspunkte[2]). Inwieweit andererseits die neuerdings geglückte analytische Bewältigung des allgemeinen Dreikörperproblems[3]) für die rechnende Astronomie fruchtbar gemacht werden kann, läßt sich heute noch kaum übersehen.

Unter diesen Umständen ist der Astronom, der den Bewegungen der Himmelskörper folgen will, beständig gezwungen, Methoden anzuwenden, deren analytische Bedeutung nicht leicht scharf erfaßt werden kann und die den sicheren Boden der strengen Analysis verlassen. Man hat sein Verfahren wohl gelegentlich als ein mathematisches Experimentieren bezeichnet und damit etwas nicht ganz Unzutreffendes gesagt, sofern man dabei auch die Vergleichung des errechneten Ergebnisses mit der Beobachtung im Auge hat.

In der Tat, wenn man eine durch ein irgendwie geartetes Verfahren vorausberechnete Naturerscheinung mit der Wirklichkeit vergleicht und Übereinstimmung findet, so wird man ein gewisses Vertrauen in dieses Verfahren setzen, mag es nun in irgendeiner Hypothese über einen Naturvorgang oder in der formellen Anwendung einer mathematischen Formel bestehen.

Das hervorragendste Beispiel eines solchen Verfahrens auf dem Gebiete der Himmelsmechanik ist KEPLERS Aufstellung seiner Gesetze der Planetenbewegung. Die Antwort auf die Frage, warum die von ihm gefundenen mathematischen Formeln gelten und ob sie wirklich streng gelten, wird er selbst wie andere nach ihm gesucht haben, bis NEWTON sie fand. Und auch diese Antwort ist keine endgültige; sie führte aber zu einem höheren Standpunkt und zu der weiteren Frage, warum das NEWTONsche Gesetz gilt und ob es streng gilt. Heute sehen wir, daß der nächste Schritt in einer Weiterbildung unserer hergebrachten Vorstellungen von Raum und Zeit besteht.

Freilich handelt es sich selten um Fragen von so gewaltiger Bedeutung wie diese; aber es läßt sich in den meisten Fällen nicht von vornherein übersehen, welche Bedeutung ihnen einmal zukommen wird. Andererseits wird man sich fragen, wie denn das erwähnte Beispiel in

---

[1]) HILL, BOHLIN, DARWIN. — Um den Aufsatz nicht mit Literaturangaben zu überladen, sind an den betreffenden Stellen nur die Namen einiger Verfasser in Fußnoten als Anhaltspunkte genannt; im übrigen sei besonders auf den VI. Band der Enzyklopädie der mathematischen Wissenschaften hingewiesen.

[2]) POINCARÉ.

[3]) SUNDMAN.

Vergleich zu stellen ist mit den rechnerischen Methoden der Himmelsmechanik; denn bei diesen setzt man die Differentialgleichungen der
Bewegung voraus und leitet daraus, immerhin deduktiv, durch Reihenentwicklungen und Integrationsverfahren die Ergebnisse ab; man versucht nicht etwa eine empirische Formel für die Bewegungen aufzustellen, wie es die KEPLERschen Gesetze zunächst waren. Es handelt
sich aber bei den genannten Methoden eben nicht um eine strenge
Deduktion; um den Anforderungen einer solchen gerecht zu werden
oder um ihnen geradezu aus dem Wege zu gehen, kann man die Sache
auch anders auffassen:

Der rechnende Astronom arbeitet mit außerordentlich verwickelten
Reihenentwicklungen nach vielen Veränderlichen, und schon darum
kann er sich über die strenge analytische Bedeutung dieser Entwicklungen keine Rechenschaft geben. Sein Verfahren ist wesentlich ein
formelles. Betrachtet man es vom allgemeinsten Standpunkt, so besteht
es im Grunde darin, daß man eine oder mehrere Größen, Konstante
oder Veränderliche, aufsucht, die hinreichend klein sind, um eine genügend starke Abnahme der ersten Glieder der Reihen zu bedingen,
wobei sich einzelne Entwicklungen allerdings als konvergent in einem
bestimmten Bereich erweisen. Ohne auf Einzelheiten einzugehen, kann
man wohl sagen, daß diese Entwicklungsmethoden darauf hinauslaufen,
eine Größe oder Funktion durch eine andere zu ersetzen, die die Form
einer endlichen oder abbrechenden Reihe hat. Das Wesentliche ist
dabei, daß die endliche Reihe einen Näherungswert geben soll. Die Ausdrücke, mit denen wir rechnen, sind nämlich immer endlich, und in
manchen Fällen, wie bei gewissen divergenten Prozessen, erhält man
gerade deswegen einen Näherungswert, weil man sie rechtzeitig abbricht.
Hierbei kann übrigens kein Zweifel darüber herrschen, daß der Grad
der Annäherung mit dem Zeitintervall zusammenhängt, auf das die
Rechnungen ausgedehnt werden. Und dieses hängt wiederum nicht allein
von der Methode ab, sondern auch von dem Grade der Genauigkeit, mit
dem die Konstanten der Bewegung — der Anfangszustand oder die
Bahnelemente — bekannt sind; dieser ist immer ein beschränkter.

Es ist nicht der Zweck dieser Zeilen, auf die analytische Bedeutung
der rechnerischen Methoden der Himmelsmechanik näher einzugehen.
Das Verfahren sollte nur dahin gekennzeichnet werden, daß man mit
Näherungswerten arbeitet, ohne über den Grad der Annäherung, den
Rest der Reihen, etwas Bestimmtes vorauszusagen. Es soll weiter unten
auf die besonderen Umstände eingegangen werden, die dem Rechner
entgegentreten; zuvor wollen wir uns darüber klar werden, welches die
Aufgaben sind, deren Lösung — oder deren vorbereitende Untersuchung — vom rechnenden Astronomen erwartet wird.

2. Aufgaben der rechnenden Himmelsmechanik. Wenn man die Bewegungen der Himmelskörper durch Rechnung und Beobachtung ver-

folgt, so ist der Zweck nicht, wie bei einer Reihe laufender Geschäfte, die Uhren richtig zu stellen, geographische Lagen zu bestimmen oder den Kalender zu redigieren, sondern es handelt sich darum, etwas Ungewöhnliches festzustellen und eine Erklärung dafür zu suchen. — Was heißt aber etwas Ungewöhnliches? Wenn wir darunter das verstehen, was noch nicht auf ein grundlegendes Prinzip zurückgeführt oder vulgo „erklärt" ist, so gibt es schon in der alten Himmelsmechanik vieles, das wir von alters her gewohnt sind, das im obigen Sinne aber ungewöhnlich ist. Die Verteilung der Bahnen der großen Planeten nach dem BODEschen Gesetz, die Teilung ihrer Rotationszeiten in zwei oder drei Gruppen mit Perioden von einem ganzen, einem halben Tage oder einem Umlauf sind solche altgewohnten Erscheinungen; über ihr Zustandekommen wissen wir nicht nur nichts, sondern sehen auch kaum einen Weg, etwas zu finden, wenn wir die Frage der Gleichheit von Rotations- und Umlaufszeit ausnehmen. Die der astronomischen Rechnung nicht entsprechende Bewegung des Merkurperihels und die Sekularbeschleunigung des Mondes sind weitere Fragen, bei denen die Hoffnung auf Aufklärung schon größer ist; insbesondere ist die Möglichkeit einer Aufklärung der ersteren in ein ganz neues Stadium getreten[1]). Freilich besteht dafür eine ganze Reihe von Möglichkeiten, was eben kein Vorteil ist, wenn sich eine von ihnen behaupten soll[2]).

Noch viele solcher Fragen ließen sich nennen, von denen manche in die Stellarastronomie übergreifen; wir wollen aber im Rahmen dieses Aufsatzes besonders auf diejenigen eingehen, die an die Bewegungen der kleinen Planeten anknüpfen.

Wenn wir einmal schon als gegeben hinnehmen, daß die Verteilung der großen Planeten mit merklichen Massen entweder von Anfang an oder im Laufe der Zeiten durch irgendwelche noch unaufgeklärte Bedingungen sich so gestaltet haben, wie wir sie beobachten, mögen sie nun dauernd oder für beschränkte Zeiten ihre heutige Stabilität behalten, so fragt sich, wie müssen sich die kleinen Massen verhalten?

Es liegt nahe, diese kleinen, kleineren und kleinsten Massen in drei Klassen zu teilen, in „ursprüngliche", die ähnlich wie die großen Planeten anscheinend einen Urbestandteil unseres Sonnensystems bilden, in „eingefangene" und in „fremde". Gewiß bildet eine solche Einteilung in bezug auf ihre scharfe Abgrenzung manche Schwierigkeit. Sollen wir die große Hauptgruppe der kleinen Planeten als ursprünglich oder als eingefangen bezeichnen? Daß die Gestalt ihrer Bahnen im wesentlichen durch den Einfluß des Jupiter bedingt ist, steht fest. Daß bei gewissen Anfangsbedingungen ganz bestimmte, zum Teil der Wirklichkeit entsprechende oder ihr nahekommende Bahnen entstehen müssen, zeigen

---

[1]) EINSTEIN, DE SITTER, GROSSMANN.

[2]) LEVERRIER, HARZER, v. HÄRDTL, v. SEELIGER, ANDING, POYNTING, NEWCOMB, C. NEUMANN, TISSERAND.

die Untersuchungen über periodische und benachbarte Lösungen[1]). Warum aber gewisse mögliche Bahnen realisiert sind und andere nicht, ist eine kosmogonische Frage, für die eine Aufklärung über die tatsächliche Verteilung der vorhandenen Bahnen durch Entdeckung und Beobachtung besonders wichtig ist.

Im Vordergrund steht die Frage nach der Entstehung der sog. Lücken im System der kleinen Planeten. Hierbei darf man nicht vergessen, wohl zu definieren, was man unter eine Lücke versteht. Der Begriff in seiner vollen Allgemeinheit ist vielleicht noch nicht ganz klar herausgeschält. Versteht man darunter das Vorhandensein einer materieleeren oder materiearmen Gegend oder das Nichtvorkommen gewisser Werte von Bahnelementen und welcher?

Sicherlich kann es kein Gebiet geben, in dem die Existenz von Materie unmöglich ist, und man kann wohl behaupten, daß in jedem Punkte des Raumes ein materielles Teilchen nicht nur existieren, sondern auch eine vorgeschriebene Geschwindigkeit haben kann, abgesehen von den Singularitäten bei einem Zusammenstoß. Wir dürfen also die Lücken nicht als verbotenes Gebiet betrachten.

Die Beobachtung zeigt zweierlei: die Lücken in den oskulierenden mittleren Bewegungen der kleinen Planeten und die Teilungen des Saturnsringes, zwei scheinbar ganz verschiedene Erscheinungen. Ihre Beziehungen zueinander scheinen mir aber unverkennbar. Die Untersuchungen vieler Astronomen, insbesondere die über periodische Bahnen, zeigen, daß an den Kommensurabilitätsstellen durch die Einwirkung des Jupiter ursprüngliche Kreisbahnen in stärker exzentrische Bahnen ausarten. Würde ein Planet eine streng elliptische Bahn mit einer in eine dieser Lücken fallenden mittleren Bewegung beschreiben, so würde er sich nur während eines kurzen Teils seiner Umlaufzeit in der Lückenzone aufhalten. Die Rechnung ergibt unter Annahme der von Klose[2]) angegebenen Dimensionen der Lücken, daß ein Planet, dessen mittlere Bewegung genau das Doppelte der Jupiterbewegung (Hekubatypus) beträgt, bei einer Exzentrizität von $0°\!.68$ die Hälfte seiner Umlaufzeit, bei einer solchen von $2°$ weniger als $1/_6$, bei einer solchen von $5°$ nur $1/_{18}$ und bei einer solchen von $10°$ nur $1/_{36}$ in der Lückenzone zubringen würde. Für die Kommensurabilitätsstelle $2/_5$ beträgt die Durchgangszeit bei einer Exzentrizität von $0°\!.68$ ebenfalls die halbe Umlaufszeit, bei größeren Exzentrizitäten nur die Hälfte der eben für den Hekubatypus genannten Zahlen.

Gibt es also einen Schwarm kleiner Körper, die unter der Einwirkung eines störenden Körpers stehen und nur an den Kommensurabilitätsstellen ausgeprägte Ellipsen-, im allgemeinen aber nahezu Kreisbahnen

---

[1]) Lagrange, Poincaré, Burrau, Thiele, Darwin, Charlier, Whittaker, Hill, Plummer, Perchot, Schwarzschild, Wilkens, Klose.

[2]) A. N. Bd. 218.

beschreiben, so muß sich an diesen Stellen eine materiearme, aber nicht materieleere, Zone zeigen. Ein solches Bild zeigt uns der Saturnsring, dessen Hauptteilungen oder Ränder im Umlaufsverhältnis von $^3/_1$, $^2/_1$ und $^3/_2$ zum innersten Monde Mimas stehen. Im Systeme der kleinen Planeten liegt die Sache allerdings so, daß auch die Planeten außerhalb der Lückenzonen keineswegs Kreise beschreiben, und nichtkommensurable Planeten werden auch in diese Zone geraten, welcher Umstand berücksichtigt werden müßte.

Auffallend ist, daß bei dieser Auslegung die der mittleren Bewegung entsprechende strenge Integrationskonstante gar keine Lücke aufzuweisen braucht. Die Beobachtungen zeigen aber sicher festgestellte Lücken in den *oskulierenden* mittleren Bewegungen, und auf diese erstrecken sich die theoretischen Untersuchungen vieler Astronomen. Mir scheint ein Umstand hier noch Berücksichtigung zu verdienen: ich habe in meiner Theorie der kleinen Planeten bewiesen, daß die Planeten an den Kommensurabilitätsstellen eine Sekularbeschleunigung erleiden, die, obwohl von der dritten Ordnung der störenden Masse, dennoch nicht unbeträchtlich ist. Diese Sekularbeschleunigung verschwindet, wenn die Perihellängen von Jupiter und Planet entweder zusammenfallen oder um 180° verschieden sind.

Ein abweichendes Verhalten zeigt sich in der Zone, deren Umlaufszeit der $^2/_3$- und $^3/_4$fachen des Jupiter entspricht; hier sind freilich die Planeten viel weniger häufig als in dem dichteren Gürtel. Es wirft sich dabei die weitere Frage auf, warum der Schwarm der kleinen Planeten überhaupt sich gerade zwischen Mars und Jupiter befindet. Vor der Entdeckung des ersten kleinen Planeten, der Ceres, hatte man an jener Stelle nach dem BODEschen Gesetz einen Planeten vermißt. Als die Ceres entdeckt wurde, glaubte man ihn gefunden zu haben, und es erregte großes Aufsehen, als man darauf einen zweiten, die Pallas, fand, der sich in einer fast gleichen Bahn bewegte. Wie sehr diese und weitere Entdeckungen die astronomische Welt, und nicht nur diese, beschäftigte, zeigt sich besonders deutlich in den zu jener Zeit zwischen GAUSS und OLBERS gewechselten Briefen.

Die sich immer stärker vermehrende Anzahl der bekannten kleinen Planeten war gewiß im oben angewandten Sinne etwas Ungewöhnliches; das bisher von der Natur gebotene Material zum Studium der Aufgaben der Himmelsmechanik im Anschluß an die Beobachtung war recht kärglich, und die Ordnung, in der die bis dahin allein bekannten großen Planeten ihre Bahnen verfolgen, gab kaum Anregung zu irgend welchen Fortschritten.

So leiteten die kleinen Planeten eine ganz neue Zeit ein; das Problem der Bahnbestimmung, das vorher nur mit ganz bestimmten Einschränkungen behandelt worden war, gewann unter den Händen eines GAUSS ein ganz neues Aussehen, und auf die neuen Entdeckungen gründeten sich bald die wichtigsten Untersuchungen bis in die neueste Zeit. Die Früchte, die die Wissenschaft hieraus geerntet hat, sind reich an Zahl.

Wenn daher schon die Entdeckung eines jeden neuen Planeten in der eigentlichen Zone zwischen Mars und Jupiter an sich begrüßt werden muß, weil sie besseren Aufschluß über deren Verteilung gibt, sei es nun zur Bearbeitung mechanischer oder kosmogonischer Fragen, so wird die Auffindung eines Körpers mit einer nach den bisherigen Erfahrungen ungewöhnlichen Bahn besonders wichtig sein.

Solche Entdeckungen beginnen mit Hilfe der Vervollkommnung der beobachtenden Astronomie auch schon sich zu mehren; wir kennen die Gruppe der sechs Jupiterplaneten, die sich nahe in der Bahn Jupiters bewegen, die Planeten Eros und 1924 QN, die die Marsbahn kreuzen, und außer den beiden Planeten Albert und Alinda mit großen Exzentrizitäten den kürzlich aufgefundenen Planeten (944) Hidalgo mit einer kometenartigen Bahn (Exzentrizität = 0,65, Neigung = 43°).

Schwerlich können wir uns ein richtiges Bild von der Verteilung dunkler Massenteilchen im Weltraume machen; doch sollte man meinen, daß ihr Vorhandensein sehr viel häufiger ist als das der leuchtenden Fixsterne, wenn auch ihre Massen im einzelnen vielleicht sehr klein im Verhältnis zu der jener Körper sein mögen, was aber keineswegs ausgemacht ist. Wissen wir doch, daß es der kleinsten Körper, der Meteore, eine so große Menge gibt, daß nach Schätzungen ihre stündliche Durchschnittshäufigkeit 4 bis 6 für den einzelnen Erdort beträgt[1]). Man fragt sich, ob es hier auch bevorzugte Bahnen gibt außer denen, in denen sich die mit Kometen zusammenhängenden Schwärme bewegen. Wenn wir von diesen absehen, so bilden noch immer Planeten, Kometen und Meteore drei deutlich voneinander getrennte Gruppen. Man könnte aber wohl erwarten, daß zwischen diesen alle denkbaren Zwischenformen existieren müssen. Wenn man nur die Form der Bahnen in Betracht zieht, so hat sich die Kluft zwischen Planeten und Kometen außerordentlich verengert. Wir kennen Kometen, z. B. 1892 III, die sich ganz in der Zone zwischen Mars und Jupiter bewegen, und wir kennen den oben erwähnten Planeten, dessen Bahn von jenseits der Marsbahn bis nahe an die Saturnsbahn reicht. Ist der Unterschied zwischen Kometen und Planeten nur durch ihre Bahnform bedingt und erklärt sie sich aus dem ungeheuren Wechsel in der Sonnenstrahlung, dem die ersteren ausgesetzt sind? Die Tatsache, daß das nebelartige Aussehen auch der schweiflosen Kometen in scharfem Gegensatz zu dem der selbst am stärksten exzentrischen Planeten steht, spricht dagegen. Ob dies jedoch ein entscheidender Beweis ist, ob nicht bei den Planeten ein Gleichgewichtszustand eingetreten ist, der die Verschiedenheit bedingt, mag die Zukunft entscheiden. Die Eigenart des Spektrums der Kometen

---

[1]) Newcomb-Engelmann: Populäre Astronomie.

dürfte keinen durchschlagenden Beweis geben; denn das Spektrum, das ein Planet unter gleichen Umständen aussenden würde, kennen wir nicht. Zwischen der Natur der kleinen Planeten und der der großen Meteore scheint der Gegensatz weit geringer, wenn überhaupt einer vorhanden ist und der Zusammenhang der letzteren mit den Kometen ist bekannt. Will man also einen durchgreifenden Unterschied zwischen den verschiedenen Gruppen der uns bekannten dunklen Himmelskörper machen, so ist man wohl in erster Linie darauf angewiesen, die Entscheidung danach zu treffen, ob man sie als „ursprüngliche", „eingefangene" oder „fremde" Körper zu betrachten hat, und hierüber liegt die Entscheidung bei der Mechanik des Himmels.

3. Notwendigkeit einer Organisation der rechnerischen Arbeiten auf dem Gebiete der Himmelsmechanik. Welche Vorstellungen man sich auch von der Natur und der Herkunft der verschiedenen Himmelskörper machen mag, so müssen wir vor allem versuchen, unsere Kenntnisse von den sich in unserem Sonnensystem bewegenden kleinen und kleinsten Körpern möglichst zu erweitern und zu vertiefen. Dazu gehört, daß die Beobachter nicht müde werden, den Himmel nicht nur nach Neuem zu durchmustern, sondern auch die bereits entdeckten Körper, wo es nötig ist, weiter zu beobachten, und daß die Rechner in der mühsamen Arbeit, die Bahnen zu verfolgen und die Beobachtungen zu identifizieren, nicht nachlassen. Bei der großen Menge des Materials ist hier eine gute Organisation ebenso nötig, wie solche bei anderen großen Aufgaben der Astronomie, im besonderen zur Beobachtung des Fixsternhimmels durch mehrere große Unternehmungen, geschaffen sind.

Eine solche Organisation für die kleinen Planeten, die wenigstens den größten Teil der Entdeckungen vor dem Verlorengehen retten konnte, ist im Berliner Astronomischen Recheninstitut durch BAUSCHINGER begonnen und durch FRITZ COHN möglichst dem immer dringender werdenden Bedürfnis angepaßt worden. Noch in frischer Erinnerung ist die bedeutende Arbeitsleistung, die BERBERICH auf diesem Gebiete bewältigte und die neuerdings von STRACKE fortgesetzt wird. Das Berliner Institut läßt sich die Sammlung und Identifizierung der Planetenbeobachtungen, eine möglichst schnelle Bahnbestimmung der neu entdeckten, sowie die Bahnverbesserung der älteren Planeten, teils mit teils ohne Berücksichtigung der Störungen angelegen sein. Ohne diese grundlegende Organisation, zu der im besonderen auch die Vorausberechnung und Veröffentlichung von Ephemeriden gehört, sind weitergehende Untersuchungen über die Bewegung der Planeten gar nicht denkbar. Dem Berliner Institut und seinen Leitern und Mitarbeitern gebührt daher der besondere Dank der Astronomen. Man wird im Anschluß an die eben genannte Organisation und mit dem durch sie und die Beobachter vorbereiteten Material weitergehen müssen, um dem Ziel, das die Beherrschung der Mechanik unseres Sonnensystems bedeutet, näherzukommen.

Das große Problem, weitgehende Aufschlüsse über die wirklich vorkommenden und über die überhaupt möglichen Bewegungen in unserem Sonnensystem zu erhalten, muß von irgendwelchen Angriffspunkten beginnend möglichst systematisch durchgeführt werden.

Einen solchen Angriffspunkt bietet die Zone der kleinen Planeten. Die große Menge der zu berechnenden Bahnen verbietet es, die Rechnung für jeden einzelnen Planeten nach einer besonderen Theorie auszuführen; man muß daher zur Gruppenrechnung greifen, und dies nicht nur wegen der großen Anzahl der Planeten, sondern auch, weil man die möglichen Bahnen kennenlernen will, in denen kein wirklicher Planet läuft.

In meiner Theorie der kleinen Planeten habe ich allgemeine Tafeln für die Störungen gegeben, die die Exzentrizitäten und Neigungen als unbestimmte Faktoren enthalten und aus denen durch Einsetzen der entsprechenden Werte für diese Größen die Störungsausdrücke direkt entnommen werden können.

In erster Linie kommt es bei allen Rechnungen natürlich darauf an, mit welcher Genauigkeit man die geozentrischen Örter der Planeten darstellen will. Ich habe für die praktische Rechnung drei Genauigkeitsgrade unterschieden, je nachdem man eine Darstellung innerhalb 20′, 1′ oder innerhalb der Beobachtungsfehler erstrebt. In meinen erwähnten Tafeln sind alle Störungsglieder berücksichtigt, die von den zweiten Potenzen der Exzentrizitäten und Neigungen abhängen und bei einem Exzentrizitätswinkel von $18°$ und einer Neigung von $28^1/_2°$ den Betrag von $8''$ überschreiten. Die Ergänzung der Tafeln durch Berücksichtigung der dritten und vierten Potenzen ist im Frankfurter Institut in Bearbeitung; die Entwicklung der Störungsfunktion ist bis zu den achten Potenzen der Exzentrizität des gestörten Körpers durchgeführt. Für kleine Exzentrizitäten und Neigungen werden die bisher berechneten Tafeln für eine Darstellung innerhalb des ersten Genauigkeitsgrades ausreichen, für größere wird die Ergänzung durch den dritten und vierten Grad nötig sein. Für solche Werte der Exzentrizitäten und Neigungen, die die oben angegebenen Grenzen überschreiten, wird man gezwungen sein, von einer eigentlichen Gruppenrechnung abzusehen und die Rechnung für einzelne Worte, insbesondere der Exzentrizität, durchzuführen, indem man eine generelle Entwicklung nach Potenzen dieser Größe vermeidet. Hierzu bieten sich verschiedene Methoden, die in einem gewissen Zusammenhang mit den bereits von Gauss und Hansen angewandten stehen. Ebenso können sich hierbei die Untersuchungen über periodische und benachbarte Bahnen nützlich erweisen.

Im übrigen erstrecken sich unsere Tafeln zunächst nur auf die Planeten, deren mittlere Bewegung größer als $709''$ ist; für die dem Jupiter näheren werden sich andere Integrationsmethoden empfehlen, über die weiter unten, S. 191, Andeutungen gemacht sind. Eine Zählung zeigt unter den ersten 911 Planeten 581 mit einer mittleren Bewegung über 709″, und darunter 17 mit einem Exzentrizitätswinkel über 18° und 4 mit einer Neigung über $28^1/_2$°. Unsere durch Berücksichtigung der Glieder dritten und vierten Grades ergänzten Tafeln werden also für mehr als die Hälfte der bekannten Planeten für den ersten und nahezu für den zweiten Genauigkeitsgrad ausreichen.

Über die Einzelheiten der im Frankfurter Institut geplanten Arbeiten habe ich im ersten Stück der Mitteilungen der Frankfurter Sternwarte berichtet. Sie erstrecken sich zunächst auf den ersten Genauigkeitsgrad, und zwar für einen Zeitraum von 100 Jahren. Bearbeitet wurden bisher etwa 230 Planeten, deren Störungsausdrücke im einzelnen aufgestellt, und etwa 130, für die mittlere Elemente abgeleitet wurden; daneben laufen einzelne Untersuchungen über besonders schwierige Planeten. Unter normalen Verhältnissen hätten gegenwärtig die Tafeln für alle Planeten der dichteren Zone vollendet und für den größten Teil auch die Störungen und die mittleren Elemente fertiggestellt sein können; doch haben die Arbeiten des Frankfurter Instituts in den letzten Jahren sehr schwer unter den Zeitverhältnissen gelitten.

Von einigen anderen Astronomen, insbesondere von Herrn BOHLIN, liegen ebenfalls vorbereitende Untersuchungen und fertige Tafeln vor, die wohl geeignet sind, ebenfalls als Grundlage für die Bearbeitung der kleinen Planeten im großen zu dienen. Es war von Anfang an beabsichtigt, im Frankfurter Institut auch diese Methoden anzuwenden; doch mußte auch dies wegen der Zeitverhältnisse aufgeschoben werden.

4. Verschiedene Methoden der Störungsrechnung. Es mag nun auf die Eigenart der verschiedenen Methoden der Störungsrechnung und auf ihre Entwicklungsmöglichkeiten etwas näher eingegangen werden, obwohl sie, besonders die ersteren, den Astronomen wohlbekannt sind.

Wir erinnern an die beiden großen klassischen Gruppen von Methoden, von denen die eine die Störungen der Koordinaten, die andere die der elliptisch gedachten Bahnelemente herzustellen sucht. Im ersten Falle ist eine Bewegung in einer festen, sog. mittleren Ellipse, zugrunde gelegt, aus der die „ungestörten" Koordinaten des Planeten berechnet werden; die Störungen dieser Koordinaten werden durch irgendein Verfahren zur genäherten Lösung der Bewegungsgleichungen abgeleitet, da die strenge nicht ausführbar ist, und den ungestörten Werten hinzugefügt. Der so vorausberechnete Ort kann mit der Beobachtung verglichen werden. Bei der zweiten Gruppe von Methoden werden die Konstanten der elliptisch gedachten Bahn als veränderlich angenommen; es wird ebenfalls eine feste Ellipse zugrunde gelegt und an deren Elemente die Elementenstörungen angebracht, womit man eine nur für den Augenblick geltende Ellipse erhält, aus der der wahre Ort des Planeten nach den Formeln der rein elliptischen Bewegung berechnet wird. Für diese mit der Zeit veränderliche Ellipse wird üblicherweise die oskulierende Ellipse gewählt, deren sechs Elemente so definiert sind, daß durch die Ellipse nicht nur die Koordinaten selbst, sondern auch die Komponenten der Geschwindigkeit dargestellt werden. Man hat diese Form gewählt, weil sich die Differentialgleichungen der Elemente als Funktionen der Zeit in besonders einfacher und übersichtlicher Weise herstellen lassen. Ich habe schon früher darauf hingewiesen, daß man allgemeiner verfahren und die veränderliche Ellipse so wählen kann, daß sie nur die Koordinaten des Planeten, nicht aber auch die Geschwindigkeiten darstellt. Bei einer solchen „instantanen" Ellipse hat man eine gewisse Freiheit, indem man mit gewissen Einschränkungen drei der Elemente beliebig annehmen oder drei beliebige Bedingungen zwischen den sechs Elementen einführen kann.

Man kann aber noch anders verfahren, indem man statt der erwähnten mittleren Ellipse eine andere Bahn zur genäherten Darstellung der Koordinaten wählt und an dieser die „Störungen" anbringt, entweder als Koordinaten- oder als Elementenstörungen. Eine solche wird in der Regel ellipsenähnlich sein, etwa eine Ellipse mit Apsidenbewegung oder noch anderen mit der Zeit langsam fortschreitenden Änderungen. Man hat eine solche Bahn nach einer von GYLDÉN, allerdings in einem spezielleren Fall, gebrauchten Bezeichnung „intermediär" genannt; sie ist ein Zwischending zwischen der ungestörten Ellipse und der wahren Bahn. Dagegen verstehen wir unter instantaner Bahn immer eine solche, die die wahren Koordinaten darstellt.

Die oskulierende Ellipse ist also ein Sonderfall der instantanen Bahn, und die Wahl der letzteren beruht mehr auf formellen Erwägungen als auf einer analytischen Definition. Ob es von Vorteil wäre, die eine oder andere Art von instantaner Bahn auch analytisch zu definieren und die Differentialgleichungen für ihre Elemente aufzustellen, ist eine offene Frage, die zu bejahen man kaum geneigt sein wird. Der Verfasser hat sie auch bisher nur zur formellen Darstellung der Bewegung, im besonderen beim Übergang von Koordinaten- zu Elementenstörungen und umgekehrt, gebraucht.

Die rechnerischen Methoden zur Auffindung der Koordinaten- und der Elementenstörungen sind nicht wesentlich verschieden, und man kann, wenn die einen gefunden sind, die anderen daraus ableiten. Der Form nach hat man für die ersteren drei Differentialgleichungen von der zweiten, für die letzteren sechs solche von der ersten Ordnung. Wir brauchen, wenn wir von besonderen Eigenschaften der Rechnungsmethoden handeln, beide Gruppen nicht scharf auseinanderzuhalten, wenn sich auch die eine oder andere Methode besser oder ausschließlich für die eine oder andere Gruppe eignet.

Bei den beiden genannten Gruppen von Methoden zur Ermittlung der Störungen kann man wiederum zwei Rechnungsarten unterscheiden, die der speziellen und die der allgemeinen Störungen. Die Methoden der ersteren Art kommen aber für Gruppenrechnungen nicht in Betracht, weil sie ihre Anwendung bei der Berechnung der speziellen Bahnen einzelner Körper finden, indem, von bestimmt vorgegebenen Anfangskoordinaten und -geschwindigkeiten ausgehend, die Bewegung sukzessive durch numerische Integration verfolgt wird. Dagegen wird diese Methode mit großem Vorteil angewandt beim Studium periodischer und anderer Bahnen und bei besonderen, auch in unserem Sonnensystem nicht realisierten Massenverteilungen; auch dient sie zur Verfolgung und Sicherung solcher Planeten, für die noch keine allgemeinen Störungen bekannt sind.

5. Die praktischen Schwierigkeiten der rechnerischen Methoden. Der grundlegende Gedanke bei den Methoden der allgemeinen Störungsrechnungen ist, eine Reihenentwicklung herzustellen, die, wenn es sich um das Dreikörperproblem handelt, nach Potenzen von 5 Größen fortschreitet, nämlich dem Verhältnis α der mittleren Entfernungen beider Planeten (des gestörten und des störenden) von der Sonne, der Exzentrizitäten e und e' und der Neigungen der Bahnen i und i' gegen die Ekliptik. An die Stelle der beiden letzteren kann die gegenseitige Neigung beider Bahnebenen treten, wodurch die Anzahl der genannten Größen auf 4 reduziert wird. Die Entwicklung nach Potenzen dieser Größen zieht eine zusammengesetzte trigonometrische Reihenentwicklung nach sich, die bei geeigneter Umformung nach Vielfachen von mehreren Argumenten fortschreitet. Diese Argumente sind ursprünglich 6 an Zahl, nämlich die Längen-, die Anomalien- und die Breitenargumente, d. h. die Winkel, die die Radienvektoren mit dem Anfangspunkt der Zählung (dem Frühlingspunkt), den Perihelen und den Knotenlinien der Bahnebenen bilden; durch Einführung der Differenzen der Längen kann die Zahl der Argumente auf 5 reduziert werden. Man kann für die genannten Winkelargumente entweder ihre „wahren" oder ihre „mittleren" oder ihre „exzentrischen" Werte gebrauchen. Ohne hier auf die Unterschiede dieser verschiedenen Entwicklungsarten eingehen zu können, bemerken wir, daß die Anwendung der wahren Längen entschiedene Vorzüge besitzt; das in der Enzyklopädie der mathematischen Wissenschaften Bd. VI. 2, S. 595 hierüber Gesagte könnte zu einem Mißverständnis führen.

Die genannten Reihenentwicklungen werden *vor der Integration* vorgenommen und bestehen im wesentlichen in der Entwicklung der „Störungsfunktion"; dazu tritt eine Entwicklung nach Potenzen oder anderen Funktionen der störenden Masse, die wir nicht hier, sondern weiter unten bei unseren Betrachtungen über die Konvergenz der sukzessiven Annäherungen zur Integration der Gleichungen behandeln.

Wir beschäftigen uns nur mit einigen allen diesen Entwicklungsarten gemeinsamen Eigentümlichkeiten, auf die an dieser Stelle etwas näher eingegangen werden soll, weil sie das wichtigste Moment für die Wahl und Anordnung der rechnerischen Methoden, insbesondere bei Gruppenrechnungen, sind. Da wir sie jedoch nicht mit allen Einzelheiten und mit ihren Komplikationen, schon des beschränkten Raumes wegen, betrachten können, so wählen wir einige einfache Beispiele, die das Verhalten der rechnerisch angewandten Entwicklungen grundsätzlich zeigen und im Anschluß an die wir untersuchen wollen, in welchem Umfange gewisse rechnerische Methoden brauchbar sind.

Nehmen wir ein System von Differentialgleichungen von der Form:

$$\frac{d x_i}{d t} = G_i (x_1, x_2, \ldots, x_n; t), \quad i = 1 \text{ bis } n \tag{1}$$

wo die G_i analytische und eindeutige Funktionen der Variablen sind, die wir ebenso wie die Variablen selbst, der Natur des Problems nach, als reell ansehen können. Wenn auch die Differentialgleichungen für die Koordinaten eines Planeten von der zweiten Ordnung sind, so können wir doch die Gleichungen (1) betrachten, wenn wir neben den Koordinaten noch die Komponenten der Geschwindigkeiten als Unbekannte einführen. Sind die x_i die Koordinaten und Geschwindigkeitskomponenten oder die Bahnelemente eines einzelnen Planeten mit verschwindender Masse, so können die Bewegungen der auf ihn wirkenden anderen Planeten durch bekannte Funktionen der Zeit t ausgedrückt gedacht und $n = 6$ gesetzt werden.

Man setze:

$$x_i = x_i^{(0)} + \xi_i$$

und nehme an, daß die Funktionen G_i sich ebenfalls teilen lassen, so daß:

$$G_i = G_i^{(0)} + H_i$$

wo die $G_i^{(0)}$ nur die Variablen $x_i^{(0)}$ enthalten und übrigens auch die Zeit t enthalten können. Durch die Lösung der Gleichungen:

$$\frac{d x_i^{(0)}}{d t} = G_i^{(0)} \tag{2}$$

seien die Funktionen $x_i^{(0)}$ eindeutig bestimmt, und man denke sich, daß diese Lösung die sog. ungestörte oder eine sonstwie genäherte Bewegung gibt, die entweder elliptisch ist oder in einer „intermediären" Bahn vor sich geht. Weiter soll angenommen werden, daß die $x_i^{(0)}$, wenigstens für einen beschränkten Zeitraum, Näherungswerte der x_i sind.

Zur Bestimmung der ξ_i ergebe sich:

$$\frac{d \xi_i}{d t} = H_i (x_1^{(0)}, x_2^{(0)}, \ldots, x_n^{(0)}; \xi_1, \xi_2, \ldots, \xi_n; t) \tag{3}$$

und die H_i seien entwickelbar nach ganzen positiven Potenzen der ξ_i. Man kann annehmen, daß die H_i und die ξ_i einen kleinen Parameter als Faktor enthalten, als den man sich die Masse des störenden Körpers denken kann, so daß wir eine Entwicklung nach Potenzen dieser Masse erhalten. Indessen binden wir uns nicht an diese Annahme, weil wir zunächst nicht die Entwicklung nach Potenzen der Masse im Auge haben, die wir weiter unten besprechen und die für numerische Rechnung im allgemeinen weniger in Betracht kommt. Wir setzen nur voraus, daß die ξ_i kleine Größen sind, die übrigens immerhin, wie die H_i, mit der störenden Masse verschwinden mögen, aber kompliziertere Funktionen dieser sein können.

Eine erste Annäherung zur Lösung der Gleichungen (3) bestehe darin, daß wir entweder die ξ_i zunächst ganz unterdrücken, d. h. gleich Null setzen, oder sie nur teilweise berücksichtigen. Es genügt für das, was wir hier zeigen wollen, eine solche erste Annäherung zu betrachten. Die $x_i^{(0)}$ sind Funktionen der Zeit t, und sie mögen einige Parameter enthalten, als die wir uns, unter der vereinfachten Annahme eines einzigen störenden Körpers, die Exzentrizitäten e und e', die Neigungen i und i' der Bahnebenen und das Verhältnis der mittleren Entfernungen von der Sonne $\alpha = \dfrac{a}{a'}\left(\text{oder } \dfrac{a'}{a}, \text{ falls } a > a'\right)$ denken und unter denen die Integrationskonstanten begriffen sein sollen. Die Neigungen wollen wir bei unseren Betrachtungen ausschließen, da die Entwicklungen nach ihnen zwar umfangreich werden können, aber keine besonderen Schwierigkeiten bieten gegenüber den hier mit Bezug auf die anderen Parameter zu besprechenden.

In den Funktionen H_i ersetzen wir die $x_i^{(0)}$ durch die Parameter e, e', α und die Zeit t und entwickeln formell nach Potenzen der drei ersten. Um leichter übersehbare Ausdrücke zu erhalten, beschränken wir uns zunächst auf die nähere Betrachtung der Entwicklung nach e. Die Entwicklung habe folgende Form:

$$H_i = a_0 + a_1 \cos(\lambda_1 t + B_1) + a_2 \cos(\lambda_2 t + B_2) + \ldots \tag{4}$$

wo:

$$a_n = b_n\, e^n (1 + c_n\, e^2 + c_n'\, e^4 + \ldots) \tag{5}$$

und a_n, b_n, c_n, c_n', λ_n, B_n konstante Größen sein mögen.

Wir wollen nun zwei mögliche Fälle betrachten, nämlich, daß die Entwicklung (4) entweder konvergiere oder von der Natur der Reihen ist, die POINCARÉ asymptotisch genannt hat. Die Schwierigkeiten, die bei der numerischen Rechnung auftreten, treffen wir in beiden Fällen. Sie bestehen darin, daß hier die Konvergenz der Reihen nicht genügt; vielmehr muß die Abnahme der Glieder so stark sein, daß eine Berechnung von nicht allzu vielen ersten Gliedern zu einem genügenden Näherungswert führt.

Wir nehmen zunächst einmal an, daß die Entwicklung (4) konvergiere, und zwar so, daß die Reihe der absoluten Beträge der a_n konvergent sei, und wenden uns zur Betrachtung der Entwicklungen (5).

Es seien, indem ich ein Beispiel wähle, ähnlich dem, das Poincaré zur Untersuchung der asymptotischen Reihen gebraucht, die a_n von der Form:

$$a_n = \frac{b_n\,e^n}{1 + n\,e^2} \tag{6}$$

Entwickelt man, um die Form (5) zu erhalten, die a_n nach Potenzen von e^2, so konvergiert diese Entwicklung nur, solange $n e^2 < 1$ ist. Setzt man trotzdem die Potenzreihen nach e^2 in (4) ein und ordnet nach Potenzen von e, so erhält man eine ähnliche Reihe, wie sie Poincaré untersucht. Hat man aber die Reihe (4) bei einem Werte von a_n abgebrochen, für den $n e^2 < 1$ ist,. indem man die nachfolgenden Koeffizienten a_n *entweder* vernachlässigt (wenn sie klein genug sind) *oder* sie sich nicht nach Potenzen von e^2 entwickelt denkt, so operiert man mit konvergenten Ausdrücken. Damit scheinen die Schwierigkeiten der Divergenz·vermieden; aber selbst in den Fällen, wo ein solches Vorgehen berechtigt ist, genügt es häufig nicht für die numerische Rechnung. Ist e sehr klein, so stößt man bei der Rechnung auf keine erheblichen Schwierigkeiten; wohl stellen sich aber solche schon bei mäßig großen Werten von e ein. Denn wir haben zwar gesehen, daß wir konvergente Ausdrücke erhalten, wenn wir die a_n für alle n-Werte, die größer als $\dfrac{1}{e^2}$ sind, fortlassen können; aber selbst wenn dies zulässig ist, so tritt für den Fall, daß $n e^2$ nicht sehr erheblich kleiner als 1 gewählt werden kann, eine so schwache Konvergenz ein, daß die numerische Rechnung teils schwierig, teils unausführbar werden kann.

Wir hatten angenommen, daß die Koeffizienten a_n von der Form (6) seien, und stießen bei der Entwicklung der Funktionen H_i nach Potenzen von e auf divergente Entwicklungen. Wir nehmen jetzt an, diese Koeffizienten haben die Form:

$$a_n = \frac{b_n\,e^n}{(1 - e^2)^n} \tag{7}$$

und e sei klein genug, damit die Reihe der a_n in (4) konvergiere. Dies ist der Fall, wenn $e < \tfrac{1}{2}(\sqrt{5} - 1)$, und auch noch für größere Werte von e, falls die b_n nach einem gewissen Gesetz abnehmen.

Entwickeln wir hier wieder nach Potenzen von e, so erhalten wir für die a_n konvergente Reihen, und die Entwicklung der H_i nach Potenzen von e konvergiert ebenfalls. Die numerische Rechnung stößt aber hier auf die gleichen, und unter Umständen auf noch größere, Schwierigkeiten, wie im vorigen Beispiel. Für größere n-Werte wachsen nämlich die Glieder der Binomialentwicklung von (7) bis zu dem Gliede, das die m-te Potenz von e enthält, wo die ganze Zahl m durch die Ungleichungen:

$$\frac{n e^2 - 1}{1 - e^2} < m < \frac{n e^2 - 1}{1 - e^2}. + 1$$

bestimmt ist. In der Theorie der kleinen Planeten treten solche Unannehmlichkeiten häufig auf, und zwar bei niedrigeren n-Werten als in dem als Beispiel gewählten Ausdruck (7), das ja nur gewählt wurde, um diese Verhältnisse in ganz elementarer Weise zu zeigen.

In der Entwicklung nach Potenzen von e', das der Jupiterexzentrizität entspricht, ist dieser Umstand weniger störend, weil e' sehr klein ist.

Dagegen zeigt sich eine ähnliche Erscheinung auch bei der Entwicklung nach Potenzen des Verhältnisses der mittleren Entfernungen α oder nach solchen von Funktionen von α, wie α^2 oder $\dfrac{\alpha^2}{1 + \alpha^2}$. Auch hier nehmen die Glieder der Entwicklungen nicht immer genügend stark ab oder anfangs sogar zu; dies ist auch der Grund, warum meine Tafeln für die Planeten bei einer mittleren Bewegung von $709''$ abbrechen. Eine Berechnung der periodischen, dem Planeten Thule entsprechenden Bahn nullten Grades, dessen mittlere Bewegung rund $400''$, d. h. das $^4/_3$fache der Jupiterbewegung, beträgt, zeigte, daß das 10. Glied der Entwicklung noch fast $2'$ und das 20. noch $4''$ in der heliozentrischen Länge ausmacht.

Es muß hier ein Mittel gesucht werden, um die Abnahme der Glieder zu verstärken. Ein solches scheint sich zu bieten, indem man statt der Zeit oder der Länge in der Bahn andere Veränderliche einführt, wobei die Beziehungen zwischen den Veränderlichen durch elliptische Funktionen ausgedrückt werden[1]). Vorläufige dahingehende Versuche lassen ein solches Verfahren aussichtsvoll erscheinen; doch sind die hier zur Rechnung nötigen Formeln noch nicht in handlicher Form hergestellt. Naturgemäß wird auch ein solches Verfahren seine natürliche Grenze haben und desto mehr zu wünschen übriglassen, je näher die Bahn der des Jupiters liegt.

Neben den Reihenentwicklungen, die wir soeben besprochen, handelt es sich noch um eine andere Gattung unendlicher Prozesse, die bei der Integration der Gleichungen auftreten, nämlich die Prozesse der sukzessiven Annäherungen. Sie sind der Analysis weit schwerer zugänglich. Die Rechnung beschränkt sich auch hier auf einen endlichen Prozeß, und zwar meist auf eine recht geringe Anzahl von Annäherungen. Andererseits bietet sich hier ein besonderer Vorteil: wenn die numerische Rechnung auch nicht den Beweis der Konvergenz oder Divergenz erbringen kann, so kann sie doch durch Verifizierung des Ergebnisses feststellen, daß dieses innerhalb der gewünschten Genauigkeit richtig ist. Die Rechnung zeigt unmittelbar, ob bei Wiederholung des Prozesses das Ergebnis innerhalb der gesetzten Grenze ungeändert bleibt und ob die Einsetzung der gefundenen Lösung in die Gleichungen diese befriedigt, wobei nur zu beachten ist, ob eine eindeutige Lösung existiert

---

[1]) Hierauf hat schon GYLDÉN hingewiesen.

oder, falls mehrere möglich sind, ob unter diesen die gewünschte gefunden wird. Wir haben hier, falls die Lösung durch Versuche nach der Methode der unbestimmten Koeffizienten gesucht wird, eine vollständige Analogie mit der Lösung gewöhnlicher Gleichungen. Es sei nur an die Keplersche Gleichung erinnert. Wenn eine andere Integrationsmethode angewandt wird, so sind die Verhältnisse ähnlich. Auch hier wollen wir den Verlauf des Prozesses an einem einfachen Beispiel zeigen.

Wir greifen zurück auf das Gleichungssystem (3), S. 188, nämlich:

$$\frac{d\xi_i}{dt} = H_i \tag{3}$$

und nehmen an, daß dieses bis zur gewünschten Genauigkeit durch den formellen endlichen Ansatz:

$$\left.\begin{array}{l} \xi_i = c_i^{(0)} + c_i^{(1)} \cos(\lambda_1 t + B_1) + c_i^{(2)} \cos(\lambda_2 t + B_2) + \cdots \\ \qquad + c_i^{(n)} \cos(\lambda_n t + B_n) \end{array}\right\} \tag{8}$$

befriedigt sei, in dem c, λ, B Konstante bedeuten. Weiter nehmen wir an, daß sich durch Einsetzen dieser Ausdrücke in (3) folgende Bestimmungsgleichungen für die c ergeben:

$$\lambda_s c_i^{(s)} = \alpha_i^{(0)} + \alpha_i^{(1)} X_i^{(1)} + \alpha_i^{(2)} X_i^{(2)} + \cdots \tag{9}$$

wo die $X_i^{(n)}$ ganze rationale Funktionen der c_i vom $n$ten Grade und die c_i wie auch die bekannten α_i von der Größenordnung der störenden Masse m sind. Die Konvergenz oder wenigstens die Brauchbarkeit der Reihen zur Bestimmung der c_i innerhalb der gewünschten Grenze setzen wir voraus.

Je nach der angewandten Methode können diese Bestimmungsgleichungen mehr oder weniger von der vorstehenden Normalform abweichen; jedoch werden unsere Betrachtungen im allgemeinen auch dann gelten.

Um den Prozeß an einem einfachen, leicht übersehbaren Beispiel zu zeigen und um zu sehen, welche Umstände sich beim numerischen Rechnen einstellen, betrachten wir statt des Gleichungssystems (9) eine einzige Gleichung mit einer Unbekannten c. Die Gleichung sei:

$$\lambda c = \alpha_0' + \alpha_1' c + \alpha_2' c^2 + \cdots \tag{10}$$

Nachdem die Größe λ und die Koeffizienten α' numerisch berechnet sind, wird der Rechner die Größe c durch Annäherungen bestimmen, wobei wir voraussetzen, daß λ weder verschwindet noch eine sehr kleine Größe ist, daß also λ von der nullten, die α', wie c selbst, von der ersten Ordnung der Masse sind. Der Fall sehr kleiner Werte von λ ist der der kleinen Integrationsdivisoren bei den Kommensurabilitätsstellen. Obwohl gerade auch in diesem Fall das im folgenden geschilderte Verfahren in Betracht kommt, gehen wir hier auf ihn nicht ein, weil dies umfangreichere Auseinandersetzungen erfordern würde und hier nicht ausführlich auf Einzelfragen eingegangen werden kann.

Der Rechner wird, falls er nur mit Zahlen rechnet, eines der beiden folgenden Näherungsverfahren einschlagen, wenn $c_1, c_2, c_3, \ldots$ die sukzessiven Näherungswerte sind und in Klammern die Größenordnung der vernachlässigten Teile angedeutet wird:

entweder:

$$c_1 = \frac{\alpha_0'}{\lambda} + (m^2)$$

$$c_2 = \frac{\alpha_0' + \alpha_1' c_1}{\lambda} + (m^3)$$

$$c_3 = \frac{\alpha_0' + \alpha_1' c_2 + \alpha_2' c_2^2}{\lambda} + (m^4)$$

usw.

oder:

$$c_1 = \frac{\alpha_0'}{\lambda - \alpha_1'} + (m^3)$$

$$c_2 = \frac{\alpha_0' + \alpha_2' c_1^2}{\lambda - \alpha_1'} + (m^4)$$

$$c_3 = \frac{\alpha_0' + \alpha_2' c_2^2 + \alpha_3' c_2^3}{\lambda - \alpha_1'} + (m^5)$$

usw.

$$\left. \right\} \quad (11)$$

Die analytische Annäherung sieht etwas anders aus, wie wir gleich sehen werden. Wir wollen die Konvergenz des Näherungsverfahrens untersuchen und sehen, ob und wie dieses unter Umständen modifiziert werden kann. Indem wir dazu das zweite der eben genannten Verfahren benutzen, bringen wir die Gleichung (10) auf die Form:

$$c = \alpha_0 + \alpha_2 c^2 + \alpha_3 c^3 + \ldots \tag{12}$$

wo c und die $\alpha_n = \dfrac{\alpha_n'}{\lambda - \alpha_1'}$ von der Größenordnung der Masse m sind. Die Reihe sei konvergent für $|c| \leqq R \neq 0$, und auch R sei von der Größenordnung m.

Wir setzen:

$$\frac{c}{R} = y, \qquad \frac{\alpha_0}{R} = x, \qquad \alpha_n R^{n-1} = a_n$$

und erhalten die Gleichung:

$$y = x + a_2 y^2 + a_3 y^3 + \cdots \tag{13}$$

in der nunmehr x und y von der Ordnung der Einheit, die a_n dagegen von der Ordnung m^n sind. Die Reihe konvergiert für $|y| \leqq 1$, und es soll y aus den bekannten Werten von x und den a_n berechnet werden.

Wie schon bemerkt, haben wir numerische und analytische Annäherung zu unterscheiden; die letztere ordnet streng nach den steigenden Potenzen von m, und die Näherungswerte sind die folgenden, indem zuerst m^2, dann m^3 usw. vernachlässigt wird,

$$\left. \begin{aligned} y_1 &= x + (m^2) \\ y_2 &= x + a_2 x^2 + (m^3) \\ y_3 &= x + a_2 x^2 + (2 a_2^2 + a_3) x^3 + (m^4) \end{aligned} \right\} \tag{14}$$

usw.

Das Verfahren führt direkt, wie man sieht, zu Lagranges Theorem der Reihenumkehrung. Die numerische Annäherung verfährt etwas anders, die Näherungswerte sind die folgenden:

$$\left.\begin{aligned}
y_1 &= x + (m^2)\\
y_2 &= y_1 + a_2\, y_1^2 = x + a_2\, x^2 + (m^3)\\
y_3 &= y_2 + a_2\, y_2^2 + a_3\, y_2^3 = x + a_2(x + a_2\, x^2)^2 + a_3(x + a_2\, x^2)^3 + (m^4)
\end{aligned}\right\}(15)$$

usw.

In beiden Methoden unterscheiden sich also die dritte und die folgenden Annäherungen, indem die numerische Methode mehr Glieder berücksichtigt, woraus aber nicht notwendig folgt, daß sie schneller zum gewünschten Ergebnis führt. Sie hat aber jedenfalls den Vorteil, daß sie bequemer ist und daß man aus den sukzessiven Näherungswerten ersieht, wie stark der Grad der Konvergenz ist, während die analytische Annäherung den Vorteil hat, y analytisch als Funktion von x zu geben. Beide Verfahren kommen auf eine Entwicklung nach Potenzen von m heraus. Die Konvergenz dieser Entwicklung folgt aus einem bekannten Satz: Es sei die Reihe, durch die y als Funktion von x dargestellt wird,

$$y = x + b_2\, x^2 + b_3\, x^3 + \cdots \tag{16}$$

so findet man die b_n durch Rekursion als Funktionen der a_n, und zwar ist b_n eine ganze rationale Funktion der $a_2, a_3, \ldots$ bis a_n mit nur positiven Vorzeichen. Es sei A der größte absolute Betrag der a_n; ein solcher muß immer vorhanden sein, weil die Reihe für $y = 1$ noch konvergiert. Setzt man:

$$b_n = \varphi_n(a_2, a_3, \ldots a_n)$$

und ersetzt man jedes a_n durch A, so ist:

$$\beta_n = \varphi_n(A) > |\, b_n\, |$$

Wenn nun die Reihe:

$$y = x + \beta_2\, x^2 + \beta_3\, x^3 + \ldots \tag{17}$$

konvergiert, so konvergiert auch die Reihe (16). Die vorstehende Reihe ist aber die Umkehrung von:

$$x = y - A(y^2 + y^3 + \ldots) = y - y^2\, \frac{A}{1 - y}$$

die Reihe (17) ist also die Entwicklung von:

$$y = \frac{1}{2(1 + A)}\left\{1 + x - \sqrt{1 - (2 + 4A)\, x + x^2}\right\}$$

wo die Wurzel negativ genommen ist, weil y mit x verschwinden soll. Die Reihe (17) konvergiert also für:

$$|x| < 1 + 2A - 2\sqrt{A + A^2} > 0$$

folglich konvergiert auch die Reihe (16) mindestens in diesem Bereich. Das vorstehende Theorem ist wohlbekannt. Es fragt sich nun, wie man vorgehen kann, wenn y einen Wert hat, für den die Reihe (16) divergiert, oder wie man die schon vorhandene Konvergenz verstärken kann.

Dazu kann man ein anderes Näherungsverfahren gebrauchen, das ich mit Vorteil in der Mondtheorie angewandt habe. Hierzu nehmen wir wieder die Gleichung (13):

$$y = x + a_2 y^2 + a_3 y^3 + \cdots \tag{13}$$

Man bestimme den ersten Näherungswert z aus der Gleichung:

$$z - a_2 z^2 = x$$

Da x klein von der Ordnung m und a_2 von der Ordnung der Einheit ist, so sind die Wurzeln reell, und man nehme diejenige als ersten Näherungswert, die für $x = 0$ verschwindet. Die analytische Annäherung kommt jetzt auf eine Reihenentwicklung nach Potenzen von z heraus; man hat:

$$y_1 = z + (m^3)$$
$$y_2 = z + a_3 z^3 + (m^4)$$
$$y_3 = z + a_3 z^3 + (2 a_2 a_3 + a_4) z^4 + (m^5)$$
$$\text{usw.}$$

Die numerische Annäherung lautet:

$$y_1 = z + (m^3)$$
$$y_2 = z - a_2 z^2 + a_2 y_1^2 + a_3 y_1^3 = z + a_3 z^3 + (m^4)$$
$$y_3 = z - a_2 z^2 + a_2 y_2^2 + a_3 y_2^3 + a_4 y_2^4 = z + 2 a_2 a_3 z^4 + a_2 a_3^2 z^6$$
$$\quad + a_3 (z + a_3 z^3)^3 + a_4 (z + a_3 z^3)^4 + (m^5)$$
$$\text{usw.}$$

Auch hier werden bei der letzteren bereits höhere Glieder berücksichtigt; die schließliche Reihe ist eine solche nach Potenzen von z; diese konvergiert sicher in einem bestimmten Bereich, etwa für $|z| \leq \varrho$. Für die Rechnung wichtig ist nun, für welche reellen Werte von y sowohl die Reihe nach Potenzen von x als auch die nach solchen von z konvergiert, und im besonderen, ob diese stärker konvergiert als jene. Der Beweis für das letztere dürfte nicht leicht zu führen sein; es scheint aber in den vorkommenden Fällen tatsächlich einzutreten.

Man kann das Verfahren erweitern und den ersten Näherungswert aus der Gleichung

$$z - a_2 z^2 - a_3 z^3 = x$$

oder aus

$$z - a_2 z^2 - a_3 z^3 - \cdots - a_n z^n = x$$

bestimmen und erhält dann y als eine Reihe, in der die zweiten bis $n$ten Potenzen von z ausfallen; man kann hier von einer Reihenumkehrung $n$ter Ordnung sprechen. Da ein allgemeiner Beweis nicht gelingt, so mag ein einfaches Beispiel das Verfahren erläutern. Es sei y aus der Gleichung:

$$y = x - \tfrac{1}{2}\,y^2 - \tfrac{1}{3!}\,y^3 - \tfrac{1}{4!}\,y^4 - \cdots \qquad (x = e^y - 1)$$

zu bestimmen, wobei bekannt ist, daß die Reihe rechter Hand in der ganzen y-Ebene konvergiert. Die Reihenumkehrungen verschiedener Ordnung lauten:

a) lineare Umkehrung:

$$y = x - \tfrac{1}{2}\,x^2 + \tfrac{1}{3}\,x^3 \mp \cdots \qquad\qquad |x| < 1$$

b) Umkehrung 2. Ordnung:

$$y = z - \tfrac{1}{6}\,z^3 + \tfrac{1}{8}\,z^4 - \tfrac{1}{20}\,z^5 + \tfrac{1}{56}\,z^7 - \tfrac{1}{64}\,z^8 \pm \cdots \qquad |z| < \sqrt{2} = 1{,}414$$

c) Umkehrung 3. Ordnung:

$$y = z - \tfrac{1}{24}\,z^4 + \tfrac{1}{30}\,z^5 - \tfrac{1}{72}\,z^6 + \tfrac{1}{252}\,z^7 - \tfrac{1}{576}\,z^8 \pm \cdots \qquad |z| < 1{,}596$$

d) Umkehrung 4. Ordnung:

$$y = z - \tfrac{1}{120}\,z^5 + \tfrac{1}{144}\,z^6 - \tfrac{1}{336}\,z^7 + \tfrac{1}{1152}\,z^8 \mp \cdots \qquad |z| < 1{,}945$$

e) Umkehrung 5. Ordnung:

$$y = z - \tfrac{1}{720}\,z^6 + \tfrac{1}{840}\,z^7 - \tfrac{1}{1920}\,z^8 \pm \cdots \qquad\qquad |z| < 2{,}181$$

Die reellen Konvergenzbereiche von x und y sind:

$$
\begin{aligned}
&\text{a)} & -\quad 1 &< x < +1 & -\,\infty\quad &< y < \log 2 = +0{,}693 \\
&\text{b)} & -\quad \tfrac{1}{2} &< x < +2{,}414 & -\,0{,}693 &< y < +1{,}228 \\
&\text{c)} & -\quad 1 &< x < +3{,}547 & -\,\infty\quad &< y < +1{,}515 \\
&\text{d)} & -\,0{,}730 &< x < +5{,}656 & -\,1{,}308 &< y < +1{,}896 \\
&\text{e)} & -\quad 1 &< x < +7{,}639 & -\,\infty\quad &< y < +2{,}156
\end{aligned}
$$

Die Einschränkung der unteren Grenze für die Konvergenz bei den Umkehrungen gerader Ordnung rührt daher, daß x als Funktion von z im Reellen ein Minimum hat. Es bleibt eine offene Frage, welche Eigenschaften die Koeffizienten a_n haben müssen, damit das Verfahren gelingt. Auf die interessanten funktionentheoretischen Verhältnisse dieser Transformationen kann hier nicht eingegangen werden.

Bemerkung zum dritten Keplerschen Gesetz.

Von **G. Herglotz**, Leipzig.

Für die Lagrangeschen Differentialgleichungssysteme:

$$\frac{d}{dt}\frac{\partial L}{\partial \dot{q}_i} - \frac{\partial L}{\partial q_i} = 0, \qquad i = 1, 2, \ldots, n \left.\right\} \quad (1)$$
$$L = L(\dot{q}_1 \ldots \dot{q}_n \mid q_1 \ldots q_n \mid t)$$

ist bekanntlich diese Tatsache grundlegend: Wird für eine Schar von Lösungen mit den Integrationskonstanten $c_1 \ldots c_k$:

$$q_i = q_i(t\, c_1 \ldots c_k) \tag{2}$$

gesetzt:

$$\sum_i p_i\, d q_i = d\Omega + H\, dt + \sum_j C_j\, d c_j \tag{3}$$

wobei:

$$p_i = \frac{\partial L}{\partial \dot{q}_i} \tag{4}$$

$$H = \sum_i p_i \dot{q}_i - L \tag{5}$$

$$\Omega = \int^t L\, dt \tag{6}$$

bedeuten soll, so sind die Größen C_j allein von den Constanten $c_1 \ldots c_k$, aber nicht von t abhängig.

Nicht nur kann auf dieselbe die Theorie der Lagrangeschen sowie der kanonischen Differentialgleichungssysteme gegründet werden, sie liefert auch in den Anwendungsgebieten dieser Systeme die gemeinsame und einfachste Quelle allgemeiner Sätze, wie z. B. der Cauchy-Helmholtzschen Wirbelsätze der Hydrodynamik oder der Kirchhoff-Abbeschen Abbildungssätze der geometrischen Optik. In dieser Richtung auf eine einfache, vielleicht nicht allgemein bemerkte Folgerung über die Perioden einer Schar periodischer Lösungen hinzuweisen, ist der Zweck dieser Zeilen.

Seien also unter Voraussetzung einer von t freien Lagrangeschen Funktion L die Lösungen der Schar Gleichung (2) periodisch mit den Perioden:

$$\tau = \tau(c_1 \ldots c_k) \tag{7}$$

und sei für sie die Konstante h des „Energieintegrales" $H = h$:

$$h = h(c_1 \ldots c_k) \tag{8}$$

so ergibt Gleichung (3) für die von t unabhängige Größe:

$$\int_t^{t+\tau} L\, dt = \omega(c_1 \ldots c_k) \tag{9}$$

in leicht erkennbarer Weise:

$$d\omega = -h\,d\tau \qquad (10)$$

Sei nun weiter h nicht etwa von $c_1 \ldots c_k$ überhaupt unabhängig, so folgt hieraus:

$$\left.\begin{array}{r}\omega + h\tau = \varphi(h) \\ \tau = \varphi'(h)\end{array}\right\} \qquad (11)$$

und damit zunächst die Tatsache, daß in der gedachten Schar von Lösungen die Periode τ allein durch die „Energiekonstante" h bestimmt wird.

Unter geeigneten besonderen Voraussetzungen über die Lagrangesche Funktion L kann nun aber auch die Art der Abhängigkeit der Periode τ von der Konstanten h näher bestimmt werden.

Zerfällt nämlich L, wie dies in der Mechanik zutrifft, in:

$$L = T(\dot{q}_1 \ldots \dot{q}_n) + U(q_1 \ldots q_n) \qquad (12)$$

und sind hier insbesondere T und U homogene Formen der $\dot{q}_i$ und der q_i bzw. von den Graden $\alpha \neq 0$ und $\beta \neq 0$, so wird:

$$H = (\alpha - 1)\,T - U = h \qquad (13)$$

$$\frac{d}{dt}\sum_i p_i q_i = \alpha T + \beta U \qquad (14)$$

woraus umgekehrt folgt[1]):

$$(\alpha\beta + \alpha - \beta)\,T = \frac{d}{dt}\Big[\ \sum_i p_i q_i + \beta h t\Big] \qquad (15\mathrm{a})$$

$$(\alpha\beta + \alpha - \beta)\,U = \frac{d}{dt}\Big[(\alpha - 1)\sum_i p_i q_i - \alpha h t\Big] \qquad (15\mathrm{b})$$

$$(\alpha\beta + \alpha - \beta)\,L = \frac{d}{dt}\Big[\ \alpha\sum_i p_i q_i \pm (\alpha - \beta) h t\Big] \qquad (15\mathrm{c})$$

und sich somit findet:

$$(\alpha\beta + \alpha - \beta)\,\omega = -(\alpha - \beta) h\tau \qquad (16)$$

Dies aber liefert mit Rücksicht auf Gleichung (10):

$$\alpha\beta h\,d\tau = (\alpha - \beta)\tau\,dh \qquad (17)$$

also unter der Voraussetzung $h \neq 0$ schließlich:

$$\tau = \mathrm{Const}\,|h|^{\frac{\alpha - \beta}{\alpha\beta}} \qquad (18)$$

wo nun die Konstante für alle Lösungen der Schar dieselbe ist.

Der Ausnahmefall periodischer Lösungen mit $h = 0$ tritt, da Gleichung (15a) zufolge:

$$(\alpha\beta + \alpha - \beta)\int_t^{t+\tau} T\,dt = \beta h\tau \qquad (19)$$

---

[1]) Über die mit diesen Gl. (15) in engster Beziehung stehende Integralinvariante vgl. Poincaré: Les méthodes nouvelles de la mécanique céleste, Bd. 3 Nr. 256.

wird, stets ein, falls $\alpha\beta + \alpha - \beta = 0$ ist, und wenn $\int\limits_t^{t+\tau} T\, dt \neq 0$, also

z. B. T eine definitive Form ist, auch nur dann ein.

Für die Bewegungsgleichungen in rechtwinkligen Koordinaten eines Systems von Massenpunkten, die sich umgekehrt proportional der λ^{ten} Potenz ($\lambda \neq 1$) der Entfernung anziehen, treffen offenbar die über L gemachten Voraussetzungen zu mit den Werten:

$$\alpha = 2, \quad \beta = 1 - \lambda \tag{20}$$

so daß hier für die Periode in einer Schar periodischer Bahnen ($h \neq 0$) gilt:

$$\tau = \text{Const} \,|\, h \,|^{\frac{1}{2}\frac{1+\lambda}{1-\lambda}} \tag{21}$$

Für das NEWTONsche Gesetz ($\lambda = 2$) und die Schar der elliptischen Bahnen des Zweikörperproblems geht diese Beziehung, wegen $h = -\dfrac{\mu}{2a}$, in das dritte KEPLERsche Gesetz über.

Der Ausnahmefall periodischer Bahnen mit $h = 0$ trifft stets und nur dann ein, falls $\alpha\beta + \alpha - \beta = 3 - \lambda = 0$ ist, d. i. bei einer Anziehung umgekehrt proportional dem Kubus der Entfernung[1]).

In dem oben ausgeschlossenen Falle $\lambda = 1$ endlich tritt an die Stelle der Homogeneitätseigenschaft von $U(q_1 \ldots q_n)$ die andere, daß:

$$\sum_i \frac{\partial U}{\partial q_i}\, q_i = c \tag{22}$$

von $q_1 \ldots q_n$ unabhängig ist, wonach sich Gleichung (14) abändert in:

$$\frac{d}{dt}\sum_i p_i q_i = \alpha T + c \tag{23}$$

daher in der früheren Weise, und zwar bei beliebig gelassenem $\alpha \neq 0$ folgt:

$$\omega = -(h + c)\tau \tag{24}$$

Ist also $c \neq 0$, so wird gelten:

$$\tau = \text{Const}\, e^{-\frac{h}{c}} \tag{25}$$

während sich bei $c = 0$ für eine Schar periodischer Lösungen $h = \text{const}$ aber keine Aussage über τ ergibt.

____

[1]) Daß bei diesem Anziehungsgesetz für periodische Bahnen $h = 0$ sein muß, bemerkt bereits JACOBI auf Grund seiner Relation:

$$\frac{d^2}{dt^2} T(q_1 \ldots q_n) + (\lambda - 3)\, U(q_1 \ldots q_n) = 2h$$

die auch sofort aus den Gl. (15) folgt. (JACOBI: Vorlesungen über Dynamik. 4. Vorlesung.)

Untersuchungen über die Figur der Himmelskörper.

Fünfte Abhandlung.

Neue Beiträge zur Maxwellschen Theorie der Saturnringe[1]).

Von **Leon Lichtenstein**, Leipzig.

Mit 5 Abbildungen.

Die Frage nach der Konstitution der Saturnringe, die jahrhundertelang die Astronomen beschäftigte, ist seit den berühmten theoretisch-photometrischen Untersuchungen von Herrn VON SEELIGER in Verbindung mit den sorgfältigen Beobachtungen von Herrn MÜLLER in Potsdam im wesentlichen als gelöst zu betrachten[2]). Wir wissen heute, daß die Saturnringe, im Einklang mit der Annahme von CASSINI, der sich später MAXWELL anschloß, aus einer Anhäufung unzählig vieler Meteoriten, dem kosmischen Staub u. dgl. bestehen. Die Ringe bilden ein „vollkommen inkohärentes" dynamisches System, d. h. ein System von Teilchen, die durch Gravitationskräfte aneinander sowie an den Zentralkörper gebunden, jedoch keinerlei Druck- oder Zugspannungen unterworfen sind.

Wenn wir somit über die physikalische Beschaffenheit der Ringe einigermaßen unterrichtet sind, so bedeutet dies keinesfalls, daß das Saturnsystem für uns kein Problem mehr birgt. Ganz im Gegenteil, gilt es doch jetzt die bekannte Konstitution der Ringe in Einklang mit der GALILEI-NEWTONschen Dynamik zu bringen, sie zu „erklären". Als erster hat sich MAXWELL dieser Aufgabe unterzogen. In seinen bekannten Untersuchungen betrachtet MAXWELL das folgende dynamische

---

[1]) Vgl. L. LICHTENSTEIN: Untersuchungen über die Figur der Himmelskörper. Erste Abhandlung. Math. Zeitschr. Bd. 10, S. 130—159. 1921. Zweite Abhandlung, ebda. Bd. 12, S. 201—218. 1922. Dritte Abhandlung, a. a. O. Bd. 13, S. 82—118. 1922. Vierte Abhandlung, a. a. O. Bd. 17, S. 62—110. 1923. Zum Verständnis der vorliegenden Arbeit ist die Kenntnis der vier vorgenannten Aufsätze nicht erforderlich.

[2]) Vgl. namentlich H. SEELIGER: Zur Theorie der Beleuchtung der großen Planeten, insbesondere des Saturn. Abh. der k. bayer. Ak. II Cl. Bd. 16, II Abt., S. 38—114. 1887; Theorie der Beleuchtung staubförmiger kosmischer Massen, insbesondere des Saturnringes, ebda. Bd. 1, Abt. I, S. 3—72. 1893.

Modell[1]). Auf einem Kreise um den Mittelpunkt des Saturn als Zentrum ist eine endliche Anzahl unter sich gleicher, äquidistanter punktförmiger Massen angebracht. Das Ganze rotiert gleichförmig um den Zentralkörper. Maxwell findet, indem er bei seinen Rechnungen nur Glieder niedrigster Ordnung berücksichtigt, daß periodische, „stabile" Zustände nicht mehr möglich sind, sobald die Gesamtmasse des Ringes eine Schranke von der Form $c\,\dfrac{M}{p^2}$ übersteigt, unter M die Masse des Saturn, c eine Konstante, p die Anzahl der Satelliten verstanden. Offenbar konvergiert die zulässige Masse des Ringes für $p \to \infty$ gegen Null. Die Maxwellschen Ergebnisse werden illusorisch. Es erscheint darum als notwendig, der Betrachtung ein Modell zugrunde zu legen, das sich der Wirklichkeit näher anschließt. Ein erster dahingehender Versuch ist vom Verfasser in einer vor einem Jahre veröffentlichten größeren Arbeit unternommen worden[2]). Dort handelt es sich um ein ebenes Problem, das wie folgt gefaßt werden kann.

In dem Ursprung eines ebenen kartesischen Koordinatensystems $\mathfrak{x}$, $\mathfrak{y}$ befindet sich eine punktförmige Masse M. Auf der Kreislinie $\mathfrak{x}^2 + \mathfrak{y}^2 = R^2$ ist eine Massenbelegung konstanter Dichte μ für die Längeneinheit verteilt. Die einzelnen Teilchen können sich, unbehindert durch die Nachbarteilchen, nach allen Richtungen frei bewegen und üben keinerlei Spannkräfte aufeinander aus.

Die einzelnen Massen sind durch Gravitationskräfte aneinander geknüpft. Die zwischen zwei in einer Entfernung r_{12} befindlichen Massen m_1 und m_2 wirkende Kraft soll den Wert $-\varkappa\,\dfrac{m_1\,m_2}{r_{12}}$, der auf das logarithmische Potential führt, haben. Ein statisches Gleichgewicht kann natürlich nicht bestehen. Nimmt man aber an, daß das ganze System mit der Winkelgeschwindigkeit $\dfrac{\varkappa^{\frac{1}{2}}}{R}\,(M + \pi\,\mu\,R)^{\frac{1}{2}}$ um den Koordinatenursprung rotiert, so wird ein dynamisches Gleichgewicht wegen der hinzutretenden Zentrifugalkräfte möglich. Dieser Bewegungszustand wird gestört, sobald dem System eine weitere punktförmige Masse $\mathfrak{M}$ hinzugefügt wird, die um den Körper M, der festgehalten wird, in der Ebene $\mathfrak{x}$, $\mathfrak{y}$ gleichförmig rotiert. Die Entfernung der Massen $\mathfrak{M}$ und M heiße $\mathfrak{R}$. Wie a. a. O. gezeigt wurde, ist für hinreichend kleine Werte

---

1) Vgl. The scientific papers of James Clerk Maxwell, Vol. 1, On the stability of the motion of Saturns Rings, S. 288—377.

2) Vgl. L. Lichtenstein: Untersuchungen über die Figur der Himmelskörper. Vierte Abhandlung. Zur Maxwellschen Theorie der Saturnringe. Math. Zeitschr. Bd. 17, S. 62—110. 1923. Siehe auch meine Schrift: Astronomie und Mathematik in ihrer Wechselwirkung. Mathematische Probleme in der Theorie der Figur der Himmelskörper, Leipzig 1923, woselbst auf S. 81—89 die Integro-Differentialgleichungen des Problems abgeleitet werden.

von $\dfrac{R}{\mathfrak{R}}$ und μ ein periodischer Bewegungszustand des Systems möglich. Des weiteren gibt es bei hinreichend kleinem μ periodische Bewegungszustände, die sich als fortschreitende Wellen charakterisieren lassen. Die Dauer einer Periode hängt von dem maximalen Ausschlag aus der Gleichgewichtslage ab, — diese Abhängigkeit tritt freilich erst in den Gliedern höherer Ordnung zum Vorschein.

Die vorstehenden Resultate sind durch völlig strenge mathematische Deduktionen gewonnen worden. Darüber hinaus sind zur Gewinnung von Stabilitätskriterien unendlich kleine Schwingungen bei nicht notwendig kleiner Dichte μ untersucht worden. Das Resultat, das, wie die Methode der kleinen Schwingungen überhaupt, nur einen bedingten Wert beanspruchen kann, lautet wie folgt. Sei M^r die Gesamtmasse des Ringes. Einer jeden natürlichen Zahl $m < m_0 = \dfrac{4\,M}{M^r} + 2$ entsprechen vier „stabile" Bewegungszustände des Ringes, die in zwei verschiedene Gruppen zerfallen und als fortschreitende Wellen aufgefaßt werden können. Für kleine m ist die Dauer einer vollen Schwingung der ersten Gruppe mit der Umlaufsdauer des Ringes vergleichbar, die der zweiten Gruppe ist wesentlich länger. Die Schwingungen der zweiten Gruppe sind auch für $m \geqq m_0$ „stabil". Nach H. Struve ist das Verhältnis der Masse des Saturn zur Gesamtmasse seiner Ringe größer als 100 000. Nehmen wir in dem hier betrachteten „ebenen" Modell $M : M^r = 100\,000$ an, so finden wir, daß die ersten $4 \cdot 10^5$ Schwingungen der ersten Gruppe „stabil" sind. Der Ring ist darum gewiß „als praktisch stabil" zu betrachten.

Es sei aber noch einmal darauf hingewiesen, daß den Ergebnissen der Methode kleiner Schwingungen grundsätzlich nur ein bedingter Wert beigemessen werden kann.

Diese Untersuchungen werden in der vorliegenden Arbeit einen Schritt weiter geführt. Wir legen unseren Betrachtungen ein Modell zugrunde, das insofern über das im vorstehenden besprochene hinausgeht, als die um M kreisende Masse diesmal über eine Anzahl koaxialer Rotationskörper rechteckigen Querschnitts verteilt ist und als das Attraktionsgesetz das Newtonsche angenommen wird. Die genaue Beschreibung des Modells findet sich in § 1 des ersten Kapitels.

Von dem Einfluß der (sehr kleinen) Neigung der Bahnen einzelner Teilchen gegen die Mittelebene der Ringe wie auch von den Zusammenstößen der Teilchen wird abgesehen. Da die Geschwindigkeitsdifferenzen nur gering sind, so ist nach einer Überschlagsrechnung von Maxwell (siehe weiter unten) die Bedeutung der Zusammenstöße sicher ganz geringfügig. *Es gelingt die Existenz eines permanenten Bewegungszustandes bei Vorhandensein einer störenden um den Zentralkörper rotierenden Masse streng zu beweisen, sofern angenommen wird, daß der Abstand des störenden Satel-*

liten (oder aber seine Masse) und die Dichte der Ringe, μ_0, klein sind. Mathematisch handelt es sich auch diesmal um den Existenzbeweis periodischer Lösungen eines Systems von zwei nicht linearen Integro-Differentialgleichungen[1]).

Für die Frage nach der Einordnung des mechanischen Verhaltens des Saturnsystems in die GALILEI-NEWTONsche Dynamik ist hiermit nur ein erster einleitender Schritt getan. Was eine Weiterführung dieser Untersuchungen betrifft, so sei vor allem bemerkt, daß es möglich sein dürfte, sich von der Einschränkung, die Dichte μ_0 sei klein, freizumachen. Des weiteren ist es leicht möglich, die Neigung der Bahn des störenden Körpers gegen die Mittelebene der Ringe zu berücksichtigen. Schwieriger erscheint demgegenüber die Aufgabe, der Neigung der Bahn verschiedener Teilchen gegen die Mittelebene Rechnung zu tragen, ohne den Boden der klassischen Dynamik zu verlassen. Da die Bahnen einzelner Teilchen im wesentlichen geordnet sind, so dürften freilich statistische Methoden bei diesen Betrachtungen wohl nur eine geringe Rolle spielen.

Bei dem vorstehenden handelte es sich um die Existenz eines permanenten Bewegungszustandes bei Vorhandensein eines störenden Körpers. Für die Kernfrage der Theorie, das Stabilitätsproblem, ist damit an sich noch nicht viel gewonnen. Die Untersuchung der Stabilität des hier betrachteten vollkommeneren Modells ist wesentlich schwieriger als diejenige des in der ersten Arbeit behandelten einfachen Systems, da die Methode der kleinen Schwingungen jetzt gar nicht oder zumindest nicht mehr in der üblichen Form angewendet werden kann. Von Wichtigkeit wird sich dabei jedenfalls die Behandlung des Problems bei beliebigem μ_0 erweisen.

Es erscheint erwünscht, einmal die Bewegung der Saturnringe für lange Zeiträume mit den Mitteln der klassischen Störungstheorie, jedoch unter Abschätzung der höheren, sonst vernachlässigten Glieder zu bestimmen[2]). Vielleicht würde sich dabei mancher Aufschluß betreffend die tatsächlich vorhandene „praktische Stabilität" der Ringe ergeben.

---

[1]) Die Bewegungsgleichungen des Problems sind von mir bereits in der loc. cit. [2]) S. 201 an zweiter Stelle genannten Schrift angegeben worden. Der Existenzbeweis ist neu.

[2]) Auf solche Abschätzungen in der Theorie der Planeten ist neuerdings von Herrn SUNDMAN hingewiesen worden. Vgl. K. F. SUNDMAN: Über die Richtungslinien für fortgesetzte Untersuchungen in den Komet- und Trabanttheorien. Wissenschaftliche Vorträge, gehalten auf dem V. Kongreß der Skandinavischen Mathematiker in Helsingfors vom 4. bis 7. Juli 1922, Helsingfors 1923, S. 48—56.

Erstes Kapitel.

§ 1. Problemstellung.

In dem Ursprung eines kartesischen Koordinatensystems $\mathfrak{x}$, $\mathfrak{y}$, $\mathfrak{z}$ möge sich eine punktförmige Masse M befinden. Über die Kreisringflächen

$$\mathfrak{z} = 0 , \quad R_{k0}^2 \leqq \mathfrak{x}^2 + \mathfrak{y}^2 \leqq (R_k^0)^2 , \quad (k = 1 , \ldots , m) \atop R_{10} < R_1^0 < R_{20} < \cdots \quad \left.\right\} \tag{1}$$

sei ferner eine Massenbelegung konstanter Dichte μ_0 verteilt (Abb. 1).

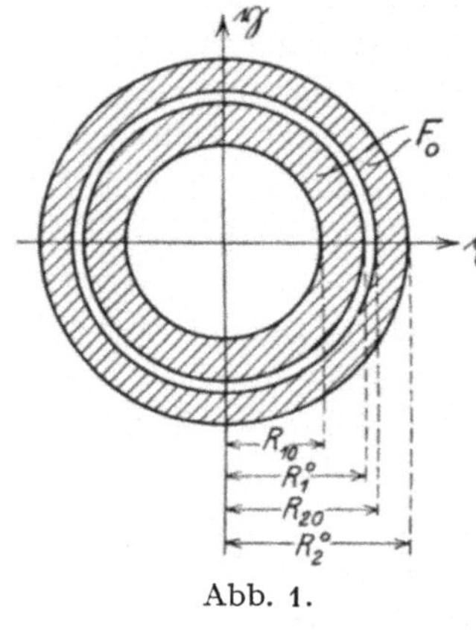

Abb. 1.

Die Gesamtmasse der Ringe ist also

$$\pi \mu_0 \sum_1^m \left\{ (R_k^0)^2 - R_{k0}^2 \right\} .$$

Die einzelnen Teilchen können sich, unbehindert durch die Nachbarteilchen, nach allen Richtungen hin frei bewegen und üben keinerlei Spannkräfte aufeinander aus.

Betrachten wir irgendeine Verschiebung der mit Masse belegten Flächenstücke, bei der der Punkt $P_0(\mathfrak{x}, \mathfrak{y}, 0)$ in $\dot{P}_0(\mathfrak{x} + \triangle \mathfrak{x}, \mathfrak{y} + \triangle \mathfrak{y}, 0)$ übergeht, unter $\triangle \mathfrak{x}$, $\triangle \mathfrak{y}$ Funktionen verstanden, die stetige Ableitungen erster Ordnung haben. Geht dabei ein P_0 enthaltendes Flächenelement $d\mathfrak{f}_0$ in $d\dot{\mathfrak{f}}$ über, so wird die Dichte $\dot{\mu}$ der Masse in $\dot{P}_0$ der Gleichung

$$\dot{\mu} \, d\dot{\mathfrak{f}} = \mu_0 \, d\mathfrak{f}_0 \tag{2}$$

gemäß festgesetzt. Die Dichte $\dot{\mu}$ ist, als Funktion von $\mathfrak{x}$ und $\mathfrak{y}$ aufgefaßt, stetig.

Bei der Anwendung auf das astronomische Problem der Saturnringe hat man sich vorzustellen, daß die einzelnen Massenteilchen einander nicht berühren, und μ_0 die mittlere Dichte der Belegung darstellt. Die Beziehung (2) ist die Kontinuitätsgleichung.

Die einzelnen Massenteilchen der Ringe sind durch Gravitationskräfte aneinander geknüpft. Die zwischen zwei in den Punkten Q_1, Q_2 in einer Entfernung r_{12} befindlichen Massenelementen dm_1 und dm_2 wirkende Kraft soll den Wert

$$- \varkappa \frac{dm_1 \, dm_2}{r_{12} \sqrt{r_{12}^2 + h^2}} , \tag{3}$$

entsprechend einem Potential

$$\frac{\varkappa}{h} \, dm_1 \, dm_2 \log \frac{\sqrt{r_{12}^2 + h^2} + h}{r_{12}} , \tag{4}$$

haben. Auf dieses Anziehungsgesetz wird man durch folgende Betrachtungen geführt.

Die Masse der Saturnringe, die wir uns mit CASSINI und MAXWELL als aus unzähligen Einzelkörpern — kleinen Satelliten, Meteorsteinen oder kosmischem Staub — bestehend vorstellen, ist tatsächlich über eine Anzahl dreidimensionaler Ringkörper vom rechteckigen Querschnitt

$$R_{k0}^2 \leqq \mathfrak{x}^2 + \mathfrak{y}^2 \leqq (R_k^0)^2, \quad -h \leqq \mathfrak{z} \leqq h \tag{5}$$

verteilt. Die Ebene $\mathfrak{x}$-$\mathfrak{y}$ ist die Mittelebene aller Ringkörper. Bewegt sich ein Teilchen der in den Ringen verteilten Materie unter der Einwirkung der Gravitationskraft des Saturn allein, so beschreibt es eine Ellipse, deren Ebene durch den Mittelpunkt des Planeten hindurchgeht. Die Bahnebenen verschiedener Teilchen schneiden einander; darum sind Zusammenstöße unvermeidlich, sofern, wie wir vorausgesetzt haben, Spannkräfte nicht vorhanden sind. Die Wirkung der Zusammenstöße ist freilich nur geringfügig, da die einzelnen Teilchen einander unter einem sehr spitzen Winkel treffen und die Geschwindigkeitsdifferenzen auch nur ganz klein sind. Nach einer von MAXWELL durchgeführten überschläglichen Berechnung dürfte der Verlust an kinetischer Energie durch Zusammenstöße (innere Reibung) im Jahre nicht mehr als den $\frac{1}{40} \cdot 10^{-12}$-ten Teil der Gesamtenergie betragen[1]). In dem folgenden soll darum von den Zusammenstößen ganz abgesehen und angenommen werden, daß alle Massenteilchen, die sich in denselben Punkt der Ebene $\mathfrak{x}$-$\mathfrak{y}$ hineinprojizieren, allemal die gleichen Bahnen in parallelen Ebenen beschreiben. Da nun die Anziehungskraft, die das den Punkt Q_2 enthaltende Elementarstäbchen von der Höhe $2\,h$ und der Gesamtmasse dm_2 auf die Masseneinheit im Punkte Q_1 ausübt, wie man leicht berechnet, den Wert

$$-\frac{\varkappa\, dm_2}{r_{12} \sqrt{r_{12}^2 + h^2}}$$

hat, so kommt man zu dem eingangs angegebenen mechanischen Modell. Die Kraft, mit der dieselbe Masseneinheit von dem Zentralkörper angezogen wird, ist natürlich die NEWTONsche.

Ein statisches Gleichgewicht des Systems ist selbstverständlich nicht denkbar. Ebensowenig kann das System wie ein starrer Körper rotieren. Nimmt man indessen an, daß die Teilchen, die sich auf der Peripherie eines Kreises vom Radius R um den Koordinatenursprung $(R_{k0} \leqq R \leqq R_k^0, \ k = 1, \ldots, m)$ befinden, mit der Winkelgeschwindigkeit ω,

$$\omega^2 = \frac{\varkappa}{R^3} [M + \mu_0 R^2 J_0(R)], \tag{6}$$

---

[1]) Vgl. The scientific papers of JAMES CLERK MAXWELL, Vol. 1, On the stability of the motion of Saturns Rings, S. 288—377, insbes. S. 354—356.

unter $J_0(R)$ eine gewisse angebbare Funktion verstanden, um den Zentralkörper rotieren, so wird das System sich im dynamischen Gleichgewicht befinden[1]). Die Winkelgeschwindigkeit nimmt mit wachsender Entfernung von dem Zentralkörper monoton ab. Diese Bewegung wird gestört, sobald dem System eine weitere punktförmige Masse $\mathfrak{M}$ hinzugefügt wird. Wir nehmen an, daß $\mathfrak{M}$ um den Körper M, der festgehalten wird, in der Ebene $\mathfrak{z} = 0$ gleichförmig rotiert. Die Entfernung der Masse $\mathfrak{M}$ von dem Koordinatenursprung heißt $\mathfrak{R}$, ihre Winkelgeschwindigkeit sei β. Wie im zweiten Kapitel gezeigt wird, *ist für hinreichend kleine Werte von* $\dfrac{\mathfrak{M}R_m^0}{\mathfrak{R}}$ *und* μ_0 *ein permanenter Bewegungszustand des Systems möglich. Die Konfiguration des Ganzen bleibt unverändert, wenn auch die relative Lage eines bestimmten Teilchens des Ringes gegen* $\mathfrak{M}$ *sich fortwährend ändert.*

§ 2. Die Integro-Differentialgleichungen der Bewegung.

Wir gehen von dem dynamischen Gleichgewichtszustande bei Abwesenheit eines störenden Körpers aus und beziehen die Lage der Teilchen, die den Kreis

$$\mathfrak{z} = 0\,, \quad \mathfrak{x}^2 + \mathfrak{y}^2 = R^2\,, \quad R_{k0} \leqq R \leqq R_k^0 \quad (k = 1,\ldots,m) \tag{7}$$

erfüllen, auf ein kartesisches Koordinatensystem x, y, z, dessen Ursprung mit dem Ursprung des Systems $\mathfrak{x}$, $\mathfrak{y}$, $\mathfrak{z}$, dessen x-y-Ebene mit der Ebene $\mathfrak{z} = 0$ zusammenfällt, und das um die z-Achse mit der Winkelgeschwindigkeit ω gleichförmig rotiert. Wir nehmen an, daß zur Zeit $t = 0$ alle die unendlich vielen rotierenden Achsenkreuze zusammenfallen.

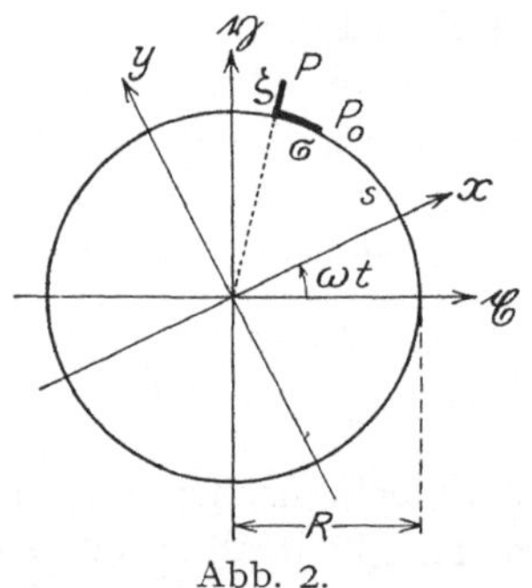

Abb. 2.

Die Lage eines bestimmten Massenteilchens P_0 des Kreises (7) ist durch seine Polarkoordinaten R, $\dfrac{s}{R}$ in bezug auf das rotierende Achsenkreuz vollkommen bestimmt. Hier ist s die von dem Punkte $(R, 0, 0)$ bis zum gerade betrachteten Punkte gerechnete Bogenlänge des Kreises $x^2 + y^2 = R^2$, $z = 0$. Tritt eine störende Kraft hinzu, so wird das Teilchen P_0 im Laufe der Zeit seine Lage gegen die rotierenden Achsen, allgemein zu reden, ändern. Seine Polarkoordinaten werden die Werte

$$R + \zeta\,, \quad \frac{s + \sigma}{R} \tag{8}$$

annehmen, unter ζ und σ gewisse Funktionen des Halbmessers R, des Polarwinkels $\dfrac{s}{R}$ und der Zeit t verstanden (Abb. 2).

---

[1]) Vergl. die Fußnote [2]) auf S. 212.

Im Laufe der weiteren Betrachtungen wird es sich stets um kleine Bewegungen handeln. Wir wollen ein für allemal $|\zeta| < R$ voraussetzen.

Es gilt jetzt

$$\mathfrak{x} = (R + \zeta) \cos\left(\omega t + \frac{s + \sigma}{R}\right), \qquad \mathfrak{y} = (R + \zeta) \sin\left(\omega t + \frac{s + \sigma}{R}\right), \qquad (9)$$

mithin

$$\left.\begin{aligned}
\frac{\partial \mathfrak{x}}{\partial t} &= -(R + \zeta)\sin\left(\omega t + \frac{s + \sigma}{R}\right)\left(\omega + \frac{1}{R}\frac{\partial \sigma}{\partial t}\right) + \frac{\partial \zeta}{\partial t}\cos\left(\omega t + \frac{s + \sigma}{R}\right), \\
\frac{\partial \mathfrak{y}}{\partial t} &= (R + \zeta)\cos\left(\omega t + \frac{s + \sigma}{R}\right)\left(\omega + \frac{1}{R}\frac{\partial \sigma}{\partial t}\right) + \frac{\partial \zeta}{\partial t}\sin\left(\omega t + \frac{s + \sigma}{R}\right),
\end{aligned}\right\} \quad (10)$$

so daß die kinetische Energie $\mathbf{T}$ den Wert

$$\mathbf{T} = \frac{1}{2}\left\{\left(\frac{\partial \mathfrak{x}}{\partial t}\right)^2 + \left(\frac{\partial \mathfrak{y}}{\partial t}\right)^2\right\} = \frac{1}{2}(R + \zeta)^2\left(\omega + \frac{1}{R}\frac{\partial \sigma}{\partial t}\right)^2 + \frac{1}{2}\left(\frac{\partial \zeta}{\partial t}\right)^2 \quad (11)$$

hat. Wir finden weiter leicht die Formeln

$$\left.\begin{aligned}
\frac{d}{dt}\left(\frac{\partial \mathbf{T}}{\partial\left(\frac{\partial \zeta}{\partial t}\right)}\right) - \frac{\partial \mathbf{T}}{\partial \zeta} &= \frac{\partial^2 \zeta}{\partial t^2} - (R + \zeta)\left(\omega + \frac{1}{R}\frac{\partial \sigma}{\partial t}\right)^2, \\
\frac{d}{dt}\left(\frac{\partial \mathbf{T}}{\partial\left(\frac{\partial \sigma}{\partial t}\right)}\right) - \frac{\partial \mathbf{T}}{\partial \sigma} &= (R + \zeta)^2\frac{1}{R^2}\frac{\partial^2 \sigma}{\partial t^2} + 2(R + \zeta)\left(\omega + \frac{1}{R}\frac{\partial \sigma}{\partial t}\right)\frac{1}{R}\frac{\partial \zeta}{\partial t}.
\end{aligned}\right\} \quad (12)$$

Die LAGRANGEschen Bewegungsgleichungen des Massenpunktes R, $\dfrac{s}{R}$ sind demnach

$$\left.\begin{aligned}
\frac{\partial^2 \zeta}{\partial t^2} - (R + \zeta)\left(\omega + \frac{1}{R}\frac{\partial \sigma}{\partial t}\right)^2 &= Q_\zeta, \\
(R + \zeta)^2\frac{1}{R^2}\frac{\partial^2 \sigma}{\partial t^2} + 2(R + \zeta)\left(\omega + \frac{1}{R}\frac{\partial \sigma}{\partial t}\right)\frac{1}{R}\frac{\partial \zeta}{\partial t} &= Q_\sigma.
\end{aligned}\right\} \quad (13)$$

Hier bezeichnen $Q_\zeta\,\delta\zeta$ und $Q_\sigma\,\delta\sigma$ die virtuellen Arbeiten, welche die auf eine Masseneinheit in $\left(R + \zeta, \dfrac{s + \sigma}{R}\right)$ wirkenden Gravitationskräfte bei ihrem Übergang in den Punkt $\left(R + \zeta + \delta\zeta, \dfrac{s + \sigma}{R}\right)$ bzw. $\left(R + \zeta, \dfrac{s + \sigma + \delta\sigma}{R}\right)$ leisten würden. Betrachten wir die Gesamtheit der bewegten Massenpunkte im Augenblick t. Die Gesamtheit der von ihnen bedeckten Flächenstücke in der Ebene x-y heiße F_*, die Dichte in einem Flächenelement $d\mathfrak{f}_*$ sei μ_* (Abb. 3). Die von der Eigengravitation der Ringe herrührenden Beiträge zu Q_ζ und Q_σ sind

$$\left.\begin{aligned}
&\frac{\varkappa}{h}\int_{F_*}\mu'_*\,d\mathfrak{f}'_*\,\frac{\partial}{\partial \nu}\log\frac{\sqrt{\varrho^2 + h^2} + h}{\varrho} \\[2ex]
\text{und}\quad &\frac{\varkappa}{h}\left(1 + \frac{\zeta}{R}\right)\int_{F_*}\mu'_*\,d\mathfrak{f}'_*\,\frac{\partial}{\partial \tau}\log\frac{\sqrt{\varrho^2 + h^2} + h}{\varrho}.
\end{aligned}\right\} \quad (14)$$

Hier bezeichnet ϱ den Abstand des betrachteten Punktes $\left(R + \zeta,\ \dfrac{s + \sigma}{R}\right)$ von dem Flächenelemente $d\mathfrak{f}'_*$. Ferner bezeichnen die Symbole $\dfrac{\partial}{\partial \nu}$ und $\dfrac{\partial}{\partial \tau}$ die Ableitungen in der Richtung der äußeren Normale an den Kreis (7) und in der darauf senkrechten Richtung in $\left(R + \zeta,\ \dfrac{s + \sigma}{R}\right)$. Ist die Entfernung des störenden Körpers $\mathfrak{M}$ von dem Punkte $\left(R + \zeta,\ \dfrac{s + \sigma}{R}\right)$ gleich P, so sind die von seiner Anziehung herrührenden Beiträge zu Q_ζ und Q_σ entsprechend gleich

$$\left.\begin{array}{l} \varkappa \mathfrak{M}\, \dfrac{\partial}{\partial \nu}\dfrac{1}{P} \\[2ex] \text{und} \\[2ex] \varkappa \mathfrak{M}\left(1 + \dfrac{\zeta}{R}\right)\dfrac{\partial}{\partial \tau}\dfrac{1}{P}\,. \end{array}\right\} \tag{15}$$

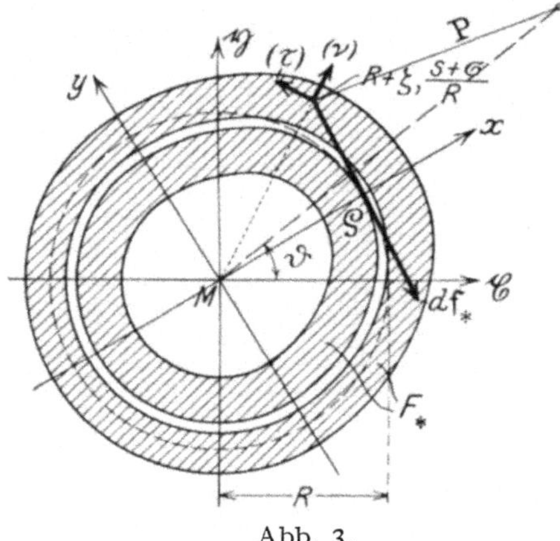

Die Anziehung des Zentralkörpers gibt nur zu Q_ζ einen Beitrag. Dieser hat den Wert

$$-\frac{M\varkappa}{(R + \zeta)^2}\,. \tag{16}$$

Die Lagrangeschen Bewegungsgleichungen lauten demnach

$$\frac{\partial^2 \zeta}{\partial t^2} - (R + \zeta)\left(\omega + \frac{1}{R}\frac{\partial \sigma}{\partial t}\right)^2 = -\frac{M\varkappa}{(R + \zeta)^2}$$

$$+ \frac{\varkappa}{h}\int\limits_{F_*}\mu'_*\,d\mathfrak{f}'_*\,\frac{\partial}{\partial \nu}\log\frac{\sqrt{\varrho^2 + h^2} + h}{\varrho} + \varkappa \mathfrak{M}\,\frac{\partial}{\partial \nu}\frac{1}{P}\,,$$

$$(R + \zeta)^2\frac{1}{R^2}\frac{\partial^2 \sigma}{\partial t^2} + 2(R + \zeta)\left(\omega + \frac{1}{R}\frac{\partial \sigma}{\partial t}\right)\frac{1}{R}\frac{\partial \zeta}{\partial t}$$

$$= \frac{\varkappa}{h}\left(1 + \frac{\zeta}{R}\right)\int\limits_{F_*}\mu'_*\,d\mathfrak{f}'_*\,\frac{\partial}{\partial \tau}\log\frac{\sqrt{\varrho^2 + h^2} + h}{\varrho} + \varkappa\left(1 + \frac{\zeta}{R}\right)\mathfrak{M}\,\frac{\partial}{\partial \tau}\frac{1}{P}\,. \tag{17}$$

Sei ϑ der Polarwinkel des Körpers $\mathfrak{M}$ in der festen $\mathfrak{x}$-$\mathfrak{y}$-Ebene zur Zeit t. Offenbar ist

$$\vartheta = \vartheta_0 + \beta t\,. \qquad (\vartheta_0 \text{ konstant.}) \tag{18}$$

Die Winkelgeschwindigkeit der relativen Bewegung von $\mathfrak{M}$ gegen das zu dem Kreise $\mathfrak{x}^2 + \mathfrak{y}^2 = R^2$, $\mathfrak{z} = 0$ gehörige rotierende Achsenkreuz x, y, z ist gleich $\beta - \omega = \gamma$. Offenbar ist γ für verschiedene Achsenkreuze verschieden.

Wir suchen jetzt Lösungen der Integro-Differentialgleichungen (17) von der Form

$$\zeta = \zeta(s, t; R) = Z(s - R\gamma t) = Z(u),$$
$$\sigma = \sigma(s, t; R) = S(s - R\gamma t) = S(u) \qquad (u = s - R\gamma t) \qquad (19)$$

zu bestimmen, unter $Z(u)$ und $S(u)$ periodische Funktionen mit der Periode $2\pi R$ verstanden. Sie werden im allgemeinen auch von R abhängen.

Sei Γ der geometrische Ort der Punkte $R + Z(s)$, $\dfrac{s + S(s)}{R}$ in der Ebene x-y. Auf der Kurve Γ liegen zur Zeit $t = 0$ alle diejenigen Teilchen, die sich im relativen Gleichgewicht auf dem Kreise (7) befanden. Die Lage aller dieser Teilchen, bezogen auf das rotierende Achsenkreuz zur Zeit t, erhält man wegen (19), wenn man Γ in der positiven Richtung durch den Winkel γt dreht. Bezogen auf das feste Koordinatensystem ist der Drehungswinkel gleich

$$\gamma t + \omega t = \beta t, \qquad (20)$$

mithin für alle Kreise (7) gleich groß, und zwar gleich dem von dem störenden Körper beschriebenen Winkel. *Die Gesamtkonfiguration des Systems ändert sich nicht mit der Zeit.* Das Ganze rotiert, vom festen Koordinatensystem aus beurteilt, mit der Winkelgeschwindigkeit β um die z-Achse. Irgendeine bestimmte Lage innerhalb dieser Konfiguration nehmen freilich im Laufe der Zeit immer wieder andere Massenteilchen ein. Das auf den Punkt R, $\dfrac{s}{R}$ des Kreises (7) bezogene Massenteilchen hat zur Zeit t in der Ebene x-y die Polarkoordinaten

$$R + Z(s - R\gamma t) = R + Z(u), \qquad \frac{s + S(s - R\gamma t)}{R} = \frac{s + S(u)}{R}. \qquad (21)$$

Seine Bahn in der x-y-Ebene erhält man, wenn man in (21) s konstant hält. Sie ist eine geschlossene Kurve, die in der Zeit $\dfrac{2\pi}{\gamma}$ einmal voll beschrieben wird.

Bevor wir die speziellen Funktionen (19) in (17) einsetzen, wollen wir den geometrischen Ort der Punkte

$$R + Z(u), \qquad \frac{u + S(u)}{R} \qquad (22)$$

in der festen Koordinatenebene $\mathfrak{x}$, $\mathfrak{y}$ einführen (Abb. 4). Dies ist, da zur Zeit $t = 0$ alle rotierenden x, y, z-Systeme und das System $\mathfrak{x}, \mathfrak{y}, \mathfrak{z}$ zusammenfallen, die vorstehend mit Γ bezeichnete Kurve. Die Gesamtheit der Kurven Γ, entsprechend allen R in (1), heiße F. Offenbar gibt F die Lage des Gebietes F_* zur Zeit $t = 0$, auf das feste Koordinatensystem $\mathfrak{x}, \mathfrak{y}, \mathfrak{z}$ bezogen, wieder. Sei $\mathfrak{M}_0$ der Punkt mit den Polarkoordinaten $\mathfrak{R}$, ϑ_0.

Betrachten wir jetzt wieder das Teilchen, das auf den Punkt $R, \dfrac{s}{R}$ des Kreises (7) bezogen ist. Es gilt zunächst

$$\frac{\partial \zeta}{\partial t} = - R\gamma \frac{dZ}{du}, \qquad \frac{\partial^2 \zeta}{\partial t^2} = R^2 \gamma^2 \frac{d^2 Z}{du^2}, \left.\begin{array}{c} \\ \\ \\ \\ \end{array}\right\}$$
$$\frac{\partial \sigma}{\partial t} = - R\gamma \frac{dS}{du}, \qquad \frac{\partial^2 \sigma}{\partial t^2} = R^2 \gamma^2 \frac{d^2 S}{du^2}. \tag{23}$$

Die Lage des betrachteten Teilchens gegen F_* und $\mathfrak{M}$ zur Zeit t ist dieselbe wie die Lage des Punktes

$$R + Z(u), \left.\begin{array}{c} \\ \\ \end{array}\right\}$$
$$\frac{u + S(u)}{R} \tag{24}$$

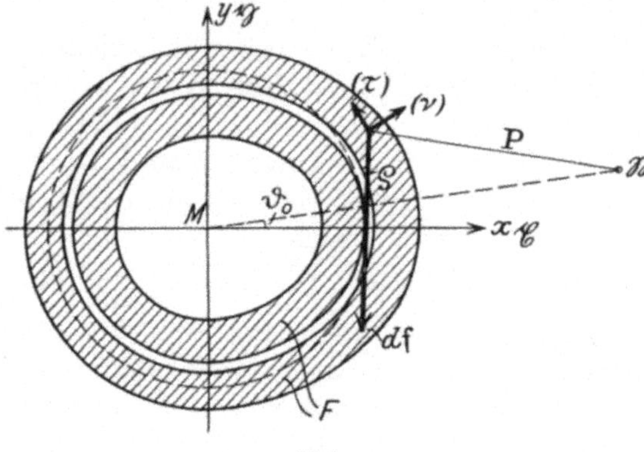

Abb. 4.

gegen F und $\mathfrak{M}_0$; F und $\mathfrak{M}_0$ entstehen aus F_* und $\mathfrak{M}$ durch eine Drehung durch den Winkel $-\gamma t$. Um nicht neue Buchstaben einführen zu müssen, wollen wir unter $\varrho_0, \varrho, P, \ldots$ Größen verstehen, die aus den vorhin ebenso bezeichneten Größen durch die vorerwähnte Drehung entstehen. Es ist also z. B. ϱ jetzt der Abstand der Punkte

$$R + Z(u), \quad \frac{u + S(u)}{R} \quad \text{und} \quad R' + Z(u'), \quad \frac{u' + S(u')}{R'},$$

unter R und R' beliebige Werte in den Intervallen (1) verstanden. Aus (17), (19) und (23) folgen nunmehr die weiteren Gleichungen

$$R^2 \gamma^2 \frac{d^2 Z}{du^2} - (R + Z)\left(\omega - \gamma \frac{dS}{du}\right)^2 = - \frac{M\varkappa}{R^2}\left(1 - \frac{2Z}{R} + \frac{3Z^2}{R^2} - \cdots\right) \left.\begin{array}{c} \\ \\ \\ \end{array}\right\}$$
$$+ \frac{\varkappa}{h}\int_F \mu' d\mathfrak{f}' \frac{\partial}{\partial \nu} \log \frac{\sqrt{\varrho^2 + h^2} + h}{\varrho} + \varkappa \mathfrak{M} \frac{\partial}{\partial \nu} \frac{1}{P}, \tag{25}$$

$$(R + Z)^2 \gamma^2 \frac{d^2 S}{du^2} - 2(R + Z)\gamma\left(\omega - \gamma \frac{dS}{du}\right)\frac{dZ}{du} \left.\begin{array}{c} \\ \\ \\ \end{array}\right\}$$
$$= \frac{\varkappa}{h}\left(1 + \frac{Z}{R}\right)\int_F \mu' d\mathfrak{f}' \frac{\partial}{\partial \tau} \log \frac{\sqrt{\varrho^2 + h^2} + h}{\varrho} + \varkappa \mathfrak{M}\left(1 + \frac{Z}{R}\right)\frac{\partial}{\partial \nu}\frac{1}{P}. \tag{26}$$

Wir bezeichnen mit F_0 das Ringsystem in der Ebene $\mathfrak{z} = 0$ bei Abwesenheit des störenden Körpers, d. h. im Zustande des dynamischen Gleichgewichtes, mit $d\mathfrak{f}'$ das Flächenelement im Punkte $R' + Z', \dfrac{u' + S(u')}{R'}$

von F, mit μ' die zugehörige Dichte. Sei schließlich $d\mathfrak{f}_0'$ das $d\mathfrak{f}'$ entsprechende Flächenelement in F_0, d. h. das Flächenelement im Punkte $R', \dfrac{u'}{R'}$. Es ist

$$\mu' d\mathfrak{f}' = \mu_0 d\mathfrak{f}_0' . \tag{27}$$

Die Integralausdrücke auf der rechten Seite der Gleichungen (25) und (26) können wir wegen (27) entsprechend durch die Ausdrücke ersetzen:

$$\int_{F_0} \mu_0 d\mathfrak{f}_0' \frac{\partial}{\partial \nu} \log \frac{\sqrt{\varrho^2 + h^2} + h}{\varrho} = \int_{F_0} \mu_0 d\mathfrak{f}_0' \left[\frac{\partial}{\partial \nu} \log \frac{\sqrt{\varrho^2 + h^2} + h}{\varrho} \right.$$
$$\left. - \frac{\partial}{\partial \nu_0} \log \frac{\sqrt{\varrho_0^2 + h^2} + h}{\varrho_0} \right] - \mu_0 h J_0(R) , \tag{28}$$

$$\int_{F_0} \mu_0 d\mathfrak{f}_0' \frac{\partial}{\partial \tau} \log \frac{\sqrt{\varrho^2 + h^2} + h}{\varrho} = \int_{F_0} \mu_0 d\mathfrak{f}_0' \left[\frac{\partial}{\partial \tau} \log \frac{\sqrt{\varrho^2 + h^2} + h}{\varrho} \right.$$
$$\left. - \frac{\partial}{\partial \tau_0} \log \frac{\sqrt{\varrho_0^2 + h^2} + h}{\varrho_0} \right] . \tag{29}$$

Hier bezeichnet ϱ_0 die Entfernung der Punkte $R, \dfrac{u}{R}$ und $R', \dfrac{u'}{R'}$; (ν_0) ist die äußere Normale an den Kreis (7) im Punkte $R, \dfrac{u}{R}$ und (τ_0) die Richtung, die man erhält, wenn man (ν_0) im positiven Sinne um 90° dreht (Abb. 5). Man überzeugt sich leicht, daß $J_0(R) > 0$ ist.

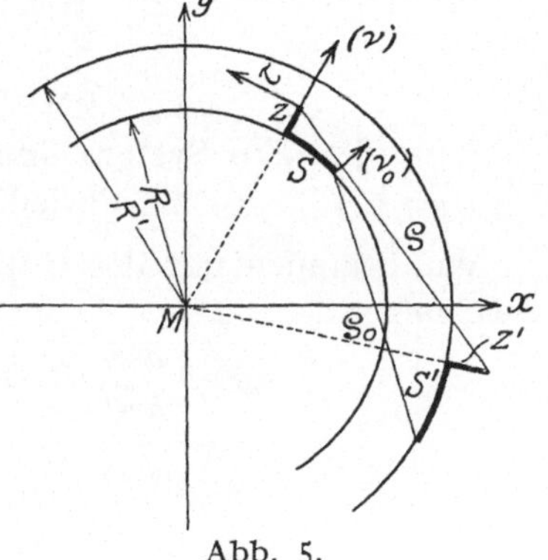

Abb. 5.

Die nur von R abhängigen Glieder in (25) sind, wenn man von $\varkappa \mathfrak{M} \dfrac{\partial}{\partial \nu} \dfrac{1}{P}$ absieht:

$$- R\omega^2 + \frac{M\varkappa}{R^2} + \varkappa \mu_0 J(R) . \tag{30}$$

Wir setzen

$$\omega^2 = \frac{M\varkappa}{R^3} + \frac{\varkappa \mu_0}{R} J_0(R) , \tag{31}$$

$$v = \omega^2 + \frac{2M\varkappa}{R^3} = 3\omega^2 - \frac{2\varkappa \mu_0}{R} J_0(R)$$

und erhalten nach einigen leichten Umformungen die endgültigen Beziehungen

$$
\begin{aligned}
L_1(Z,S) &\equiv R^2\gamma^2\frac{d^2 Z}{du^2} + 2R\gamma\omega\frac{dS}{du} - vZ = R\gamma^2\left(\frac{dS}{du}\right)^2 - 2\gamma\omega Z\frac{dS}{du} \\
&\quad + \gamma^2 Z\left(\frac{dS}{du}\right)^2 - \frac{M\varkappa}{R^2}\left(\frac{3Z^2}{R^2} - \frac{4Z^3}{R^3} + \cdots\right) \\
&\quad + \frac{\varkappa}{h}\int_{F_0}\mu_0\,d\mathfrak{f}_0'\left[\frac{\partial}{\partial\nu}\log\frac{\sqrt{\varrho^2+h^2}+h}{\varrho} - \frac{\partial}{\partial\nu_0}\log\frac{\sqrt{\varrho_0^2+h^2}+h}{\varrho_0}\right] \\
&\quad + \varkappa\,\mathfrak{M}\,\frac{\partial}{\partial\nu}\frac{1}{P} \equiv \varLambda_1(Z,S),
\end{aligned} \right\} \quad (32)
$$

$$
\begin{aligned}
L_2(Z,S) &\equiv R^2\gamma^2\frac{d^2 S}{du^2} - 2R\gamma\omega\frac{dZ}{du} = -2\gamma^2 RZ\frac{d^2 S}{du^2} - \gamma^2 Z^2\frac{d^2 S}{du^2} \\
&\quad - 2\gamma^2 R\frac{dS}{du}\frac{dZ}{du} + 2\gamma Z\omega\frac{dZ}{du} - 2\gamma^2 Z\frac{dS}{du}\frac{dZ}{du} + \varkappa\,\mathfrak{M}\left(1+\frac{Z}{R}\right)\frac{\partial}{\partial\tau}\frac{1}{P} \\
&\quad + \frac{\varkappa}{h}\left(1+\frac{Z}{R}\right)\int_{F_0}\mu_0'\,d\mathfrak{f}_0'\left[\frac{\partial}{\partial\tau}\log\frac{\sqrt{\varrho^2+h^2}+h}{\varrho} - \frac{\partial}{\partial\tau_0}\log\frac{\sqrt{\varrho_0^2+h^2}+h}{\varrho_0}\right] \\
&\equiv \varLambda_2(Z,S).
\end{aligned} \right\} \quad (33)
$$

Dies ist ein System nichtlinearer Integro-Differentialgleichungen zur Bestimmung von Z und S [1]) [2]).

Zweites Kapitel.

§ 1. Ein System linearer Differentialgleichungen. Periodische Lösungen.

Wir beginnen mit der Untersuchung des Systems der Differentialgleichungen:

$$
\left.\begin{aligned}
R^2\gamma^2\frac{d^2 Z}{du^2} + 2R\gamma\omega\frac{dS}{du} - vZ &= F_1(u), \\
R^2\gamma^2\frac{d^2 S}{du^2} - 2R\gamma\omega\frac{dZ}{du} &= F_2(u),
\end{aligned}\right\} \quad (1)
$$

unter $F_1(u)$ und $F_2(u)$ stetige, periodische, reelle oder komplexe Funktionen des Argumentes u mit der Periode $2\pi R$ verstanden. Gesucht werden die etwa vorhandenen nebst ihren Ableitungen erster und zweiter Ordnung stetigen periodischen Lösungen mit derselben Periode.

Das allgemeine Integral der zu (1) gehörigen homogenen Differentialgleichungen ($F_1 = F_2 = 0$) hat, wie man leicht verifiziert, die Form

---

[1]) Es ist zu beachten, daß Z und S auch noch von R abhängen, so daß für $\dfrac{dZ}{du}$, $\dfrac{dS}{du}$ usw. eigentlich $\dfrac{\partial Z}{\partial u}$, $\dfrac{\partial S}{\partial u}$ usw. geschrieben werden sollte.

[2]) Die Gleichungen (32) und (33) sind für $\mathfrak{M} = 0$ durch $Z \equiv 0$, $S \equiv 0$ erfüllt. Damit ist die Behauptung am Schluß der Seite 205 begründet.

$$Z(u) = B + C\cos\frac{pu}{\gamma R} + D\sin\frac{pu}{\gamma R},$$

$$S(u) = A + B\frac{vu}{2R\gamma\omega} + Cq\sin\frac{pu}{\gamma R} - Dq\cos\frac{pu}{\gamma R}, \qquad (2)$$

$$p = \sqrt{4\omega^2 - v} = \sqrt{\omega^2 + \frac{2\varkappa\mu_0}{R}J_0(R)}, \qquad q = \frac{2\omega}{p}.$$

Ist $p = \gamma$, so haben die zu (1) gehörigen homogenen Differentialgleichungen drei linear unabhängige Systeme periodischer Lösungen mit der Periode $2\pi R$:

$$0, \quad R; \quad R\cos\frac{u}{R}, \quad \frac{2R\omega}{\gamma}\sin\frac{u}{R}; \quad R\sin\frac{u}{R}, \quad -2R\frac{\omega}{\gamma}\cos\frac{u}{R}. \quad (3)$$

Ist, wie wir voraussetzen wollen, $\mathfrak{M}$ ein Satellit des Zentralkörpers M, so ist für alle in Betracht kommenden R gewiß $\beta < \omega$, mithin $\gamma < 0$. Es gibt demnach nur ein System periodischer Lösungen mit der Periode $2\pi R$, nämlich 0, R.

Sei jetzt $\mathfrak{G}(\xi, u)$ die, wie man sich leicht überzeugt, stets vorhandene periodische GREENsche Funktion der Differentialgleichung:

$$\frac{d^2 Z}{du^2} - \frac{v}{\gamma^2 R^2}Z = 0, \qquad (4)$$

$$\left(\mathfrak{G}(\xi,0) = \mathfrak{G}(\xi,2\pi R), \quad \frac{\partial}{\partial u}\mathfrak{G}(\xi,0) = \frac{\partial}{\partial u}\mathfrak{G}(\xi,2\pi R)\right).$$

Das allgemeine Integral von (4) ist $C_1 e^{\frac{vu}{\gamma R}} + C_2 e^{-\frac{vu}{\gamma R}}$. Periodische Lösungen mit der Periode $2\pi R$ sind nicht vorhanden. Also existiert in der Tat die GREENsche Funktion. Sie hängt offenbar auch von R ab.

Mit Rücksicht auf eine spätere Verwendung sei bemerkt, daß $\mathfrak{G}(\xi, u)$ eine für alle R in (1) I, alle ξ in $< 0, 2\pi R >$ und $\xi - 2\pi R \leqq u \leqq \xi$, $\xi \leqq u \leqq \xi + 2\pi R$ analytische und reguläre Funktion der drei Veränderlichen darstellt. Sie ist überdies für alle u und ξ stetig, während ihre partiellen Ableitungen in bezug auf ξ und u sich auf den Geraden $u = \xi$, $u = \xi \pm 2\pi R$ sprungweise ändern können[1].

Die Funktion $\mathfrak{G}_1(\xi, u) = \mathfrak{G}(\xi + \alpha, u + \alpha)$ (α konstant) ist, als Funktion von u aufgefaßt, offenbar periodisch mit der Periode $2\pi R$, genügt der Differentialgleichung (4) und verhält sich für $u = \xi$ wie $\mathfrak{G}(\xi, u)$. Da die GREENsche Funktion der Gleichung (4) vollkommen bestimmt ist, so gilt

$$\mathfrak{G}(\xi, u) = \mathfrak{G}(\xi + \alpha, u + \alpha)$$

und für $\alpha = -u$

$$\mathfrak{G}(\xi, u) = \mathfrak{G}(\xi - u, 0).$$

---

[1] Die partielle Ableitung $\frac{\partial}{\partial u}\mathfrak{G}(\xi, u)$ erleidet definitionsgemäß für $u = \xi$ einen Sprung, $\frac{\partial^2}{\partial u^2}\mathfrak{G}(\xi, u)$ ist wegen (4) durchweg stetig.

Die Funktion $\mathfrak{G}(\xi, u)$ hängt also nur von $\xi - u$ ab. Also ist

$$\frac{\partial}{\partial \xi} \mathfrak{G}(\xi, u) = - \frac{\partial}{\partial u} \mathfrak{G}(\xi, u).$$ (6)

Da $\mathfrak{G}(\xi, u)$ periodisch ist, so ist gewiß

$$\int_0^{2\pi R} \frac{\partial}{\partial u} \mathfrak{G}(\xi, u)\, du = 0,$$ (7)

also wegen (6) auch

$$\frac{\partial}{\partial \xi} \int_0^{2\pi R} \mathfrak{G}(\xi. u)\, du = 0, \qquad \int_0^{2\pi R} \mathfrak{G}(\xi, u)\, du = \text{Const.}$$ (8)

Übrigens folgt aus (4) fast unmittelbar

$$\frac{v}{\gamma^2 R^2} \int_0^{2\pi R} \mathfrak{G}(\xi, u)\, du = 1.$$ (9)

Schreibt man die Gleichungen (1) in der Form

$$\left.\begin{aligned}
\frac{d^2 Z}{d u^2} - \frac{v}{\gamma^2 R^2} Z &= - 2 \frac{\omega}{\gamma R} \frac{d S}{d u} + \frac{1}{\gamma^2 R^2} F_1(u), \\
\frac{d^2 S}{d u^2} - \frac{v}{\gamma^2 R^2} S &= 2 \frac{\omega}{\gamma R} \frac{d Z}{d u} - \frac{v}{\gamma^2 R^2} S + \frac{1}{\gamma^2 R^2} F_2(u),
\end{aligned}\right\}$$ (10)

so findet man zur Bestimmung der etwa vorhandenen periodischen Lösungen nach bekannten Sätzen die Formeln

$$\left.\begin{aligned}
Z(\xi) &= 2 \frac{\omega}{\gamma R} \int_0^{2\pi R} \mathfrak{G}(\xi, u) \frac{dS}{du}\, du - \frac{1}{\gamma^2 R^2} \int_0^{2\pi R} \mathfrak{G}(\xi, u) F_1(u)\, du, \\
S(\xi) &= - 2 \frac{\omega}{\gamma R} \int_0^{2\pi R} \mathfrak{G}(\xi, u) \frac{dZ}{du}\, du + \frac{v}{\gamma^2 R^2} \int_0^{2\pi R} \mathfrak{G}(\xi, u) S(u)\, du \\
&\quad - \frac{1}{\gamma^2 R^2} \int_0^{2\pi R} \mathfrak{G}(\xi, u) F_2(u)\, du,
\end{aligned}\right\}$$ (11)

oder, nach einer teilweisen Integration, die linearen Integralgleichungen

$$\left.\begin{aligned}
Z(\xi) &= - 2 \frac{\omega}{\gamma R} \int_0^{2\pi R} \frac{\partial}{\partial u} \mathfrak{G}(\xi, u) S(u)\, du - \frac{1}{\gamma^2 R^2} \int_0^{2\pi R} \mathfrak{G}(\xi, u) F_1(u)\, du, \\
S(\xi) &= 2 \frac{\omega}{\gamma R} \int_0^{2\pi R} \frac{\partial}{\partial u} \mathfrak{G}(\xi, u) Z(u)\, du + \frac{v}{\gamma^2 R^2} \int_0^{2\pi R} \mathfrak{G}(\xi, u) S(u)\, du \\
&\quad - \frac{1}{\gamma^2 R^2} \int_0^{2\pi R} \mathfrak{G}(\xi, u) F_2(u)\, du.
\end{aligned}\right\}$$ (12)

Wie man weiß, können die Gleichungen (12) durch eine einzige Integralgleichung von der Form

$$Z(\xi) = \int_E K(\xi,\,u)\,Z(u)\,du + \varPhi(\xi) \tag{13}$$

ersetzt werden, wenn man als Integrationsgebiet E das zweimal durchlaufene Intervall $(0,\,2\pi R)$ annimmt. Die Funktion $Z(\xi)$ ist beim erstmaligen Durchlaufen der Strecke $(0,\,2\pi R)$ gleich $Z(\xi)$, bei dem darauffolgenden Passieren desselben Intervalls gleich $S(\xi)$ zu setzen. Die entsprechenden Werte von $\varPhi(\xi)$ sind:

$$\left.\begin{aligned}
\varPhi^{(1)}(\xi) &= -\frac{1}{\gamma^2 R^2}\int_0^{2\pi R} \mathfrak{G}(\xi,\,u)\,F_1(u)\,du, \\[2ex]
\varPhi^{(2)}(\xi) &= -\frac{1}{\gamma^2 R^2}\int_0^{2\pi R} \mathfrak{G}(\xi,\,u)\,F_2(u)\,du.
\end{aligned}\right\} \tag{14}$$

Der Kern $K(\xi,u)$ ist der folgenden Tabelle zu entnehmen:

	Der erste Umlauf von u	Der zweite Umlauf von u
Der erste Umlauf von ξ	0	$-2\dfrac{\omega}{\gamma R}\dfrac{\partial}{\partial u}\mathfrak{G}(\xi,\,u)$
Der zweite Umlauf von ξ	$2\dfrac{\omega}{\gamma R}\dfrac{\partial}{\partial u}\mathfrak{G}(\xi,\,u)$	$\dfrac{v}{\gamma^2 R^2}\mathfrak{G}(\xi,\,u)$

Betrachten wir die zu (13) adjungierte homogene Integralgleichung

$$\mathsf{U}(\xi) = \int_E K(u,\,\xi)\,\mathsf{U}(u)\,du \tag{15}$$

Sie faßt die folgenden simultanen Integralgleichungen

$$\left.\begin{aligned}
\tilde{Z}(\xi) &= 2\frac{\omega}{\gamma R}\int_0^{2\pi R}\frac{\partial}{\partial\xi}\mathfrak{G}(\xi,\,u)\,\tilde{S}(u)\,du, \\[2ex]
\tilde{S}(\xi) &= -2\frac{\omega}{\gamma R}\int_0^{2\pi R}\frac{\partial}{\partial\xi}\mathfrak{G}(\xi,\,u)\,\tilde{Z}(u)\,du + \frac{v}{\gamma^2 R^2}\int_0^{2\pi R}\mathfrak{G}(\xi,\,u)\,\tilde{S}(u)\,du
\end{aligned}\right\} \tag{16}$$

zusammen.

Wegen (6) sind (16) mit den zu (12) gehörigen homogenen Integralgleichungen identisch. Das System (12) ist sich selbst adjungiert.

Die Integralgleichung

$$Z(\xi) = \int_E K(\xi,\,u)\,Z(u)\,du \tag{17}$$

hat, wie bereits erwähnt, eine Nullösung $W(\xi)$, gleich 0 in dem ersten, gleich R in dem zweiten Intervall $(0,\,2\pi R)$. Damit die nicht homogene

Gleichung (13), oder, was dasselbe ist, die Gleichungen (12) Lösungen haben, ist notwendig und hinreichend, daß eine Integralbeziehung erfüllt ist. Diese lautet:

$$\int_E \Phi(\xi) W(\xi)\, d\xi = R \int_0^{2\pi R} \Phi^{(2)}(\xi)\, d\xi = -\frac{1}{\gamma^2 R} \int_0^{2\pi R} d\xi \int_0^{2\pi R} \mathfrak{G}(\xi, u) F_2(u)\, du = 0, \quad (18)$$

oder wegen (9), mit Rücksicht auf $\mathfrak{G}(\xi, u) = \mathfrak{G}(u, \xi)$, einfacher:

$$\int_0^{2\pi R} F_2(u)\, du = 0. \quad (19)$$

Ist diese Bedingung erfüllt, so findet man nach bekannten Sätzen für die Lösungen $Z(\xi)$, $S(\xi)$, wenn man mit c eine willkürliche Konstante bezeichnet, die Formeln von der Form

$$\left.\begin{aligned}
Z(\xi) &= \int_0^{2\pi R} H_1(\xi, u)\,\Phi^{(1)}(u)\, du + \int_0^{2\pi R} H_2(\xi, u)\,\Phi^{(2)}(u)\, du, \\
S(\xi) &= \int_0^{2\pi R} H_3(\xi, u)\,\Phi^{(1)}(u)\, du + \int_0^{2\pi R} H_4(\xi, u)\,\Phi^{(2)}(u)\, du + cR.
\end{aligned}\right\} \quad (20)$$

$H_1(\xi, u), \ldots, H_4(\xi, u)$ hängen natürlich auch von R ab und stellen für alle R in (1) I, alle ξ in $< 0, 2\pi R >$ und

$$\xi - 2\pi R \leq u \leq \xi, \quad \xi \leq u \leq \xi + 2\pi R$$

analytische und reguläre Funktionen ihrer drei Argumente dar.

In der Tat ist jetzt

$$Z(\xi) = \int_E H(\xi, u)\,\Phi(u)\, du + cW(\xi). \quad (21)$$

Die Formeln (20) drücken diese Beziehung lediglich in einer anderen Form aus. Wie bekannt, ist

$$\int_E H(\xi, u) W(\xi)\, d\xi = 0. \quad (22)$$

Aus (21) und (22) folgt demnach

$$\int_E Z(\xi) W(\xi)\, d\xi = c \int_E W^2(\xi)\, d\xi = 2\pi R^3 c, \quad (23)$$

mithin

$$c = \frac{1}{2\pi R^2} \int_0^{2\pi R} S(\xi)\, d\xi. \quad (24)$$

Wir spezialisieren die Lösung durch die Festsetzung

$$\int_0^{2\pi R} S(\xi)\, d\xi = 0, \quad \text{d. h.} \quad c = 0.$$

Bei den folgenden Entwicklungen werden die Funktionen $F_1(u)$, $F_2(u)$ auch noch von R abhängen. Wir tragen diesem Umstande wie üblich durch geeignete Differentiationszeichen Rechnung.

Ist in den Gebieten

$$0 \leqq u \leqq 2\pi R, \quad R_{k0} \leqq R \leqq R_k^0 \qquad (k = 1, \ldots, m) \quad (26)$$

$$|F_1(u)|, \quad |F_2(u)| \leqq \mathfrak{h}, \tag{27}$$

so folgt aus (20)

$$|Z(u)|, \quad |S(u)|, \quad \left|\frac{\partial Z(u)}{\partial u}\right|, \quad \left|\frac{\partial S(u)}{\partial u}\right| \leqq \alpha_1 \mathfrak{h}, \tag{28}$$

unter α_1 eine positive Konstante verstanden, wie später unter α_2, α_3, $\ldots$ Aus (10) ergeben sich die weiteren Ungleichheiten

$$\left|\frac{\partial^2 Z(u)}{\partial u^2}\right|, \quad \left|\frac{\partial^2 S(u)}{\partial u^2}\right| \leqq \alpha_2 \mathfrak{h}. \tag{29}$$

Haben $F_1(u)$, $F_2(u)$ stetige partielle Ableitungen erster Ordnung $\dfrac{\partial F_1}{\partial u}, \ldots, \dfrac{\partial F_2}{\partial R}$ und ist

$$\left|\frac{\partial F_1}{\partial u}\right|, \quad \ldots, \quad \left|\frac{\partial F_2}{\partial R}\right| \leqq \alpha_3 \mathfrak{h}, \tag{30}$$

so haben, wie wieder aus (14), (20) und (10) erschlossen werden kann, vor allem $Z(u)$, $S(u)$ auch noch stetige partielle Ableitungen

$$\frac{\partial Z}{\partial R}, \ \frac{\partial S}{\partial R}, \ \frac{\partial^2 Z}{\partial R \partial u}, \ \frac{\partial^2 S}{\partial R \partial u}. \tag{30*}$$

Die in (28), (29), (30*) zusammengestellten Ableitungen haben ferner ihrerseits stetige Ableitung in bezug auf u. Es gelten schließlich die weiteren Beziehungen

$$\left|\frac{\partial Z}{\partial R}\right|, \ \left|\frac{\partial^2 Z}{\partial R \partial u}\right|, \ \left|\frac{\partial^3 Z}{\partial R \partial u^2}\right|, \ \left|\frac{\partial^3 Z}{\partial u^3}\right|, \ \left|\frac{\partial S}{\partial R}\right|, \ \ldots, \ \left|\frac{\partial^3 S}{\partial u^3}\right| \leqq \alpha_4 \mathfrak{h}. \tag{31}$$

Genügen schließlich $\dfrac{\partial F_1}{\partial u}, \ldots, \dfrac{\partial F_2}{\partial R}$ einer Hölderschen Bedingung (H-Bedingung) mit dem Exponenten λ [1])

$$\left.\begin{array}{c} \left|\dfrac{\partial}{\partial u} F_1(x+h,\, y+k) - \dfrac{\partial}{\partial u} F_1(x,\, y)\right| \\ \cdots\cdots\cdots\cdots\cdots\cdots\cdots\cdots\cdots\cdots\cdots\cdots\cdots\cdots \\ \left|\dfrac{\partial}{\partial R} F_2(x+h,\, y+k) - \dfrac{\partial}{\partial R} F_2(x,\, y)\right| \end{array}\right\} \leqq \alpha_5 \mathfrak{h}\,(|h| + |k|)^\lambda, \tag{32}$$

so haben die Funktionen (28), (29), (31) die gleiche Eigenschaft. Es ist z. B.

$$\left|\frac{\partial^3 S}{\partial R \partial u^2}(x+h,\, y+k) - \frac{\partial^3 S}{\partial R \partial u^2}(x,\, y)\right| \leqq \alpha_6 \mathfrak{h}\,(|h| + |k|)^\lambda. \tag{33}$$

---

[1]) Wir fassen jetzt vorübergehend F_1, F_2, Z, S als Funktionen von x und y auf und bringen dies durch die Schreibweise $F_1(x,\, y)$, $F_2(x,\, y)$ usw. zum Ausdruck.

Ist insbesondere

$$F_1(u) = F_1(2\theta_0 - u), \quad F_2(u) = -F_2(2\theta_0 - u), \tag{34}$$

so genügen die jetzt betrachteten Lösungen $\left(\int\limits_0^{2\pi R} S(u)\,du = 0\right)$ den Beziehungen

$$Z(u) = Z(2\theta_0 - u), \quad S(u) = -S(2\theta_0 - u). \tag{35}$$

Sind nämlich $Z(u)$, $S(u)$ $\left(\int\limits_0^{2\pi R} S(u)\,du = 0\right)$ Lösungen der Gleichungen (10), so bilden, wie man leicht sieht, $Z(2\theta_0 - u)$, $-S(2\theta_0 - u)$ ein weiteres System von Lösungen der betrachteten speziellen Art. Da es aber nur ein System dieser Eigenschaft gibt, so gelten in der Tat die Formeln (35).

Betrachten wir die in (32) und (33) I vorkommenden Ausdrücke

$$J = \int\limits_{F_0} d\mathfrak{f}_0' \left[\frac{\partial}{\partial \nu} \log \frac{\sqrt{\varrho^2 + h^2} + h}{\varrho} - \frac{\partial}{\partial \nu_0} \log \frac{\sqrt{\varrho_0^2 + h^2} + h}{\varrho_0} \right] \tag{36}$$

und

$$K = \int\limits_{F_0} d\mathfrak{f}_0' \left[\frac{\partial}{\partial \tau} \log \frac{\sqrt{\varrho^2 + h^2} + h}{\varrho} - \frac{\partial}{\partial \tau_0} \log \frac{\sqrt{\varrho_0^2 + h^2} + h}{\varrho_0} \right], \tag{37}$$

unter Z und S diesmal irgendwelche in den Bereichen F_0, (1) I, erklärte Funktionen von R und u verstanden, die stetige, der H-Bedingung genügende partielle Ableitungen erster Ordnung haben. Die Integrale J und K sind in F_0 erklärte Funktionen von R und u, die daselbst (den Rand inbegriffen) stetige, der H-Bedingung genügende partielle Ableitungen erster Ordnung haben. Dies ist bewiesen, sobald gezeigt ist, daß die Ausdrücke

$$J^{(1)} = \int\limits_{F_0} \frac{\partial}{\partial \nu} \log \frac{\overline{R}}{\varrho}\, d\mathfrak{f}_0' \quad \text{und} \quad K^{(1)} = \int\limits_{F_0} \frac{\partial}{\partial \tau} \log \frac{\overline{R}}{\varrho}\, d\mathfrak{f}_0' \tag{38}$$

$$(\overline{R} \text{ konstant})$$

die behauptete Eigenschaft haben. Wegen

$$\mu'\, d\mathfrak{f}' = \mu_0\, d\mathfrak{f}_0' \tag{39}$$

können $J^{(1)}$ und $K^{(1)}$ auch in der Form geschrieben werden:

$$J^{(1)} = \frac{1}{\mu_0} \int\limits_{F} \mu'\, d\mathfrak{f}' \frac{\partial}{\partial \nu} \log \frac{\overline{R}}{\varrho}, \quad K^{(1)} = \frac{1}{\mu_0} \int\limits_{F} \mu'\, d\mathfrak{f}' \frac{\partial}{\partial \tau} \log \frac{\overline{R}}{\varrho}. \tag{40}$$

Wie sich in bekannter Weise zeigen läßt, ist

$$d\mathfrak{f}' = d\mathfrak{f}_0' \left(1 + \frac{\partial Z'}{\partial R'} + \frac{\partial S'}{\partial u'} \right), \tag{41}$$

mithin

$$\mu' = \frac{\mu_0}{1 + \dfrac{\partial Z'}{\partial R'} + \dfrac{\partial S'}{\partial u'}}. \tag{42}$$

Als Ortsfunktion in F_0 betrachtet, erfüllt demnach μ' die H-Bedingung. Die gleiche Eigenschaft hat offenbar μ', wenn man es als Ortsfunktion in F auffaßt. Nach bekannten Sätzen haben $J^{(1)}$ und $K^{(1)}$ als Ortsfunktionen in F, also auch als Ortsfunktionen in F_0, stetige Ableitungen erster Ordnung, die einer H-Bedingung genügen.

Es mögen jetzt die Funktionen Z und S den Ungleichheiten genügen:

$$\left.\begin{aligned}
&|Z|,\ |S|,\ \left|\frac{\partial Z}{\partial R}\right|,\ \ldots,\ \left|\frac{\partial S}{\partial u}\right| \leqq \Omega^0, \\[2mm]
&\left|\frac{\partial}{\partial R}Z(x+h,\ y+k) - \frac{\partial}{\partial R}Z(x,y)\right|,\ \ldots, \\[2mm]
&\left|\frac{\partial}{\partial u}S(x+h,\ y+k) - \frac{\partial}{\partial u}S(x,y)\right| \leqq \Omega^0(|h|+|k|)^\lambda.
\end{aligned}\ \right\} \quad (43)$$

Es ist im vorstehenden stillschweigend angenommen worden, daß dem Kreisringsystem F_0 ein System von kreisringähnlichen Bereichen F umkehrbar eindeutig und stetig zugeordnet ist. Dies ist gewiß der Fall, wenn Ω^0 hinreichend klein ist. Der Rand von F besteht dann aus geschlossenen Kurven mit stetiger Tangente, wobei die Richtungskosinus der Normale einer H-Bedingung genügen.

Es mögen nunmehr $\dot{Z}$ und $\dot{S}$ zwei, was ihre Stetigkeitseigenschaften betrifft, wie Z und S beschaffene Funktionen bezeichnen, die zu (43) ganz analoge Ungleichheiten erfüllen. Es möge ferner

$$\left.\begin{aligned}
&|\dot{Z}-Z|,\ |\dot{S}-S|,\ \left|\frac{\partial}{\partial R}(\dot{Z}-Z)\right|,\ \ldots,\ \left|\frac{\partial}{\partial u}(\dot{S}-S)\right| \leqq \Omega, \\[2mm]
&\left|\frac{\partial}{\partial R}[\dot{Z}(x+h,\ y+k) - Z(x+h,\ y+k)] - \frac{\partial}{\partial R}[\dot{Z}(x,\ y) - Z(x,\ y)\right| \\[2mm]
&\hspace{5cm} \leqq \Omega(|h|+|k|)^\lambda, \\[2mm]
&\cdots\cdots\cdots\cdots\cdots\cdots\cdots\cdots\cdots\cdots\cdots \\[2mm]
&\left|\frac{\partial}{\partial u}[\dot{S}(x+h,y+k) - S(x+h,y+k)] - \frac{\partial}{\partial u}[\dot{S}(x,y) - S(x,y)]\right| \\[2mm]
&\hspace{4cm} \leqq \Omega(|h|+|k|)^\lambda \hspace{2cm} (\Omega \leqq \tfrac{1}{2}\Omega^0)
\end{aligned}\ \right\} \quad (44)$$

sein.

Die zu $\dot{Z}$ und $\dot{S}$ gehörigen Werte der Integrale J und K mögen $\dot{J}$ und $\dot{K}$ heißen. Die Differenzen $J - \dot{J},\ K - \dot{K}$ haben, *als Ortsfunktionen in F_0 aufgefaßt*, stetige partielle Ableitungen erster Ordnung, die der H-Bedingung genügen. Wie in dem vierten Kapitel gezeigt werden wird, gelten nun die folgenden Ungleichheiten:

$$|\dot{J}-J|,\ |\dot{K}-K|,\ \left|\frac{\partial}{\partial R}(\dot{J}-J)\right|,\ \ldots,\ \left|\frac{\partial}{\partial u}(\dot{K}-K)\right| \leqq \alpha_7\Omega, \quad (45)$$

$$\left| \frac{\partial}{\partial R} [\dot{J}(x+h, y+k) - J(x+h, y+k)] - \frac{\partial}{\partial R} [\dot{J}(x,y) - J(x,y)] \right|$$
$$\leq \alpha_7 \, \Omega \, (|h| + |k|)^\lambda ,$$

$$\dots\dots\dots\dots\dots\dots\dots\dots\dots\dots\dots\dots\dots\dots\dots\dots\dots$$

$$\left| \frac{\partial}{\partial u} [\dot{K}(x+h, y+k) - K(x+h, y+k)] - \frac{\partial}{\partial u} [\dot{K}(x,y) - K(x,y)] \right|$$
$$\leq \alpha_7 \, \Omega \, (|h| + |k|)^\lambda .$$

Der Faktor α_7 gilt für alle den Beziehungen (43) und (44) genügenden Z, S, $\dot{Z}$, $\dot{S}$ gleichmäßig.

Wir haben uns vorhin eingehend mit der Bestimmung periodischer Lösungen der Differentialgleichungen (1) beschäftigt.

Handelt es sich speziell um die Gleichungen (32) I und (33) I, so ist

$$
\left.
\begin{aligned}
F_1(u) &= R\gamma^2 \left(\frac{dS}{du}\right)^2 - 2\gamma\,\omega\,Z\frac{dS}{du} + \gamma^2 Z \left(\frac{dS}{du}\right)^2 - \frac{M\varkappa}{R^2}\left(\frac{3Z^2}{R^2} - \frac{4Z^3}{R^3} + \dots\right) \\
&\quad + \frac{\varkappa}{h}\int_{F_0} \mu_0\, d\mathfrak{f}_0' \left[\frac{\partial}{\partial \nu}\log\frac{\sqrt{\varrho^2 + h^2} + h}{\varrho} - \frac{\partial}{\partial \nu_0}\log\frac{\sqrt{\varrho_0^2 + h^2} + h}{\varrho_0}\right] \\
&\quad + \varkappa\mathfrak{M}\frac{\partial}{\partial \nu}\frac{1}{P}, \\[4pt]
F_2(u) &= -2\gamma^2 R Z \frac{d^2 S}{du^2} - \gamma^2 Z^2 \frac{d^2 S}{du^2} - 2\gamma^2 R \frac{dS}{du}\frac{dZ}{du} + 2\gamma Z \omega \frac{dZ}{du} \\
&\quad - 2\gamma^2 Z \frac{dS}{du}\frac{dZ}{du} + \frac{\varkappa}{h}\left(1 + \frac{Z}{R}\right)\int_{F_0} \mu_0'\, d\mathfrak{f}_0'\left[\frac{\partial}{\partial \tau}\log\frac{\sqrt{\varrho^2 + h^2} + h}{\varrho}\right. \\
&\quad \left. - \frac{\partial}{\partial \tau_0}\log\frac{\sqrt{\varrho_0^2 + h^2} + h}{\varrho_0}\right] + \varkappa\mathfrak{M}\left(1 + \frac{Z}{R}\right)\frac{\partial}{\partial \tau}\frac{1}{P}.
\end{aligned}
\right\} \quad (46)
$$

Es mögen jetzt vorübergehend Z und S irgendwelche Funktionen mit der Periode $2\pi R$ bezeichnen, die folgende Eigenschaften haben. Die in (28), (29), (31) zusammengestellten Ableitungen sind vorhanden und stetig; die vier partiellen Ableitungen dritter Ordnung genügen der Hölder-schen Bedingung (33). Es gelten die Beziehungen (35)[1]. Führt man nunmehr in (46) rechts die betrachteten Funktionen ein, so findet man, daß $F_1(u)$ und $F_2(u)$ stetig sind und stetige, der Hölderschen Bedingung genügende partielle Ableitungen erster Ordnung haben[2]. Ferner ist, wie man ohne Schwierigkeiten sieht,

$$F_1(u) = F_1(2\theta_0 - u), \qquad F_2(u) = -F_2(2\theta_0 - u) \qquad (47)$$

und darum auch

$$\int_0^{2\pi R} F_2(u)\, du = 0. \qquad (48)$$

---

[1] Die unter Zugrundelegung der Funktionen $Z(u)$, $S(u)$ konstruierte Fläche F_0 ist in bezug auf den Strahl $y = x\,\mathrm{tang}\,\theta_0$ symmetrisch.

[2] Vgl. weiter unten S. 221.

Drittes Kapitel.

§ 1. Bestimmung periodischer Lösungen der Integro-Differentialgleichungen (32) I und (33) I durch sukzessive Approximationen.

Wir versuchen jetzt die in § 2 I näher charakterisierten periodischen Lösungen der Gleichungen (32) I und (33) I wie folgt durch sukzessive Approximationen zu gewinnen.

Wir setzen

$$Z_0 = 0, \quad S_0 = 0 \tag{1}$$

und bestimmen nacheinander die periodischen Lösungen mit der Periode $2\pi R$ der Gleichungen

$$\left.\begin{aligned}
L_1(Z_1, S_1) &= \Lambda_1(Z_0, S_0) = F_{10}(u), \\
L_2(Z_1, S_1) &= \Lambda_2(Z_0, S_0) = F_{20}(u),
\end{aligned} \quad \int\limits_0^{2\pi R} S_1(u)\,du = 0, \right\} \tag{2}$$

$$\left.\begin{aligned}
L_1(Z_2, S_2) &= \Lambda_1(Z_1, S_1) = F_{11}(u), \\
L_2(Z_2, S_2) &= \Lambda_2(Z_1, S_1) = F_{21}(u),
\end{aligned} \quad \int\limits_0^{2\pi R} S_2(u)\,du = 0, \right\} \tag{3}$$

$$\left.\begin{aligned}
L_1(Z_3, S_3) &= \Lambda_1(Z_2, S_2) = F_{12}(u), \\
L_2(Z_3, S_3) &= \Lambda_2(Z_2, S_2) = F_{22}(u),
\end{aligned} \quad \int\limits_0^{2\pi R} S_3(u)\,du = 0, \right\} \tag{4}$$

. . . .

Es ist nicht schwer zu sehen, daß der Bestimmung der Wertfolge $Z_1(u)$, $S_1(u)$; $Z_2(u)$, $S_2(u)$; ... nichts im Wege steht. Es ist zunächst

$$F_{10}(u) = \varkappa\,\mathfrak{M}\,\frac{\partial}{\partial \nu}\,\frac{1}{P}\,, \qquad F_{20}(u) = \varkappa\,\mathfrak{M}\,\frac{\partial}{\partial \tau}\,\frac{1}{P}\,. \tag{5}$$

Die Funktionen $F_{10}(u)$, $F_{20}(u)$ haben gewiß partielle Ableitungen erster Ordnung $\dfrac{\partial F_{10}}{\partial u}$, ..., $\dfrac{\partial F_{20}}{\partial R}$, die einer HÖLDERschen Bedingung

$$\left|\frac{\partial F_{10}}{\partial u}(x+h,\ y+k) - \frac{\partial F_{10}}{\partial u}(x,\ y)\right| \le \alpha_8(|h|+|k|)^\lambda \quad (0<\lambda<1) \tag{6}$$

. . . .

genügen[1]). Aus (6) folgert man den Schlußausführungen des § 2 II gemäß, daß die Funktion Z_1 stetige Ableitungen $\dfrac{\partial Z_1}{\partial R}$, $\dfrac{\partial Z_1}{\partial u}$, $\dfrac{\partial^2 Z_1}{\partial R\,\partial u}$, $\dfrac{\partial^2 Z_1}{\partial u^2}$, $\dfrac{\partial^3 Z_1}{\partial R\,\partial u^2}$, $\dfrac{\partial^3 Z_1}{\partial u^3}$ hat und alle diese Ableitungen einer H-Bedingung mit dem Exponenten λ genügen. Die Funktion S_1 hat dieselben Eigenschaften.

Man sieht jetzt leicht, daß die Funktionen $F_{11}(u)$ und $F_{21}(u)$ gewiß stetige, die H-Bedingung erfüllende Ableitungen erster Ordnung be-

---

[1]) Vgl. die Fußnote S. 217.

sitzen. Dies hat zur Folge, daß Z_2 und S_2 die gleichen Stetigkeitseigenschaften wie Z_1 und S_1 haben. Die Funktionen F_{12} und F_{22} verhalten sich wie F_{11} und F_{21}, so daß man ungehindert zu weiteren Approximationen gelangen kann.

Sei jetzt
$$Z_n - Z_{n-1} = \mathfrak{z}_n, \qquad S_n - S_{n-1} = \mathfrak{s}_n \qquad (n \geqq 1) \quad (7)$$

$$\text{Max}\left\{|\mathfrak{z}_n|, \left|\frac{\partial \mathfrak{z}_n}{\partial u}\right|, \left|\frac{\partial \mathfrak{z}_n}{\partial R}\right|, \left|\frac{\partial^2 \mathfrak{z}_n}{\partial u^2}\right|, \left|\frac{\partial^3 \mathfrak{z}_n}{\partial u^3}\right|, \left|\frac{\partial^2 \mathfrak{z}_n}{\partial R \partial u}\right|, \left|\frac{\partial^3 \mathfrak{z}_n}{\partial R \partial u^2}\right|, |\mathfrak{s}_n|, \ldots, \left|\frac{\partial^3 \mathfrak{s}_n}{\partial R \partial u^2}\right|\right\} = \mathfrak{v}_n, \quad (8)$$

$$\text{Max}\left\{|Z_m|, \left|\frac{\partial Z_m}{\partial u}\right|, \left|\frac{\partial Z_m}{\partial R}\right|, \ldots, \left|\frac{\partial^3 Z_m}{\partial R \partial u^2}\right|, |S_m|, \ldots, \left|\frac{\partial^3 S_m}{\partial R \partial u^2}\right|\right\} = \mathfrak{w}_n \quad (m = 1, \ldots, n) \quad (9)$$

gesetzt. Wir bezeichnen ferner zur Vereinfachung die obere Grenze der Größen

$$\left.\begin{array}{c} \dfrac{1}{(|h| + |k|)^\lambda} |\mathfrak{z}_n(x+h, y+k) - \mathfrak{z}_n(x, y)|, \ldots, \\[2mm] \dfrac{1}{(|h| + |k|)^\lambda} \left|\dfrac{\partial^3 \mathfrak{s}_n}{\partial R \partial u^2}(x+h, y+k) - \dfrac{\partial^3 \mathfrak{s}_n}{\partial R \partial u^2}(x, y)\right| \qquad (n \geqq 1) \end{array}\right\} \quad (10)$$

mit $\mathfrak{D}_n$, die obere Grenze der Größen

$$\left.\begin{array}{c} \dfrac{1}{(|h| + |k|)^\lambda} |Z_m(x+h, y+k) - Z_m(x, y)|, \ldots, \\[2mm] \dfrac{1}{(|h| + |k|)^\lambda} \left|\dfrac{\partial^3}{\partial R \partial u^2} S_m(x+h, y+k) - \dfrac{\partial^3}{\partial R \partial u^2} S_m(x, y)\right| \quad (m = 1, \ldots, n) \end{array}\right\} \quad (11)$$

mit $\mathfrak{E}_n$. Aus (1), (2), (3) ergibt sich zunächst

$$\begin{aligned} L_1(\mathfrak{z}_n, \mathfrak{s}_n) &= \Lambda_1(Z_{n-1}, S_{n-1}) - \Lambda_1(Z_{n-2}, S_{n-2}), \\ L_2(\mathfrak{z}_n, \mathfrak{s}_n) &= \Lambda_2(Z_{n-1}, S_{n-1}) - \Lambda_2(Z_{n-2}, S_{n-2}). \qquad (n \geqq 2) \end{aligned} \quad (12)$$

Aus (32) I und (33) I folgt, wenn man $\mathfrak{w}_n \leqq \Omega^0$, $\mathfrak{E}_n \leqq \Omega^0$, $\mathfrak{v}_n \leqq \frac{1}{2}\Omega^0$, $\mathfrak{D}_n \leqq \frac{1}{2}\Omega^0$ annimmt, wegen (7), (8), (9), (43) II, (44) II und (45) II, wie man leicht findet,

$$\left.\begin{aligned} |\mathsf{M}_{n-1}| &= |\Lambda_1(Z_{n-1}, S_{n-1}) - \Lambda_1(Z_{n-2}, S_{n-2})| \\ &\leqq \alpha_9\left(\mathfrak{w}_{n-1} + \mathfrak{w}_{n-1}^2 + \mu_0 + \frac{\varkappa \mathfrak{M}}{\mathfrak{R}^3}\right)(\mathfrak{v}_{n-1} + \mathfrak{D}_{n-1}), \\ |\mathsf{N}_{n-1}| &= |\Lambda_2(Z_{n-1}, S_{n-1}) - \Lambda_2(Z_{n-2}, S_{n-2})| \\ &\leqq \alpha_9\left(\mathfrak{w}_{n-1} + \mu_0 + \frac{\varkappa \mathfrak{M}}{\mathfrak{R}^3}\right)(\mathfrak{v}_{n-1} + \mathfrak{D}_{n-1}). \end{aligned}\right\} \quad (13)$$

In der Tat kommt nach (32) I in M_{n-1} zunächst die Differenz $R\gamma^2\left\{\left(\dfrac{dS_{n-1}}{du}\right)^2 - \left(\dfrac{dS_{n-2}}{du}\right)^2\right\}$ vor. Es gilt

$$\left|R\gamma^2\left\{\left(\frac{dS_{n-1}}{du}\right)^2 - \left(\frac{dS_{n-2}}{du}\right)^2\right\}\right| = R\gamma^2\left|\frac{dS_{n-1}}{du} + \frac{dS_{n-2}}{du}\right|\left|\frac{dS_{n-1}}{du} - \frac{dS_{n-2}}{du}\right|$$

$$\leqq 2R\gamma^2\,\mathfrak{w}_{n-1}\,\mathfrak{z}_{n-1} \leqq 2R\gamma^2\,\mathfrak{w}_{n-1}\,\mathfrak{v}_{n-1}.$$

Das zweite und das vierte Glied rechts in (32) I führt zu einem analogen Ausdruck, das dritte liefert eine Schranke $3\,\gamma^2\,\mathfrak{w}_{n-1}^2\,\mathfrak{v}_{n-1}$, das letzte eine Schranke von der Form const. $\varkappa\,\dfrac{\mathfrak{M}}{\mathfrak{R}^3}\,\mathfrak{v}_{n-1}$. Was das über F_0 erstreckte Integral betrifft, so läßt sich nach (43) II, (44) II, (45) II für die Differenz der zu $\Lambda_1\,(Z_{n-1},\,S_{n-1})$ und $\Lambda_1\,(Z_{n-2},\,S_{n-2})$ gehörigen Werte dieses Integrals eine Schranke angeben, sobald für alle in (44) II links angegebenen Differenzen, mit $Z_{n-1},\,\ldots,\,S_{n-2}$ für $\dot{Z},\,\ldots,\,S$, eine gemeinsame obere Schranke (dort mit Ω bezeichnet) vorliegt. Eine solche ist gewiß in der größeren der beiden Zahlen $\mathfrak{v}_{n-1}$ und $\mathfrak{D}_{n-1}$ gegeben. Die gesuchte obere Schranke kann also auch gleich const. $\mu_0\,(\mathfrak{v}_{n-1}+\mathfrak{D}_{n-1})$ gesetzt werden. Es ist jetzt klar, daß die erste Ungleichheit (13) gilt, unter α_9 eine hinreichend große feste Zahl verstanden. In analoger Weise gelangt man zu der Formel (13) für $|N_{n-1}|$. Man findet ferner die weiteren Ungleichheiten

$$
\left.
\begin{aligned}
&\left|\frac{\partial}{\partial R}M_{n-1}\right|,\ \left|\frac{\partial}{\partial u}M_{n-1}\right|\leqq\alpha_{10}\left(\mathfrak{w}_{n-1}+\mathfrak{w}_{n-1}^2+\mu_0+\frac{\varkappa\mathfrak{M}}{\mathfrak{R}^4}\right)(\mathfrak{v}_{n-1}+\mathfrak{D}_{n-1}),\\[2mm]
&\left|\frac{\partial}{\partial R}N_{n-1}\right|,\ \left|\frac{\partial}{\partial u}N_{n-1}\right|\leqq\alpha_{10}\left(\mathfrak{w}_{n-1}+\mathfrak{w}_{n-1}^2+\mu_0+\frac{\varkappa\mathfrak{M}}{\mathfrak{R}^4}\right)(\mathfrak{v}_{n-1}+\mathfrak{D}_{n-1}),\\[2mm]
&\frac{1}{(|h|+|k|)^\lambda}\left|\frac{\partial}{\partial R}M_{n-1}(x+h,\,y+k)-\frac{\partial}{\partial R}M_{n-1}(x,\,y)\right|\\[2mm]
&\qquad\qquad\leqq\alpha_{10}\left(\mathfrak{w}_{n-1}+\mathfrak{w}_{n-1}^2+\mu_0+\frac{\varkappa\mathfrak{M}}{\mathfrak{R}^4}\right)(\mathfrak{v}_{n-1}+\mathfrak{D}_{n-1}),
\end{aligned}
\right\}\tag{14}
$$

Nimmt man $\mathfrak{w}_{n-1}\leqq 1$ an und setzt zur Vereinfachung

$$\frac{1}{2}\left(\mu_0+\frac{\varkappa\mathfrak{M}}{\mathfrak{R}^3}\right)=\widehat{\beta},\tag{15}$$

so kann man für (13) und (14) auch schreiben:

$$
\left.
\begin{aligned}
&|M_{n-1}|,\ |N_{n-1}|,\left|\frac{\partial}{\partial R}M_{n-1}\right|,\ \ldots,\ \left|\frac{\partial}{\partial u}N_{n-1}\right|\leqq\alpha_{11}(\mathfrak{w}_{n-1}+\widehat{\beta})(\mathfrak{v}_{n-1}+\mathfrak{D}_{n-1}),\\[2mm]
&\frac{1}{(|h|+|k|)^\lambda}\left|\frac{\partial}{\partial R}M_{n-1}(x+h,\ y+k)-\frac{\partial}{\partial R}M_{n-1}(x,\,y)\right|\\[2mm]
&\qquad\qquad\leqq\alpha_{11}(\mathfrak{w}_{n-1}+\widehat{\beta})(\mathfrak{v}_{n-1}+\mathfrak{D}_{n-1}),
\end{aligned}
\right\}\tag{16}
$$

Aus (2), (27) II bis (33) II folgt zunächst

$$\mathfrak{v}_1=\mathfrak{w}_1\leqq\alpha_{12}\frac{\mathfrak{M}}{\mathfrak{R}^2},\quad\mathfrak{D}_1\leqq\alpha_{12}\frac{\mathfrak{M}}{\mathfrak{R}^2}.\tag{17}$$

Aus (12), (13), (14), (27) II bis (33) II ergibt sich weiter

$$\mathfrak{v}_n\leqq\alpha_{13}(\mathfrak{w}_{n-1}+\widehat{\beta})(\mathfrak{v}_{n-1}+\mathfrak{D}_{n-1})\ \text{und}\ \mathfrak{D}_n\leqq\alpha_{13}(\mathfrak{w}_{n-1}+\widehat{\beta})(\mathfrak{v}_{n-1}+\mathfrak{D}_{n-1}),\tag{18}$$

darum mit

$$\mathfrak{v}_n + \mathfrak{D}_n = \theta_n \tag{19}$$

einfacher

$$\theta_n \leqq 2\,\alpha_{13}(\mathfrak{w}_{n-1} + \hat{\beta})\,\theta_{n-1}. \tag{20}$$

Ist etwa

$$\mathfrak{w}_l \leqq \beta_0 \leqq \Omega^0, \quad \mathfrak{E}_l \leqq \beta_0 \leqq \Omega^0, \tag{21}$$

so folgt aus (32) I, (33) I gewiß

$$\left.\begin{array}{l} |\Lambda_1(Z_l,\,S_l)|,\ |\Lambda_2(Z_l,\,S_l)|, \\[4pt] \left|\dfrac{\partial}{\partial R}\Lambda_1(Z_l,\,S_l)\right|,\dots,\left|\dfrac{\partial}{\partial u}\Lambda_2(Z_l,\,S_l)\right|, \\[8pt] \dfrac{1}{(|h|+|l|)^{\lambda}}\left|\dfrac{\partial}{\partial R}\Lambda_1(Z_l,\,S_l)_{x+h,\,y+k} - \dfrac{\partial}{\partial R}\Lambda_1(Z_l,\,S_l)\right| \end{array}\right\} \begin{array}{l} \leqq \alpha_{14}\Big(\beta_0^2 + \beta_0^3 \\[6pt] \quad + \dfrac{\varkappa\mathfrak{M}}{\mathfrak{R}^2} + \mu_0\Big), \end{array} \tag{22}$$

. . . .

demnach wegen (27) II bis (33) II, sofern $\Omega^\circ \leqq 1$ vorausgesetzt wird,
wie man leicht sieht,

$$\mathfrak{w}_{l+1} \leqq \alpha_{15}\,(\beta_0^2 + \hat{\beta}), \qquad \mathfrak{E}_{l+1} \leqq \alpha_{15}(\beta_0^2 + \hat{\beta}). \tag{23}$$

Gelingt es zu erreichen, daß

$$\alpha_{15}\,(\beta_0^2 + \hat{\beta}) \leqq \beta_0 \tag{24}$$

gilt, so wird sich auch

$$\mathfrak{w}_{l+1} \leqq \beta_0, \qquad \mathfrak{E}_{l+1} \leqq \beta_0 \tag{25}$$

ergeben.

Wir wählen β_0 und $\hat{\beta}$ den folgenden Ungleichheiten gemäß

$$\beta_0 \leqq \tfrac{1}{2}\,\Omega^0 \leqq \tfrac{1}{2}, \quad \beta_0 < \frac{1}{\alpha_{15}}, \quad \hat{\beta} \leqq \frac{\beta_0(1-\alpha_{15}\beta_0)}{\alpha_{15}} = \beta_0\Big(\frac{1}{\alpha_{15}} - \beta_0\Big). \tag{26}$$

Sei weiter

$$\alpha_{12}\,\frac{\mathfrak{M}}{\mathfrak{R}^2} \leqq \beta_0. \tag{27}$$

Aus (26) folgt, wie man sofort sieht, (24). Da jetzt wegen (17)
und (27)

$$\mathfrak{w}_1 \leqq \beta_0, \qquad \mathfrak{E}_1 = \mathfrak{D}_1 \leqq \beta_0 \tag{28}$$

ist, so ist für *alle* l

$$\mathfrak{w}_l \leqq \beta_0 \leqq \tfrac{1}{2}\,\Omega^0, \qquad \mathfrak{E}_l \leqq \beta_0 \leqq \tfrac{1}{2}\,\Omega^0. \tag{29}$$

Aus (20) folgt jetzt nach (26)

$$\theta_n \leqq 2\,\alpha_{13}(\beta_0 + \hat{\beta})\,\theta_{n-1} \leqq 2\,\alpha_{13}\,\theta_{n-1}\,\beta_0\Big(1 + \frac{1}{\alpha_{15}} - \beta_0\Big). \tag{30}$$

Ist schließlich, was wir annehmen wollen,

$$2\,\alpha_{13}\beta_0\Big(1 + \frac{1}{\alpha_{15}}\Big) \leqq q \leqq \tfrac{1}{2}, \tag{31}$$

d. h.

$$\beta_0 \leqq \frac{q}{2\,\alpha_{13}\left(1 + \dfrac{1}{\alpha_{15}}\right)} \qquad (32)$$

so sind die Reihe $\sum \theta_n$ und darum auch die beiden Reihen $\sum \mathfrak{v}_n$ und $\sum \mathfrak{D}_n$ gleichmäßig konvergent[1]).

Sei jetzt

$$Z = Z_1 + \sum_{k=2}^{\infty}(Z_k - Z_{k-1})\,, \qquad S = S_1 + \sum_{k=2}^{\infty}(S_k - S_{k-1})\,. \qquad (33)$$

Die Funktionen Z und S sind stetig, periodisch mit der Periode $2\pi R$ und haben stetige, den Beziehungen

$$\left| \frac{\partial}{\partial R} Z(x+h,\, y+k) - \frac{\partial}{\partial R} Z(x,y) \right| \leqq \beta_0\,(|h| + |k|)^\lambda \qquad (34)$$

. . . .

genügende partielle Ableitungen erster Ordnung. Sie erfüllen, wie sich ohne er stliche Schwierigkeiten durch Grenzübergang zeigen läßt, die Integro-Differentialgleichungen (32) I, (33) I. Es ist schließlich

$$\int_0^{2\pi R} S(u)\, du = 0\,. \qquad (35)$$

Damit ist die eingangs behauptete Existenz eines permanenten Bewegungszustandes des betrachteten vollkommen inkohärenten Kontinuums unter der gleichzeitigen Wirkung der Eigengravitation, der Zentrifugalkräfte und der Anziehung durch den Zentralkörper sowie den störenden Körper $\mathfrak{M}$ *bewiesen.* Vorausgesetzt wurde dabei, daß die Dichte der stetig verteilten Materie hinreichend klein, der Abstand des störenden Körpers von dem Zentralkörper hinreichend groß oder aber seine Masse genügend klein ist.

Viertes Kapitel.

§ 1. Der Hilfssatz.

Wir haben nur noch den Beweis des in § 2 I angegebenen Hilfssatzes nachzutragen.

Wir betrachten hierzu diejenige von einem Parameter ε abhängende stetige Schar topologischer Abbildungen der Bereiche F auf Bereiche F_ε, bei der dem Punkte $R + Z,\ \dfrac{u+S}{R}$ der Punkt

$$R + Z_\varepsilon = R + Z + \frac{\varepsilon}{\Omega}(\dot Z - Z)\,, \quad \frac{u + S_\varepsilon}{R} = \frac{1}{R}\left(u + S + \frac{\varepsilon}{\Omega}(\dot S - S)\right) \qquad (1)$$

zugeordnet wird. Dem Wert $\varepsilon = 0$ entsprechen die Bereiche F selbst (identische Abbildung), dem Werte $\varepsilon = \Omega$ die Bereiche $\dot F$.

___

[1]) Aus (30) folgt $\theta_n < q\,\theta_{n-1}$, darum gewiß für alle $n > 1 \ldots \theta_n \leqq \tfrac{1}{2}\Omega^0$, $\mathfrak{v}_n \leqq \tfrac{1}{2}\Omega^0$, $\mathfrak{D}_n \leqq \tfrac{1}{2}\Omega^0$.

Sei

$$J_\varepsilon = \int\limits_{F_0} d\mathfrak{f}_0' \left[\frac{\partial}{\partial \nu} \log \frac{\sqrt{\varrho_\varepsilon^2 + h^2} + h}{\varrho_\varepsilon} - \frac{\partial}{\partial \nu_0} \log \frac{\sqrt{\varrho_0^2 + h^2} + h}{\varrho_0} \right],$$

$$K_\varepsilon = \int\limits_{F_0} d\mathfrak{f}_0' \left[\frac{\partial}{\partial \tau} \log \frac{\sqrt{\varrho_\varepsilon^2 + h^2} + h}{\varrho_\varepsilon} - \frac{\partial}{\partial \tau_0} \log \frac{\sqrt{\varrho_0^2 + h^2} + h}{\varrho_0} \right], \qquad (2)$$

unter ϱ_ε der Abstand der Punkte

$$R + Z_\varepsilon, \quad \frac{u + S_\varepsilon}{R} \qquad \text{und} \qquad R' + Z_\varepsilon', \quad \frac{u' + S_\varepsilon'}{R'} \qquad (3)$$

verstanden. Die Ausdrücke (2) haben für alle dem absoluten Betrage nach hinreichend kleinen *komplexen* Werte von ε eine bestimmte Bedeutung. Für reelle Z und S findet man unmittelbar beispielsweise

$$K^{(1)} = \int\limits_{F_0} d\mathfrak{f}_0' \frac{\partial}{\partial \tau} \log \frac{\bar{R}}{\varrho} = \int\limits_{F_0} d\mathfrak{f}_0' \frac{R' + Z'}{\varrho^2} \sin\left(\frac{u' + S'}{R'} - \frac{u + S}{R} \right),$$

$$\varrho^2 = (R' + Z')^2 + (R + Z)^2 - 2(R + Z)(R' + Z') \cos\left(\frac{u' + S'}{R'} - \frac{u - S}{R} \right). \qquad (4)$$

Durch die gleiche Formel wird $K^{(1)}$ für komplexe Z und S *definiert*. Die Argumente R und u sind nach wie vor reell.

Es gilt also

$$K_\varepsilon^{(1)} = \int\limits_{F_0} d\mathfrak{f}_0' \frac{\partial}{\partial \tau} \log \frac{\bar{R}}{\varrho_\varepsilon} = \int\limits_{F_0} d\mathfrak{f}_0' \frac{R' + Z_\varepsilon'}{\varrho_\varepsilon^2} \sin\left(\frac{u' + S_\varepsilon'}{R'} - \frac{u + S_\varepsilon}{R} \right),$$

$$\varrho_\varepsilon^2 = (R' + Z_\varepsilon')^2 + (R + Z_\varepsilon)^2 - 2(R + Z_\varepsilon)(R' + Z_\varepsilon') \cos\left(\frac{u' + S_\varepsilon'}{R'} - \frac{u + S_\varepsilon}{R} \right). \qquad (5)$$

Wir wählen Ω^0 *so klein, daß für alle* $|\varepsilon| \leqq \Omega^0$ *und alle* $\left(R, \dfrac{u}{R} \right)$ *und* $\left(R', \dfrac{u'}{R'} \right)$ *in* F_0 *gewiß* $\left| \dfrac{\varrho_\varepsilon}{\varrho_0} \right| \geqq q > 0$ *ausfällt.*

Wie man leicht sieht, sind J_ε und K_ε für alle $|\varepsilon| \leqq \Omega^0$ analytische und reguläre Funktionen von ε. Ferner ist

$$\dot{J} - J = \int\limits_0^\Omega \frac{\partial J_\varepsilon}{\partial \varepsilon} d\varepsilon. \qquad (6)$$

Sei weiter C_0 der Kreis vom Radius Ω^0 um den Koordinatenursprung in der Ebene der komplexen Variablen ε.

Der Cauchysche Integralsatz liefert für $\dfrac{\partial J_\varepsilon}{\partial \varepsilon}$ $(|\varepsilon| \leqq \Omega)$ den Ausdruck

$$\frac{\partial J_\varepsilon}{\partial \varepsilon} = \frac{1}{2\pi i} \int\limits_{C_0} \frac{d\delta}{(\delta - \varepsilon)^2} J_\delta, \qquad (\delta = \Omega^0 e^{i\psi}) \quad (7)$$

In (7) bezeichnet J_δ den Wert, den das erste der Integrale (2) erhält, wenn man dem Parameter ε den Zahlwert $\delta = \Omega^0 e^{i\psi}$ erteilt. Durch eine sinngemäße Übertragung der bekannten, auf Herrn Hölder zurückgehenden Überlegungen der Theorie des logarithmischen Potentials einer ebenen Flächenbelegung läßt sich zeigen, daß auch jetzt noch J_δ und K_δ im Innern und auf dem Rande von F_0 stetige, der H-Bedingung genügende partielle Ableitungen erster Ordnung haben. Es gelten für alle den Beziehungen (43) II und (44) II genügenden Z, S; $\dot{Z}$, $\dot{S}$ und alle $|\delta| = \Omega^0$ gleichmäßig

$$\left.\begin{aligned} &|J_\delta|, \quad |K_\delta|, \quad \left|\frac{\partial J_\delta}{\partial x}\right|, \quad \left|\frac{\partial J_\delta}{\partial y}\right| \leqq \alpha_{16}, \\[2mm] &\left|\frac{\partial}{\partial x} J_\delta(x+h, y+k) - \frac{\partial}{\partial x} J_\delta(x, y)\right|, \\[2mm] &\left|\frac{\partial}{\partial y} J_\delta(x+h, y+k) - \frac{\partial}{\partial y} J_\delta(x, y)\right| \leqq \alpha_{16}\,(|h| + |k|)^\lambda. {}^1) \end{aligned}\right\} \tag{8}$$

Nach (7) und (8) finden wir zunächst für alle $|\varepsilon| \leqq \Omega \leqq \dfrac{\Omega^0}{2}$

$$\left|\frac{\partial J_\varepsilon}{\partial \varepsilon}\right| \leqq \frac{4}{\Omega^0}\,|J_\delta| \leqq \alpha_{17}, \tag{9}$$

mithin wegen (6)
$$|\dot{J} - J| \leqq \alpha_{17}\,\Omega. \tag{10}$$

Auf (7) folgt weiter
$$\frac{\partial^2 J_\varepsilon}{\partial x\,\partial \varepsilon} = \frac{1}{2\pi i}\int\limits_{C_0} \frac{d\delta}{(\delta - \varepsilon)^2}\,\frac{\partial J_\delta}{\partial x}\ {}^2). \tag{11}$$

Aus (8), (6) und (11) folgt darum
$$\left|\frac{\partial}{\partial x}(\dot{J} - J)\right| \leqq \alpha_{18}\,\Omega. \tag{12}$$

Analog ist
$$\left|\frac{\partial}{\partial y}(\dot{J} - J)\right| \leqq \alpha_{18}\,\Omega. \tag{13}$$

In ähnlicher Weise finden wir die Ungleichheiten

$$\left.\begin{aligned} &\left|\frac{\partial}{\partial x}[\dot{J}(x+h, y+k) - J(x+h, y+k)] - \frac{\partial}{\partial x}[\dot{J}(x, y) - J(x, y)]\right| \\[2mm] &\left|\frac{\partial}{\partial y}[\dot{J}(x+h, y+k) - J(x+h, y+k)] - \frac{\partial}{\partial y}[\dot{J}(x, y) - J(x, y)]\right| \\[2mm] &\qquad\qquad \leqq \alpha_{19}\,\Omega\,(|h| + |k|)^\lambda. \end{aligned}\right\} \tag{14}$$

Die gleichen Überlegungen führen zu entsprechenden Formeln für K. Die Ungleichheiten (45) II sind eine unmittelbare Folge der Beziehungen (10), (12), (13) und (14).

---

^{1}) Wir führen jetzt vorübergehend kartesische Koordinaten ein. Der Punkt (x, y) ist mit dem Punkt $\left(R, \dfrac{u}{R}\right)$ in F_0 identisch. Man überzeugt sich leicht, daß $|Z_\varepsilon|$, $|S_\varepsilon|$, $\left|\dfrac{\partial Z_\varepsilon}{\partial u}\right|, \ldots \leqq 2\Omega^0$ sind.

^{2}) In (11) werden J_ε und J_δ als Ortsfunktionen in F_0 aufgefaßt.

Zur Durchmusterung des Problème restreint.

Asymptotisch-periodische Lösungen.

Von **Elis Strömgren**, Kopenhagen.

Mit 11 Abbildungen.

Den Anregungen folgend, die in den von Lagrange und Hill erreichten Resultaten über Librationspunkte einerseits und über periodische Lösungen des Dreikörperproblemes anderseits lagen, hat Poincaré in seinen „Méthodes nouvelles de la Mécanique Céleste" (1892 bis 1899) das Problem der periodischen Lösungen studiert und die Existenz solcher Lösungen im Dreikörperproblem mit einer überwiegenden Masse nachgewiesen.

In den Arbeiten über periodische Lösungen, die wir in Kopenhagen in den letzten Jahrzehnten ausgeführt haben[1]), ist das erweiterte „Problème restreint" behandelt worden, das von einem Gesichtspunkte allgemeiner ist als das Poincarésche Problem, indem wir auf die überwiegende eine Masse verzichtet haben. Unsere Methode ist, ebenso wie in dem Darwinschen Werk „Periodic Orbits", eine gemischte: Theorie und numerische Rechnung in stetiger Zusammenarbeit. Unsere Hilfsmittel und die Ausgangspunkte bei der Arbeit waren: die Thielesche Transformation (die die Singularitäten in den endlichen Massenpunkten aufhebt); die seit langem bekannten infinitesimalen periodischen Lösungen um die Librationspunkte; die von Herrn Fischer-Petersen definierten „charakteristischen Kurven"[2]); die von Hill und Bohlin in das Problem eingeführte Aufteilung der Bahnebene in erlaubte und unerlaubte Gebiete, nebst einigen allgemeinen Sätzen über Spitzen und

---

[1]) Zusammenfassende Übersichten der erhaltenen Resultate sind in den folgenden zwei Arbeiten gegeben: „Forms of periodic motion in the restricted problem and in the general problem of three bodies, according to researches executed at the Observatory of Copenhagen" (a lecture delivered at the inter-Scandinavian congress of mathematicians at Helsingfors, july 1922), Publ. der Kopenhagener Sternwarte Nr. 39, und „Tre Aartier Celest Mekanik paa Köbenhavns Observatorium" (Festschrift der Universität Kopenhagen, Nov. 1923).

[2]) Siehe z. B. Publ. der Kopenhagener Sternwarte Nr. 27, S. 20.

Schleifen im Problème restreint, die von dem Unterzeichneten im Jahre 1907 in den Astr. Nachr. Nr. 4155 mitgeteilt sind[1]); zuletzt — und vor allen Dingen — die mächtige Methode der numerischen Integration.

———

Die Hauptaufgabe, die wir uns gestellt hatten, war diese: eine möglichst vollständige Durchmusterung des Gebietes des Problème restreint nach einfach-periodischen Bahnen, und dies Ziel ist jetzt in weitem Umfange erreicht.

Ein Hauptresultat aus den letzten Jahren ist die Erkenntnis, daß wenigstens die meisten der in unserem Problem existierenden Klassen periodischer Bahnen in irgendeiner Weise einen natürlichen Abschluß finden, und zwar nicht so, wie BURRAU sich die Sache zurechtgelegt hatte — und so, wie wir es auch lange auffaßten —, daß der natürliche Abschluß einer Klasse einfach-periodischer Bahnen mit der Ejektionsbahn identisch sei, zu der wir allmählich geführt werden, wenn die periodische Bahn einer der zwei endlichen Massen immer näherrückt.

Eine solche Ejektionsbahn gleitet von selbst in eine kompliziertere Bahn hinüber, eine mit Schleife um die endliche Masse versehene Bahn. Wenn wir uns, was ja tatsächlich der Fall war, ausschließlich auf einfach-periodische Bahnen beschränken wollten, wäre man anscheinend berechtigt, bei der Ejektionsbahn abzubrechen und diese Ejektionsbahn als den natürlichen Abschluß der Klasse aufzufassen. Nun stellte es sich aber allmählich heraus, daß es Fälle gab, wo die Klasse, über die Ejektionsbahn weitergeführt, schließlich wieder zu einfach-periodischen Bahnen zurückkehrte, in einer solchen Weise, daß wir mit einer in sich geschlossenen Klasse zu tun hatten.

Der erste hierher gehörende Fall ist die in der Kopenhagener Publ. Nr. 39 (S. 26) mit o bezeichnete Klasse. In derselben Publikation ist auch angedeutet worden, daß wir für mehrere andere Klassen eine ähnliche Entwicklung erwarteten: Klasse a (die THIELE-BURRAUsche Klasse)—und damit natürlich auch Klasse b — Klasse k (die direkten Bahnen um die beiden endlichen Massen, mit zwei Unterklassen) und Klasse l (retrograde Bahnen um beide Massen, absolute Bewegung direkt). Inzwischen ist für alle diese Klassen der Beweis dafür gebracht worden, daß unsere Vermutung richtig war: diese Klassen bilden, jede für sich, in sich selbst geschlossene Bahnklassen, die uns nur vorübergehend in das trostlose — weil endlose — Gebiet der komplizierteren Bahnformen führten.

Die Entwicklung der Klasse a bzw. b wird an anderer Stelle gegeben werden. Für die Klassen k_1, k_2 und l werden wir im folgenden die Hauptresultate wiedergeben.

———

[1]) In der Hauptsache dieselben Sätze hat F. R. MOULTON, S. 496, in seinem 1920 herausgegebenen Werke „Periodic Orbits" gegeben, anscheinend ohne meine Arbeit aus dem Jahre 1907 zu kennen.

Außer den *periodischen* Bahnen im Dreikörperproblem hat POINCARÉ in seinen „Méthodes nouvelles" den von ihm entdeckten *asymptotischen* Bahnen große Aufmerksamkeit gewidmet. Die Untersuchungen fußen auf der Behandlung der sog. Variationsgleichungen des Problems. Von einer in der Zeit t periodischen Lösung für eine Anzahl Veränderliche

$$x_i = \varphi_i(t), \qquad (i = 1, 2, \ldots n)$$

ausgehend, setzt man in die Differentialgleichungen des Problems

$$x_i = \varphi_i(t) + \xi_i$$

ein. Man entwickelt nach Potenzen von ξ und behält nur die Glieder ersten Grades in bezug auf ξ. Dadurch erhält man die sog. Variationsgleichungen des Problems, die aus einem System homogener, linearer Differentialgleichungen mit periodischen Koeffizienten bestehen. Die Lösung ergibt sich — von einem Spezialfall abgesehen — in der Form:

$$\xi_1 = e^{\alpha_p \cdot t} S_{1,p} \ldots \bar{\xi}_n = e^{\alpha_p \cdot t} S_{n,p} \qquad (p = 1, 2, \ldots n),$$

wo α_p Konstanten sind und $S_{i,p}$ periodische Funktionen mit derselben Periode wie $\varphi_i(t)$ bedeuten.

Die Konstanten α_p sind die sog. charakteristischen Exponenten. Die Form der Lösung des Problems hängt von diesen Exponenten ab.

Wenn der reelle Teil eines Exponenten nicht gleich Null ist, erhalten wir für die ξ-Werte einen periodischen Ausdruck, mit dem Faktor e^a, oder e^{-at} versehen (wo a eine reelle, positive Größe bedeutet), also Lösungen, die sich für $t = -\infty$ bzw. $t = +\infty$ einer periodischen Lösung unbegrenzt nähern.

POINCARÉ hat auch die Existenz von „doppelt-asymptotischen" Lösungen bewiesen, d. h. Bewegungsformen, wo der Körper für sowohl $t = -\infty$ wie für $t = +\infty$ sich einer periodischen Lösung asymptotisch nähert.

––––––––

Wenn wir überhaupt die Gesichtspunkte zusammenstellen wollten, die für die Untersuchung der Bewegungsverhältnisse im Problème restreint maßgebend gewesen sind, könnte es in der folgenden Weise getan werden. Man hat den folgenden Klassen von Bahnen besondere Aufmerksamkeit geschenkt:

1. den periodischen Bahnen;
2. hiervon besonders den einfach-periodischen Bahnen;
3. den asymptotischen Bahnen;
4. hiervon besonders den doppelt-asymptotischen Bahnen;
5. Bahnen, die einen natürlichen Abschluß einer ganzen Klasse periodischer Bahnen bilden.

Die Bahnen, die den Hauptgegenstand der vorliegenden Untersuchung bilden, besitzen alle diese Merkmale auf einmal.

––––––––

In der auf S. 228 zitierten Abhandlung „Tre Aartieı Celest Mekanik paa Köbenhavns Observatorium" sind in dänischer Sprache die Überlegungen wiedergegeben, die uns zu dem vorliegenden Problem führten. Die im Jahre 1918 veröffentlichte Untersuchung der retrograden Bahnen um die beiden endlichen Massen (absolute Bewegung direkt)[1] gab eine Entwicklung, die durch die Abb. 1 dargestellt wird. Die Entwick-

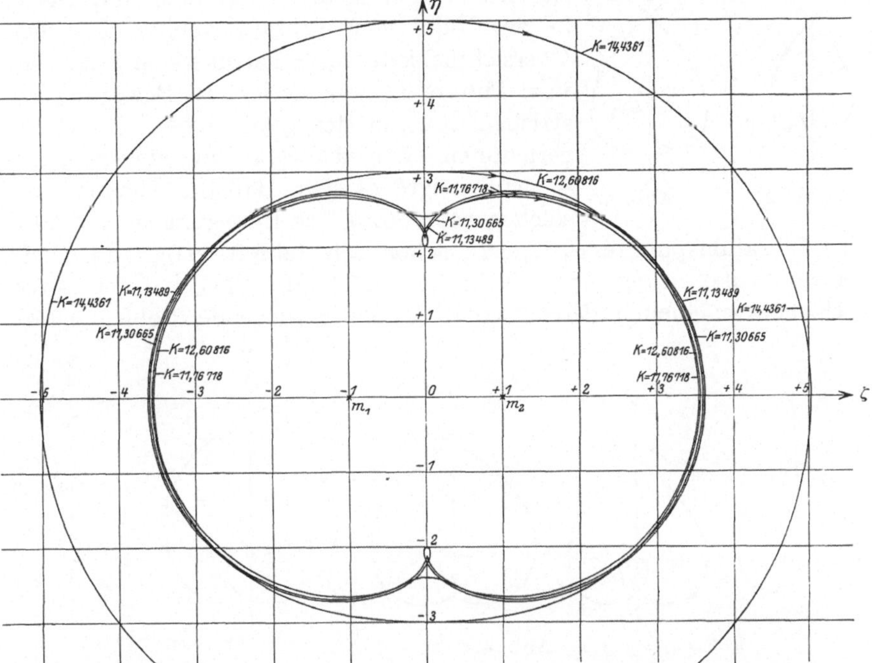

Abb. 1. Klasse *l*. Retrograde Bahnen um beide Massen, absolute Bewegung direkt.

lung fängt mit zirkulären Bahnen mit unendlich großem Radius an. Wenn wir nach innen gehen, platten sich die Bahnen allmählich ab; auf einem gewissen Stadium bemerkt man auf der η-Achse eine nach innen gerichtete Einbuchtung, die später zu einer Spitze (Bahngeschwindigkeit Null) und danach zu einer Schleife (Bahngeschwindigkeit sehr klein) wird; alles nach innen, d. h. oberhalb der ξ-Achse nach unten (Abb. 2), unterhalb der ξ-Achse nach oben gerichtet, und alles sich in der Nähe der Librationspunkte L_4 und L_5 abspielend. Eine weitergeführte verfeinerte Untersuchung zeigte, daß die kleine Schleife in Abb. 2 später auf der unteren Seite eine kleine, nach oben gerichtete Einbuchtung erhält (Abb. 3). Hiernach lag es nahe, sich die weitere Entwicklung in der folgenden Weise vorzustellen. Die kleine Einbuch-

---

[1] Kopenhagener Publ. Nr. 30.

tung unten in der Abb. 3 entwickelt sich zu einer Spitze, dann zu einer Schleife. Und, wenn wir weiter so fortsetzen, ergibt sich schließlich die in Abb. 4 dargestellte Entwicklung: eine in zwei Spiralen asymptotisch, d. h. mit gegen Null konvergierender Geschwindigkeit, in den Librationspunkt L_4 hinein bzw. von demselben Punkte wegführende Bewegung. Ganz analog entwickelt sich selbstverständlich die Sache in der Nähe vom Librationspunkt L_5. Der Übersichtlichkeit wegen haben wir in Abb. 5 die vom Punkte L_4 weg gerichtete Bewegung gestrichen und nur den gegen diesen Punkt hin gerichteten Zweig der Bewegung stehen lassen.

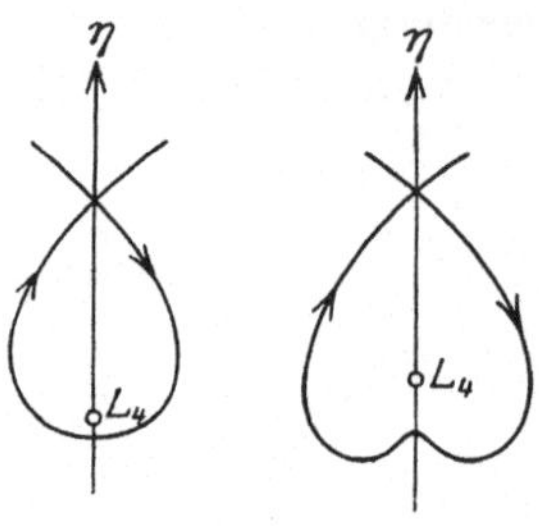

Abb. 2.   Abb. 3.

Nun zeigte es sich, daß die Theorie einer solchen asymptotischen Bewegung in der Nähe der Librationspunkte L_4 und L_5 schon seit Jahren fertig vorlag. In einem Artikel in den Astr. Nachr. Nr. 4015[1]) (Mai 1905) hat der Unterzeichnete zu einem anderen Zwecke das hier vorliegende Problem gelöst.

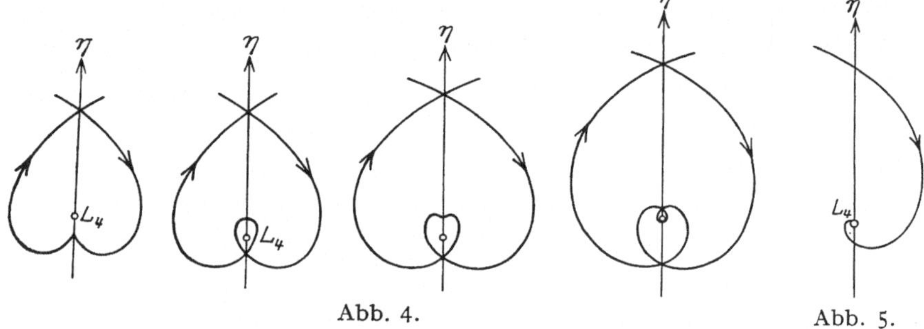

Abb. 4.   Abb. 5.

Die Bewegungsgleichungen der unendlich kleinen Masse in der unmittelbaren Nähe vom Librationspunkt L_4 lauten (siehe z. B. Moultons Celestial Mechanics, 1. Aufl., S. 210):

$$\left.\begin{aligned}
\frac{d^2 x}{d t^2} - 2\frac{dy}{dt} &= \frac{3}{4}x + \frac{3\sqrt{3}}{4}(1 - 2\mu)y\,,\\[2mm]
\frac{d^2 y}{d t^2} + 2\frac{dx}{dt} &= \frac{3\sqrt{3}}{4}(1 - 2\mu)x + \frac{9}{4}y\,,
\end{aligned}\right\} \tag{1}$$

wo μ die kleinere der zwei endlichen Massen bedeutet, $1 - \mu$ die größere, und die Einheiten sonst die in dem zitierten Werke benutzten sind (diese Einheiten sind nicht mit denjenigen identisch, die wir sonst in den Kopenhagener Publikationen gebrauchen; der Unterschied spielt doch für das Folgende keine Rolle).

---

[1]) Ein asymptotischer Fall im Dreikörperproblem.

Wie sofort ersichtlich, vereinfachen sich die Glieder rechts in den Gleichungen (1) wesentlich, wenn wir uns gleich auf das in den Kopenhagener Arbeiten immer benutzte Massenverhältnis $\mu = \frac{1}{2}$ spezialisieren.

Die Hauptresultate der zitierten Abhandlung lauten folgenderweiße: Wir erhalten für die Koordinaten der unendlich kleinen Maße, x und y, die auf L_4 als Origo und auf ein Achsensystem bezogen sind, das sich in üblicher Weise mit derselben Geschwindigkeit dreht wie das System der zwei endlichen Massen:

$$\left.\begin{array}{l} x = 2\,A e^{at} \cos bt + 2B e^{at} \sin bt\,, \\[4pt] y = 2\,C e^{at} \cos bt + 2D e^{at} \sin bt\,, \end{array}\right\} \qquad (2)$$

wo A, B, C, D Integrationskonstanten bedeuten (wovon nur zwei unabhängig sind). Die Buchstaben a und b stellen reelle Größen dar, und a hat einen positiven Wert, so lange wie wir uns auf die Glieder beschränken, die für $t = -\infty$ die asymptotische Lösung geben.

In unserem Probleme ($\mu = \frac{1}{2}$) haben a und b die folgenden Werte:

$$\left.\begin{array}{ll} a = \dfrac{\sqrt{23}}{2\sqrt{2}\,\sqrt{2 + 3\sqrt{3}}}\,, & \log a = 9.80077 - 10\,, \\[18pt] b = \dfrac{\sqrt{2 + 3\sqrt{3}}}{2\sqrt{2}}\,, & \log b = 9.97700.5 - 10\,. \end{array}\right\} \qquad (3)$$

Die ausführliche Behandlung des Problems soll in einer späteren größeren Arbeit gegeben werden. Wir beschränken uns hier auf die Hauptresultate, die in der folgenden Weise ausgedrückt werden können:

1. Die Bewegung geht in einer Spirale vor sich, nach den obigen Festlegungen von L_4 weg.

2. Weil a und b beide von Null verschiedene Werte besitzen, gibt es nur gemischt exponentiell-trigonometrische Glieder, keine rein exponentielle und keine rein trigonometrische (bekanntlich gibt es in der Umgebung von L_4 und L_5 periodische Bewegungen, wenn $1 - 27\mu(1 - \mu) \geqq 0$, d. h. wenn $\mu \leqq 0.0385 \ldots$). Rein exponentielle Glieder gibt es, wie später besprochen werden wird, in den entsprechenden Problemen für L_1, L_2 und L_3.

3. Es gibt unendlich viele Spiralbewegungen von den betreffenden Librationspunkten weg (und auf sie zu) gerichtet.

4. Für die Umlaufszeit (T) in den Spiralen erhalten wir bei unserer Wahl der Einheiten den für alle Spiralen gültigen Wert:

$$T = \frac{2\pi}{b} = 6.624\,83 \ldots$$

5. Der Faktor e^{at} erhält für einen ganzen Umlauf in der Spirale den Wert:

$$e^{aT} = 65.850 \ldots$$

und für die halbe Umlaufszeit den Wert:

$$e^{a\frac{T}{2}} = 8.1148\ldots$$

Hier haben wir also das Verhältnis der Stücke der η-Achse von L_4 aus gerechnet, die von den aufeinander folgenden Spiralwindungen abgeschnitten werden.

6. Der konstante Winkel (Θ), den die Bahntangente in einem Punkte auf der η-Achse mit dieser Achse macht, ist durch folgende Gleichung gegeben:

$$\mathrm{Tg}\,\Theta = \frac{1}{a},$$

woraus sich der für alle Spiralen gültige Wert

$$\Theta = 57°42'.25$$

ergibt.

Die jetzt angedeuteten Sätze haben nur so lange strenge Gültigkeit, als wir die Entfernung von dem Librationspunkte als differentiell behandeln dürfen. In einer Abhandlung aus dem Jahre 1919[1]) hat Buchanan in diesem Probleme höhere Potenzen mitgenommen und ein spezielles numerisches Beispiel bis auf die zweite Potenz durchgeführt. Für unsere Zwecke würde die Berechnung auf diesem Wege allzu umständlich werden, wenn überhaupt möglich, und wir haben deshalb einen anderen Weg eingeschlagen: durch numerische Integration der exakten Differentialgleichungen des Problems haben wir die Bahnbewegung von einem Punkte $\hat{\xi} = 0$, $\eta = \eta_0$ an verfolgt, der dem betreffenden Librationspunkte sehr nahe liegt. Dem Anfang der Rechnung, also vom Librationspunkt bis zum Punkte 0, η_0, liegt die differentielle Theorie zugrunde. Mit Hilfe des Jacobischen Integrals ließ sich der zu dem Punkte 0, η_0 gehörende Wert der Bahngeschwindigkeit aus dieser Theorie berechnen; ebenso, mit Hilfe des oben angegebenen Wertes des Winkels Θ, die Komponenten dieser Geschwindigkeit auf die ξ- bzw. η-Achse. Durch Anwendung der oben gegebenen Sätze 4 bis 6 über die Spiralbewegung ließ sich die Genauigkeit der Ingangsetzung der Rechnung immer kontrollieren.

Bei der außerordentlich umfassenden Rechenarbeit haben mich die folgenden Mitarbeiter unterstützt: Dr. Burmeister und Herr Stammhammer (München), Dr. Stobbe und Frl. Dr. Güssow (Neubabelsberg), Dr. Noteboom (Rathenow), Admiral Korsakoff, Assistent Johannsen und Herr Lökkegaard (Kopenhagen). An der Organisation der ganzen Arbeit hat Herr Assistent J. P. Möller teilgenommen.

---

[1]) Asymptotic Satellites near the equilateral-triangle equilibrium points in the problem of three bodies (Transactions of the Cambridge Philosophical Society).

Wie oben (S. 229) angedeutet, soll auch die Klasse k der Publ. 39, mit ihren zwei Unterklassen k_1 und k_2, hier behandelt werden. Das Wesentliche der Entwicklung dieser Klasse wird in den Abb. 6—8

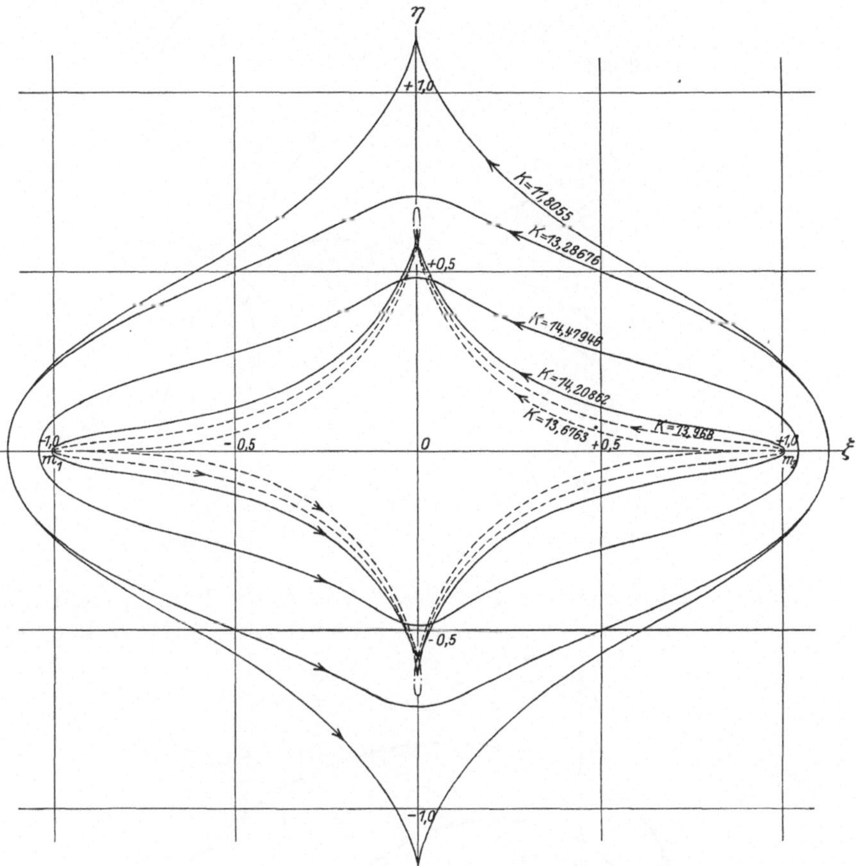

Abb. 6.   Klasse k. Direkte Bahnen um beide Massen.

dargestellt: die Entwicklung der direkten Bahnen um die beiden end-lichen Massen, die sich in zwei Unterklassen aufteilen lassen (nähere Details s. Publ. 32 und 39). Die Bahnen jeder Unterklasse besitzen große, um die η-Achse symmetrische Schleifen in der Nähe von den Librationspunkten L_4 und L_5, und die weitere Entwicklung der Bahnen geht ganz analog der Entwicklung der Klasse l vor sich, bis die Bahnen asymptotisch in L_4 und L_5 enden.

Als diese Entwicklung der Bahnklassen k_1, k_2 und l uns klar war, wurde die Arbeit mit den asymptotischen Bahnen in Gang gesetzt.

Die Resultate werden durch die drei Abb. 9 bis 11 dargestellt. Das ganze Gebiet der gegen L_4 und L_5 asymptotischen Bahnen sollte durch

Variation des Anfangswertes von η durchmustert werden. Abb. 9 gibt
den einen Quadranten der bis jetzt erhaltenen fünf asymptotischen

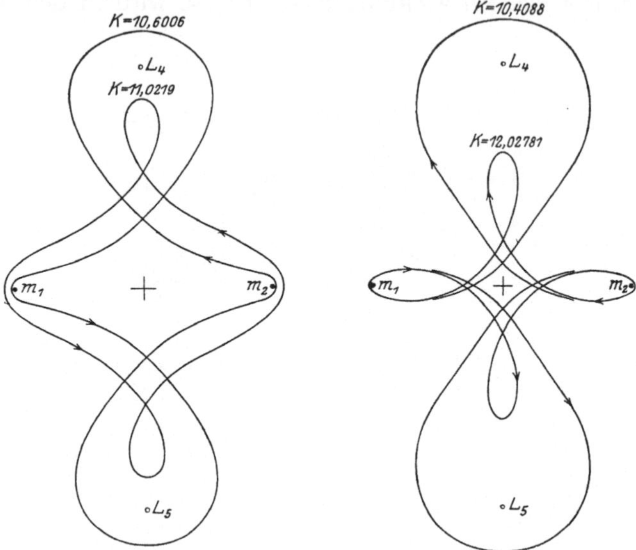

Abb. 7.   Unterklasse k_1.   Abb. 8. Unterklasse k_2.

Bahnen, *die gleichzeitig periodisch sind*. Von diesen Bahnen stellt die
Bahn I die Begrenzungsbahn der Klasse k_1 (vgl. Abb. 7), II die Begren-

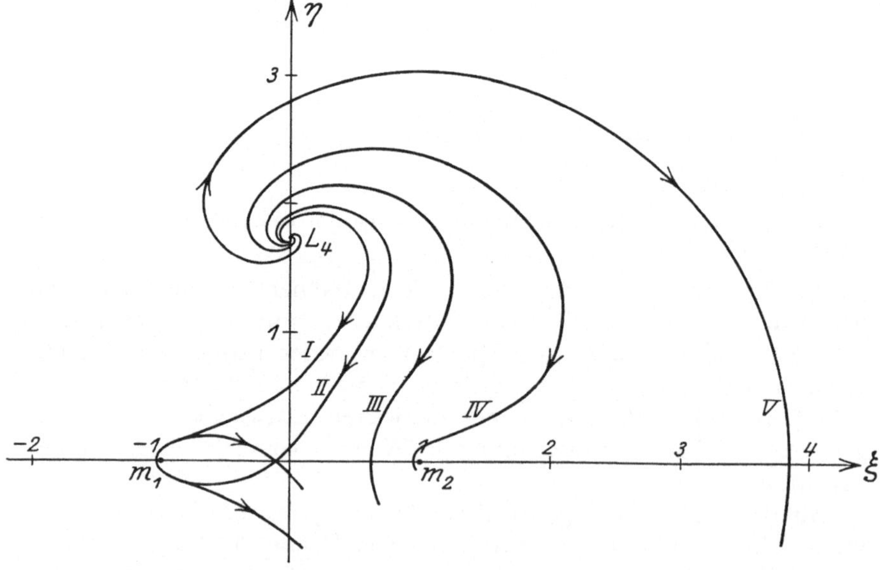

Abb. 9.   Asymptotisch-periodische Bahnen. Bahnen I, II und V sind die Be-
grenzungsbahnen der Klassen k_1 bzw. k_2 und l.

zungsbahn der Klasse k_2 (vgl. Abb. 8) und V die Begrenzungsbahn der Klasse l dar (vgl. Abb. 1). Abb. 10 gibt in größerem Maßstabe den Anfang dieser Bahnen, und Abb. 11 stellt die Bahn V dar, durch die entsprechenden Teile der Bewegung in den drei anderen Quadranten zu einem vollen Bilde der Bewegungsverhältnisse ergänzt (vgl. auch Abb. 1).

Die Entwicklung der drei Bahnklassen k_1, k_2 und l liegt also jetzt vollständig klar: die Klassen k und l bilden beide in sich geschlossene Klassen. Die Untersuchung hat uns als Nebenprodukte noch die asymptotisch-periodischen Bahnen III und IV gegeben. Und wir sind damit noch nicht zu Ende. So ist es durch Untersuchungen, die nach dem Abschluß der Rechnungen ausgeführt sind, die dem vorliegenden Artikel zugrunde liegen, mit ziemlicher Sicherheit festgestellt worden, daß eine zwischen IV und V liegende asymptotisch-periodische Bahn

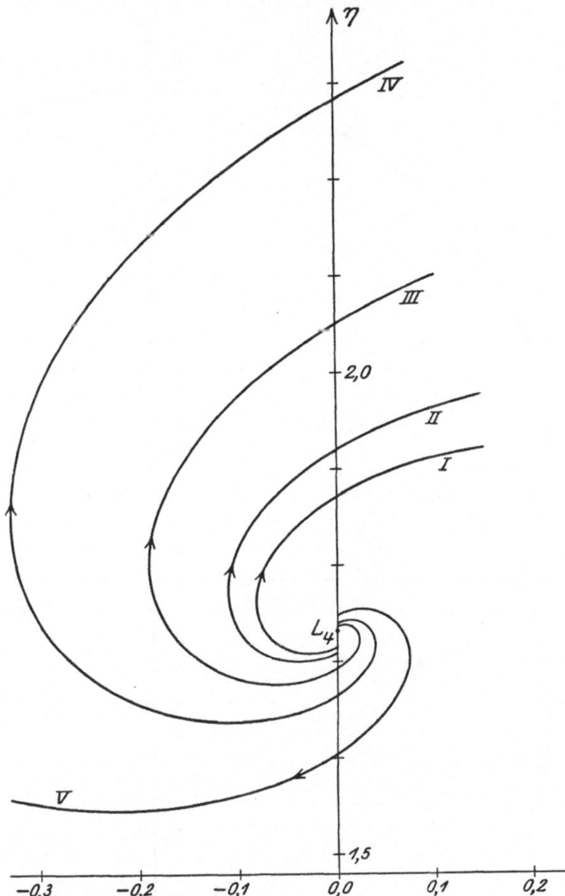

Abb. 10. Asymptotisch-periodische Bahnen. Anfang der Bahnen in vergrößertem Maßstabe.

mit Schleifen um die zwei endlichen Massen die Grenzbahn der (über die Doppelejektionsbahn hinaus weitergeführten) Librationen um L_1 (Gruppe c der Kopenhagener Publikation 39) ausmacht. Darüber, sowie auch über die Stellung der zwei Bahnen III und IV im ganzen Problem, Näheres in einer späteren Arbeit.

Die POINCARÉschen Untersuchungen über asymptotische Lösungen bezogen sich ganz allgemein auf Bewegungen, die sich *einer periodischen Lösung* asymptotisch nähern. In den Problemen, die wir hier studiert haben, war (im rotierenden Achsensystem) die generierende periodische Lösung auf einen Punkt — einen Librationspunkt — zusammengeschrumpft. Wir haben gesehen, daß es in diesen Fällen (bei

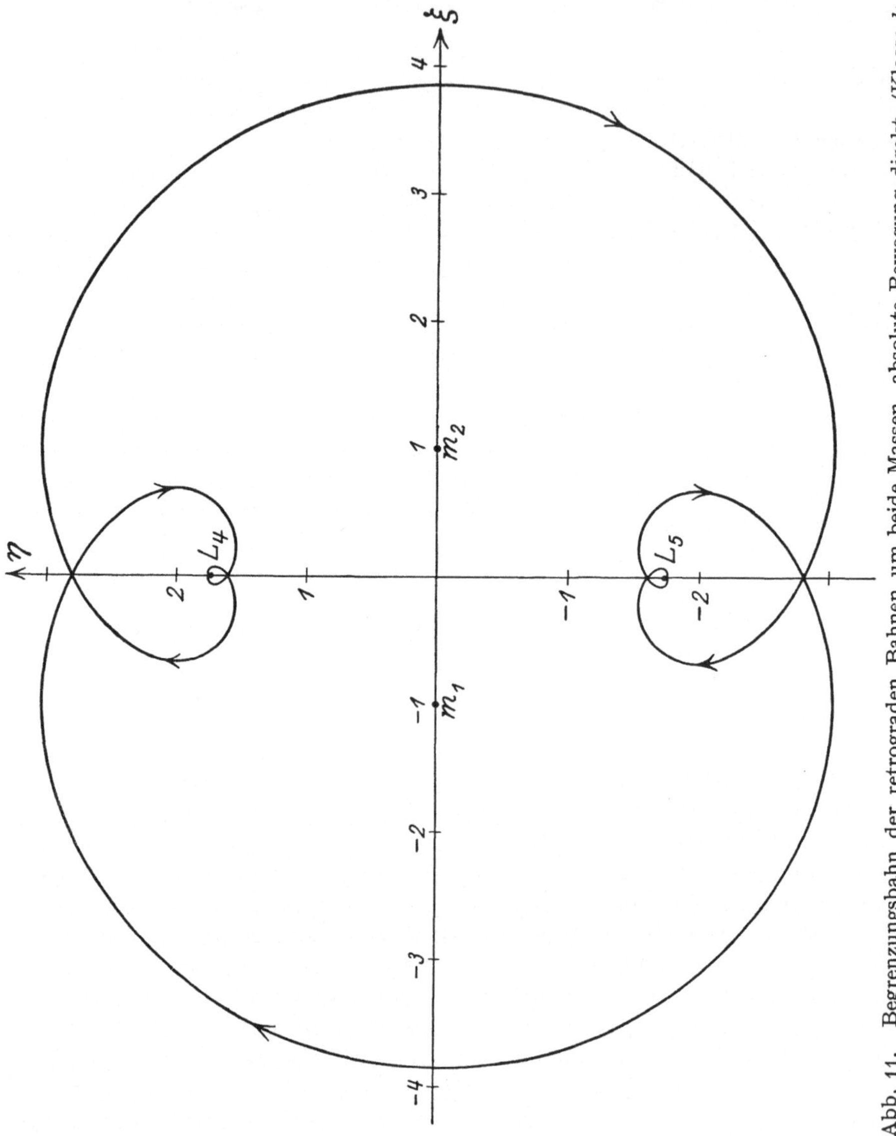

Abb. 11. Begrenzungsbahn der retrograden Bahnen um beide Massen, absolute Bewegung direkt, (Klasse l; Bahn V in den Abb. 9 und 10).

L_4 und L_5) unendlich viele asymptotische Lösungen gab, darunter eine endliche Zahl einfach-periodische. In den Fällen L_1, L_2 und L_3 dagegen liegen die Verhältnisse ganz anders, indem es hier nur rein exponentielle Lösungen gibt, und zwar nur je zwei Lösungen gegen den Librationspunkt bzw. von ihm weggerichtet[1]). Diese zuletzt erwähnten Fälle bieten deshalb ein verhältnismäßig geringes Interesse. Und es ist leicht zu zeigen, daß es keine Lösungen gibt, die zu anderen im rotierenden Koordinatensystem festen Punkten asymptotisch sind, als gerade zu den fünf Librationspunkten.

––––––

Das jetzt mühsam aufgebaute Gerüst von periodischen und asymptotisch-periodischen Bahnen im Problème restreint hätte sich ohne die Methoden der numerischen Integration wohl nie konstruieren lassen. Wie schwierig es ist, ohne solche handgreifliche Grundlagen zu richtigen Resultaten zu kommen, zeigen mit aller Deutlichkeit verschiedene Versuche, die MOULTON im zusammenfassenden Schlußkapitel seiner ,,Periodic Orbits" gemacht hat, die Entwicklung gewisser Bahnklassen vorherzusagen.

So wie die Verhältnisse jetzt liegen, darf es wohl gesagt werden, daß für das Problème restreint ein so umfassendes, übersichtliches, konkretes Material vorliegt, wie es mit keinem zweiten mathematisch ungelösten Problem der Fall ist.

––––––

[1]) WARREN, L. A. H.: A Class of asymptotic orbits in the Problem of three Bodies (American Journal of Mathematics 1916).

Zur Weiterentwicklung der Weltgeometrie (Relativitätstheorie).

Von **A. Kopff**, Heidelberg-Königstuhl.

1. Einleitung. Durch die letzten Arbeiten vor allem von WEYL, EDDINGTON und EINSTEIN hat die Relativitätstheorie eine Richtung eingeschlagen, die vielleicht zuerst den Eindruck erweckt, als seien die Probleme rein formaler Natur, und ihre Verknüpfung mit den Ergebnissen der Beobachtung drohe mehr und mehr sich zu verlieren. Vertieft man sich jedoch etwas weiter in die Gedankengänge, wie sie besonders EDDINGTON in seinem Lehrbuch[1]) dargelegt hat, so ergibt sich doch, daß die Weiterentwicklung, die notwendig aus dem bisher Erreichten hervorgeht, eines tieferen physikalischen Sinnes nicht entbehrt. Der Weg führt nicht von der Natur ab; nur muß man daran denken, daß man nicht fragen kann: Wie ist die Natur?, sondern fragen muß: Wie stellt die Natur sich unseren Sinnen und unserer Vernunft dar? Betrachtet man, wie es im folgenden geschehen soll, ausgehend von der klassischen Physik, die Entwicklung der Relativitätstheorie unter diesem Gesichtspunkt, so zeigt sich, daß der Sinn der physikalischen Naturforschung hierbei nicht verdunkelt wird, sondern sich nur immer klarer heraushebt.

2. Die klassische Physik. In der Einleitung zu seinen Prinzipien der Mechanik sagt HEINRICH HERTZ: „Wir machen uns innere Scheinbilder oder Symbole der äußeren Gegenstände, und zwar machen wir sie von solcher Art, daß die denknotwendigen Folgen der Bilder stets wieder die Bilder seien von den naturnotwendigen Folgen der abgebildeten Gegenstände." Damit ist das Ziel der Physik klar umschrieben. Diese Scheinbilder sind die physikalischen Begriffe, von denen unser Denken ausgeht.

Die klassische Physik kennt zwei solche Begriffssysteme, das der Mechanik und das der Elektrodynamik, aus denen alle Erscheinungen sich darstellen lassen. Die Mechanik baut sich auf den Begriffen Bewegung (Geschwindigkeit, Beschleunigung), Kraft und Masse auf; die Elektrodynamik auf elektrischer und magnetischer Kraft, Elektrizitätsmenge und elektrischem Strom, und dazu kommt bei der Darstellung der mechanischen Vorgänge elektrodynamischen Ursprungs der Begriff

---

[1]) EDDINGTON, A. S.: The Mathematical Theory of Relativity. Cambridge University Press 1923.

der ponderomotorischen Kraft. Die Begriffe Raum und Zeit bestehen unabhängig voneinander und von beiden Systemen; das physikalisch Reale wird in Raum und Zeit eingeordnet, die beide euklidische Struktur besitzen.

Die beiden Begriffssysteme enthalten jedoch — und darin liegt ihre augenfälligste Schwäche — Absolutes, was sich der sinnlichen Wahrnehmung entzieht. Man wird EINSTEIN[1]) schon recht geben müssen, wenn er von einem Begriffssystem in der Physik verlangt: „Begriffe und Unterscheidungen sind nur insoweit zulässig, als ihnen beobachtbare Tatbestände eindeutig zugeordnet werden können." Diese Forderung ist jedoch vom Absoluten nicht erfüllt.

Das Absolute in der Mechanik ist der absolute Raum oder das Inertialsystem, worauf sich der Begriff der Trägheitsbewegung aufbaut. Das Absolute ist hierbei durch die Formulierung der NEWTONschen Bewegungsgesetze bedingt, und man hat die Berechtigung dieses Absoluten durch das Auftreten der Zentrifugalkräfte bei der Rotation zu begründen versucht. Wahrnehmbar ist jedoch nur die Rotation als Bewegung in bezug auf die äußeren Massen, und auf die Existenz dieser letzteren müßte man das Zentrifugalfeld zurückzuführen suchen. In der Elektrodynamik ist das Absolute der Äther im Sinne der MAXWELL-LORENTZschen Theorie, und für diese Theorie ist er identisch mit dem an ein bevorzugtes Koordinatensystem gebundenen elektromagnetischen Kraftfeld. Alle Versuche, dieses durch die Gültigkeit der Grundgleichungen ausgezeichnete Koordinatensystem nachzuweisen, sind gescheitert. In Mechanik und Elektrodynamik erhebt sich also die Forderung, das Begriffssystem so aufzustellen, daß es von dem bevorzugten Koordinatensystem des absoluten Raumes bzw. des ruhenden Äthers frei wird, daß es unabhängig von irgendeinem besonderen Bezugssystem gilt.

3. Die spezielle Relativitätstheorie. Die spezielle Relativitätstheorie baut sich auf der Invarianz der Grundgleichungen der Elektrodynamik gegenüber der Lorentz-Transformation auf. Das *eine* bevorzugte System des ruhenden Äthers wird dadurch illusorisch. Das Absolute bleibt aber, ebenso wie in der klassischen Mechanik, als die dreifach unendliche Schar von Koordinatensystemen bestehen, für welche die Grundgleichungen gelten.

Dadurch, daß aber in der Physik die Galilei-Transformation durch die Lorentz-Transformation ersetzt wird, treten in der physikalischen Begriffsbildung grundlegende Änderungen ein, durch die das Wesen der physikalischen Naturerkenntnis erst in voller Deutlichkeit hervortritt. Die zeitlichen Größen werden ebenso wie die räumlichen abhängig von

---

[1]) EINSTEIN, A.: Grundgedanken und Probleme der Relativitätstheorie. (Nord. Naturforschervers. Gotenburg 11. VII. 1923.) Lex prix Nobel en 1921—22. Stockholm 1923.

der Wahl des bewegten Koordinatensystems. Raum und Zeit verlieren damit ihre Selbständigkeit, und ihre Struktur ist durch die physikalischen Begriffe bedingt, deren Verknüpfung in Raum und Zeit erfolgt. Geometrisch werden beide gleichartig; sie lassen sich zu einem vierdimensionalen Raum-Zeit-Kontinuum, der *Welt*, zusammenfassen. Die Geometrie der Welt ist pseudoeuklidisch. Alle physikalischen Größen und Begriffe sind geometrisch aufgefaßt Skalare, Vektoren oder Tensoren (allgemein: Tensoren) in diesem vierdimensionalen Kontinuum, und zwar kontravariante Tensoren, die sich transformieren wie Strecken, oder kovariante Tensoren, die sich transformieren wie die durch die Differentialquotienten eines Potentials dargestellten Kraftkomponenten. Wie die realen Dinge selbst, bestehen die Tensoren begrifflich unabhängig vom Koordinatensystem; nur die numerischen Werte ihrer Komponenten sind durch die Koordinatenwahl bedingt. Andere Begriffe, als die durch solche Tensoren darstellbaren, dürften infolge der Gebundenheit unserer sinnlichen Wahrnehmung an Raum und Zeit in der Physik nicht vorkommen. Jedes physikalische Begriffssystem ist zugleich ein System von Tensoren im vierdimensionalen Raum-Zeit-Kontinuum.

Die physikalischen Naturgesetze, worunter hier die Grundgesetze der theoretischen Physik verstanden seien, sind geometrische Beziehungen zwischen Tensoren[1]); es sind also denknotwendige Gesetze zugeordnet den Dingen in der Natur. Insofern gehören diese Naturgesetze nicht der Natur, sondern unserem Denken an. Nicht jeder geometrische Satz ist ein Naturgesetz, aber jedes Naturgesetz ist ein geometrischer Satz. Wo uns eine geometrische Beziehung zwischen irgendwelchen Tensoren fehlt, fehlt uns das physikalische Naturgesetz.

Wesentlich für unsere Naturerkenntnis ist nicht das Gesetz, sondern die Zuordnung zu einem Tensorsystem. Die wahren, inneren Zusammenhänge in der Natur vermögen wir niemals zu erkennen. Unsere geistige Struktur läßt uns die physikalischen Dinge und Begriffe immer nur als Tensoren in der „Welt" auffassen; alle Beziehungen in der Natur können also immer nur als geometrische Beziehungen begreifbar sein. Unser Fortschreiten in der physikalischen Erkenntnis besteht in einer immer vollkommeneren Zuordnung der sinnlich wahrnehmbaren Dinge und physikalischen Begriffe zu einem System von Tensoren. Diese Zuordnung ist eine richtige, wenn die geometrischen Beziehungen zwischen den Tensoren denen unserer Beobachtung in der Natur nach bestehenden Zusammenhängen entsprechen. Die geometrischen Beziehungen sind dann naturnotwendig.

Beim Aufsuchen der naturnotwendigen Beziehungen unter den geometrisch überhaupt möglichen werden wir in der Physik vor allem von

---

[1]) Vgl. hierzu auch die Aufsätze von A. Haas in Die Naturwissensch. Bd. 7. 1919 und Bd. 8. 1920, die allerdings in der Auffassung von der hier gegebenen Darstellung stark abweichen.

einer Forderung geleitet: die Beziehungen sollen einfach sein, einfacher als andere denkmögliche Beziehungen. Allem Anschein nach ist diese Aufstellung des Kriteriums der Einfachheit identisch mit der Aufstellung eines HAMILTONschen Prinzips, dessen Brauchbarkeit zur Auffindung naturnotwendiger Beziehungen mehr und mehr hervortritt.

Im Rahmen der speziellen Relativitätstheorie treten diese Zusammenhänge noch nicht mit derselben Deutlichkeit wie bei der allgemeinen Relativitätstheorie hervor; sie sind aber doch überall vorhanden und erkennbar. Durch die Zuordnung der elektrischen und magnetischen Feldstärke zu einem auf ein Vektorpotential zurückführbaren antisymmetrischen Tensor II. Ranges und durch die Zuordnung der Stromdichte und Ladungsdichte zu einem Vektor ist der Zusammenhang der elektrodynamischen Erscheinungen dargestellt. Ein HAMILTONsches Prinzip führt unmittelbar zu einem Ausdruck für die ponderomotorische Kraft, die auf den Energie-Impuls-Tensor zurückgeführt werden kann.

Daß unsere fortschreitende Naturerkenntnis gerade durch die Art der Zuordnung zum Ausdruck kommt, zeigt sich u. a. darin, daß Energie und Materie durch denselben Tensor, den Energie-Impuls-Tensor, dargestellt werden. Auch unsere Auffassung vom Äther läßt die Bedeutung der Zuordnung erkennen. Der das elektromagnetische Kraftfeld darstellende Feldtensor II. Ranges besteht unabhängig von den einzelnen durch die Lorentz-Transformation festgelegten Bezugssystemen. Bezeichnet man diesen Feldtensor als Äther (bzw. Lichtäther), so entspricht dies in viel höherem Sinn als bei allen Ätherhypothesen der Forderung einer restlosen begrifflichen Erfassung der in der Natur wahrnehmbaren Erscheinungen und Zusammenhänge.

Die Versuche, die Grundgedanken der speziellen Relativitätstheorie auf die Mechanik auszudehnen, brachte andererseits in der Darstellung der Gravitationserscheinungen keinen Erfolg, weil es hier nicht gelang, die Zuordnung der trägen und schweren Masse zu demselben Tensor auf innere Gründe zurückzuführen. Hier hat erst die allgemeine Relativitätstheorie weitergeholfen.

4. **Die allgemeine Relativitätstheorie.** Die Forderung der Kovarianz der Naturgesetze für beliebig bewegte Koordinatensysteme und damit der Unabhängigkeit der Naturgesetze von jedem Koordinatensystem war nur durch eine Erweiterung der Weltgeometrie möglich. An Stelle der pseudoeuklidischen Geometrie trat die allgemeine Differentialgeometrie RIEMANNS für das Raum-Zeit-Kontinuum. Der Tensor g_{ik} legt die Geometrie in der Umgebung jedes Punktes fest.

Das im allgemeinen Relativitätsprinzip enthaltene Äquivalenzprinzip ermöglicht nun die Zuordnung der *zwei* in der Natur auftretenden Erscheinungsformen der Masse zu *einem* Tensorbegriff und führt zugleich zur Zuordnung der Komponenten des Gravitationsfeldes zum Tensorfeld der g_{ik}. Die Gravitation wird damit vom GALILEIschen und NEWTONschen

Kraftbegriff losgelöst. Das Gravitationsfeld wird zum Führungsfeld (WEYL); neben den Lichtäther tritt das Tensorfeld der g_{ik} als Gravitationsäther. Die Weltgeometrie ist also durch das Gravitationsfeld festgelegt.

Wiederum gelangt man mittels eines HAMILTONschen Prinzipes einerseits zu den mechanischen Gleichungen, andererseits zu den Gravitationsgleichungen, den Beziehungen zwischen dem Materie-Tensor T_{ik} und dem Gravitationstensor g_{ik}. Das HAMILTONsche Prinzip besteht unabhängig von jedem Koordinatensystem; dieselbe Unabhängigkeit gilt auch für alle daraus hergeleiteten Naturgesetze, die damit vom Absoluten der klassischen Physik frei sind. Ihre Kovarianz ist eine Folge des HAMILTONschen Prinzips. Das allgemeine Relativitätsprinzip wird demnach durch die zu Tensoren einer allgemeinen nichteuklidischen Geometrie hinführende Begriffsbildung in der Physik und durch die Forderung der Einfachheit der Naturgesetze im Sinne eines HAMILTONschen Prinzips zu einer notwendigen Folgerung. Zugleich aber erhält diese Zuordnung und diese Forderung ihre tiefere Berechtigung erst durch die Tatsache der empirischen Gültigkeit des Relativitätsprinzips. Daß jedoch der Weg von der klassischen Physik zur allgemeinen Weltgeometrie über das spezielle und allgemeine Relativitätsprinzip gegangen ist, erscheint uns heute bereits als historische Zufälligkeit in ähnlichem Sinn, wie dies MACH von der klassischen Mechanik betont.

Die Zuordnung der physikalischen Erfahrung zu einem System von Tensoren im allgemeinen vierdimensionalen Raum-Zeit-Kontinuum und eine möglichst einfache geometrische Verknüpfung dieser Tensoren (im Sinne eines HAMILTONschen Prinzips) tritt uns so als das Ziel der physikalischen Forschung entgegen.

5. Die Weiterentwicklung der Weltgeometrie. Von diesem letzteren Gesichtspunkt aus muß man auch die weitere Entwicklung der Relativitätstheorie oder, wie man jetzt wohl besser sagt, der Weltgeometrie betrachten. Die spezielle Relativitätstheorie hatte die Elektrodynamik auf eine neue Grundlage gestellt, die allgemeine Relativitätstheorie erreichte dasselbe für die Mechanik. Jedoch konnte die Elektrodynamik nur lose in die allgemeine Relativitätstheorie eingereiht werden; die innere Verknüpfung von Elektrodynamik und Mechanik fehlte zunächst. Das Tensorfeld des Lichtäthers und das des Gravitationsäthers waren unvermittelt nebeneinandergestellt. Es entsteht die Forderung, Elektrizität und Gravitation auf eine gemeinsame Wurzel zurückzuführen.

Freilich fehlt es an Erfahrungstatsachen, welche über die Art der Verknüpfung bestimmten Aufschluß geben könnten. Aber die Forderung selbst ist berechtigt. Sie bestand schon vor der Relativitätstheorie und besteht jetzt um so mehr, wo nun auch angenommen werden muß, daß die Gravitation sich mit Lichtgeschwindigkeit fortpflanzt. Der einzig mögliche Weg, eine solche Verknüpfung herzustellen, ist augenblicklich der, eine geometrische Beziehung zwischen dem Tensor des elektromagnetischen und des Gravitationsfeldes aufzusuchen, ohne daß man zu Widersprüchen mit den Beobachtungen geführt wird. Dieser Weg birgt natürlich die Gefahr leerer Spekulation in sich. Aber es scheint doch, daß vor allem die von EDDINGTON gefundene Lösung in einfachster Weise die gestellte Forderung erfüllt.

Zwei andere Versuche sind vorher noch hervorzuheben. HILBERT[1]) hat, ausgehend von einem HAMILTONschen Prinzip, gezeigt, daß zwischen den 10 Gravitationsgleichungen und den 4 Zustandsgleichungen der Elektrodynamik wegen der freien Wahl des Koordinatensystems vier Bedingungen bestehen, und hat diesen Zusammenhang dahin gedeutet, daß die elektrodynamischen Erscheinungen durch die Gravitation verursacht sind. Dieser Schluß ist *nicht notwendig*. Man kann auch, wie dies z. B. bei v. LAUE[2]) geschieht, lediglich vier der die Metrik bestimmenden g_{ik} durch die Wahl des Koordinatensystems als vorgeschrieben ansehen.

Da die Gravitation schon allein die durch die RIEMANNsche Geometrie geforderten Zusammenhänge festlegt, so ist es wahrscheinlich, daß eine Verknüpfung von Gravitation und Elektrizität nur durch eine nochmalige Erweiterung der Weltgeometrie über die RIEMANNsche hinaus erreicht werden kann. WEYL[3]) erweitert die Metrik der RIEMANNschen Geometrie und setzt die Koeffizienten φ_i der metrischen Fundamentalform den elektromagnetischen Potentialen gleich. Die dadurch erreichte Verknüpfung ist jedoch rein formaler Natur, und es bedarf außerdem noch, um mit der Erfahrung in Einklang zu bleiben, der Einführung des Begriffes der Einstellung zum Gegensatz desjenigen der Beharrung.

Demgegenüber ist es natürlicher, auf dem von EDDINGTON eingeschlagenen Weg den inneren Zusammenhang zwischen Elektrizität und Gravitation herzustellen. Es handelt sich darum, den Lichtäther und Gravitationsäther zu einem einzigen Weltäther zu vereinigen. Das heißt aber nichts anderes, als *den Tensor des elektromagnetischen Feldes und den des Gravitationsfeldes auf einen gemeinsamen Tensor zurückführen*. EDDINGTON erreicht dies, indem er von der allgemeinen affinen Differentialgeometrie ausgeht. Ein unendlich kleiner Tensor A^μ erfährt danach bei der Parallelverschiebung von einem Punkt P zu einem unendlich benachbarten Punkt P' die Änderung $dA^\mu = - \Gamma^\mu_{\nu\alpha} A^\alpha \, dx^\mu$. Aus den 40 Komponenten des affinen Zusammenhanges $\Gamma^\mu_{\nu\alpha}$ lassen sich die Tensoren (In-Tensoren) $^*G^\varepsilon_{\mu\nu\sigma}$ und $^*B_{\mu\nu}$ herleiten, welche die geometrische Struktur der Welt festlegen. Den symmetrischen Tensor des

---

[1]) Nachr. d. Kgl. Ges. d. Wiss., Göttingen, Math.-physik. Klasse 1915, S. 395; und 1917, S. 53.

[2]) LAUE, M. v.: Die Relativitätstheorie. II. Bd., 2. Aufl. 1923, S. 170.

[3]) WEYL, H.: Raum, Zeit, Materie. 4. Aufl. 1921, § 35.

Gravitationsfeldes und den antisymmetrischen Tensor des elektromagnetischen Feldes können wir als die beiden Teile des einzigen Tensors $*G_{\mu\nu}$ auffassen; dieser stellt den *Weltäther* dar. Gravitation und Elektrizität sind Teile desjenigen Realen, auf das alle mechanischen und elektrodynamischen Erscheinungen zurückgeführt werden können, und das zugleich die innere geometrische Struktur der Welt festlegt. Aber der symmetrische Tensor der Gravitation allein bestimmt, wie die Bildung des Linienelementes zeigt, genau wie in der allgemeinen Relativitätstheorie die Metrik in der Umgebung jedes Punktes im Sinne der Riemannschen Geometrie. Einstein[1]) hat mit Hilfe eines Hamiltonschen Prinzips dann im einzelnen gezeigt, daß die Elektrodynamik und die Gravitationstheorie des Relativitätsprinzips in der Weltgeometrie Eddingtons enthalten ist. Der Anschluß an die physikalische Erfahrung ist damit vollständig hergestellt.

Die Gesamtheit der physikalischen Erscheinungen, soweit die Vorgänge außerhalb des Atoms in Betracht kommen, läßt sich also begrifflich durch Zuordnung zu einem einzigen Tensorfeld darstellen; aus ihm gewinnt man durch geometrische Verknüpfung mittels eines Hamiltonschen Prinzips zugleich die Gesetze der Mechanik und der Elektrodynamik.

---

[1]) Sitzungsber. d. preuß. Akad. d. Wiss., Physik.-math. Klasse 1923, S. 32, 76 und 137.

Die Verteilung der Leuchtkräfte der Sterne, besonders des M-Typus.

Von **P. J. van Rhijn**, Groningen.

Mit 1 Abbildung.

Kurzer Inhalt: Das Häufigkeitsgesetz für Sterne von allen Spektral-klassen zusammen ist abgeleitet worden. Die Kurve ist bis zur absoluten Größe $+6$ noch immer steigend.

Das Häufigkeitsgesetz der M-Sterne konnte zwischen den absoluten Größen -9 bis -3 und zwischen $+3.5$ bis $+6.5$ ermittelt werden. Die Häufigkeit der zwischenliegenden absoluten Größen ist unbekannt. Wahrscheinlich ist sie etwas kleiner als die größte Häufigkeit der Giganten; es gibt keine Tatsache, welche der Verteilung 11 der Abb. 1 auf S. 259 widerspricht, für welche die Häufig-keit der Zwischenglieder ungefähr gleich der mittleren Häufigkeit der Giganten ist. Es ist daher nicht gestattet, die Existenz dieser Zwischenglieder zu leugnen.

Die Verteilung der absoluten Helligkeiten der Sterne ist eine der wichtigsten Funktionen der Stellarstatistik. Nicht nur weil man mit ihrer Hilfe die Dichtigkeit des Sternsystems aus bloßen Zählungen der Sterne jeder scheinbaren Größe ableiten kann, sondern auch weil die Verteilung der absoluten Helligkeiten für *jede einzelne Spektralklasse* zur Evolution der Sterne in naher Beziehung steht.

Vor einigen Jahren haben KAPTEYN und ich eine vorläufige Be-stimmung der Verteilung der absoluten Leuchtkräfte für die Sterne zwischen den galaktischen Breiten $\pm 40°$ und $\pm 90°$ durchgeführt[1]). In den letzten Monaten habe ich dieselbe Methode auf die Sterne in den sonstigen galaktischen Zonen angewendet; weiter sind die Sterne des M-Typus gesondert untersucht worden. Ich möchte in diesem Auf-satz die Hauptresultate dieser Untersuchung mitteilen.

1. Sterne aller Spektralklassen zusammen. An erster Stelle habe ich die Verteilung der Leuchtkräfte aller Spektralklassen zusammen bestimmt nach der Methode, welche KAPTEYN schon 1902 angewendet hat[2]). Diese Methode wird im folgenden mit I bezeichnet werden. Die Beobachtungsdaten, auf welchen diese Bestimmung beruht, sind die folgenden:

a) die Zahlen von Sternen bestimmter Größe m und totaler Eigen-bewegung μ;

---

[1]) Contributions from the Mount Wilson Observatory Bd. 9, S. 289. 1920.
[2]) Publications of the Astronomical Laboratory at Groningen 1902, Nr. 11.

b) die mittleren Parallaxen von Sternen bestimmter Größe und Eigenbewegung;

c) die Streuung der wahren Parallaxen unter den Sternen bestimmter Größe und Eigenbewegung.

Will man, wie in dem gegenwärtigen Falle, die Verteilung der Leuchtkräfte für jede von drei galaktischen Zonen[1]) ableiten, so müssen natürlich erst die Funktionen 1 bis 3 für jede dieser Zonen ermittelt werden. Die Bestimmung dieser Funktionen ist in den Groningen Publications Nr. 30 und 34 durchgeführt worden; Nr. 30, Tab. 19, enthält die Zahlen von Sternen bestimmter Größe und Eigenbewegung, während die mittleren Parallaxen in Tab. 19 von Nr. 34 gegeben worden sind. Die Streuung der wahren Parallaxen für jede Gruppe ist nach der Formel (42) der Groningen Publications Nr. 34, S. 43, berechnet worden.

Die Resultate dieser Untersuchungen werden sofort mitgeteilt werden. Es sei aber gleich erwähnt, daß ich die Verteilung der Leuchtkräfte an zweiter Stelle nach einer anderen Methode (II) abgeleitet habe, welche als einzige Beobachtungsdaten die trigonometrischen Parallaxen voraussetzt. Die Methode ist die folgende[2]):

Wir werden die Verteilung der Leuchtkräfte aus der Verteilung der trigonometrischen Parallaxen für Sterne der scheinbaren Größen 3, 4, 5... bis 9 ableiten. Wie findet man aber die Verteilung der Parallaxen z. B. für Sterne der scheinbaren Größe 7? Es genügt selbstverständlich nicht, die gemessenen Parallaxen einfach abzuzählen; denn die Sterne in der Nähe der Sonne sind vollständiger beobachtet worden als die mehr entfernten Objekte. Wir haben diese Unvollständigkeit berücksichtigt, indem die Voraussetzung gemacht wurde, daß die *nichtbeobachteten* Sterne einer gewissen Eigenbewegung im Mittel dieselben Parallaxen haben wie die *beobachteten* Sterne derselben Eigenbewegungsgruppe. Ein Beispiel für die galaktische Zone $\pm 40° \pm 90°$ und die scheinbare Größe 7 wird das Verfahren erläutern:

Wir zählen die beobachteten Parallaxensterne zwischen den Grenzen der Eigenbewegung, welche in der ersten Spalte der Tab. 1 angegeben sind.

In der dritten Spalte findet man die totalen Zahlen per 10 000 Quadrat-Grade nach Groningen Publications Nr. 30, Tab. 19. Der Quotient der zweiten und dritten Spalte ergibt die Häufigkeit für jede Eigenbewegungsgruppe; die reziproken Werte dieser Häufigkeit sind in der vierten Spalte eingesetzt worden. Bei der Ableitung der Verteilung der

---

[1]) Die Grenzen der drei galaktischen Zonen sind die folgenden:

I. $- 20°$ bis $+ 20°$.

II. $- 20°$ bis $- 40°$ und $+ 20°$ bis $+ 40°$.

III. $- 40°$ bis $- 90°$ und $+ 40°$ bis $+ 90°$.

[2]) Kapteyn (Astr. Nachr. Bd. 183, S. 313. 1910) und Hess (Astr. Nachr. Bd. 220, S. 65. 1923) haben ungefähr dieselbe Methode verwendet.

Parallaxen für die betrachtete Gruppe wird nun jeder Parallaxenstern der Eigenbewegung $0''.200$ bis $0''.400$ für 13 Sterne gezählt, jeder Parallaxenstern der Eigenbewegung $0''.400$ bis $0''.800$ für 3.7 Sterne usw. Auf diese Weise kann man der Unvollständigkeit der Parallaxenbeobachtungen Rechnung tragen.

Tabelle 1.

Beispiel zur Ableitung der Häufigkeiten der Parallaxen.
Galaktische Zone $\pm 40° \pm 90°$, Größe 7.

Grenzen μ	Zahl der π-Sterne	Zahl G. P. 30	$\dfrac{1}{\text{Häufigkeit}}$
$0''.200$ bis $0''.400$	8	102	13
.400 bis .800	9	33	3.7
.800 bis 2.00	7	7.2	1.0
> 2.00	3	1.0	0.3

Man sieht aus der Tab. 1, daß die Sterne mit einer Eigenbewegung kleiner als $0''.200$ gar nicht berücksichtigt worden sind. Die Häufigkeiten in der Gruppe $\mu < 0''.200$ sind so klein, daß man mit ihrer Hilfe kaum zuverlässige Resultate ableiten kann. Die hier verwendete Methode gibt daher *nicht* die Verteilung der Parallaxen für *alle* Sterne der Größe 7, sondern nur für diejenigen Sterne, deren Eigenbewegung größer ist als $0''.200$. Die erhaltenen Zahlen von Sternen in jeder Parallaxengruppe müssen also noch mit einem Proportionalitätsfaktor K multipliziert werden, wo:

$$K = \frac{\text{Zahl aller Sterne in der Parallaxengruppe } \pi_1 \text{ bis } \pi_2}{\text{Zahl der Sterne } \mu > 0''.200 \text{ in der Parallaxengruppe } \pi_1 \text{ bis } \pi_2}$$

Dieser Faktor K ist gleich:

$$K = \frac{\text{Zahl aller Sterne in der Parallaxengruppe } \pi_1 \text{ bis } \pi_2}{\text{Zahl der Sterne, deren transversale lineare Bewegung } > 4 \cdot 74 \dfrac{0.200}{\pi}}$$

wo $\bar\pi$ die mittlere Parallaxe der Gruppe ist.

Man kann daher die Faktoren K aus der Verteilung der linearen transversalen Geschwindigkeiten berechnen. Diese letztere Verteilung wurde in Groningen Publications Nr. 34, S. 10, aus den bekannten radialen Geschwindigkeiten abgeleitet.

Die Verteilung der Parallaxen für jede scheinbare Größe wurde weiter für den Einfluß der zufälligen Beobachtungsfehler korrigiert. Aus den erhaltenen Resultaten für die Zahlen von Sternen von jeder Größe und Parallaxe kann man leicht die Verteilung der Leuchtkräfte ableiten.

Das sind wohl die wichtigsten Punkte dieser zweiten Methode, nach welcher die Verteilung der Leuchtkräfte bloß mittels der trigonometrischen Parallaxen bestimmt wird. Ein systematischer Fehler in diesen Parallaxen wird natürlich die Genauigkeit der Resultate besonders für die absolut hellen Sterne verringern. Ich habe daher diese systematischen Fehler besonders für Sterne mit kleiner Eigenbewegung[1]) durch eine Vergleichung mit den mittleren Parallaxen von Groningen Publications Nr. 34 zu bestimmen versucht. Meiner Ansicht nach sind diese letzteren Parallaxen für Sterne mit kleiner Eigenbewegung, was die systematischen Fehler anbelangt, besser als die trigonometrischen Parallaxen. Die systematischen Unterschiede zwischen den verschiedenen trigonometrischen Parallaxenbestimmungen sind doch immer noch größer als $0''004$; Fehler dieser Größenordnung in den mittleren Parallaxen der Sterne mit kleiner Eigenbewegung sind nach meiner Ansicht ausgeschlossen.

Ich habe in Tab. 2 die Parallaxenbestimmungen von Allegheny, Mc Cormick, Mount Wilson und Yerkes für Sterne, deren Eigenbewegung kleiner ist als $0''200$, mit den mittleren Parallaxen von Groningen Publications Nr. 34 verglichen.

Tabelle 2.

Mittlere Differenzen zwischen den absoluten trigonometrischen Parallaxen und Groningen Publications Nr. 34 für Sterne, deren Eigenbewegung kleiner ist als $0''200$

Sternwarte	G. P. 34 — trig. π
Allegheny	$0''0000$
Mc Cormick	— 0.0015
Yerkes	— 0.0035
Mount Wilson	— 0.0045
Dearborn	— 0.0170
Greenwich ($\mu > 0''200$) . .	— 0.0000

Einen starken Grund für die Überzeugung, daß die Groninger Parallaxen systematisch richtig sind, findet man in der Tatsache, daß von den zwei meist umfassenden Parallaxenreihen die eine (Allegheny) genau mit dem Groninger System übereinstimmt, während die andere (Mc Cormick) einen relativ kleinen Unterschied gegen die Groninger Parallaxen zeigt. Die Abweichungen der sonstigen Parallaxen von dem Groningen-Alleghenysystem sind größer; besonders der Betrag für Dearborn ist beträchtlich, aber zweifellos reell.

Die meisten Eigenbewegungen der Greenwicher Parallaxensterne sind größer als $0''200$; die Vergleichung mit dem Groninger System hat für diese Sterne nicht dieselbe Bedeutung wie für Sterne mit kleiner Eigenbewegung. Denn erstens sind die Groninger Parallaxen für Sterne mit großer Eigenbewegung auf die trigonometrischen Parallaxen ge-

---

[1]) Für Sterne mit größerer Eigenbewegung wird das Groninger System selbstverständlich mit dem mittleren trigonometrischen System übereinstimmen, weil es auf das letztere gegründet ist.

gründet worden und ist Übereinstimmung daher selbstverständlich; zweitens aber sind die Abweichungen der individuellen Parallaxen von den mittleren Parallaxen beträchtlicher als für die Sterne mit kleiner Eigenbewegung, und ist daher der wahrscheinliche Fehler dieser mittleren Parallaxen größer.

Ich habe die systematischen Differenzen der Tab. 2 als Korrektionen an die Parallaxen angebracht. Nur die Dearborn-Parallaxen sind fortgelassen worden, weil es nicht möglich ist, ihre Korrektion mit einer genügenden Genauigkeit zu bestimmen.

Die Resultate für die Verteilung der Leuchtkräfte findet man in der Tab. 3, in welcher die Logarithmen der Zahlen von Sternen per 1000 Kubik-parsec gegeben sind. Ich habe nur das Endresultat für alle galaktischen Zonen zusammen mitgeteilt, weil der Unterschied zwischen den für die drei Zonen gefundenen Funktionen gering ist; wahrscheinlich gibt es in der Milchstraße relativ etwas mehr absolut helle Sterne, als in der Nähe der Milchstraßenpole.

Tabelle 3.

Logarithmen der Zahlen von Sternen per 1000 Kubik-parsec in der Nähe der Sonne.
$$M = m + 5 \log \pi.$$

M	Methode I	Methode II	K	I—II	I—K	II—K
— 8.5 bis — 7.5	6.76	7.30	6.91	— 0.54	— 0.15	+ 0.39
— 7.5 bis — 6.5	7.46	7.60	7.61	— .14	— .15	— .01
— 6.5 bis — 5.5	8.09	8.34	8.25	— .25	— .16	+ .09
— 5.5 bis — 4.5	8.74	8.90	8.81	— .16	— .07	+ .09
— 4.5 bis — 3.5	9.28	9.38	9.31	— .10	— .03	+ .07
— 3.5 bis — 2.5	9.71	9.74	9.74	— .03	— .03	.00
— 2.5 bis — 1.5	0.04	0.04	0.09	.00	— .05	— .05
— 1.5 bis — 0.5	0.29	0.27	0.39	+ .02	— .10	— .12
— 0.5 bis + 0.5	0.49	0.40	0.61	+ .09	— .12	— .21
+ 0.5 bis + 1.5	0.62	0.47	0.76	+ .15	— .14	— .29
+ 1.5 bis + 2.5	0.67	0.31	0.84	+ .36	— .17	— .53
+ 2.5 bis + 3.5	0.65	0.48	0.85	+ .17	— .20	— .37
+ 3.5 bis + 4.5	0.57	0.78	0.80	— .21	— .23	— .02
+ 4.5 bis + 5.5	0.44	0.38	0.67	+ .06	— .23	— .29
+ 5.5 bis + 6.5	0.27	0.92	0.48	— .65	— .21	+ .44

Tabelle 3a.

Logarithmen der Zahlen von Sternen per 1000 Kubik-parsec in der Nähe der Sonne, abgeleitet mittels der korrigierten Parallaxen von Groningen Publ. Nr. 34.

M	$\log \varphi (M)$	*Kapteyn*
6.4	0.82	0.39
7.4	0.80	0.10
8.4	0.81	9.75
9.4	0.71	9.30

Die zweite und dritte Spalte der Tab. 3 geben die Resultate der ersten resp. der zweiten Methode dieses Aufsatzes; K bedeutet die Verteilung der Leuchtkräfte nach der oben angeführten vorläufigen Lösung[1]). Man sieht aus der Tabelle, daß die drei Resultate I, II und K für Sterne heller als $M = 0$ gut übereinstimmen; die etwas größeren Unterschiede für die allerhellsten Sterne sind wohl der kleinen Zahl der beobachteten Sterne zuzuschreiben.

Für die schwächeren Sterne ist die Übereinstimmung nicht so gut: die Zahlen I und II sind kleiner als die Kapteynschen Werte und die Kurve II zeigt nicht die Abnahme für absolute Größen schwächer als $+3$, welche für das Häufigkeitsgesetz K charakteristisch ist. Nach meiner Ansicht müssen die Werte II für die schwächsten Sterne den beiden anderen vorgezogen werden und ist die Abnahme der Häufigkeitskurve für absolute Größen schwächer als $+3$ nicht reell. Denn die Verteilungen I und K beruhen für schwache Sterne hauptsächlich auf scheinbar schwachen Sternen. Die mittleren Parallaxen aber dieser Sterne sind in Tab. 19 von Groningen Publications Nr. 34 einfach mittels der bekannten Parallaxen der helleren Sterne extrapoliert, weil zur Zeit keine trigonometrische Parallaxen von diesen schwachen Sternen bekannt waren. Nachher sind die Parallaxen von einigen dieser Objekte beobachtet worden, und es hat sich gezeigt, daß die mittleren Parallaxen von Groningen Publications Nr. 34 zu klein sind. Die mittlere trigonometrische Parallaxe aller beobachteten Sterne schwächer als die Größe 10 mit großer Eigenbewegung ist nämlich:

$$\pi = 0\overset{''}{.}087 \quad \text{für} \quad \mu = 1\overset{''}{.}34 \quad \text{und} \quad m = 12.0 \ (12 \text{ Sterne}),$$

während der entsprechende Wert von Groningen Publications Nr. 34 $0\overset{''}{.}038$ beträgt. Falls man Korrektionen von dieser Ordnung an die mittleren Parallaxen von der betrachteten scheinbaren Größe und Eigenbewegung anbringt, werden die berechneten Zahlen von absolut schwachen Sternen zweifellos größer werden. Um das zu zeigen, habe ich die Verteilung der Parallaxen für die scheinbare Größe 14 nach der zweiten oben beschriebenen Methode abgeleitet in der Annahme, daß die mittleren Parallaxen von Groningen Publications Nr. 34, Tab. 19, für Sterne $m = 14$, $\mu > 0\overset{''}{.}200$ mit 2 multipliziert werden müssen. Dieser Faktor ist dem Resultat für $m = 12.0$, $\mu = 1\overset{''}{.}34$ gemäß genommen worden. Mittels dieser Verteilung findet man die Häufigkeitskurve φ von Tab. 3a.

Der Faktor 2, mit welchem die mittleren Parallaxen von Groningen Publications Nr. 34 multipliziert worden sind, ist natürlich sehr unsicher, und die abgeleiteten Zahlen φ sind wenig zuverlässig. Aber sie zeigen doch, daß die Anbringung einer nicht unwahrscheinlichen Korrektion an die mittleren Parallaxen der schwachen Sterne mit großer Eigenbewegung die Zahlen per 1000 Kubik-parsec zwischen $M = +6.4$

---

[1]) Contributions from the Mount Wilson Observatory Bd. 9, S. 289. 1920.

und $+9.4$ dermaßen vergrößert, daß von einer Abnahme nach $M = +3$ keine Rede mehr ist. Weiter werden wir in dem folgenden Abschnitt finden, daß 1000 Kubik-parsec 10 M-Sterne zwischen den absoluten Größen 5.5 und 6.5 enthalten. Dieses Resultat ist ziemlich unsicher, aber es zeigt doch, daß die Zahl *aller* Sterne per 1000 Kubik-parsec wohl größer sein muß als der KAPTEYNsche Wert 3.

Jedenfalls sind die genauen Zahlen von Sternen per Volumeneinheit für diese schwachen Objekte ziemlich unsicher, und für $M > 6.5$ wissen wir gar nichts über den Verlauf der Häufigkeitskurve; wir wissen sogar nicht, ob sie noch immer steigt oder wieder abnimmt. Und doch liegt die Lösung dieses Problems gewiß innerhalb der Grenzen der gegenwärtigen Hilfsmittel. Man hat zu diesem Zweck nur die Parallaxen von scheinbar schwachen Sternen mit großer Eigenbewegung zu bestimmen; z. B. Sterne $m = 9.5$ bis 13.5 und Eigenbewegung $\mu > 0''.300$. Die mittleren Parallaxen aller dieser Sterne sind wahrscheinlich größer als $0''.020$ und können daher ziemlich genau bestimmt werden. Hoffentlich werden die Parallaxenbeobachter diese Objekte in ihr Programm aufnehmen; eine Liste der bekannten Sterne mit Eigenbewegung größer als $0''.500$ ist von LUYTEN zusammengestellt worden[1]), und einen Hinweis auf die Sterne $\mu = 0''.200$ bis $0''.500$ findet man auf S. 80 von Groningen Publications Nr. 30.

Die Zahl der jetzt schon bekannten absolut schwachen Sterne genügt nicht zu einer genauen statistischen Untersuchung der Verteilung der Leuchtkräfte für absolute Größen schwächer als $M = +5$. Von großer Wichtigkeit für diese Untersuchungen sind daher die Beobachtungen, welche das Auffinden von scheinbar schwachen Sternen mit großer Eigenbewegung beabsichtigen, wie z. B. die Arbeit von WOLF, INNES, TURNER usw.; denn die große Eigenbewegung ist eben ein Anzeichen geringer Leuchtkraft. Die Tab. 4 gibt die mutmaßlichen Zahlen

Tabelle 4.

Zahlen der Sterne zwischen den scheinbaren Größen 9.5 und 13.5, deren Eigenbewegung größer ist als $0''.300$.

$m + 5 \log \pi$	Zahl
$+ 1.0$ bis $+ 2.0$	330
$+ 2.0$ bis $+ 3.0$	640
$+ 3.0$ bis $+ 4.0$	911
$+ 4.0$ bis $+ 5.0$	1560
$+ 5.0$ bis $+ 6.0$	850
$+ 6.0$ bis $+ 7.0$	350
$+ 7.0$ bis $+ 8.0$	160
$+ 8.0$ bis $+ 9.0$	33
$+ 9.0$ bis $+ 10.0$	16

---

[1]) LUYTEN, W. J.: Lick Observatory Bulletin 1923, Nr. 344.

von Sternen am Himmel zwischen den scheinbaren Größen 9.5 und 13.5, deren Eigenbewegung $0''.300$ übertrifft.

Man sieht aus dieser Tabelle, daß das Absuchen von 10 000 Quadratgrad mit dem Blinkmikroskop bis zur visuellen Größe 13.5 noch ungefähr 40 Sterne zwischen den absoluten Größen 7 und 8 liefern wird, falls alle Eigenbewegungen größer als $0''.300$ per annum aufgefunden werden; die Verteilung der Leuchtkräfte könnte daher in diesem Falle bis zur Größe 8 genau bestimmt werden, während bis zur Größe 10 eine genäherte Kenntnis in Aussicht gestellt werden könnte. Die Auffindung aller Sterne $\mu > 0''.300$, $m = 9.5$ bis 13.5 per 10 000 Quadratgrade stellt gar keine unmögliche Aufgabe dar; man hat bis jetzt schon 2000 Quadratgrade untersucht für die Sterne heller als die scheinbare Größe 12 und 1000 Quadratgrade für die Sterne der Größen 12 bis 15[1])

2. Sterne des Ma- bis Mc-Typus. Die Verteilung der Leuchtkräfte der M-Sterne ist von großer Bedeutung für die Theorie der Riesen- und Zwergsterne, welche zum Teil auf einer vermeintlichen Lücke in dieser Verteilung gegründet worden ist.

Die Md-Sterne sind in dieser Untersuchung fortgelassen worden, weil sie wohl besser als eine gesonderte Gruppe behandelt werden sollten.

Die Verteilung der Leuchtkräfte für die absolut hellen Sterne ist nach der ersten in dem vorhergehenden Paragraphen beschriebenen Methode bestimmt worden. Von den erforderlichen Daten findet man die Verteilung der Eigenbewegungen in Groningen Publications Nr. 30[2]); diese wurde für die Größen 7 und 8 und $\mu < 0''.100$ noch verbessert mittels einiger kürzlich veröffentlichten Eigenbewegungen[3]); die Größe 9 könnte jetzt auch hinzugefügt werden. Für dieselben Größenklassen 7, 8 und 9 und $\mu = 0''.100$ bis $0''.500$ wurde die Verteilung der Eigenbewegungen ganz neu gerechnet mittels der Daten aus den folgenden Katalogen:

Boss' Preliminary General Catalogue;
Publications of the Cincinnati Observatory Nr 18;
Greenwich Catalogue 1910, S. B XXXII;
Greenwich Astrographic Catalogue Vol. 4, S. A 19.

Wir bestimmten die Zahlen von Sternen in diesen Katalogen für jede Spektralklasse zwischen gewissen Grenzen von Eigenbewegung und Größe; diese Zahlen wurden auf die Gesamtzahl per 10 000 Quadratgrade reduziert durch Multiplikation mit den Faktoren:

---

[1]) Siehe Groningen Publications 1920, Nr. 30, S. 80 und 81.

[2]) Groningen Publications Nr. 30. S. 91 für die Größen 3 und 4; S. 98 für die Größen 5 bis 8.

[3]) Wilson, R. E.: Astronom. Journ. Bd. 34, S. 183. 1923, und Bd. 35, S. 125. 1923. — Prager, R.: Veröffentl. der Univ.-Sternwarte zu Berlin-Babelsberg Bd. 4. 1923.

Gesamtzahl der Sterne aller Spektralklassen per 10 000 Quadratgrade der Eigenbewegung μ und Größe m

Gezählte Zahl der Sterne aller Spektralklassen der Eigenbewegung μ und Größe m

Die Zähler dieser Faktoren sind in Groningen Publications Nr. 30, Tab. 19 zusammengestellt worden.

Die also gefundenen Zahlen von Sternen $\mu = 0\rlap{.}''100$ bis $0\rlap{.}''500$ und $m = 7$ und 8 beruhen auf einem sehr viel größeren Material als die Zahlen von Groningen Publications Nr. 30. Denn zur Zeit der Veröffentlichung dieser letzten Schrift waren nur die ersten 4 Stunden des HENRY-DRAPER-Katalogs erschienen, während jetzt alle Sterne bis zur 19 Stunde berücksichtigt worden sind.

Die Zahlen von Sternen des M-Typus für $\mu > 0\rlap{.}''500$ sind nach derselben Methode aus LUYTENS Schrift abgeleitet worden[1]); die Spektralbestimmungen von L. O. B. Nr. 351 sind hierbei auch verwendet worden. Die benutzte Verteilung der Eigenbewegungen ist in der Tab. 5 enthalten. $m = 3$ bedeutet die Größe zwischen 2.45 und 3.44 usw.

Tabelle 5.

Zahl der Ma- bis Mc-Sterne am ganzen Himmel.

— bedeutet, daß unter den beobachteten Sternen dieser Eigenbewegung und Größe keine M-Sterne gefunden sind; eine offene Stelle bedeutet, daß die Spektralklassen der Sterne dieser Eigenbewegung und Größe überhaupt unbekannt sind.

μ \ m	3	4	5	6	7	8	9	10	11
$0\rlap{.}''000$ bis $0\rlap{.}''020$	0.0	3.0	23	63	158	545	917		
.020 bis .040	1.0	4.0	23	52	140	413	1028		
.040 bis .060	0.0	2.0	25	46	63	144	264		
.060 bis .080	1.0	3.0	7	35	41	42	—		
.080 bis .100	1.0	0.0	5	11	18	7	—		
.100 bis .150	2.0	2.0	4	16	5	1	—		
.150 bis .200	1.0	2.0	2	2	2	3	—		
.200 bis .300	3.0	1.0	—	—	—	—	—		
.300 bis .400	—	—	—	—	—	—	4		
.400 bis .500	—	1.0	—	—	—	—	—		
.500 bis .600	—	—	—	—	—	—	4	25	
.600 bis .700	—	—	—	—	—	—	7	14	
.700 bis .800	—	—	—	—	—	—	2		
.800 bis .900	—	—	—	—	—	—	7	} 51	} 82
.900 bis 1.00	—	—	—	—	—	—	3		
1.00 bis 1.50	—	—	—	—	—	—	14	20	41
1.50 bis 2.00	—	—	—	—	—	—	3	4	16
> 2.00	—	—	—	—	3	7	5	8	8

Außer der Verteilung der Eigenbewegungen braucht man zur Ableitung des Häufigkeitsgesetzes der Leuchtkräfte noch die mittleren

---

[1]) L. O. B. Nr. 344.

Parallaxen von Sternen bestimmter Eigenbewegung und Größe. Dieselben sind in Groningen Publications Nr. 34, S. T 24, gegeben worden; sie sind aber auf Grund nachher veröffentlichter Daten um 10% zu vermindern.

Die Verteilung der Leuchtkräfte habe ich nun nach der Methode I des vorhergehenden Paragraphen abgeleitet. Die Resultate für $M = -9.0$ bis -3.0 enthält die Tab. 6.

Tabelle 6.

Logarithmen der Zahlen von M-Sternen per 1000 Kubik-parsec in der Nähe der Sonne.

M	$\log \varphi (M)$	M	$\log \varphi (M)$
$- 9.0$ bis $- 8.0$	(4.53)	$+ 3.5$ bis $+ 4.5$	0.36
$- 8.0$ bis $- 7.0$	6.012	$+ 4.5$ bis $+ 5.5$	0.79
$- 7.0$ bis $- 6.0$	6.915	$+ 5.5$ bis $+ 6.5$	1.12
$- 6.0$ bis $- 5.0$	7.502		
$- 5.0$ bis $- 4.0$	7.766		
$- 4.0$ bis $- 3.0$	7.740		

Wir finden auf diese Weise das Häufigkeitsgesetz bloß für die absolut hellen Sterne. Das hätte man erwarten können; denn die benutzten Sterne der Tab. 5 ($m = 3$ bis 8, $\mu < 0''.500$) sind scheinbar helle Sterne mit kleiner Eigenbewegung und haben daher eine große absolute Leuchtkraft; die Sterne der scheinbaren Größen 9, 10 und 11 habe ich fortgelassen, weil deren mittlere Parallaxen unbekannt sind.

Die Bestimmung des Häufigkeitsgesetzes für absolut schwache Sterne mittels scheinbar schwacher Sterne mit großer Eigenbewegung ($m = 6.5$ bis 10.5, $\mu > 0''.500$) kann am besten nach der zweiten in Abschnitt 1 beschriebenen Methode ausgeführt werden; denn die trigonometrischen Parallaxen dieser Sterne sind groß, und ihre absoluten Größen können daher sehr genau bestimmt werden.

Die Tab. 7 zeigt die Zahlen von M-Sternen mit bekannten Parallaxen zwischen bestimmten Grenzen von Eigenbewegung und Größe.

Tabelle 7.

Zahl von Sternen mit bekannten Parallaxen. Relative Häufigkeit dieser Sterne.

μ \ m	Zahl der Parallaxensterne					Häufigkeit der Parallaxensterne				
	7	8	9	10	11	7	8	9	10	11
$0''.500$ bis $0''.600$			2					0.50		
.600 bis .700			1	1				0.17	0.071	
.700 bis 1.00			2	1				0.17	0.020	
1.00 bis 1.50			}	}	}			}	}	}
1.50 bis 2.00			} 6	} 2	} 4			} 0.36	} 0.062	} 0.062
> 2.00	2	4	2	}	}	0.67	0.59	0.40	}	}

Die zweite Hälfte der Tabelle gibt die Häufigkeiten, definiert durch:

$$\frac{\text{Zahl der } \textit{gezählten} \text{ Parallaxensterne des M-Typus der Eigenbewegung } \mu \text{ und der Größe } m}{\textit{Gesamtzahl} \text{ der M-Sterne derselben Eigenbewegung und Größe am ganzen Himmel}}$$

Wir bestimmen nun die Zahl der Sterne mit bekannter Parallaxe zwischen bestimmten Parallaxengrenzen für die Größenklassen 7, 8, 9, 10 und 11. Ein Parallaxenstern der Eigenbewegung $0\rlap{.}{''}500$ bis $0\rlap{.}{''}600$ und der Größe 9 zählt dabei für $\dfrac{1}{0.50}$ Sterne, weil die relative Häufigkeit für diese Sterne 0.50 beträgt usw.

Die Sterne mit Eigenbewegung kleiner als $0\rlap{.}{''}500$ sind auf die in dem ersten Abschnitt beschriebene Weise berücksichtigt worden. Wir erhalten daher die Gesamtzahl der Sterne bestimmter scheinbarer Größe, deren Parallaxe zwischen bestimmten Grenzen liegt. Aus diesen Daten kann die Verteilung der Leuchtkräfte abgeleitet werden. Das Resultat ist in die Tab. 6 aufgenommen worden. Man sieht, daß die Zahl der absolut schwachen Sternen diejenige der absolut hellen Sterne weitaus übertrifft.

3. Theorie der Riesen- und Zwergsterne. Die bisher verwendeten Daten reichen offenbar nur aus zur Ableitung des Häufigkeitsgesetzes für die absoluten Größen $- 9$ bis $- 3$ und $+ 3.5$ bis $+ 6.5$; die zwischenliegenden Größen sind noch unbestimmt. Kann man diese Lücke nicht ausfüllen? Ein Vertreter der Riesen- und Zwergtheorie wird antworten, daß das unmöglich ist, weil eben keine M-Sterne der absoluten Größen $- 3$ oder $- 2$ bis $+ 3$ existieren. Ich glaube, daß diese Antwort verfrüht sein würde. Wir haben bisher zwei Arten von M-Sternen untersucht: die scheinbar hellen Sterne und die scheinbar schwachen Sterne mit großer Eigenbewegung. Es ist nun leicht ersichtlich, daß die erste Art die sog. Riesensterne enthalten wird, die zweite Art aber die Zwerge. Könnte diese zweifache Wahl nicht die Ursache für das Fehlen der M-Sterne von mittleren absoluten Helligkeiten sein? Eine Antwort auf diese Frage wird man nur geben können, indem man nicht wie oben aus den gegebenen Zahlen $N_{m,\,\mu}$[1]) das Häufigkeitsgesetz, sondern umgekehrt aus verschiedenen Annahmen über das Häufigkeitsgesetz zwischen $M = -2$ und $+3$ die Zahlen $N_{m,\,\mu}$ ableitet und diese mit den beobachteten Zahlen von Tab. 5 vergleicht.

Die Zahl von Sternen zwischen den scheinbaren Größen $m - \tfrac{1}{2}$ und $m + \tfrac{1}{2}$ und den Eigenbewegungen μ_1 und μ_2 ist gegeben durch die Formel:

$$N(m, \mu_1, \mu_2) = 4\pi \int_0^\infty r^2 \varphi(m - 5\log r) D(r)\{\psi(4.74\mu_1 r) - \psi(4.74\mu_2 r)\}\,dr \quad (1)$$

---

[1]) $N_{m,\,\mu} =$ Zahl von Sternen zwischen bestimmten Grenzen von Größe (m) und Eigenbewegung (μ).

wo $D(r)$ = Dichte als Funktion der Entfernung: Einheit = Dichte in der Nähe der Sonne;

$\varphi(M)$ = Zahl der Sterne in der Volumeneinheit in der Nähe der Sonne zwischen den absoluten Größen $M - \frac{1}{2}$ und $M + \frac{1}{2}$;

$\psi(v)$ = Prozentsatz der Sterne, deren Geschwindigkeit relativ zur Sonne größer ist als v. Es handelt sich dabei nicht um die ganzen Geschwindigkeiten, sondern nur um ihrer Projektionen auf die Sphäre.

Die Berechnung der Zahlen $N(m, \mu_1, \mu_2)$ setzt also die Kenntnis von D, φ und ψ voraus. Wir werden kurz auseinandersetzen, welche Funktionen wir für diese drei Größen benutzt haben.

a) *Dichte $D(r)$.* Bei der Ableitung des Häufigkeitsgesetzes nach der ersten Methode findet man die Zahlen von Sternen jeder absoluten Größe, welche zwischen bestimmten Grenzen der Entfernung liegen. Aus diesen Daten ist die Dichte als Funktion der Entfernung leicht abzuleiten[1]). Das Resultat findet man in der Tab. 8.

Tabelle 8.

Logarithmus der Dichte.

π	$(\log D)_{\text{Beob.}}$	$(\log D)_{\text{Rechn.}}$
$> 0''\!.0118$	0.00	
.0118	9.91	9.94
.0074	9.85	9.83
.0047	9.69	9.68
.0030	9.46	9.47
.0019	9.21	9.22
.0012	8.92	8.91

Die Logarithmen der dritten Spalte sind mittels der Formel:

$$\log D = -1.652 - 2.040 \log \pi - 0.631 \log^2 \pi$$

berechnet worden; man sieht, daß diese Formel die beobachteten Zahlen sehr gut wiedergibt. In einigen Fällen sind die Dichten für Entfernungen größer als 1000 Parsec mittels dieser Formel extrapoliert worden.

b) *Geschwindigkeitsverteilung $\psi(v)$.* Die Funktion $\psi(v)$ ist nach der in Groningen Publications Nr. 34, S. 10, beschriebenen Methode aus den Radialgeschwindigkeiten abgeleitet worden; die Abhängigkeit der Verteilung der Geschwindigkeiten von der absoluten Helligkeit der Sterne wurde berücksichtigt.

Wir kennen jetzt in der Formel (1) die Funktionen $D(r)$ und $\psi(v)$; die Berechnung der Zahlen $N(m, \mu_1, \mu_2)$ könnte daher ausgeführt werden, falls das Häufigkeitsgesetz $\varphi(M)$ ebenfalls bekannt wäre. Ich habe dieses Gesetz, soweit es in dem vorhergehenden Abschnitt berechnet

---

[1]) Siehe z. B. Groningen Publications. 1902, Nr. 11.

worden ist, in der Figur durch eine ausgezogene Linie dargestellt[1]). Der Teil zwischen $M = -3$ und $+3$ ist unbekannt; ich habe daher fünf Annahmen über den Verlauf des Häufigkeitsgesetzes zwischen diesen Grenzen gemacht, welche in der Abbildung durch punktierte Linien dargestellt worden sind. Die Zahlen $N(m, \mu_1, \mu_2)$ sind mittels der Formel (1) nach jeder dieser fünf Annahmen berechnet worden.

Die fünf Häufigkeitsgesetze findet man in der Tab. 9; die berechneten Zahlen N sind in der Tab. 10 zusammengestellt worden.

Tabelle 9.

Die fünf Annahmen für das Häufigkeitsgesetz zwischen $M = -3$ und $+3$.

M	I	II	III	IV	V
-2.5	7.93	7.63	7.49	7.12	7.78
-1.5	8.21	7.41	7.00	6.04	7.78
-0.5	8.60	7.13	6.06	4.05	7.78
$+0.5$	9.03	7.02	5.00	3.00	7.78
$+1.5$	9.44	7.55	6.05	4.20	8.32
$+2.5$	9.82	9.16	8.89	8.55	9.35

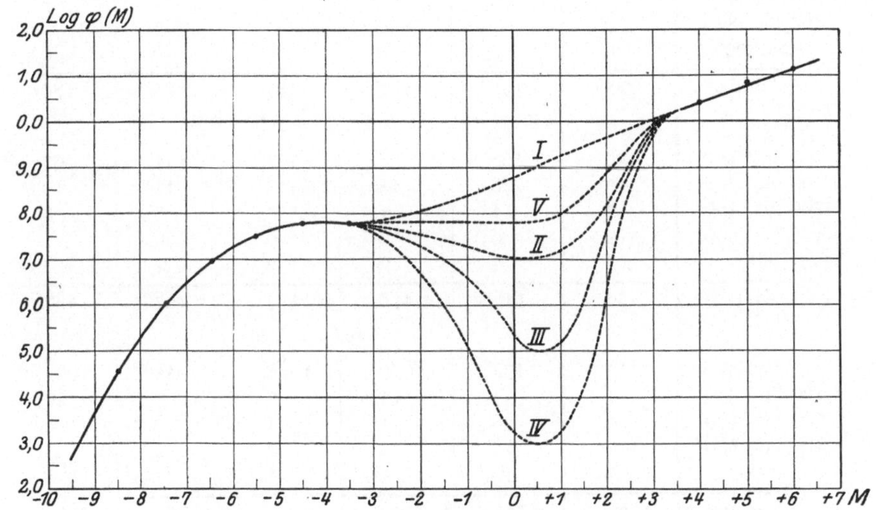

Abb. 1. Häufigkeitsgesetz der M-Sterne.

Man sieht aus der Tab. 10, daß das Häufigkeitsgesetz I, welches den Riesen- und Zwergenast durch eine stetig steigende Kurve verbindet[2]), Zahlen $N(m, \mu_1, \mu_2)$ liefert, die *nicht* mit den beobachteten Zahlen übereinstimmen. Die berechneten Zahlen von Sternen $m = 8$, $\mu = 0''.100$ bis $0''.400$ sind viel größer als die beobachteten Zahlen;

---

[1]) Siehe Tabelle 6.
[2]) Siehe die Abb.

Tabelle 10.

Berechnete Zahlen $N\,(m, \mu_1, \mu_2)$ von M-Sternen zwischen bestimmten Grenzen der scheinbaren Größe und Eigenbewegung.

μ	Beobacht.	Berechnete Zahlen mit verschiedenen φ				
		I	II	III	IV	V
		Größe 2.5 bis 4.5				
alle	26	28	28	28	28	28
		Größe 4.5 bis 5.5				
$< 0\overset{''}{.}100$	83	52	52	52	52	52
.100 bis .200	6	13	13	13	13	13
.200 bis .400	0	2	2	2	2	2
.400 bis .800	0	1	0	0	0	0
$>$ .800	0	0	0	0	0	0
Summe	89	68	67	67	67	67
		Größe 5.5 bis 6.5				
$< 0\overset{''}{.}100$	207	171	169	169	169	169
.100 bis .200	18	19	16	16	15	17
.200 bis .400	0	5	2	2	1	3
.400 bis .800	0	2	0	0	0	0
$>$ .800	0	1	0	0	0	0
Summe	225	198	187	187	185	189
		Größe 6.5 bis 7.5				
$< 0\overset{''}{.}100$	420	468	458	457	454	462
.100 bis .200	7	28	15	13	11	19
.200 bis .400	0	11	1	1	0	3
.400 bis .800	0	6	0	0	0	1
$>$ .800	3	4	2	2	2	2
Summe	430	517	476	473	467	487
		Größe 7.5 bis 8.5				
$< 0\overset{''}{.}100$	1151	1138	1079	1068	1055	1097
.100 bis .200	4	54	11	7	3	20
.200 bis .400	0	28	1	1	0	4
.400 bis .800	0	15	2	2	2	3
$>$ .800	7	10	7	7	7	8
Summe	1162	1245	1100	1085	1067	1132
		Größe 8.5 bis 9.5				
$< 0\overset{''}{.}100$	2209	2536	2262	2222	2174	2335
.100 bis .200	0	127	9	4	2	23
.200 bis .400	4	75	7	5	5	13
.400 bis .800	13	35	14	13	12	17
$>$ .800	32	26	25	24	24	25
Summe	2258	2799	2317	2268	2217	2413
		Größe 9.5 bis 10.5				
$0\overset{''}{.}400$ bis $0\overset{''}{.}800$	70	88	66	64	62	69
$>$ .800	64	65	65	65	65	65

doch sind diese letzteren zuverlässig, weil für $\mu = 0''\!.100$ bis $0''\!.200$ wenigstens ein fünftel Teil der Gesamtzahl und für $\mu = 0''\!.200$ bis $0''\!.800$ etwas mehr als die Hälfte der Gesamtzahl aller Sterne beobachtet worden ist.

Die Verteilungen der Leuchtkräfte II bis IV geben alle ungefähr dieselben Zahlen von Sternen zwischen den Grenzen von Eigenbewegung und Größe, welche in der Tab. 10 berücksichtigt worden sind; die Unterschiede liegen wenigstens innerhalb der Unsicherheitsgrenzen, welche den Zahlen anhaften. Es ist daher unmöglich, auf Grund dieser Zahlen zwischen den Annahmen II bis IV eine Wahl zu treffen.

Man sieht weiter, daß die berechneten Zahlen II bis IV gut mit den beobachteten Werten übereinstimmen.

Nachdem diese Rechnung gezeigt hatte, daß sogar eine Kurve mit einer solchen kleinen Einsenkung wie die Kurve II auf Grund der bisherigen Daten nicht verworfen werden kann, habe ich untersucht, ob die Kurve V, welche ganz zwischen I und II liegt, den Beobachtungen widerspricht. Es zeigt sich wirklich, daß die mit V berechneten Zahlen für Eigenbewegungen $0''\!.200$ bis $0''\!.800$ etwas zu groß sind.

Das Resultat dieser Untersuchung kann auf die folgende Weise zusammengestellt werden: Wir wissen auf Grund der beobachteten Zahlen von Sternen bestimmter Eigenbewegung und Größe, daß das Häufigkeitsgesetz der M-Sterne zwischen den absoluten Größen -3 und $+3$ eine Einsenkung zeigt oder wenigstens nicht stetig steigt. Aber wir können aus den bisherigen Daten unmöglich den Schluß ziehen, daß diese M-Sterne überhaupt nicht existieren; denn die Verteilung der absoluten Leuchtkräfte II, welche für die Sterne $M = -3$ bis $M = +3$ eine Häufigkeit gibt gleich der mittleren Häufigkeit der sog. Riesensterne, widerspricht in keiner Hinsicht den beobachteten Zahlen.

Das angebliche Fehlen der Zwischenglieder ist wohl einer der Anhaltspunkte der Theorie der Riesen- und Zwergsterne gewesen. So sagt RUSSELL[1]): There seem to be no stars of class M which are closely comparable with the Sun in brightness, they are either much brighter or much fainter. More than 100 stars of large proper motion have been investigated for parallax without any knowledge of their spectra and it has later been found, when the spectra were investigated that all those of about the sun's brightness were fairly similar to the sun in spectrum, so that it seems pretty certain that the absence of stars of spectrum M similar in brightness to the sun cannot be result of the process of selection but must be a real phenomenon."

Die Trennung der M-Sterne in zwei Klassen kann, nach meiner Ansicht, auf diese Weise nicht bewiesen werden. In den folgenden Betrachtungen werde ich nachweisen, daß es unmöglich ist, das Fehlen

---

[1]) The Observatory Bd. 36, S. 325. 1913.

der Zwischenglieder zu beweisen mittels der bisherigen Daten, und daß
es schwer sein wird, diese Zwischenglieder, falls sie überhaupt existieren,
aus den Tausenden von Sternen, unter welche sie gemischt sind, aus-
zuwählen.

Ich habe in der folgenden Tabelle die Zahlen von Sternen jeder
scheinbaren Größe zwischen den Grenzen der absoluten Größe $< - 2.0$
(Riesen), $- 2.0$ bis $+ 3.0$ (Zwischenglieder), und $> + 3.0$ (Zwerge) zusammengestellt. Die Zahlen sind nach der Formel:

$$4\pi \int r^2 \varphi (m - 5 \log r) D(r) dr$$

gerechnet worden, wo φ das Häufigkeitsgesetz II bedeutet und $D(r)$ die
Dichte nach S. 258; die Grenzen des Integrals sind leicht aus den gegebenen Grenzen der absoluten Größe abzuleiten.

Tabelle 11.
Zahl von M-Sternen, Häufigkeitsgesetz II.

Grenzen m	Zahl			
	Riesen	Zwischen-glieder	Zwerge	Summe
4.5 bis 5.5	68	0	0	68
5.5 bis 6.5	187	0	0	187
6.5 bis 7.5	473	2	2	477
7.5 bis 8.5	1081	10	9	1100
8.5 bis 9.5	2241	34	41	2316
9.5 bis 10.5	4202	104	174	4480
10.5 bis 11.5	7226	315	714	8255

Die Sterne, deren scheinbare visuelle Helligkeit größer ist als 7^m5,
enthalten praktisch gar keine Zwischenglieder, obwohl deren Häufigkeit
im Raume bei der Rechnung ungefähr gleich der mittleren Häufigkeit der
Giganten (Kurve II) genommen worden ist. Man kann daher unmöglich die
Nichtexistenz der Zwischenglieder mittels der Sterne heller als 7.5 beweisen: denn diese scheinbar hellen Sterne enthalten eben keine Objekte
dieser Art, sogar nicht in dem Falle, daß ihre Häufigkeit im Raume
dieselbe wäre, wie diejenige der Riesensterne.

Zur Entdeckung der Sterne mittlerer absoluter Helligkeit können
die 10 Zwischenglieder der achten scheinbaren Größe nicht viel nützen,
denn die meisten dieser gehören zu der Gruppe $M = - 2.0$ bis $- 1.0$,
welche der Gigantengruppe doch sehr nahe ist.

So müssen wir denn zu Sternen von noch schwächeren scheinbaren Größen unsere Zuflucht nehmen.

Das Auffinden der Zwischenglieder würde sehr erleichtert werden
durch die Kenntnis der Eigenbewegungen, wie die Tab. 12 zeigt. Diese
enthält die Zahlen von M-Sternen (Riesen, Zwischenglieder und Zwerge)
zwischen bestimmten Grenzen der Eigenbewegung für die scheinbaren
Größen 9 und 10. Unter den 276 M-Sternen der scheinbaren Größe

8.5 bis 10.5 und Eigenbewegung $0''.040$ bis $0''.400$ sind ungefähr 85 Zwischenglieder vorhanden, falls das Häufigkeitsgesetz II richtig ist. Aber das Auffinden der M-Sterne $m = 8$ und 9, $\mu = 0''.040$ bis $0''.400$ unter Tausenden von anderen wird eine umfassende Arbeit sein.

Tabelle 12.

Zahlen von M-Sternen zwischen bestimmten Grenzen von scheinbarer Größe und Eigenbewegung

$$M < -2.0 = \text{Riesen.}$$
$$M = -2.0 \text{ bis } +3.0 = \text{Zwischenglieder.}$$
$$M > +3.0 = \text{Zwerge.}$$

μ	Riesen	Zwischen-glieder	Zwerge	Summe
		$m = 8.5$ bis 9.5		
$0''.000$ bis $0''.040$	2147	6	0	2153
.040 bis .100	96	13	0	109
.100 bis .200	1	7	1	9
.200 bis .400	0	3	4	7
$> .400$	0	4	36	40
		$m = 9.5$ bis 10.5.		
$0''.000$ bis $0''.040$	4165	38	0	4203
.040 bis .100	42	36	3	81
.100 bis .200	0	12	10	22
.200 bis .400	0	14	34	48
$> .400$	0	5	126	131

Diese Betrachtungen zeigen die großen Schwierigkeiten, welche dem Auffinden der Zwischenglieder anhaften; weil man sich aber bis jetzt nicht besonders bemüht hat, Sterne dieser Art zu finden, befremdet es uns nicht, daß sie nicht entdeckt worden sind.

Man kann fragen, was die Ursache davon ist, daß die Zwischenglieder unter den Sternen bis zur neunten Größe so selten vorkommen, obwohl für ihre Häufigkeit im Raume bei unseren Rechnungen ungefähr derselbe Wert angenommen ist wie für die Giganten. Die Antwort lautet: Die Giganten sind unter den Sternen bis zur neunten Größe zahlreicher, weil wir sie bis in viel größere Entfernungen beobachten können. Die Zwerge haben eine größere Häufigkeit im Raume als die Zwischenglieder und kommen daher in größerer Frequenz in der nächsten Umgebung der Sonne vor, wo sie durch ihre große Eigenbewegung unsere Aufmerksamkeit auf sich ziehen. Die Zwischenglieder aber sind in der unglücklichen Lage, daß sie in kleinerer Zahl ohne auffallendes Merkmal unter Tausende von anderen Sternen gemischt sind; und weil wir sie aus diesem Grunde nicht haben finden können, haben wir ihre Existenz überhaupt geleugnet.

Der Versuch, die Trennung der M-Sterne in zwei Gruppen zu beweisen, indem man die absoluten Helligkeiten einer gewissen Zahl scheinbar heller Sterne beobachtet und zeigt, daß diese eine Lücke aufweisen, muß zweifellos als verfehlt angesehen werden.

Adams und Joy haben auf diese Weise die Trennung in Riesen- und Zwergsterne zu beweisen versucht[1]). Die M-Sterne ihrer Liste schwächer als die Größe 8.5 sind aber ohne Ausnahme nach großer Eigenbewegung ausgewählt worden; ihr Programm enthält daher nur Sterne heller als die Größe 8.5, welche nach Tab. 11 notwendig Riesen sind, und Sterne schwächer als die Größe 8.5 mit großer Eigenbewegung, welche notwendig Zwerge sind. Die Verteilungskurve der absoluten Helligkeiten der M-Sterne in der Abb. 1 ihres Aufsatzes gibt gar keine Auskunft über die Verteilung der Leuchtkräfte im Raume; sie ist nur ein Maß für die Auswahl der M-Sterne, welche Adams und Joy getroffen haben.

Jackson und Furner[2]) und Haas[3]) haben die Verteilung der Leuchtkräfte durch eine Betrachtung der Sterne innerhalb einer gewissen Entfernung von der Sonne abgeleitet. Die Raumvolumina, welche diese Autoren berücksichtigt haben, sind aber notwendig so klein, daß darin gar keine Zwischenglieder des M-Typus erwartet werden können.

---

[1]) Mount Wilson Contributions Bd. 7, S. 319. 1917.
[2]) Monthly Notices of the Royal Astronomical Society Bd. 81, S. 13—16. 1920.
[3]) Veröffentl. der Univ.-Sternwarte zu Berlin-Babelsberg Bd. 3, H. 3. 1923.

Die Verteilungsfunktion der absoluten Helligkeiten in ihrer Abhängigkeit vom Spektrum.

Von R. Hess, München.

Mit 1 Abbildung.

1. Daß die Verteilung der Leuchtkräfte der Sterne vom Spektrum abhängt, ist längst bekannt. HERTZSPRUNG[1]) und RUSSELL[2]) unterscheiden zwei Gruppen von Sternen: Die Riesen, deren Leuchtkraft oder absolute Helligkeit mit dem Spektraltypus wenig veränderlich ist, und die Zwerge, die mit zunehmendem Spektraltypus einen rapiden Helligkeitsabfall zeigen. Es sei an dieser Stelle auch auf das allbekannte RUSSELLsche Diagramm verwiesen. Man hat dann auch versucht, für die einzelnen Spektraltypen die mittlere absolute Helligkeit[3]) der Riesen wie der Zwerge aufzustellen. Ich erwähne z. B. die Bestimmungen von ADAMS und JOY[4]), P. DOIG[5]) sowie meine eigenen[6]). Ganz besonders aber möchte ich auf eine interessante Arbeit von LUNDMARK und LUYTEN[7]) aufmerksam machen, die den Zusammenhang zwischen absoluter Helligkeit und Spektraltypus aus Doppelsternkomponenten ableiten.

Die bisher besprochenen Bestimmungen der mittleren absoluten Helligkeiten von Riesen und Zwergen verschiedener Spektren sind im wesentlichen Bestimmungen der Lage der Maxima der Verteilungsfunktion $\varphi(M)$ der absoluten Helligkeiten M, getrennt nach Spektraltypen. Wir haben also damit bei weitem nicht alle Daten gewonnen, die die Verteilung der Sterne nach Leuchtkraft und Spektrum bestimmen. Wir müssen vielmehr erst für jeden Spektraltypus den Verlauf von $\varphi(M)$ aufstellen. Dies ist in zwei oben erwähnten Arbeiten ge-

---

[1]) Zur Strahlung der Sterne. Zeitschr. f. wiss. Photographie Bd. 3.

[2]) Relations between the spectra and other characteristics of the stars. Publ. Amer. Astr. Soc. Bd. 3.

[3]) Unter absoluter Helligkeit soll hier stets die scheinbare Helligkeit verstanden werden, die ein Stern erhält, wenn man ihn in die der Parallaxe 0″1 entsprechende Entfernung versetzt.

[4]) Mt. Wilson Contr. 142.

[5]) Monthly Notices of Roy. Astr. Soc. (M. N.) 82.

[6]) Astr. Nachr. Bd. 220, S. 65—78. 1923.

[7]) Astronom. Journal Nr. 828.

schehen, das eine Mal aus 500[1]), das andere Mal aus 1646[2]) spektroskopischen Parallaxen.

Die Zusammenstellung einer Reihe von Kurven für die verschiedenen Spektraltypen bringt eine gewisse Unübersichtlichkeit mit sich. Dazu kommt noch der weitere Nachteil, daß die Kurven nur für einige wenige Spektraltypen aufgestellt werden können, während über den Verlauf von $\varphi(M)$ in den zwischenliegenden Typen manche Unklarheit herrscht. Diese beiden Nachteile lassen sich vermeiden durch Einführung einer Häufigkeitsfunktion $\Psi(M, S)$, die wir dadurch definieren, daß $\Psi(M, S)\, dM\, dS$ die Anzahl der Sterne sein soll, deren absolute Helligkeiten zwischen M und $M + dM$, und deren Spektren zwischen S und $S + dS$ liegen. Durch eine solche Funktion $\Psi(M, S)$ ist die statistische Verteilung der Sterne nach Leuchtkraft und Spektrum eindeutig festgelegt. Halten wir einen Wert S_0 von S fest, so gibt $\Psi(M, S_0) = \varphi_{S_0}(M)$ die Luminositätskurve für das Spektrum S_0.

Es empfiehlt sich, die Funktion $\Psi(M, S)$ geometrisch zu veranschaulichen; denn nur auf diese Weise wird sie uns, wenn sie einmal bekannt ist, eine klare Vorstellung von der Verteilung der Sterne nach Leuchtkraft und Spektrum vermitteln. Zu diesem Zwecke machen wir in der Horizontalebene die Abszisse zur S-Achse und tragen auf der letzteren die Spektren B 0, A 0, F 0, G 0, K 0, M in gleichen Abständen auf. Es ist wohl ganz klar, wie man die Zwischenstufen B 1, B 2, ... A 1, A 2 usw. durch Unterteilung der Abstände einschaltet. Auf der Ordinatenachse dagegen tragen wir die absoluten Helligkeiten M auf. Jedem Wertepaar M, S ist nun ein Punkt der Horizontalebene eindeutig zugeordnet und umgekehrt. Andererseits ist jedem Punkte (M, S) ein bestimmter Funktionswert $\Psi(M, S)$ zugeordnet, den wir senkrecht zur Horizontalebene (M, S-Ebene) nach oben auftragen. Selbstverständlich können negative Funktionswerte nicht auftreten. $\Psi(M, S)$ ist nun dargestellt als eine Fläche, die wir als die Verteilungsfläche der Spektren und absoluten Helligkeiten bezeichnen wollen. Halten wir wieder einen Wert S_0 von S fest, so gibt $\Psi(M, S_0) = \varphi_{S_0}(M)$ einen Querschnitt durch die Verteilungsfläche, die wir also konstruieren können, wenn genügend solcher Querschnitte bekannt sind. Diesen Weg der Bestimmung von $\Psi(M, S)$ wollen wir auch einschlagen.

Wir können uns aber an Hand der bis jetzt bekannten Tatsachen bereits ein Bild machen von dem ungefähren Aussehen der Fläche Ψ: Sie wird im wesentlichen aus zwei Erhebungen bestehen, entsprechend den beiden Gruppen: Riesen- und Zwergsterne. Der ungefähre Verlauf der Kammlinien dieser beiden Erhebungen ist durch das RUSSELLsche Diagramm oder durch die oben erwähnten Bestimmungen der mittleren

[1]) Mt. Wilson Contr. 142.
[2]) Astr. Nachr. Bd. 220, S. 65—78. 1923.

absoluten Helligkeiten von Riesen und Zwergen verschiedener Spektraltypen bekannt. Wir wissen auch, daß die Anzahl der Zwerge erheblich größer ist als die der Riesen. Wir werden deshalb erwarten, daß die den Zwergen zugehörige Häufigkeitserhebung diejenige der Riesen weit überragt. Nachdem wir also bereits eine rohe Vorstellung der Fläche Ψ gewonnen haben, gehen wir daran, $\Psi(M, S)$ so genau zu bestimmen, als es das gegenwärtig noch spärliche Material erlaubt.

2. Für Abzählungen der absoluten Helligkeiten dürfte es sich wohl empfehlen, spektroskopische Parallaxen zu verwenden. Gegenüber den trigonometrischen Parallaxen haben sie folgende Vorteile: 1. Der mittlere Fehler in M ist nicht von der Entfernung abhängig, während die absoluten Helligkeiten, die aus kleinen trigonometrischen Parallaxen berechnet sind, außerordentlich unsicher sind. 2. Bei Verwendung von trigonometrischen Parallaxen müßte man die negativen Werte streichen, oder ein ähnliches Verfahren einschlagen wie Luyten[1]), der $M = -3$ setzt bei negativen Parallaxen, was jedoch beides nicht statthaft ist.

Im folgenden gebe ich eine Zusammenstellung von vorhandenem Material an spektroskopischen Parallaxen:

Anzahl der Sterne		Spektrum	Quellenangabe
Mt.-Wilson-Parallaxen	1646	A 3 bis M	Mt. Wils. Contr. 199
	544	B 7 bis F 2	Mt. Wils. Contr. 244
	299	O 8 bis A 1	Mt. Wils. Contr. 262
	137		Mt. Wils. Contr. 228
	6	M	Publ. Astr. Soc. Pac. 35, S. 328—329
Harvard-Parallaxen	50	K	Harvard Circ. 228
	87	K	Harvard Circ. 232
	116	G, K	Harvard Circ. 246
	233	K	Harvard Circ. 243
N. Lockyer Obs.-Parall.	49	B 0 bis B 8	M. N. 83, 53—54
	500	F 0 bis M b	Mem. R. A. S. 62. IV.

Bei der Auswahl des zu benutzenden Materials muß nun untersucht werden, ob die drei Systeme: Mt. Wilson, Harvard und Norman Lockyer Observatory keine systematischen Unterschiede aufweisen. Dies ist aber leider der Fall. Nach Shapley und Miss Ames[2]) ist die systematische Differenz in M:

$$\text{Mt. Wils. Contr. 199} - \text{Harvard Circ. 243} = +0.53$$
$$\text{Harv. Circ. 228, 232} - \text{Harvard Circ. 243} = +0.39$$

Aus diesem Grunde sind die Harvard-Parallaxen in der vorliegenden Arbeit nicht verwendet worden. Günstiger dagegen scheinen die Verhältnisse zu liegen, wenn man Mt.-Wilson- und Lockyer-Parallaxen miteinander vergleicht. Von den 49 Sternen aus M. N. 83 kommen

---

[1]) Publ. Astr. Soc. Pac. Nr. 206.
[2]) Harvard Circ. 243.

17 in Mt. Wilson Contr. 262 vor. Die in den beiden Publikationen gegebenen Werte von M für diese 17 gemeinsamen Sterne wurden untersucht. Es stehen hier drei Bestimmungen zur Verfügung: 1. Aus der Linie $\lambda 4144$ (Lockyer,), 2. aus der Linie $\lambda 4388$ (Lockyer) und 3. aus dem Spektraltypus (Mt. Wilson). Es wurden die Differenzen in M gebildet: $1 - 3$ und $2 - 3$. Im Mittel aus den 17 Sternen ergab sich $1 - 3 = +0.2 \pm 0.3$ und $2 - 3 = 0.0 \pm 0.3$, so daß also keinerlei systematische Differenz nachweisbar ist. Was die Bestimmung des Spektraltypus betrifft, so ist die mittlere Differenz Lockyer—Mt. Wilson $= -0.6 \pm 0.5$, wobei das Intervall zwischen B0 und A0 gleich 10.0 gesetzt ist. Es lag also kein Grund vor, die Norman-Lockyer-Parallaxen auszuscheiden. Die 49 Sterne aus M. N. 83 wurden übernommen, und zwar wurde das Mittel der aus den beiden Linien $\lambda 4144$ und $\lambda 4388$ sich ergebenden Werte M akzeptiert. Dagegen wurden die 500 Sterne in Mem. R. A. S. 62 nicht benutzt, da sie sämtlich unter den 1646 Parallaxen in Mt. Wilson Contr. 199 enthalten sind. Endlich wurde auch unter den Mt.-Wilson-Parallaxen eine gewisse Auswahl getroffen: Von den 1646 Sternen in Contr. 199 wurden gestrichen: 1. 45 Sterne, die besser aus Contr. 244 entnommen werden, da die betreffenden absoluten Helligkeiten in Nr. 199 durch Extrapolation der Reduktionskurven des F-Typus erhalten sind. 2. 5 M-Sterne, für die ADAMS und JOY verbesserte Werte geben[1]). Ferner wurden in Contr. 262 die mit M. N. 83 gemeinsamen 17 Sterne gestrichen, da es empfehlenswerter sein dürfte, die Bestimmungen aus Spektral*linien* als solche aus dem Spektral*typus* heranzuziehen, um die Streuung innerhalb eines Spektralbereichs zu untersuchen. Bei den 95 Sternen, die in Contr. 244 und 262 gemeinsam auftreten, wurde das Mittel der absoluten Helligkeiten genommen. Schließlich waren auch die 137 Sterne aus Contr. 262 nicht geeignet für die vorliegende Untersuchung.

Für das ausgewählte Material wurde ein Zettelkatalog angelegt, enthaltend die Bezeichnung des Sterns, scheinbare Größe m (nach HARVARD), E. B., Spektrum, absolute Helligkeit M und die aus $(M - m)$ folgende Parallaxe, welche nötigenfalls aus einer zu diesem Zwecke angefertigten kleinen Tabelle entnommen wurde. Außerdem findet sich hinter M die Quellenangabe. Mit diesem Zettelkatalog wurden die weiteren Untersuchungen angestellt.

3. Bereits in A. N. 5261 wurde darauf aufmerksam gemacht, daß die Bevorzugung großer scheinbarer Helligkeiten und großer E. B. bei der Auswahl der Sterne, deren M spektroskopisch bestimmt wurden, zu Verfälschungen der Verteilungsfunktion der absoluten Helligkeiten sowie der Dichtefunktion führt. Um die Wirkung dieser Bevorzugungen auszuschalten, führte ich in A. N. 5261 Gewichte $p_{m,\mu,S} = \dfrac{N_{m,\mu,S}}{n_{m,\mu,S}}$

---

[1]) Publ. Astr. Soc. Pac. 35, S. 328—329.

ein, wobei m = scheinbare Größe, μ = E. B., S = Spektrum ist und N und n die wirkliche Anzahl bzw. die Anzahl im Material der Sterne von der Eigenschaft (m, μ, S) bedeutet. Jeder Stern wurde bei der Abzählung nicht *ein*mal gezählt, sondern p-mal. Die Bezeichnung „Gewicht" für p könnte vielleicht zu Irrtümern Veranlassung geben. p hat nichts zu tun mit Vertrauenswürdigkeit; im Gegenteil: Je größer die p sind, um so unsicherer werden die Abzählungsresultate. Wir wollen daher für die p von nun an den Ausdruck „Bevorzugungszahlen" verwenden.

Auch bei der vorliegenden Arbeit wurden Bevorzugungszahlen eingeführt. Die Berechnung der N geschah aus den Zahlen von KAPTEYN und VAN RHIJN[1]) in der Weise, wie schon in A. N. 5261 beschrieben wurde. Die n ergaben sich durch Abzählung des Materials, wobei Sterne von unbekannter E. B. von der weiteren Untersuchung ausgeschlossen wurden. Bei der Einteilung des Materials nach Spektren wurden der Typus B 5 und frühere Typen zum Spektrum B 2 gezählt, B 6 bis A 5 kam zu A, A 6 bis F 5 zu F, F 6 bis G 5 zu G, G 6 bis K 5 zu K und die späteren Typen zu M. Leider konnten die Bevorzugungszahlen meist nur bis zur scheinbaren Größe $6^{\text{M}}\!.4$ erteilt werden. Die schwächeren Helligkeiten waren nämlich in so geringen Mengen im Material vertreten, daß die p meist zu groß wurden. Die nicht mit Bevorzugungszahlen versehenen Sterne wurden aus dem weiteren Verlauf der Untersuchung ausgeschieden.

4. In A. N. 5261 wurde auch gezeigt, daß es unstatthaft ist, die Verteilungsfunktion $\varphi(M)$ durch Abzählung des ganzen Materials zu bestimmen, sondern daß man endlich begrenzte Zonen abzählen muß. Es wurde daher auch hier in Entfernungszonen (konzentrische Kugel-

Tabelle 1.

Zonenübersicht.

Zone	Parallaxe		parsecs
I		$> 0\rlap{.}''1000$	0 bis 10
II	$0\rlap{.}''1000$ bis	.0500	10 ,, 20
III	.0500 ,,	.0333	20 ,, 30
IV	.0333 ,,	.0250	30 ,, 40
V	.0250 ,,	.0200	40 ,, 50
VI	.0200 ,,	.0133	50 ,, 75
VII	.0133 ,,	.0100	75 ,, 100
VIII	.0100 ,,	.0075	100 ,, 133
IX	.0075 ,,	.0065	133 ,, 154
X	.0065 ,,	.0045	154 ,, 222
XI	.0045 ,,	.0035	222 ,, 286
XII	.0035 ,,	.0025	286 ,, 400
XIII	.0025 ,,	.0015	400 ,, 667

---

[1]) Groningen Publ. 30.

schalen um die Sonne) geschieden. Tab. 1 gibt eine Übersicht über die Zonen. Neben diesen 13 Zonen wurde zuweilen auch eine Zone (II) benutzt mit den Begrenzungen

$$0''.1000 > \pi > 0''.0625.$$

Durch die scheinbare Grenzgröße m_g, bei der die Erteilung der Bevorzugungszahlen aufhört, ist für jede Entfernung eine absolute Grenzgröße

$$M_g = m_g + 5 + 5 \log \pi$$

gegeben. M_g wurde für den äußeren Rand einer jeden Zone berechnet. In keiner Zone darf über das zugehörige M_g hinaus abgezählt werden. Um ein Beispiel für die Abzählungsresultate zu geben, greife ich den Typus K heraus, für den Tab. 2 die Summen der Bevorzugungszahlen gibt. Die Treppenlinie deutet die absolute Grenzgröße M_g an.

5. Nun handelt es sich darum, aus den Abzählungen für die einzelnen Spektraltypen die Verteilungsfunktionen zu bestimmen. Wir verfahren dabei auf folgende Weise: Die in der letzten Zone (in Tab. 2: Zone X) noch verwendbaren absoluten Helligkeiten werden über alle Zonen

Tabelle 2.

Abzählung des K-Typus.

M	−1.3	−0.8	−0.3	+0.2	+0.7	1.2	1.7	2.2	2.7	3.2	3.7	4.2	4.7	5.2	5.7	6.2	6.7	7.2
Von	−1.5	−1.0	−0.5	0.0	0.5	1.0	1.5	2.0	2.5	3.0	3.5	4.0	4.5	5.0	5.5	6.0	6.5	7.0
bis	−1.1	−0.6	−0.1	+0.4	0.9	1.4	1.9	2.4	2.9	3.4	3.9	4.4	4.9	5.4	5.9	6.4	6.9	7.4
Zone																		
I	—	—	—	—	—	1.2	1.2	—	—	1.4	—	—	2.7	—	7.8	10.9	57.2	4.0
(II)	—	—	—	—	—	(1.2)	—	—	(1.2)	(2.2)	(1.1)	—	(2.0)	(11.7)	(13.0)	(6.0)		
II	—	—	—	—	1.2	2.4	—	1.4	2.3	2.2	1.1	1.0	3.0	15.7	20.0			
III	—	—	—	3.6	7.6	2.6	6.2	2.7	4.9	1.5	—	3.0	5.0					
IV	—	—	4.4	8.0	19.9	19.4	5.2	14.0	2.7	8.2	9.2	151.7						
V	—	—	—	7.3	32.5	16.4	19.7	1.5	11.2	9.1	1.0							
VI	—	3.6	3.9	35.7	101.5	91.6	105.0	211.8	159.2									
VII	3.6	1.9	8.1	73.0	155.5	157.1	155.4	47.7										
VIII	—	—	15.8	122.7	620.8	573.6	—											
IX	—	—	18.3	31.2	1003.1	—												
X	—	3.2	166.3	471.4	898.2													

hinweg abgezählt. Beim Typus K z. B. sind dies die Helligkeiten
— 1.3, — 0.8, — 0.3 (Die Helligkeiten 0.2 und 0.7 sind nicht verwendbar, da die Summen 471.4, 898.2 und 1003.1 nur aus 8, 6 und 14 Sternen
bzw. abgeleitet sind.) Durch vertikale Addition erhalten wir also für
$M = -1.3$, — 0.8, — 0.3 die Werte 3.6, 8.7, 217 für $\varphi(M)$. Hierauf
scheiden wir die letzte Zone aus. In der Regel werden wir dadurch
imstande sein, ein oder zwei Vertikalspalten mehr aufaddieren zu
können. In Tab. 2 würden wir also Zone I—IX addieren bis zu M
$= + 0.2$. Wir gewinnen neue Werte von $\varphi(M)$, die aber mit den letzten
nicht vergleichbar sind, da ja über ein kleineres Volumen summiert
wurde. Wir müssen daher die neuen Werte erst mit einem Faktor
multiplizieren. Dieser läßt sich aus irgendeinem M berechnen, für das
wir $\varphi(M)$ sowohl bezüglich der Zonen I—X wie auch in bezug auf
I—IX kennen, indem wir einfach zwei entsprechende Werte von $\varphi(M)$
durcheinander dividieren. Im Falle des Typus K sind das die Helligkeiten $M = -1.3$, — 0.8, — 0.3. Am sichersten aber bestimmt man
den Faktor aus allen diesen Helligkeiten zusammen, indem man z. B.
hier die Anzahlen der Sterne, die absolut heller oder gleich — $0^M.1$ sind,
für Zone I—X und Zone I—IX bestimmt und durcheinander dividiert.
Mit diesem Faktor, der immer > 1 ist, werden die neuen Werte von
$\varphi(M)$ multipliziert. Durch sukzessives Ausscheiden der jeweils
letzten noch mitgenommenen Zone bekommt man $\varphi(M)$ für immer
schwächere Helligkeiten. Die verschiedenen Kurven $\varphi(M)$ der einzelnen Spektren sind aber noch nicht vergleichbar, da sie sich auf verschiedene Volumbereiche beziehen (bei K z. B. auf die Zonen I—X).
Wir multiplizieren daher jedes $\varphi(M)$ mit einem von Spektraltypus
zu Spektraltypus verschiedenen konstanten Faktor. Diese Konstanten wählen wir derart, daß $\varphi(M)\,dM$ die Anzahl der Sterne in
den Zonen I—VII, d. h. in einer Kugel vom Radius 100 parsecs um
die Sonne gibt. Diese endgültigen Werte von $\varphi(M)$ sind in Tab. 3
zusammengestellt.

Dabei zeigen sich einige Mängel: Beim B-Typus läßt sich der Verlauf
der Kurve für schwächere Helligkeiten nicht angeben. Ferner fehlt
jede Möglichkeit, die Verteilungskurve der M-Zwerge zu bestimmen.
Es scheint, daß der Anstieg dieser Kurve etwa bei 8^M beginnt. Vermutlich erreicht sie ihr Maximum bei 9 oder 10^M.

6. Wir sind jetzt imstande, vermöge der in Tab. 3 gegebenen Werte
die Fläche $\Psi(M, S)$ zu konstruieren. Die Darstellung der Fläche
wollen wir mit Hilfe von Höhenlinien, d. h. Linien konstanten Funktions-
wertes Ψ, also konstanter Häufigkeit vornehmen. Mehrere Querschnitte
$\varphi(M)$ durch die Fläche sind uns durch Tab. 3 gegeben. Es wurden also
zunächst die verschiedenen Kurven $\varphi(M)$ gezeichnet, wobei natürlich
immer eine Kurve gezeichnet wurde, die sich den bekannten Punkten
möglichst gut anschmiegte.

Tabelle 3.

Die Werte $\varphi(M)$ für die verschiedenen Spektren.

M	B2	M	A0	F0	G0	K0	M
−4	0.04	−1.8			1.3		
−3	0.98	−1.3		5.7	2.1	0.6	0.7
−2	4.4	−0.8	12	8.8	6.6	1.5	4.7
−1	32	−0.3	39	9.0	18	37	12
0	20	+0.2	160	23	54	184	69
+1	29	0.7	595	64	57	718	76
		1.2	1097	88	34	581	16
		1.7	1501	169	98	575	0.0
		2.2	2417	1990	149	258	0.0
		2.7	0	3670	36	231	0.0
		3.2	0	1610	854	310	
		3.7	247[1])	7120	4293	240	
		4.2		1650	4530	289	
		4.7			4221	771	
		5.2			4166	3122	
		5.7			4257	5528	
		6.2			0	4651	

Um die Höhenlinien zu erhalten, mußten nun etliche Parallelebenen zur (M, S)-Ebene mit der Verteilungsfläche geschnitten werden. Diese Parallelebenen schneiden die Querschnitte $\varphi(M)$ in parallelen Linien zur Abszisse (M-Achse). Es wurden also Parallele zur Abszisse in bestimmten Abständen gezogen und die Schnittpunkte mit den Kurven $\varphi(M)$ auf die Abszisse projiziert. Jede derartige Projektion gibt einen Punkt der betreffenden Höhenlinie. Man liest auf der Abszisse den Wert von M_0 ab für den projizierten Punkt. Handelt es sich dabei um die Kurve vom Spektrum S_0, so ist (M_0, S_0) ein Punkt der Höhenlinie. Jede Höhenlinie wird also gegeben durch eine Reihe von Punkten. So wurde das beigefügte Bild der Fläche $\Psi(M, S)$ erhalten Es zeigt eine gewisse Ähnlichkeit mit dem RUSSELLschen Diagramm. Die Scheidung in Riesen und Zwerge ist deutlich erkennbar. Die Trennung scheint nur beim A-Typus zu fehlen, wo die Kammlinien sich schneiden. Die Zweiteilung der B-Sterne dürfte wohl verbürgt sein. Auch EDWARDS tritt für die Zweiteilung des B-Typus ein[2]). Daß die Unterbrechung der Kammlinie für die Riesen zwischen F- und G-Typus reell ist, muß wohl bezweifelt werden Die M-Zwerge ließen sich leider nicht einzeichnen. Nur der vermutliche Verlauf der Kammlinie von den K-Zwergen zu den M-Zwergen ist angegeben. Aus dem Bilde ersehen wir auch, daß die Häufigkeit der Zwerge die der Riesen übertrifft. Wir sehen ferner, daß die B-Sterne sehr seltenes Vorkommen auf-

---

[1]) Für $M = 4.0$ gültig.
[2]) Observatory 46, S. 19—22.

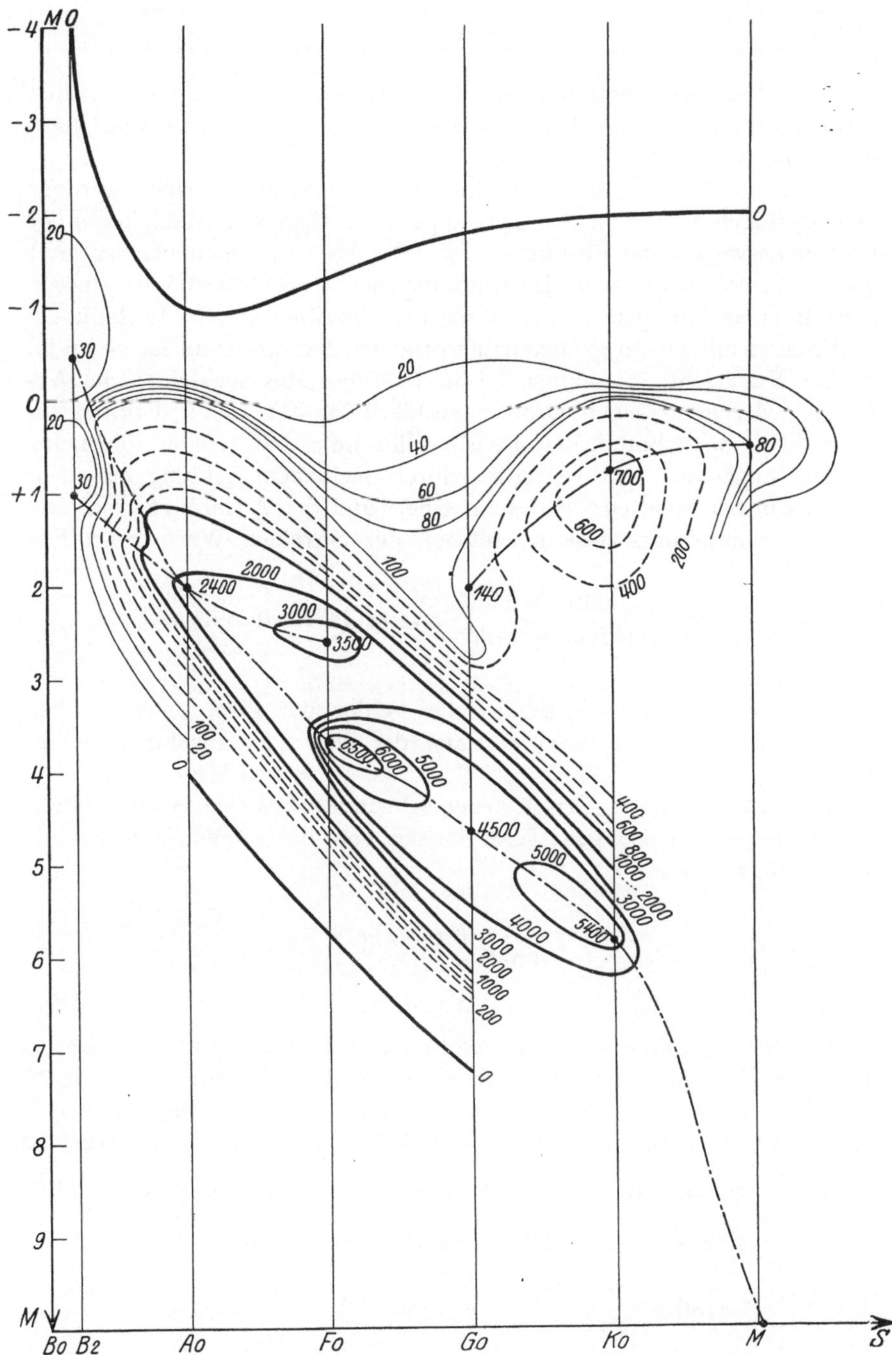

Abb. 1. *Die Verteilungsfläche* $\Psi(M, S)$ *der absoluten Helligkeiten und der Spektren.*
Die Linien parallel zur M-Achse deuten die Spektraltypen an, für die Querschnitte durch die Fläche bekannt sind. Die strichpunktierten Linien geben
den Verlauf der Kammlinien der Fläche an, also die Häufigkeitsmaxima der
einzelnen Spektraltypen. Die dickausgezogenen Linien sind Höhenlinien (Kurven
konstanter Häufigkeit) von 1000 zu 1000, die gestrichelten von 200 zu 200, die
dünnausgezogenen von 20 zu 20. Diese „Höhenzahlen" sind Sternanzahlen in
der Kugel vom Radius 100 parsecs um die Sonne.

weisen; dies gilt jedoch nur für die betrachtete Kugel vom Radius 100 parsecs. In großen Entfernungen treten die B-Sterne wohl häufiger auf.

7. Es muß noch auf einen Punkt aufmerksam gemacht werden: Die zufälligen Fehler in den Messungen der absoluten Helligkeiten M sind keineswegs ohne Einfluß auf $\varphi(M)$. Aber es schien bei der jetzt noch ziemlich ungenauen Bestimmung der $\varphi(M)$ überflüssig, diesen Einfluß in Rechnung zu setzen. Wird aber die Ableitung der Verteilungsfunktionen mit einem größeren Material wiederholt, dann ist es nötig, darauf Rücksicht zu nehmen. Daß zufällige Messungsfehler die Abzählung der gemessenen Größen verfälschen, hat zuerst EDDINGTON bemerkt[1]). Zur Elimination dieser Verfälschung hat er eine allgemeine Formel abgeleitet und weiterhin durch Reihenentwicklung eine Näherungsformel erhalten. Hat man eine Tafel der Abzählungsresultate, so ist die näherungsweise Korrektion der einzelnen Werte nach EDDINGTON:

$$= - \left(\frac{1.046 \cdot \text{w. F.}}{\text{Tafelintervall}}\right)^2 \cdot \text{zweite Tafeldifferenz}$$

Unter w. F. ist der wahrscheinliche Fehler der gemessenen Größen zu verstehen, also in unserem Falle der w. F. der absoluten Helligkeiten M.

Ist $\varphi(M)$ nicht numerisch gegeben, sondern als GAUSSsche Fehlerkurve dargestellt, so geht man besser von der allgemeinen Formel EDDINGTONS aus:

$$u(m) = \frac{h}{\sqrt{\pi}} \int\limits_{-\infty}^{+\infty} v(m+x)\, e^{-h^2 x^2}\, dx$$

in der $v(m)$ die wahre, $u(m)$ die gefälschte Verteilungsfunktion und h das Maß der Präzision der gemessenen Größen m bedeutet. KAPTEYN und VAN RHIJN haben bei der Untersuchung der Verteilung der E. B.[2]) diese Formel für die GAUSSsche Fehlerkurve spezialisiert Ihr Ergebnis ist das folgende: Ist $v(m)$ eine GAUSSsche Fehlerkurve $\dfrac{h_1}{\sqrt{\pi}}\, e^{-h_1^2(m-m_1)^2}$, so ist auch $u(m)$ eine solche und wird dargestellt durch:

$$u(m) = \frac{h'}{\sqrt{\pi}}\, e^{-h'^2(m-m_1)^2}, \quad \text{wo} \quad h' = \frac{1}{\sqrt{\dfrac{1}{h_1^2} + \dfrac{1}{h^2}}}$$

---

[1]) M. N. 73, Nr. 5.
[2]) Groningen Publ. 30.

Die letzte Beziehung ist nichts anderes als das Additionsgesetz der zufälligen Fehler. Bekannt sind h und h'. Man berechnet h_1 nach der Formel: $h_1 = \dfrac{h\,h'}{\sqrt{h^2 - h'^2}}$.

8. Zum Schluß sei darauf hingewiesen, wo das Material der absoluten Helligkeiten einer Ergänzung bedarf: Vor allem müssen die schwachen M-Sterne herangezogen werden, von der scheinbaren Größe 6.5 an, um die Verteilung der M-Zwerge zu erhalten. Ferner muß das Material der B-Sterne noch vergrößert werden. Für alle Spektraltypen wäre es empfehlenswert, auch die schwächeren Sterne (jenseits der scheinbaren Größe 6.5) bedeutend stärker heranzuziehen, als dies bisher der Fall war. Man beschränke sich dabei aber nicht auf große E. B.!

Die Grenzen des typischen Sternsystems und die Verteilungsfunktion der absoluten Leuchtkräfte.

Von **Walter Sametinger**, München.

1. **Die Integralgleichungen der Stellarstatistik bei endlicher Ausdehnung des Systems und endlichen Maximalwerten der Leuchtkräfte.** Die mathematische Behandlung des Problems, aus der scheinbaren Sternverteilung die räumliche Struktur des Sternsystems zu ermitteln, wird, wie SCHWARZSCHILD[1]) gezeigt hat, relativ einfach, wenn man annimmt, daß das System unendlich weit ausgedehnt ist und daß beliebig große Leuchtkräfte vorkommen können. Beide Voraussetzungen stehen jedoch im Widerspruch mit der Wirklichkeit. Daß die absoluten Leuchtkräfte eine ziemlich scharf betonte obere Grenze haben, geht aus dem bisher bekannten Material unzweideutig hervor. Ferner läßt sich, wenn man nur die Ansätze entsprechend macht, die Existenz einer ausgesprochenen Begrenzung des Sternsystems aus den beobachteten Sternzahlen beweisen.

Im Jahre 1898 hat SEELIGER[2]) zum erstenmal die strenge Formulierung der Beziehungen gegeben, welche grundlegend sind für die Untersuchungen über die Konstitution des Fixsternsystems. Er stellte die Integralgleichungen auf, welche die Beobachtungsdaten verbinden mit den unbekannten Funktionen: der räumlichen Dichteverteilung und der Verteilungsfunktion der Leuchtkräfte. Damit wurde das Problem erst auf eine exakte Grundlage gestellt.

SEELIGER hat seine Ansätze entsprechend der Annahme einer endlichen Ausdehnung des Systems und eines endlichen Maximalwertes der Leuchtkräfte gemacht. Es soll in den folgenden Untersuchungen gezeigt werden, daß die so gewonnenen Gleichungen es gestatten, unsere Kenntnis vom Bau des Sternsystems in manchen Punkten wesentlich zu erweitern und daß die Resultate den gewünschten Grad der Zuverlässigkeit haben.

SEELIGER führte gleich von Anfang an den Begriff des typischen Sternsystems ein. Die hervorstechendste Eigenschaft der scheinbaren

---

1) Über die Integralgleichungen der Stellarstatistik. Astr. Nachr. Bd. 185, Nr. 4422. 1910.

2) Betrachtungen über die räumliche Verteilung der Fixsterne. Abh. Akad. München 1898.

Sternverteilung ist die Symmetrie in bezug auf die Milchstraße, welcher eine Symmetrie der räumlichen Sternverteilung in bezug auf die Milchstraßenebene entspricht. Demgegenüber tritt die Abhängigkeit der Sternverteilung von der galaktischen Länge mehr in den Hintergrund. Im typischen Sternsystem wird die Abhängigkeit von der galaktischen Länge vernachlässigt, die Abhängigkeit von der galaktischen Breite wird dadurch berücksichtigt, daß die Sphäre in eine Reihe von parallel und symmetrisch zur Milchstraße gelegenen Zonen eingeteilt wird, die Beobachtungsdaten für jede dieser Zonen gemittelt und darnach die unbekannten Funktionen bestimmt werden. Man schreibt also auf diese Weise dem Sternsystem die Gestalt eines Rotationskörpers zu mit der Achse in Richtung der Milchstraßenpole. Es ist schon sehr viel gewonnen, wenn wir das so definierte ideale Modell des Systems in seiner Struktur und Ausdehnung kennen. Das Studium der einzelnen Ab weichungen von der typischen Form wird dann das Ziel spezieller Untersuchungen sein, wie sie ja auch schon, z. B. für die Milchstraßenwolken, von verschiedener Seite angestellt worden sind [EASTON[1]), PANNEKOEK[2]), KOPFF[3])].

SEELIGER hat dann in einer zweiten Arbeit vom Jahre 1909[4]) seine Ansätze auf die allgemeinste Form gebracht, indem er auch die Absorption des Lichtes im Weltraum als unbekannte Funktion einführte. Die Beziehungen, welche sich auf diese Weise aufstellen lassen, sind von ihm selbst in dieser und seinen folgenden Abhandlungen, außerdem von DEUTSCHLAND[5]), ausführlich diskutiert worden. Da das gegenwärtig verfügbare Beobachtungsmaterial nicht ausreicht, um auch die Absorption zu bestimmen, sollen die folgenden Untersuchungen so angestellt werden, als ob keine Absorption vorhanden wäre.

Die scheinbare Sternverteilung ist dann abhängig von zwei Funktionen, der Dichtefunktion D und der Verteilungsfunktion φ der absoluten Leuchtkräfte, und außerdem von der räumlichen Ausdehnung des Systems. Dabei bedeutet D die Anzahl der Sterne in der Volumeinheit und $\varphi(i)\,di$ den Bruchteil davon, dessen Leuchtkräfte zwischen i und $i + di$ liegen.

Ist H der Höchstwert der Leuchtkräfte, so soll also sein:

$$\int_0^H \varphi(i)\,di = 1 \ .$$

Wir machen weiter die Annahme, daß die Funktion φ vom Ort unabhängig ist, daß also das Mischungsverhältnis der Leuchtkräfte überall im Sternsystem das gleiche ist.

[1]) Astr. Nachr. Bd. 141, Nr. 3279.
[2]) Bull. of the Astr. Inst. of the Netherlands Nr. 11.
[3]) Über Häufigkeitsfunktion und Entfernung bei den hellen Milchstraßenwolken. Astr. Nachr. Bd. 216, Nr. 5177.
[4]) Abh. Akad. München.   [5]) Vierteljahrsschr. Astr. Ges. Bd. 54.

Die Ableitung der Grundgleichungen läßt sich leicht durchführen. Die Anzahl der Sterne in dem Volumelement $\omega \varrho^2 d\varrho$, deren Leuchtkraft zwischen i und $i + di$ oder deren scheinbare Helligkeit zwischen $h_m = \dfrac{i}{\varrho^2}$ und $h_m + dh_m = \dfrac{i + di}{\varrho^2}$ liegt, ist gegeben durch:

$$n(h_m)\,dh_m = \omega\,\varrho^2\,D(\varrho)\,d\varrho\,\varphi(i)\,di$$

Will man alle Sterne in dem betreffenden Element erhalten, deren scheinbare Helligkeit größer oder gleich h_m ist, so hat man über i zu integrieren von $h_m\varrho^2$ bis H; man bekommt so:

$$a(h_m) = \omega\,\varrho^2 D(\varrho)\,d\varrho \int\limits_{h_m\varrho^2}^{H} \varphi(i)\,di$$

Um endlich die Anzahl A_m aller Sterne von den hellsten bis zu denen von der scheinbaren Größe m, welche auf dem Areal ω stehen, zu erhalten, hat man über ϱ zu integrieren von $\varrho = r_0$, d. h. der Entfernung, innerhalb welcher keine Sterne mehr vorkommen, bis zu derjenigen Entfernung, außerhalb welcher keine Sterne mehr einen Beitrag zu dem betreffenden A_m liefern. Dabei hat man zwei Fälle zu unterscheiden; es sei r_1 die Entfernung bis zur Grenze des Systems, und wir definieren eine Sterngröße n so, daß:

$$h_n r_1^2 = H$$

Ist dann $m < n$, so hat man, um A_m zu erhalten, der oberen Grenze des Integrals über ϱ den Wert $\sqrt{\dfrac{H}{h_m}}$ zu geben. Ist jedoch $m > n$, so ist das Integral zu erstrecken bis $\sqrt{\dfrac{H}{h_n}} = r_1$.

Man erhält so für die A_m die folgenden beiden Ausdrücke:

$$A_m = \omega \int\limits_{r_0}^{\sqrt{\frac{H}{h_m}}} D(\varrho)\,\varrho^2\,d\varrho \int\limits_{h_m\varrho^2}^{H} \varphi(i)\,di, \quad \text{wenn} \quad m < n \tag{I}$$

$$A_m = \omega \int\limits_{r_0}^{\sqrt{\frac{H}{h_n}}} D(\varrho)\,\varrho^2\,d\varrho \int\limits_{h_m\varrho^2}^{H} \varphi(i)\,di, \quad \text{wenn} \quad m > n \tag{II}$$

Die Ableitung der Formeln für die mittleren Parallaxen π_m ist ebenfalls leicht auszuführen. Es sei $N_m\,dm$ die Anzahl der Sterne auf einem Areal ω, deren scheinbare Sterngröße zwischen m und $m + dm$ liegt. Dann ist:

$$N_m\,dm = -N(h_m)\,\frac{dh_m}{dm}\,dm$$

wo $N(h_m)\,dh_m$ aus $n(h_m)\,dh_m$ durch Integration über ϱ entsteht. Es ist also:

$$N(h_m)\,dh_m = \omega\,dh_m \int\limits_{r}^{\sqrt{\frac{H}{h_m}}} D(\varrho)\,\varrho^4\,\varphi(h_m\varrho^2)\,d\varrho$$

woraus:

$$N_m\,dm = \frac{0.4}{\log e}\,\omega\,dm \int\limits_{r_0}^{\sqrt{\frac{H}{h_m}}} D(\varrho)\,\varrho^4\,h_m\,\varphi(h_m\varrho^2)\,d\varrho \tag{1}$$

Beachtet man nun, daß die Parallaxe π eines Sternes $= \dfrac{1}{\varrho}$ ist, falls man als Entfernungseinheit die Sternweite ($\pi = 1''.0$) benützt, so ergibt sich sofort, wenn π_m die mittlere Parallaxe der Sterne von der Größe m ist:

$$N_m \cdot \pi_m = \frac{0.4}{\log e}\,\omega \int\limits_{r_0}^{\sqrt{\frac{H}{h_m}}} D(\varrho)\,\varrho^3\,h_m\,\varphi(h_m\varrho^2)\,d\varrho$$

oder:

$$\pi_m \cdot \int\limits_{r_0}^{\sqrt{\frac{H}{h_m}}} D(\varrho)\,\varrho^4\,\varphi(h_m\varrho^2)\,d\varrho = \int\limits_{r_0}^{\sqrt{\frac{H}{h_m}}} D(\varrho)\,\varrho^3\,\varphi(h_m\varrho^2)\,d\varrho\,, \qquad m < n \tag{III}$$

Für $m > n$ hat man die oberen Grenzen der Integrale durch $\sqrt{\dfrac{H}{h_n}}$ zu ersetzen.

Man hat nun vier Integralgleichungen für die Bestimmung der unbekannten Funktionen D und φ. In diesen Gleichungen kommen außerdem die unbekannten Größen H und h_n vor. H und h_n lassen sich jedoch, wie sich später zeigen wird, auf andere Weise bestimmen. Da die Helligkeit der h_n entsprechenden Sterngröße n in der Nähe der 8. oder 9. Größe liegt, läßt sich die zweite Gleichung für π_m gegenwärtig nicht verwerten, weil die mittleren Parallaxen für so schwache Sterne zu wenig zuverlässig sind. Man hat also schließlich für die Bestimmung von $D(\varrho)$ und $\varphi(i)$ die drei Gleichungen (I), (II) und (III).

Insbesondere ist der wichtige Umstand hervorzuheben, daß sich die Dichtefunktion D und die Verteilungsfunktion φ allein aus den Sternzahlen A_m für $m < n$ und $m > n$ bestimmen lassen, wenn man, wie wir hier getan haben, die Absorption vernachlässigt.

Andere Autoren haben später bei ihren Untersuchungen die Zahlen N_m benützt. Ein grundsätzlicher Unterschied zwischen der Verwendung der A_m und derjenigen der N_m besteht nicht. Wenn man will, kann man jederzeit von den A_m, falls man sie durch eine analytische Formel dargestellt hat, zu den N_m übergehen, indem man $\dfrac{dA_m}{dm}$ bildet.

KAPTEYN baute seine Forschungen in der Hauptsache auf die Zahlen $N_{m\mu}$, d. h. auf die Anzahl der Sterne auf, deren scheinbare Größe zwischen $m - \frac{1}{2}$ und $m + \frac{1}{2}$ liegt und deren Eigenbewegung den Wert μ hat. Mit Hilfe der mittleren Parallaxen $\pi_{m\mu}$ ermittelte KAPTEYN daraus die Anzahl aller Sterne, welche in einer bestimmten Entfernung von der Sonne stehen, geordnet nach absoluten Helligkeiten. Zu diesem Zweck machte KAPTEYN die hypothetische Annahme, daß die Größen $z = \log \dfrac{\pi}{\pi_0}$, wo π die individuelle, π_0 die wahrscheinliche Parallaxe ist, nach dem Gaußschen Fehlergesetz gestreut sind.

Man bekommt auf diese Weise verhältnismäßig ohne große Mühe den Verlauf der Dichtefunktion und der Verteilungsfunktion der absoluten Leuchtkräfte, wobei auch von KAPTEYN die Annahme gemacht wird, daß die Absorption unmerklich ist. Es muß jedoch hervorgehoben werden, daß mit der relativ geringen Rechenarbeit, welche diese Methode erfordert, eine größere Unsicherheit der Resultate in Kauf genommen wird. Während sich bei einigermaßen sorgfältigen Untersuchungen die A_m gegenwärtig bis zur 15. Größe herab recht genau bestimmen lassen, werden die $N_{m\mu}$ und die $\pi_{m\mu}$, sobald m den Betrag 8 oder 9 übersteigt, immer weniger zuverlässig. Die $\pi_{m\mu}$ haben seit ihrer ersten Bestimmung durch KAPTEYN im Laufe der Zeit sehr beträchtliche Änderungen erfahren und es ist immerhin zu vermuten, daß auch die aus ziemlich umfangreichem Material in neuerer Zeit von VAN RHIJN[1]) abgeleiteten mittleren Parallaxen in der Folge noch mehr oder weniger korrigiert werden. Außerdem geht, wie bereits oben erwähnt, in die Methode KAPTEYNS die Hypothese ein, daß die Größen:

$$z = \log \frac{\pi}{\pi_0}$$

nach dem Gaußschen Fehlergesetz gestreut sind. Allerdings hat SCHWARZSCHILD[2]) dieser Annahme eine gewisse Stütze gegeben. Es ist aber daran zu erinnern, daß die Untersuchungen SCHWARZSCHILDS auf der Voraussetzung eines unendlich ausgedehnten Sternsystems und auf der Zulassung beliebig großer Leuchtkräfte beruhen. Bei einem System von endlicher Ausdehnung wird natürlich z nie streng eine Gaußsche Fehlerkurve durchlaufen, weil ja dann überhaupt nur π bis zu einem gewissen kleinsten Wert vorkommen. Und schließlich bleibt doch immer noch die Form der Fehlerkurve, d. h. der wahrscheinliche Wert ϱ von z zu bestimmen; dies geschieht auf empirischem Weg und dadurch kommt eine neue Fehlerquelle in die Ergebnisse.

Der Hauptwert bei der Bestimmung der charakteristischen Funktionen des Sternsystems sollte also auf die Sternanzahlen A_m gelegt

---

1) Publ. Groningen Nr. 34.
2) Zur Stellarstatistik. Astr. Nachr. Bd. 190, Nr. 4557. 1912.

werden; man kann dabei außerdem die mittleren Parallaxen π_m mitbenützen. Ermittelt man jedoch die Funktionen D und φ allein aus den $N_{m\mu}$ und aus den $\pi_{m\mu}$, so wird, insbesondere für größere Entfernungen von der Sonne, der Verlauf von D und, für kleinere Leuchtkräfte, derjenige von φ nicht mehr verbürgt sein.

2. Die Beobachtungsdaten. Den folgenden Untersuchungen liegen die von VAN RHIJN in Publ. Groningen Nr. 27 abgeleiteten Sternanzahlen zugrunde. Es ist für die Beurteilung der Zuverlässigkeit der aus ihnen zu ziehenden Schlüsse von Wichtigkeit zu wissen, wie sie entstanden sind.

Die Anzahlen für die Sterngrößen $m > 10$ lieferten 65 nördliche Durchmusterungsplatten der Selected Areas. Die Helligkeitsskala ist die der photographischen Nordpolsequenz nach Harv. Ann. 71. Es mußten also auch die Anzahlen für die helleren Sterne auf diese Skala bezogen werden. Die A_m für die Größen 6.0, 6.5, 7.0 wurden aus der Göttinger Aktinometrie ermittelt, deren Skala fast vollkommen mit der in H. A. 71 übereinstimmt[1]). Die Anzahlen der Sterne heller als 6.0 wurden durch Abzählungen in der Revised Harvard Photometry gewonnen, nachdem vorher für jede halbe Spektralklasse die entsprechenden Farbenkorrektionen, d. h. die Farbenindizes nach H. A. 71, angebracht worden waren.

In einer Reihe von Zonen konnten außerdem auch noch die A_m für die Größen 6.5 bis 8.5 aus einer Arbeit von CHAPMAN und MELOTTE[2]) ermittelt werden, nachdem die dort für das Gebiet von $+69°$ Dekl. bis zum Nordpol gegebenen photographischen Größen auf das System H. A. 71 reduziert worden waren. Das Material, welches VAN RHIJN zu benützen gezwungen war, ist also nicht sehr homogen. Dieser Umstand beeinflußt natürlich die Sicherheit der daraus abgeleiteten Sternzahlen: einmal kommen nicht vollkommen kontrollierbare Skalendifferenzen herein, welche im Verlauf der A_m Unregelmäßigkeiten hervorrufen werden. Ferner, was noch wichtiger ist, es wird die Tatsache, daß die A_m für verschiedene m nicht auf gleiche Areale am Himmel bezogen sind, besonders in niederen galaktischen Breiten den Verlauf der A_m in schwer feststellbarer Weise beeinflussen. Wir wollen ja die Verhältnisse im typischen Sternsystem studieren, d. h. es sollen Mittelwerte über die einzelnen Zonen gebildet und dann soll mit diesen Mittelwerten gerechnet werden. Nun sind wohl die Sternzahlen bis zur Größe 5.5 aus der Rev. Harv. Photom. durch Abzählungen über den ganzen Himmel gewonnen, die Anzahlen der Gött. Akt. dagegen gelten nur für das Gebiet von $0°$ bis $+20°$ Dekl., diejenigen CHAPMANS und MELOTTES nur für das Gebiet von $+69°$ bis zum Nordpol, und endlich

---

[1]) Publ. Groningen Nr. 27, S. 16.
[2]) Memoirs of the Roy. Astr. Soc. Bd. 60, Teil 4. 1914.

sind die Sternzahlen für $m > 10$ aus verhältnismäßig kleinen, über den ganzen Himmel verteilten Feldern abgeleitet. Wenn also, wie besonders in Milchstraßennähe, die Sternfülle relativ stark mit dem Ort innerhalb ein und derselben Zone variiert, so kann dadurch unter Umständen eine Veränderung im Verlauf der A_m hervorgerufen werden, welche nicht vorhanden wäre, wenn man immer nur Mittelwerte über die ganze Zone betrachten würde. Bei den Zahlen für den ganzen Himmel wird sich noch ein weiterer Umstand störend bemerkbar machen, welcher bei den A_m für die einzelnen Zonen nicht so ins Gewicht fällt. Die A_m bis 5.5 sind tatsächlich Mittelwerte aus Abzählungen über den ganzen Himmel. Die Sternzahlen für die folgenden Größen bis 7.0 stammen aus der Gött. Akt. und beziehen sich nur auf einen verhältnismäßig schmalen Streifen. Selbst wenn in jeder einzelnen Zone die Sternfülle konstant wäre, so ist es klar, daß die mittlere Sternfülle innerhalb dieses Streifens keineswegs gleich der mittleren Sternfülle des ganzen Himmels sein wird, da ja der Streifen in den verschiedenen Zonen prozentual ganz verschieden große Gebiete bedeckt.

VAN RHIJN hat seine $\log A_m$ durch Kurvenzeichnen ausgeglichen und sie dann auf die photometrische Harvard-Skala reduziert. Ich habe jedoch, um mit den immerhin nicht ganz sicheren Skalenkorrektionen nicht eine neue Fehlerquelle in die Beobachtungen hereinzubringen, die auf das photographische System bezogenen A_m bei meinen Rechnungen benützt, und zwar die ursprünglichen, unausgeglichenen Zahlen, was bei gewissen Untersuchungen wichtig ist.

Leider fehlen gerade die Sternzahlen in dem Intervall von $m = 7.5$ bis $m = 10.0$, bzw. in einigen Zonen für die Größen 9.0, 9.5, 10.0. Dies ist um so bedauerlicher, als in dieser Gegend nach der Theorie SEE-LIGERS Veränderungen im Verlauf der A_m auftreten müssen, welche gestatten, die räumliche Ausdehnung des Systems zu ermitteln. Immerhin wird sich zeigen, daß es trotzdem gelingt, die Existenz dieser Veränderungen nachzuweisen.

Die Sternzahlen wurden in drei Zonen zusammengefaßt. Zone I enthält das Gebiet von $\pm 50°$ galaktischer Breite bis zu den Polen, Zone II geht von $\pm 20°$ bis $\pm 50°$ und Zone III von $0°$ bis $\pm 20°$.

Um nun die oben besprochenen systematischen Einflüsse auf die A_m der helleren Sterne wenigstens einigermaßen zu kompensieren, habe ich zunächst an die Sternzahlen der Zone III folgende drei Korrektionen angebracht:

a) Eine Korrektion für Skalenunterschied Gött. Akt.—Rev. Harv. Phot. Die Werte hierfür sind gegeben in Publ. Gron. 27, S. 13.

b) Eine Reduktion der Göttinger Zahlen auf die mittlere Sternfülle der ganzen Zone nach Angaben von NORT[1]).

---

[1]) Rech. Astr. de l'Obs. d'Utrecht VII.

c) Eine Korrektion des $\log A_m$ für Unvollständigkeit der Gött. Akt. für $m = 7.0$. Sie wurde veranlaßt durch das auffallende Absinken der $\log A_m$ für alle Zonen bei der Größe 7.0 im Vergleich mit den aus umfangreicherem Material abgeleiteten und daher genaueren Zahlen aus der Potsdamer Durchmusterung[1]). Es konnte dafür keine andere Erklärung gefunden werden als die, daß die Gött. Akt. nicht alle Sterne der photographischen Größe 7.0 enthält.

Es ist natürlich schwierig, diese Unvollständigkeit zu beseitigen. Ich wählte dazu folgenden, allerdings nicht ganz einwandfreien Weg: ich bildete die Differenzen zwischen den nach a) und b) korrigierten $\log A_m$ VAN RHIJNS und den $\log A_m$ aus der P. D.; sie ließen sich annähernd durch eine in m lineare Formel darstellen. Ermittelt man diese Formel nun aus den Zahlen für $m = 4.0$ bis $m = 6.5$ einschließlich und extrapoliert dann auf $m = 7.0$, so hat man damit sofort den richtigen $\log A_m$ der photographischen Skala. Voraussetzung ist dabei, daß die Zahlen der P. D. bis $m = 7$ richtig sind.

Die Ableitung der Korrektion c) soll hier für die Zone III kurz durchgeführt werden. Eine Ausgleichung ergab für die Darstellung der Differenzen $\log A_m$ (Publ. Gron. 27) — $\log A_m$ (P. D) folgende Beziehung:

$$\Delta_1 = -0.010 + 0.045\,(m - 5.0)$$

Dabei wurde den Differenzen für die Größen 4.0 und 4.5 geringeres Gewicht gegeben, da naturgemäß die Sternzahlen der helleren Sterne ungenauer sind. Tab. 1 enthält die Werte Δ_1 neben den beobachteten Differenzen. Es ergibt sich daraus, daß $\log A_{7.0}$ um den Betrag $+0.038$ zu korrigieren ist.

Tabelle 1.

Differenzen zwischen den $\log A_m$ in Publ. Gron. 27 und den $\log A_m$ aus der P. D für Zone III.

m	Gron. 27	P. D.	P. D. – Gron. 27	Δ_1
4.0	8.116	8.028	-0.088	-0.065
4.5	8.348	8.289	-0.059	-0.033
5.0	8.609	8.585	-0.024	-0.010
5.5	8.844	8.870	$+0.026$	$+0.013$
6.0	9.095	9.104	$+0.009$	$+0.035$
6.5	9.323	9.389	$+0.067$	$+0.058$
7.0	9.511	9.629	$+0.118$	$+0.080$

In den Zonen I und II und bei den A_m für den ganzen Himmel wurde die Korrektion b) außer acht gelassen, da in höheren galaktischen Breiten die Sternfülle sich nicht so stark innerhalb einer Zone ändert. Korrektion a) und c) wurden dagegen wie in Zone III angebracht.

---

[1]) SEELIGER: Sitz.-Ber., Akad. München 1912. MÜLLER und KEMPF: Astr. Nachr. Bd. 180, Nr. 4312.

Man darf sich nicht verhehlen, daß alle diese Reduktionen durchaus nicht sehr sicher sind. Die Angaben in Publ. Gron. 27, welche a) zugrunde liegen, sind Mittelwerte über alle Zonen; wendet man sie auf eine einzelne Zone an, so begeht man damit eine mehr oder weniger große Ungenauigkeit. Die b) zugrunde liegenden Angaben von NORT gelten für die Verteilung der Sterne von der 11. photometrischen Sterngröße. Man macht also bei ihrer Anwendung die Annahme, daß die Gesetzmäßigkeiten der scheinbaren Verteilung der Sterne der 7. Größe die gleichen seien wie die der Sterne 11. Größe. Es wird also in den A_m der helleren Sterne immer noch eine gewisse Unsicherheit zurückbleiben; man wird dies bei der Berechnung der Funktionen φ und D und bei der Diskussion der Ergebnisse berücksichtigen müssen.

Für die mittleren Parallaxen benützte ich die in Publ. Gron. 34, Tab. 40 gegebenen Werte; mittels der dem Harv. Circ. 226 entnommenen Verteilung der Spektraltypen und den Zahlen N_m wurden die π_m für alle Typen zusammen und für den ganzen Himmel bestimmt; es zeigte sich übrigens, daß die so erhaltenen π_m nicht sehr von den von SCHOUTEN[1]) gegebenen Werten abweichen. Da die auf diese Weise bestimmten π_m auf die visuelle Harvard-Skala bezogen sind, habe ich die Parallaxen für die entsprechenden photographischen Größen durch Interpolation bestimmt, wobei die in Tab. 2 gegebenen Korrektionen verwendet wurden. Sie sind aus einem Vergleich der Gött. Akt. mit dem Henry Draper Catalog in den Rektaszensionsstunden 0^h bis 19^h gewonnen und so durch eine Kurve ausgeglichen, daß diese sich für die schwächeren m an diejenige anschließt, welche durch die in Publ. Gron. 27, S. 42, gegebenen Differenzen Harv. phot. — Harv. vis. bestimmt ist.

Tabelle 2.

Differenzen zwischen der Skala der Göttinger Aktinometrie und der visuellen Harv.-Skala. (Mittlere Farbenindices.)

Harv. vis.	Gött. Akt. — Harv. vis.
4.00	+ 0.60
5.00	+ 0.58
6.00	+ 0.55
7.00	+ 0.52
8.00	+ 0.50

Es besteht übrigens ein auffallender Unterschied zwischen den dort angeführten Skalendifferenzen für die helleren Sterne und den von mir aus einer Anzahl von insgesamt 2659 Sternen bestimmten. Es gibt zwei Erklärungen dafür. Entweder sind die Korrektionen in Publ. Gron. 27 für die helleren Sterne unrichtig, oder die Skala der Gött. Akt. schließt

---

[1]) On the Determination of the Principal Laws of Statistical Astronomy. Amsterdam 1918.

sich doch schlechter an die photographische Harvard-Skala an als bisher angenommen.

3. Die Grenzen des typischen Sternsystems. SEELIGER hat den Weg vorgezeichnet[1]), auf welchem wir zu einer Kenntnis der räumlichen Ausdehnung des Sternsystems gelangen können. Bildet man die ersten und zweiten Differentialquotienten der A_m nach h_m nach den Gleichungen (I) und (II) des 1. Kapitels, so hat man für $m = n$:

$$\frac{1}{\omega}(A_n)_\mathrm{I} - \frac{1}{\omega}(A_n)_\mathrm{II} = 0 \quad \text{und} \quad \frac{1}{\omega}\left(\frac{dA_n}{dh_n}\right)_\mathrm{I} = \frac{1}{\omega}\left(\frac{dA_n}{dh_n}\right)_\mathrm{II} = 0$$

$$\frac{1}{\omega}\left[\left(\frac{d^2A_n}{dh_n^2}\right)_\mathrm{I} - \left(\frac{d^2A_n}{dh_n^2}\right)_\mathrm{II}\right] = \frac{1}{\omega}\,\triangle\left(\frac{d^2A_n}{dh_n^2}\right) = \frac{1}{2H}\left(\frac{H}{h_n}\right)^{\frac{1}{2}} D\left(\sqrt{\frac{H}{h_n}}\right)\varphi(H) \quad \text{(IV)}$$

Also: während die A_m selbst und ihre ersten Differentialquotienten an der Stelle $m = n$ stetig verlaufen, besitzt der zweite Differentialquotient an dieser Stelle eine Unstetigkeit, deren Betrag durch den Ausdruck (IV) gegeben ist. Voraussetzung für die Existenz eines solchen Sprunges ist, daß die Grenze des Systems bei einem von Null verschiedenen Wert der Dichte erreicht wird und daß $\varphi(H) \neq 0$ ist. Die erste Bedingung versteht sich von selbst aus dem Wesen des Begriffes Grenze heraus; denn sobald die Dichtefunktion sich asymptotisch dem Wert 0 nähert, läßt sich ja keine Grenze mehr definieren. Die zweite Bedingung bedeutet eine gewisse Erweiterung unserer früheren Definition der oberen Grenze H der Leuchtkräfte. Wir nennen nun H diejenige größte absolute Helligkeit, welche noch so viele Sterne besitzen, daß deren Anzahl im statistischen Sinne noch Bedeutung hat. Kommt man zu noch größeren Werten der Leuchtkräfte, so soll die Funktion φ entweder Null sein, oder, allgemeiner gesprochen, so kleine Werte annehmen, daß dieser Teil von φ auf statistische Untersuchungen keinen Einfluß mehr ausübt. Man müßte also eigentlich, um diesen allgemeineren Fall zu berücksichtigen, wie dies SEELIGER zuletzt getan hat, eine Funktion $\Phi(i)$ so definieren, daß $\Phi(i) = \varphi(i)$, solange $i < H$, und daß $\Phi(i) = \chi(i)$, wenn $i > H$; der Übergang an der Stelle H soll unstetig erfolgen. Führt man diese Definition in die Ausdrücke für die A_m ein, so findet sich, daß schließlich in (IV) an Stelle von $\varphi(H)$ die Differenz $\varphi(H) - \chi(H)$ steht. Es genügt jedoch für unsere gegenwärtigen Zwecke anzunehmen, daß $\chi(i)$ Null ist; dann gilt nach wie vor die Form (IV).

SEELIGER hat die Größe H aus den mittleren Parallaxen KAPTEYNS bestimmt[2]). Er findet für die H entsprechende Sterngröße den Wert -4.3, wobei Entfernungseinheit die Siriusweite ist. Bezieht man die

[1]) Abh. Akad. München 1898 und Sitz.-Ber. 1920 (Untersuchungen über das Sternsystem).

[2]) Über die räumliche Verteilung der Sterne im schematischen Sternsytem. Sitz.-Ber. Akad. München 1911.

absoluten Helligkeiten auf das parsec als Entfernungseinheit, so ergibt sich $M_H = -7.8$.

Es ist natürlich kein Beweis für die Unrichtigkeit dieser Größe, wenn man, wie dies SCHOUTEN tut, fünf Sterne bekannter Parallaxen anführt, deren Leuchtkräfte diesen Wert übersteigen. Abgesehen von der Unsicherheit der Parallaxenbestimmung (M ändert sich um $+ 0.4$ Größenklassen, wenn man die Parallaxe um 20% vergrößert), spielen diese verschwindenden Ausnahmen bei statistischen Untersuchungen nicht die geringste Rolle.

Tabelle 3.

Die Verteilungsfunktion für die B0—B5-Sterne nach KAPTEYN.

M	n
— 3.5	61
— 4.0	74
— 4.5	58
— 5.0	66
— 5.5	34
— 6.0	42
— 6.5	27
— 7.0	25
— 7.5	12
— 8.0	2
— 8.5	3
— 9.0	1

KAPTEYN[1]) hat die absoluten Helligkeiten einer Reihe von B-Sternen abgeleitet; Tab. 3 enthält seine Zusammenstellung der Sterne mit bestimmter Leuchtkraft, bezogen auf das parsec als Einheit. Es zeigt sich, daß von der Größe -8.0 an nur noch sehr wenige Sterne vorkommen; der Übergang von -7.5 zu -8.0 findet sprunghaft statt. Nimmt man -7.8 als Übergangsstelle und nimmt man an, daß die B-Sterne von allen Typen die größte absolute Helligkeit erreichen können, so kommt man genau auf den Wert, welchen SEELIGER für M_H gefunden hat. In einer zweiten Arbeit bestimmte KAPTEYN[2]) für eine weitere Anzahl von Sternen der Typen B_0 bis B_5 die absoluten Helligkeiten; der unstetige Übergang ist hier nicht so ausgeprägt, läßt sich aber immerhin feststellen, wenn man die prozentuale Abnahme der Zahlen ins Auge faßt. Die Grenze H würde darnach allerdings zwischen die Größen -7.0 und -7.5 fallen. ADAMS und JOY[3]) veröffentlichen eine Zusammenstellung, welche die absoluten Helligkeiten einer größeren Anzahl von B-Sternen enthält. 39 Sterne haben absolute Helligkeiten zwischen -7.0 und -8.0; jenseits -8.0 liegen dagegen nur 5 Sterne.

---

[1]) Mt. Wils. Contr. 82.   [2]) Mt. Wils. Contr. 147.
[3]) Mt. Wils. Contr. 262.

Wenn man also für unsere Zwecke annimmt, die H entsprechende photographische Sterngröße ist — 8.0, so befindet man sich damit in guter Übereinstimmung mit dem empirischen Befund und mit dem Wert, welchen SEELIGER bereits früher gefunden hat. Man wird dann schätzungsweise einen Fehler von kaum mehr als 0.2 oder 0.3 Größenklassen begehen; wenn erst einmal das Beobachtungsmaterial erweitert sein wird, dann wird sich die Größe H wesentlich genauer bestimmen lassen.

Die Bestimmung der Grenzen des Systems stützt sich nun auf die Ermittlung des Sprunges $\varDelta \left(\dfrac{d^2 A_m}{d h_m^2} \right)_{m=n}$ und der Stelle $m = n$, bei welcher er auftritt. Man hat dann zwei Methoden, um die Größe r_1, welche die Entfernung bis zur Grenze des Systems darstellt, zu berechnen. Zunächst ist ja einfach:

$$h_n r_1^2 = H$$

woraus sich r_1 ergibt. Kennt man ferner die Funktionen $D(\varrho)$ und $\varphi(i)$, so kann man in den Ausdruck (IV) $\varphi(H)$ und $D\left(\sqrt{\dfrac{H}{h_n}} \right) = D(r_1)$ einsetzen; man kann dann die so entstehende Gleichung nach r_1 auflösen. Besonders einfach gestaltet sich dies, wenn die Dichtefunktion die ihr durch SEELIGER gegebene Gestalt:

$$D(\varrho) = \gamma \varrho^{-\lambda} - \alpha \varrho^{-\lambda_1}$$

hat, wenn also $D(r_1) = \gamma(r_1^{-\lambda} - \alpha r_1^{-\lambda_1})$ wird.

Es soll nun die Größe n und der Betrag $\varDelta$ des Sprunges im zweiten Differentialquotienten nach einem Verfahren ermittelt werden, welches SEELIGER zuerst in einem Fall angewandt hat[1]). Bekanntlich lassen sich die $\log A_m$ in einem nicht zu großen Intervall durch eine in m quadratische Interpolationsformel darstellen. Wir gehen nun so vor, daß wir zwei Formeln ansetzen, von denen die eine für $m < n$, die andere für $m > n$ gilt. Die beiden Formeln sollen folgende Gestalt haben:

$$\log A_m = \alpha + \beta(m - n) - \gamma_1(m - n)^2 \qquad m < n \qquad (2)$$
$$\log A_m = \alpha + \beta(m - n) - \gamma_2(m - n)^2 \qquad m > n \qquad (3)$$

$\log A_m$ selbst ist danach an der Stelle $m = n$ stetig, ebenso $\dfrac{d \log A_m}{d m}$; der zweite Differentialquotient $\dfrac{d^2 \log A_m}{d m^2}$ macht hier jedoch einen Sprung vom Betrag $2(- \gamma_2 + \gamma_1)$. Gelingt es also, aus den $\log A_m$ die Koeffizienten α, β, γ_1, γ_2 und die Größe n durch Ausgleichung mit genügender Sicherheit zu bestimmen und lassen sich dadurch die $\log A_m$ befriedigend darstellen, so ist die endliche Begrenzung des Systems bewiesen. Gleich-

---

[1]) Astr. Nachr. Bd. 214 (Nr. 5121).

zeitig hat man dann alle Größen, welche nötig sind, um die räumliche Ausdehnung des Systems zu bestimmen.

Ein Umstand, welcher den im folgenden benützten $\log A_m$ anhaftet, könnte vielleicht als Einwand gegen dieses Verfahren benützt werden. Die Größe n liegt zwischen $7^m\!.0$ und $9^m\!.0$, fällt also gerade in das Intervall, für welches keine A_m beobachtet sind. Das obige Verfahren kommt also in diesem Fall einer Extrapolation von beiden Seiten her in die Lücke hinein gleich. Diese Extrapolation erstreckt sich jedoch nur über zwei, höchstens drei Größenklassen und die Darstellung der $\log A_m$ durch die Formeln ist so, daß ernstere Bedenken deshalb nicht auftreten können.

Die Rechnung wurde nach der Methode der kleinsten Quadrate mit den nötigen Kontrollen für alle Zonen und für den ganzen Himmel durchgeführt. Es ergaben sich die in Tab. 4 mit ihren mittleren Fehlern enthaltenen Werte der Unbekannten. Wieweit die Formeln sich an die beobachteten $\log A_m$ anpassen, ist aus Tab. 5 ersichtlich. In den Zonen I und II wurden die Formeln für $m < n$ so angesetzt, daß γ_1 gleich von vornherein zu Null angenommen wurde; und zwar deshalb, weil in diesen Zonen der Verlauf der $\log A_m$ sehr nahe linear ist, und weil sich infolge der Unsicherheit der $\log A_m$ für die helleren Sterne der Betrag von γ_1 doch nicht verbürgen ließe.

Tabelle 4.

Die Konstanten der Interpolationsformeln für die $\log A_m$, die Grenzgröße n, der Sprung im zweiten Diff. Quot. und die Entfernung r_1 bis zur Grenze des Systems.

Zone	α	β	γ_1	γ_2	n	$\Delta\left(\dfrac{d^2 \log A_n}{d n^2}\right)$	r_1 (Einh. $\pi = 1''\!.0$)
I (Pol)	-0.569	0.5195	0	$+0.0195$ ± 0.0007	7.49 ± 0.28	0.0390 ± 0.0007	1260
II	-0.150	0.4948	0	$+0.0175$ ± 0.0011	8.16 ± 0.20	0.0350 ± 0.0011	1700
III (Milch-straße)	$+0.605$	0.4653	$+0.0021$ ± 0.0023	$+0.0132$ ± 0.0022	9.22 ± 0.98	0.0222 ± 0.0032	2820
Ganzer Himmel	-0.025	0.4568	$+0.0016$ ± 0.007	$+0.0138$ ± 0.0012	8.4 ± 2.0	0.0244 ± 0.0080	1900

Die Größen n zeigen deutlich die zu erwartende Abnahme, wenn man von der Milchstraße zur Polgegend übergeht: das System ist in der Richtung der Pole weniger ausgedehnt als in Richtung der Milchstraße. Das Achsenverhältnis ist ungefähr $1 : 3.5$. Die Δ zeigen ebenfalls eine ausgeprägte Abhängigkeit von der galaktischen Breite. Sie nehmen zu mit wachsender Breite.

Aus den mittleren Fehlern ε_Δ der Größen Δ der Tab. 4 geht hervor, daß die Beträge der Unstetigkeiten im zweiten Differentialquotienten so sicher bestimmt sind, daß an ihrer Existenz kein Zweifel mehr be-

stehen kann. Die Größen n sind in den Zonen I und II ebenfalls recht
genau bestimmt; sie könnten natürlich noch genauer festgelegt werden,
wenn die Lücke im Beobachtungsmaterial bei den Größen 8 und 9 ausgefüllt werden würde. In Zone III (Milchstraße) hat der mittlere Fehler ε_n
von n einen beträchtlich größeren Wert. Wenn es erlaubt ist, dies
physikalisch zu interpretieren, so kann man annehmen, daß am Äquator-
gürtel des Systems der Übergang vom sternerfüllten zum sternleeren
Raum langsamer erfolgt als in höheren Breiten, daß die Dichte weniger
sprunghaft den Wert 0 annimmt, d. h. mit anderen Worten, daß die
Grenze verwaschener ist.

Tabelle 5.

Die Darstellung der $\log A_m$ durch die Interpolationsformeln.

m	Zone I			Zone II			Zone III			Ganzer Himmel		
	Beob.	Rechn.	B—R	Beob.	Rechn.	B—R	Beob.	Rechn.	B—R	Beob.	Rechn.	B—R
5.0	8.131	8.137	— 0.006	8.293	8.286	+ 0.007	8.609	8.605	+ 0.004	8.408	8.404	+ 0.004
5.5	8.396	8.397	— .001	8.533	8.534	— .001	8.844	8.345	— .001	8.649	8.637	+ .012
6.0	8.655	8.657	— .002	8.810	8.781	+ .029	9.095	9.085	+ .010	8.881	8.870	+ .011
6.5	8.871	8.917	— .046	9.022	9.027	— .005	9.323	9.324	— .001	9.088	9.101	— .013
7.0	9.150	9.176	— .026	9.273	9.276	— .003	9.549	9.562	— .013	9.337	9.332	+ .005
7.5				9.519	9.523	— .004						
8.0				9.770	9.771	— .001						
8.5				0.007	0.017	— .010						
10.5	0.825	0.817	+ 0.008	0.907	0.912	— 0.005	1.198	1.179	+ 0.019	0.865	0.873	— 0.008
11.0	1.022	1.013	+ .009	1.124	1.111	+ .013	1.369	1.392	— .023	1.062	1.070	— .008
11.5	1.216	1.199	+ .017	1.332	1.308	+ .024	1.606	1.598	+ .008	1.277	1.259	+ .018
12.0	1.391	1.376	+ .014	1.519	1.491	+ .028	1.800	1.797	+ .003	1.457	1.440	+ .017
12.5	1.541	1.544	— .003	1.675	1.667	+ .008	1.982	1.990	— .008	1.618	1.616	+ .002
13.0	1.706	1.702	+ .004	1.825	1.834	— .009	2.190	2.176	+ .014	1.788	1.784	+ .004
13.5	1.842	1.849	— .007	1.966	1.993	— .027	2.350	2.356	— .006	1.933	1.945	— .012
14.0	1.978	1.987	— .011	2.126	2.142	— .016	2.514	2.529	— .015	2.087	2.100	— .013
14.5	2.122	2.114	+ .008	2.294	2.285	+ .009	2.700	2.695	+ .005	2.255	2.247	+ .008

Die Ergebnisse aus den Zahlen für den ganzen Himmel bestimmen
die Grenzen des sog. schematischen Sternsystems. Im schematischen
Sternsystem wird auch die Abhängigkeit der Sternverteilung von der
galaktischen Breite vernachlässigt, die Flächen gleicher Dichte sind
Kugeln, das System ist kugelförmig begrenzt. Das schematische Sternsystem entfernt sich also noch mehr von der Wirklichkeit als das
typische. Wenn man bedenkt, daß in die Bestimmung der Grenzen des
schematischen Systems die ganze Variation der Grenzen des typischen
Systems mit der galaktischen Breite eingeht, wird man erwarten müssen,
daß sich die Grenzen des schematischen Systems überhaupt nur sehr
unsicher aus den Sternabzählungen bestimmen lassen. In der Tat ist
auch der mittlere Fehler ε_n bei den Zahlen für den ganzen Himmel sehr
groß. Die Wahrscheinlichkeit ist $\frac{1}{2}$, daß n in den Grenzen liegt:

$$(8.4 - 1.3) < n < (8.4 + 1.3) \qquad \text{oder} \qquad 7.1 < n < 9.7$$

Mit den für die Zonen I und III erhaltenen Werten ε_n liegt dieses selbe n mit der Wahrscheinlichkeit $\frac{1}{2}$ zwischen den Grenzen:

$$(7.49 - 0.18) < n < (9.22 + 0.66)$$
$$7.3 < n < 9.9$$

Also eine ausgezeichnete Übereinstimmung, welche eine schöne Bestätigung der theoretischen Grundlagen bedeutet.

Um auch die Formel (IV) für die Bestimmung der Grenzen heranziehen zu können, muß man $\dfrac{d^2 A_m}{d\,h_m^2}$ durch $\dfrac{d^2 \log A_m}{d\,m^2}$ ausdrücken. Setzt man $p = \dfrac{0.4}{\log e}$, so wird:

$$\varDelta \left(\frac{d^2 \log A_m}{d\,m^2}\right) = + \frac{0.4}{A_m}\,\varDelta \left(\frac{d^2 A_m}{d\,h_m^2}\right) p\,h_m^2$$

Nimmt man $m = n$ und führt den Ausdruck für $\varDelta \left(\dfrac{d^2 A_n}{d\,h_n^2}\right)$ aus (IV) in diese Gleichung ein, so erhält man:

$$\varDelta \left(\frac{d^2 \log A_n}{d\,n^2}\right) = \omega p \cdot \frac{0.2}{A_n}\, D\left(\sqrt{\frac{H}{h_n}}\right)\left(\frac{H}{h_n}\right)^{\frac{3}{2}} H \cdot \varphi\,(H) \qquad\qquad \text{(V)}$$

Wir werden im nächsten Abschnitt die Funktionen φ und D aus den Zahlen für den ganzen Himmel berechnen und damit die Brauchbarkeit der Formel (V) an der Übereinstimmung des sich aus ihr ergebenden Resultates mit dem auf dem ersten Weg gefundenen nachprüfen.

Das Ergebnis der Betrachtungen des gegenwärtigen Abschnittes läßt sich dahin zusammenfassen, daß aus dem Verlauf der $\log A_m$ trotz der Lücken im Material mit ausreichender Sicherheit sich die räumliche Ausdehnung des Systems bestimmen läßt. Die Entfernung bis zur Grenze ist in Richtung der Milchstraße etwa 3.5mal so groß als in Richtung der Pole. Die Größen n, welche die Grenzentfernung bestimmen, lassen sich in höheren galaktischen Breiten schärfer fixieren als in der Milchstraßenzone.

4. Die Verteilungsfunktion der Leuchtkräfte. Den Beziehungen, welche im 1. Kapitel aufgestellt wurden, lag u. a. die Annahme zugrunde, daß die Funktion φ unabhängig vom Ort ist. Diese Annahme war deshalb notwendig, weil es gegenwärtig nicht möglich ist, die Integralgleichungen auszuwerten, wenn man voraussetzt, daß $\varphi(i)$ die Entfernung ϱ explizite enthält. Wir werden also bei der Bestimmung der Funktion φ die über den ganzen Himmel gemittelten Sternzahlen benützen; es dient uns somit das schematische Sternsystem als Hilfsmittel, um die Verteilungsfunktion φ zu bestimmen. Inwieweit sich die so gewonnene Funktion φ von der Wirklichkeit im einzelnen entfernt, hängt nur davon ab, in welchem Maße φ vom Ort unabhängig ist.

Wie im 1. Kapitel abgeleitet wurde, hat man für die Bestimmung der Funktionen $\dot{\varphi}(i)$ und $D(\varrho)$ die drei Integralgleichungen (I), (II) und (III). Es müßte φ und D so bestimmt werden, daß die beobachteten A_m und π_m befriedigend dargestellt werden. Dies ist natürlich nur dann möglich, wenn das Beobachtungsmaterial frei von systematischen Fehlern ist. Aus den Betrachtungen des 2. Abschnittes ergab sich, daß die A_m der helleren Sterne sehr wahrscheinlich systematischen Verfälschungen unterlegen sind. Es wird also zu erwarten sein, daß sich die Gleichungen (I), (II) und (III) nicht vollkommen befriedigen lassen. Dann fragt sich, welchen beiden von den drei Gleichungen man den Vorzug bei der Bestimmung von φ und D geben soll. Die A_m für die schwächeren Sterne sind einheitlich bestimmt. Sie wird man also vor allem zur Ermittlung von φ heranziehen. Weiter hielt ich es, eben aus den im 2. Kapitel besprochenen Gründen, für geboten, mehr Wert auf die Darstellung der mittleren Parallaxen zu legen als auf die der Sternzahlen für die helleren Sterne.

Die Theorie der Integralgleichungen gestattet bis jetzt noch nicht eine strenge Auflösung der Gleichungen (I), (II) und (III). Wir müssen daher den Weg eines Näherungsverfahrens einschlagen. Wir setzen die Verteilungsfunktion in der jetzt allgemein gebräuchlichen Form einer Exponentialfunktion an, und zwar so, daß:

$$\varphi(i) = \Gamma_1 e^{a\log\frac{i}{H} - b\left(\log\frac{i}{H}\right)^2} \tag{4}$$

SEELIGER hat die Konstanten dieser Funktion zum erstenmal in seiner Abhandlung vom Jahre 1911 aus den mittleren Parallaxen KAPTEYNS abgeleitet.

Das Verfahren, welches wir im folgenden anwenden werden, um die Konstanten a und b zu bestimmen, soll kurz skizziert werden: Wir bestimmen zunächst die Funktion D nach einem näherungsweise gültigen Ansatz. Ein solcher ist leicht zu beschaffen. Als SEELIGER seine ersten Untersuchungen über das Sternsystem veröffentlichte, standen ihm an Beobachtungsmaterial nur seine Abzählungen aus der B. D. zur Verfügung. Diese Sternzahlen zeigten nach ihrer Reduktion auf die Harvard-Skala die Gesetzmäßigkeit, daß sehr nahe:

$$\log A_m - \log A_{m-\frac{1}{2}} = \log \alpha = \text{konst.} \tag{5}$$

was gleichbedeutend ist mit der Beziehung $A_m = c\,h_m^{\frac{\lambda-3}{2}}$, wo c und λ Konstante sind und λ mit α in der Weise zusammenhängt, daß $\log\alpha = \dfrac{3-\lambda}{10}$. SEELIGER zeigte, daß sich, wenn (5) gilt, die Dichte D aus der Gleichung bestimmt:

$$\int\limits_{r_0}^{\sqrt{\frac{H}{h_m}}}\left(\lambda D\cdot\varrho^2 + \varrho^3\frac{dD}{d\varrho}\right)d\varrho\int\limits_{h_m\varrho^2}^{H}\varphi(i)\,di = 0 \tag{6}$$

Eine triviale Lösung dieser Gleichung bildet die von Schouten vermerkte Annahme, daß alle Sterne in der gleichen Entfernung ϱ stehen, wobei dann natürlich φ so verlaufen muß wie die Sternzahlen A_m. Sieht man hiervon ab, so ist die Gleichung nur erfüllt, wenn:

$$\lambda D(\varrho)\,\varrho^2 + \varrho^3\,\frac{d\,D\,(\varrho)}{d\,\varrho} = 0$$

woraus folgt:

$$D(\varrho) = \gamma\,\varrho^{-\lambda} \tag{7}$$

Führt man dies in den Ausdruck für die mittleren Parallaxen ein, so ergibt sich:

$$\pi_m = C \cdot h_m^{\frac{1}{2}} \tag{8}$$

wo C unabhängig von h_m ist. Nun verlaufen bekanntlich die π_m keineswegs so, wie die durch (8) definierten Parallaxen, welche Seeliger „normale" Parallaxen nennt. Diese Unvereinbarkeit der normalen mit den mittleren Parallaxen könnte zweierlei Ursachen haben:

1. es existiert eine Absorption des Lichts im interstellaren Raum.
2. die durch (7) definierte Dichtefunktion entspricht nicht dem wahren Dichteverlauf.

Da Annahme 1, wie Seeliger gezeigt hat[1]), zu unmöglich großen Beträgen der Absorption führt, so bleibt nur Annahme 2 übrig. Um den mittleren Parallaxen zu genügen, hat dann Seeliger der Dichtefunktion folgende Form gegeben[2]):

$$D(\varrho) = \gamma\,(\varrho^{-\lambda} - \alpha\,\varrho^{-\lambda_1}) \tag{9}$$

Es ist klar, daß die Formel für $D(\varrho)$ mit Hinzunahme des zweiten Gliedes interpolarischen Charakter annimmt. Es besteht jedoch kein Anlaß dazu, die Form (9) etwa als weniger begründet anzusehen wie den von Schwarzschild und Kapteyn benützten Exponentialansatz für $D(\varrho)$. Schwarzschild hat zwar gezeigt[3]), daß D die exponentielle Form hat, wenn sowohl die Zahlen N_m als auch die Funktion $\varphi(i)$ durch einen Exponentialansatz darstellbar sind. Abgesehen davon jedoch, daß es nicht erwiesen ist, daß vor allem φ in seinem *ganzen* Verlauf die Form einer Gaußschen Fehlerkurve hat, beruhen, wie immer wieder betont werden muß, die Untersuchungen Schwarzschilds auf der nicht zutreffenden Annahme eines unendlich ausgedehnten Systems und auf der Zulassung beliebig großer Leuchtkräfte. Es ist daher nicht bewiesen, daß die Dichtefunktion tatsächlich die Form haben muß, welche ihr Schwarzschild gibt. Damit nimmt jedoch der Exponentialansatz für $D(\varrho)$ ebenfalls den Charakter einer reinen Interpolationsformel an und hat als solche durchaus nicht mehr Berechtigung wie die Form (9),

---

[1]) Abh. Akad. München 1909, S. 45 ff.
[2]) Sitz.-Ber. Akad. München 1911.
[3]) Astr. Nachr. Bd. 185, Nr. 4422.

wenn nur durch diese die Beobachtungen genügend dargestellt werden. Die Form (9) hat lediglich den einen Nachteil, daß für Werte von ϱ, welche kleiner sind als die Größe $r_0 = \alpha^{\lambda - \lambda_1}$, die rechte Seite von (9) negativ, die Dichte also imaginär wird. Es wird sich jedoch zeigen, daß r_0 sich so klein ergibt, daß dieser Umstand keine Rolle spielt. Dagegen hat der Ansatz (9) den großen Vorzug, daß er sich in geschlossener Form integrieren läßt. Andrerseits wird auch der Exponentialansatz in der Nähe der Sonne unbrauchbar, da er für $\varrho = 0$ auf den Wert $D = 0$ führt, was nicht richtig sein kann.

Das Näherungsverfahren bei der Bestimmung von φ gestaltet sich nun folgendermaßen. Wir berechnen mit Hilfe des Koeffizienten β der Interpolationsformel für die Zahlen des ganzen Himmels die erste Näherung der Dichtefunktion nach (7), setzen sie in (II) ein und bestimmen die Konstanten der Funktion φ durch Versuche so, daß die A_m der schwächeren Sterne so gut als möglich dargestellt werden. Mit dem so erhaltenen φ gehen wir in (III) ein und bestimmen die Konstanten λ, λ_1 und α der Formel (9) so, daß die mittleren Parallaxen befriedigt werden. Auf diese Weise gewinnen wir eine verbesserte Dichtefunktion; wir setzen sie wiederum in (II) ein und bestimmen daraus die zweite Näherung der Verteilungsfunktion der Leuchtkräfte. Dies wird im allgemeinen genügen, um die beobachteten Größen, mittlere Parallaxen und Sternzahlen der schwächeren Sterne, befriedigend darzustellen. Man kann, wenn nötig, das Verfahren noch einmal wiederholen. Mit den schließlich erhaltenen Ausdrücken für φ und D müssen sich nun auch die Sternzahlen der helleren Sterne für $m < n$ so darstellen lassen, daß die Abweichungen durch die möglichen Beträge der im 2. Kapitel besprochenen systematischen Verfälschungen erklärt werden können.

Es sollen nun die für die Rechnung bequemen Formeln abgeleitet werden. Da die $\log A_m$ durch quadratische Interpolationsformeln dargestellt sind, lassen sich sehr leicht die Differentialquotienten $\dfrac{d\log A_m}{dm}$ bilden. Es liegt also nahe, die Formel (II) nach h zu differenzieren und damit weiterzurechnen. Es ist:

$$\frac{dA_m}{dh_m} = -\int\limits_0^{\sqrt{\frac{H}{h_m}}} D(\varrho)\, \varrho^4\, \varphi(h_m \varrho^2)\, d\varrho\,, \qquad \text{wenn} \quad m < n \quad (10)$$

Hierin ist die Näherungslösung von $D(\varrho)$ einzusetzen; man bekommt dann:

$$\frac{dA_m}{dh_m} = -\gamma \int\limits_{r_0}^{\sqrt{\frac{H}{h_m}}} \varrho^{4-\lambda}\, \varphi(h_m \varrho^2)\, d\varrho$$

Wenn man noch den Ausdruck für φ einsetzt, so läßt sich die rechte Seite durch eine Reihe von elementaren Substitutionen leicht auf die Form bringen:

$$\frac{1}{2}\left(\frac{H}{h_m}\right)^{\frac{5-\lambda}{2}}\frac{\Gamma\gamma\sqrt{\pi}}{2\sqrt{b}}\cdot e^{q^2}\cdot\frac{2}{\sqrt{\pi}}\int_\eta^\xi e^{-z^2}\,dz$$

wobei:

$$\frac{2a+5-\lambda}{4\sqrt{b}}=q\,,\qquad q+\sqrt{b}\,p\,(m-n)=-\xi$$

$$q-\sqrt{b}\log\frac{h_m\varrho^2}{H}=-\eta\,,\qquad p=\frac{0.4}{\log e}$$

Anderseits ist nun:

$$\frac{\partial A_m}{\partial m}=\frac{\partial A_m}{\partial h_m}\frac{d h_m}{d m}=-p\,e^{-pm}\frac{\partial A_m}{\partial h_m}$$

und:

$$\frac{\partial\log A_m}{\partial m}=\frac{\partial\log A_m}{\partial A_m}\cdot\frac{\partial A_m}{\partial m}=\frac{\log e}{A_m}\frac{\partial A_m}{\partial m}$$

Es ergibt sich sofort:

$$\frac{\partial A_m}{\partial h_m}=-2.5\,A_m\,e^{pm}\frac{\partial\log A_m}{\partial m}$$

also:

$$5\left(\frac{h_m}{H}\right)^{\frac{5-\lambda}{2}}A_m\,e^{pm}\frac{\partial\log A_m}{\partial m}=\frac{\gamma\,\Gamma\sqrt{\pi}}{2\sqrt{b}}\,e^{q^2}\frac{2}{\sqrt{\pi}}\int_\eta^\xi e^{-z^2}\,dz\tag{11}$$

wobei auf der linken Seite alles bekannt ist.

Aus der Interpolationsformel für die $\log A_m$ ergab sich $\lambda=0.72$. Nach einigen Versuchen erhielt ich damit aus Gleichung (11) folgende Werte für die Konstanten der Funktion φ:

$$a=-2.337,\qquad b=+0.0490$$

Führt man in die Gleichung (III) die Ausdrücke (4) und (9) ein, so läßt sich durch ähnliche Umformungen wie oben die Beziehung aufstellen:

$$\left[e^{\tau'^2}\frac{2}{\sqrt{\pi}}\int_{\sigma'}^{\tau'}e^{-z^2}\,dz-\alpha\left(\frac{h_m}{H}\right)^{\frac{\lambda_1-\lambda}{2}}e^{\tau_1'^2}\frac{2}{\sqrt{\pi}}\int_{\sigma_1'}^{\tau_1'}e^{-z^2}\,dz\right]\pi_m$$

$$=\left[e^{\tau^2}\frac{2}{\sqrt{\pi}}\int_\sigma^\tau e^{-z^2}\,dz-\alpha\left(\frac{h_m}{H}\right)^{\frac{\lambda_1-\lambda}{2}}e^{\tau_1^2}\frac{2}{\sqrt{\pi}}\int_{\sigma_1}^{\tau_1}e^{-z^2}\,dz\right]\left(\frac{h_m}{H}\right)^{\frac{1}{2}}\tag{12}$$

wo:

$$\tau'=q\,,\qquad\tau=\tau'-\frac{1}{4\sqrt{b}}\,,\qquad\tau_1'=\frac{2a+5-\lambda_1}{4\sqrt{b}}\,,$$

$$\tau_1=\tau_1'-\frac{1}{4\sqrt{b}}\quad\text{und}\quad\sigma=\tau-\sqrt{b}\log\left(\frac{h_m r_0^2}{H}\right)$$

Durch Versuche ergab sich mit den obigen Werten der Konstanten aus dieser Gleichung:

$$\lambda = 0.64, \qquad \lambda_1 = 1.14, \qquad \log \alpha = 9.800 \tag{13}$$

Die zweite Näherung von φ bestimmt sich aus der Gleichung:

$$\left.\begin{aligned}
5&\left(\frac{h_m}{H}\right)^{\frac{5-\lambda}{2}} A_m e^{pm} \frac{\partial \log A_m}{\partial_m} \\
&= \frac{\Gamma \gamma \sqrt{\pi}}{2\sqrt{b}} \left[e^{q^2} \frac{2}{\sqrt{\pi}} \int_\eta^\xi e^{-z^2} dz \quad \alpha \left(\frac{h_m}{H}\right)^{\frac{\lambda_1-\lambda}{2}} e^{q_1^2} \frac{2}{\sqrt{\pi}} \int_{\eta_1}^{\xi_1} e^{-z^2} dz \right]
\end{aligned}\right\} \tag{14}$$

wo q, ξ, η die Bedeutung wie in (11) haben und:

$$q_1 = \frac{2a + 5 - \lambda_1}{4\sqrt{b}} = \tau_1', \qquad -\xi_1 = q_1 + \sqrt{b}\, p(m - n),$$

$$-\eta_1 = q_1 - \sqrt{b}\, \log \frac{h_m r_0^2}{H}$$

ist. Der Einfluß des zweiten Terms in der Klammer ist sehr gering, die Versuche gestalten sich daher relativ einfach. Es ergaben sich für die Konstanten die Werte:

$$a = -2.279, \quad b = +0.0428 \tag{15}$$

SEELIGER fand für die Konstanten a und b aus den mittleren Parallaxen KAPTEYNS die Beträge: $a = -2.338$, $b = 0.0434$. Die Übereinstimmung ist also gut.

Tabelle 6.

Die Darstellung der $\log A_m$ für den ganzen Himmel mit den gefundenen Funktionen $D(\varrho)$ und $\varphi(i)$.

m	Beob.	Rechn.	$B-R$
4.5	8.138	8.115	+ 0.023
5.0	8.408	8.357	+ .051
5.5	8.649	8.592	+ .057
6.0	8.881	8.847	+ .034
6.5	9.088	9.079	+ .009
7.0	9.337	9.315	+ .022
10.5	0.865	0.863	+ 0.002
11.0	1.062	1.062	.000
11.5	1.277	1.271	+ .006
12.0	1.457	1.450	+ .007
12.5	1.618	1.617	+ .001
13.0	1.788	1.785	+ .003
13.5	1.933	1.949	− .016
14.0	2.087	2.104	− .017
14.5	2.255	2.261	− .006

Tabelle 7.

Darstellung der mittleren Parallaxen aus Gron. Publ. 34 mit den Funktionen $\varphi(1)$ und $D(\varrho)$.

m	π_m beob.	π_m berechn.	$B - R$
4	0$.''$0365	0$.''$0365	0$.''$0000
5	0.0231	.0244	$-$ .0013
6	0.0153	.0161	$-$ .0008
7	0.0104	.0105	$-$ .0001
8	0.0073	.0068	$+$ .0005

Mit den Konstanten (13) und (15) ergeben sich die in den Tab. 6 und 7 enthaltenen Darstellungen der π_m und der A_m. Es ergab sich für die Konstante $\log \Gamma_0$, wo $\Gamma_0 = \dfrac{\gamma \, \Gamma \sqrt{\pi}}{2\sqrt{b}}$, der Wert $-$ 10.687. Die Darstellung ist bei den A_m für $m > n$ und bei den π_m genügend. Bei den A_m der helleren Sterne zeigen sich systematische Abweichungen, welche jedoch nicht sehr groß sind. Sie können z. B. erklärt werden durch einen Skalenunterschied von $0^{m}.06$ bis $0^{m}.08$ zwischen der Skala für die helleren und derjenigen für die schwächeren Sterne.

Mit den Konstanten von φ und D läßt sich nun auch noch die Grenze des schematischen Sternsystems nach Formel (V) berechnen. Es wird:

$$\varDelta \left(\frac{d^2 \log A_n}{d n^2} \right) = 0.2 \, \omega \, p \, \Gamma \, \gamma \, \frac{H}{A_n} \, r_1^{3-\lambda} \left(1 - \frac{\alpha}{\sqrt{r_1}} \right)$$

oder:

$$\varDelta \left(\frac{d^2 \log A_n}{d n^2} \right) = 0.2 \, p \cdot \frac{H}{A_n} \, \Gamma_0 \cdot \frac{2\sqrt{b}}{\sqrt{\pi}} \, r_1^{3-\lambda} \left(1 - \frac{\alpha}{\sqrt{r_1}} \right) \qquad \text{(VI)}$$

wo:

$$\Gamma_0 = \omega \, \frac{\Gamma \, \gamma \sqrt{\pi}}{2\sqrt{b}}$$

woraus $r_1^{3-\lambda} \left(1 - \dfrac{\alpha}{\sqrt{r_1}} \right) = [7.217]$ und $r_1 = 1200$ pars. Nach den Betrachtungen des 2. Kapitels ist die Wahrscheinlichkeit $\frac{1}{2}$, daß $\log r_1$ zwischen den Grenzen liegt:

$$3.02 < \log r_1 < 3.54 \quad \text{oder} \quad 1050 < r_1 < 3470 \text{ pars.}$$

In Anbetracht der Unsicherheit der Grenzbestimmung im schematischen Sternsystem bedeutet dies eine gute Übereinstimmung der beiden auf ganz verschiedene Weise bestimmten Werte. In einer späteren Arbeit sollen die Konstanten der Dichtefunktion auch für die einzelnen Zonen abgeleitet werden. Man darf erwarten, daß dann die aus (1) und (VI) ermittelten r_1 für die einzelnen Zonen einander noch mehr bestätigen werden.

Die Funktion φ läßt sich leicht in die Form einer Gaußschen Fehlerkurve bringen. Beachtet man, daß:

$$\varphi(i)\,di = -\varphi(i)\frac{di}{dM}\,dM = p\,e^{-pM}\,\varphi(i)\,dM\,, \quad \text{wo} \quad p = \frac{0.4}{\log e}$$

so ergibt sich, da $\log \dfrac{i}{H} = -p\,(M - M_H)$, wobei $M_H = -8.0$, folgender Ausdruck:

$$\varphi(i)\,di = p\,\Gamma_1 e^{0.597\,M - 0.0363\,M^2}\,dM\,, \quad \text{wo} \quad \Gamma_1 = \Gamma e^{14.469}$$

Schließlich erhält man, wenn man noch berücksichtigt, daß $\int_0^H \varphi(i)\,di = 1$ sein soll:

$$\varphi(M)\,dM = \frac{0.1905}{\sqrt{\pi}}\,e^{-1/2\,(M-8.22)^2/3.71^2}\,dM$$

Die absoluten Sterngrößen sind also um den mittleren Wert $+ 8.2$ nach dem Gaußschen Fehlergesetze gestreut entsprechend einem mittleren Fehler von ± 3.71.

Die obige Funktion φ weicht sehr beträchtlich ab von der Verteilungsfunktion, welche KAPTEYN[1]) aus den $N_{m\mu}$ bestimmt hat. Das Maximum der KAPTEYNschen Funktion liegt bei $M = + 2.7$, und die Streuung entspricht einem mittleren Fehler vom Betrag ± 2.5 (entsprechend einem wahrscheinlichen Fehler von ± 1.69).

Es ist sehr wahrscheinlich, daß das φ KAPTEYNS für die absolut schwächeren Sterne zu kleine Werte annimmt. Nach KIENLE[2]) ergibt sich für die Umgebung der Sonne aus den Sternen mit bekannter Parallaxe, daß innerhalb einer Kugel mit 5 pars. Radius das Verhältnis der Anzahlen der Sterne mit $M < 2.7$ zu denen mit $M > 2.7 = \frac{1}{2}$ ist, statt nach KAPTEYNS Kurve $= \frac{1}{4}$. Das Verhältnis wird sich noch weiter verschieben, da hellere Sterne in diesem Raum nicht mehr dazukommen werden, sicher aber noch eine Anzahl schwächere. Es zeigt sich, daß das Maximum von φ für die bis jetzt bekannten Sterne, deren Parallaxe $> 0\rlap{.}{''}2$ ist, etwa bei der Größe 6 oder 7 liegt. Je mehr man den Radius der Kugel vergrößert, desto mehr rückt das Maximum von φ nach der Seite der kleinen M, da von den absoluten schwachen Sternen immer mehr weggelassen werden.

Die KAPTEYNsche Verteilungsfunktion ist aus den Zahlen $N_{m\mu}$ und den mittleren Parallaxen $\pi_{m\mu}$ abgeleitet. Wir wollen einmal einen Raum um die Sonne abgrenzen, der etwa gerade so groß ist, daß statistische Mittelbildungen noch mit genügender Sicherheit durchzuführen sind. Wir wählen dazu eine Kugel mit dem Radius 20 pars. = 4 Siriusweiten. Wir wollen die Bedingungen festlegen, unter welchen die Methode

---

1) Mt. Wils. Contr. 188 und Mt. Wils. Contr. 229.
2) Zur Kritik der Verteilungsfunktion der absoluten Leuchtkräfte. Astr. Nachr. Bd. 218, Nr. 5216.

Kapteyns innerhalb dieses Raumes alle Sterne mit der absoluten Größe
$+ 3^m0$ berücksichtigt. Ein Stern der absoluten Größe $+ 3.0$ erscheint
an der Grenze der Kugel von der Größe 9^m5. Wenn also alle Sterne
mit $M = +3.0$ innerhalb dieser Kugel zu dem φ einen Beitrag liefern
sollen, so müssen mindestens die $N_{m\mu}$ bis $m = 9.5$ bekannt sein. Die
Sterne, welche die Zahl $N_{m\mu}$ bilden, haben verschiedene Leuchtkräfte.
Diese nehmen zu mit abnehmender Parallaxe. Kapteyn nahm an,
daß die individuellen Parallaxen, wenn π_0 ihr wahrscheinlichster Wert
ist, derart verteilt sind, daß die Größen $\log \dfrac{\pi}{\pi_0} = z$ nach dem Gaußschen
Fehlergesetz gestreut sind, und zwar so, daß der wahrscheinliche
Fehler ϱ von z den Wert ± 0.19 hat. Die Streuung der Parallaxen um
ihren wahrscheinlichsten Wert ist nicht sehr groß; es ergibt sich $h = 2.510$,
$h^2 = 6.30$. Ist $\log \pi - \log \pi_0 = \pm 0.60$, so ist die Häufigkeit nur noch
0.026; ist $\log \pi - \log \pi_0 = \pm 0.70$, so wird sie 0.017. Da die wahrscheinliche Parallaxe π_0 sich nicht viel von der mittleren unterscheidet,
werden also von den Anzahlen $N_{m\mu}$ der schwachen Sterne nur diejenigen
in die Kugel vom Radius 20 pars. hereingestreut, welche große Eigenbewegungen μ haben. Wenn also die $N_{m\mu}$ für größere m zu wenig
Sterne mit großer EB. enthalten, dann werden die Werte von φ systematisch zu klein für absolut schwächere Sterne.

Man kann sicher annehmen, daß noch eine beträchtliche Anzahl
von schwachen Sternen mit großem μ festgestellt werden wird, so daß
gerade die $N_{m\mu}$ für die großen μ vergrößert werden. Das Absteigen
der Kurve φ nach Kapteyn können wir also nicht als reell ansehen.
Es findet seine Erklärung in der prozentual zu geringen beobachteten
Anzahl schwacher Sterne mit großer Eigenbewegung.

Es muß hervorgehoben werden, daß bei der von uns verwendeten
Methode der Bestimmung aus den Sternzahlen das φ frei von systematischen Einflüssen der eben besprochenen Art bleibt. Selbstverständlich ist auch hier für sehr schwache M die Kurve nicht mehr
verbürgt; wieweit man ihren Verlauf für gesichert halten kann,
hängt davon ab, bis zu welchen m Sternzahlen beobachtet sind. Wir
konnten die A_m bis zu $m = 14.5$ verwenden, berücksichtigen daher
in der Kugel vom Radius 20 pars. alle Sterne bis zur absoluten Größe
$M = + 8.5$.

Tab. 8 enthält die nach der Formel

$$\log \varphi (M) = - 2.034 + 0.2592\, M - 0.0158\, M^2$$

berechneten Werte der Verteilungsfunktion der Leuchtkräfte.

Zusammenfassung. Die Einführung der Grenzgrößen H und h_n in
die Integralgleichungen der Stellarstatistik nach Seeliger ermöglicht
es, die Ausdehnung des typischen Sternsystems mit hinreichender Sicherheit zu bestimmen.

Tabelle 8.

Die Verteilungsfunktion der Leuchtkräfte.

M	$\log \varphi(M)$	$\varphi(M)$
$-\ 8.0$	$4.863-10$	0.00000730
$-\ 6.0$	5.842	0.0000695
$-\ 4.0$	6.676	0.000474
$-\ 2.0$	7.385	0.00243
0.0	7.966	0.00925
$+\ 2.0$	8.421	0.0264
$+\ 4.0$	8.750	0.0563
$+\ 6.0$	8.952	0.0895
$+\ 8.0$	9.011	0.103
$+\ 10.0$	8.978	0.0951
$+\ 12.0$	$8.796-10$	0.0625

Die Verteilungsfunktion der Leuchtkräfte, welche sich aus den Sternzahlen A_m für $m > n$ ergibt, stimmt überein mit der Verteilungskurve, welche sich aus den Sternen der nächsten Umgebung der Sonne bestimmt. Im Widerspruch zu ihr steht dagegen die Verteilungsfunktion KAPTEYNS, welche für große M zu kleine Werte liefert.

Bei der Ermittlung von Sternzahlen, welche zu Untersuchungen über das typische Sternsystem dienen sollen, ist vor allem darauf zu achten, daß man die Variation der Sternfülle mit der galaktischen Länge kompensiert. Dies geschieht dadurch, daß man möglichst viele gleichmäßig über die Zone verteilte Felder abzählt, wenn man nicht, was natürlich noch besser ist, das ganze Gebiet der Zone abzählen kann. Ist man gezwungen, verschiedene Helligkeitsskalen zu benützen, so sind die Skalendifferenzen, insbesondere auch in ihrer Abhängigkeit von der galaktischen Breite, sorgfältig zu bestimmen.

Eigenbewegungen.

Von Ernst Grossmann, München.

Als W. Herschel im Jahre 1783 den kühnen Entschluß faßte, den
Aufbau des Stellarsystems zu erforschen, stand ihm außer 8 EB. keinerlei
Beobachtungsmaterial zur Verfügung; er mußte es sich erst selbst ver-
schaffen. Die nur vereinzelten Bestrebungen in den nächsten hundert
Jahren, in Herschels Sinn weiterzuarbeiten, brachten keine nennens-
werten Erfolge. Auch als Herr v. Seeliger im Jahre 1898 seine erste
Arbeit über die räumliche Verteilung der Fixsterne veröffentlichte,
mußte er noch sagen, „daß sich durch die statistische Methode manche
wichtige und wertvolle Einsicht gewinnen ließe, wenn nur erst das notwendige Material vorliegen würde". Er empfiehlt bereits die Anwendung der Photographie zu diesem Zweck und fügt hinzu: „Es ist wohl
kaum eine lohnendere und wichtigere Beobachtungsarbeit im Gebiete
der Stellarastronomie für solche Beobachter anzugeben, die für Dauer-
aufnahmen eingerichtet sind." Seine Mahnung ist auf fruchtbaren
Boden gefallen. Wenn heute die Stellarastronomie sich so großer Erfolge
rühmen darf, so verdanken wir dieses in erster Linie den langjährigen
scharfsinnigen Arbeiten v. Seeligers; wir müssen ihn als den Begründer
der Stellarstatistik ehren.

Die letzten Jahrzehnte haben uns eine wertvolle Bereicherung des
Beobachtungsmaterials gebracht, besonders dank der organisatorischen
Tätigkeit Kapteyns; es weiter zu vervollständigen, muß unsere nächste
Aufgabe sein, aber auch es weiter auszufeilen und zu verfeinern; hierauf
möchte ich an dieser Stelle das Augenmerk lenken.

Das Fundamentalproblem der Stellarstatistik besteht in der Herstellung einer Beziehung zwischen mittleren Parallaxen einerseits, der
Anzahl der Sterne, ihrer Helligkeit und ihrer Geschwindigkeit andererseits; in zweiter Annäherung treten Spektraltypus und galaktische
Breite, in dritter galaktische Länge hinzu. Unsere Aufgabe besteht
darin, die räumliche Verteilung der Sterne als Funktion dieser Elemente
zu untersuchen. Soweit diese als charakteristische Merkmale der Sterne
zu Klassifizierungszwecken dienen, dürften sie bei dem heutigen Stande
unseres Problems mit hinreichender Genauigkeit bekannt sein; hier ist
Vervollständigung die Hauptsache. Aber zum Teil müssen sie mehr
leisten: Unseren Methoden der Parallaxenbestimmung, der trigonometrischen sowohl wie der spektroskopischen, sind Grenzen gesteckt;

wir müssen deshalb zur Ableitung der mittleren Parallaxen die EB. heranziehen, die uns in mehrfacher Weise, besonders in Gestalt ihrer in den größten Kreis Stern-Apex fallenden Komponente hierbei wertvolle Dienste leistet. Leider aber ist auch sie es, die uns die größten Schwierigkeiten bereitet; es ist nicht möglich, sie lediglich durch ad hoc angestellte Beobachtungen auf den zu erstrebenden höchsten Grad der Genauigkeit zu bringen. Besonders mißlich liegt die Sache bei den schwachen Sternen.

Wir haben zu unterscheiden zwischen EB., die aus absoluten Positionsbestimmungen hervorgegangen sind und solchen, die auf Anschlußbeobachtungen an diese, sei es mit Hilfe des Meridiankreises, sei es durch mikrometrische Messung, beruhen. Dazu treten heute noch vielfach EB., die zunächst nur relativer Natur sind, weil ein Anschlußstern auf der photographischen Platte fehlt; nur auf hypothetischem Wege können sie in absolute verwandelt werden.

Die erste Kategorie geben uns zur Hauptsache unsere FC. Daß diese heute kein uneingeschränktes Vertrauen mehr verdienen, unterliegt keinem Zweifel. In meinem Aufsatz: „Zur Revision unseres Bezugssystems"[1]) habe ich die Abweichungen der FC. gegen neuere Beobachtungsreihen in AR. und ihren EB., den Äquinoktialfehler und die Ungenauigkeit der NEWCOMBschen Präzessionskonstanten eingehend besprochen und die Notwendigkeit einer Neubeobachtung der FC. dargetan, und zwar besonders nach einer zur Elimination des von mir diskutierten Fehlers $\Delta \alpha_\delta$ geeigneten Anordnung des Beobachtungsprogramms. Ich halte diese Arbeit für um so dringlicher, als nach Herrn F. RENZ[2]) der Katalog Pulkowa 1915 den Fehler nicht bestätigt hat. In München ist im vorigen Jahre von Herrn KIENLE und mir eine entsprechende Beobachtungsreihe begonnen worden, aber noch nicht abgeschlossen. Ich möchte an dieser Stelle ein Versehen berichtigen, welches mir in obiger Arbeit unterlaufen ist: Meine Behauptung, daß für Pulkowa 1842—44 Mittelwerte des Azimuts aus längeren Zeiträumen benutzt seien, trifft nicht zu; es sind vielmehr kurzperiodische Schwankungen streng berücksichtigt.

Auf einen Fehler $\Delta \delta_\delta$ der FC. bin ich in meiner Arbeit „Über die astronomische Refraktion"[3]) näher eingegangen. Mehrere neuere Beobachtungsreihen weisen Differenzen gegen die FC. auf, die für die Sterne vom Äquator bis $\delta = +50°$ bis zu einer Sekunde ansteigen, und ich habe damals die Vermutung ausgesprochen, daß wahrscheinlich auch die EB. mit mehr oder weniger starken Fehlern behaftet seien. Auch diese Differenz bedarf dringend der Aufklärung. Durch gleich-

---

[1]) Astr. Nachr. Bd. 215. 1922.   [2]) Astr. Nachr. Bd. 217. 1923.
[3]) Abhandl. d. Kgl. bayr. Akad. d. Wissensch., mathem.-physik. Kl. Bd. 23, 9. Abh. 1917.

zeitige Beobachtungen in Heidelberg [KOPFF[1])] und München (GROSS-MANN und KIENLE) konnte sie nicht gewonnen werden, und rätselhaft ist noch die große Differenz zwischen den beiden Pulkowaer Vertikalkreisen von ERTEL und REPSOLD, von der ich bereits in obiger Arbeit gesprochen und die nunmehr von BONSDORFF[2]) eingehend, jedoch ohne Erfolg, untersucht ist.

Nunmehr hat KAPTEYN[3]) auf Grund einer Vergleichung von fünf neueren Katalogen mit Boss PGC durch ROY[4]) gefolgert, daß die Korrektion des PGC für die Epoche 1906 $+ 0\rlap{.}''56$ für $\delta = 0°$ betrüge, und daß daher, da zur Zeit der mittleren Epoche des PGC (1873.9) $\varDelta\delta_\delta = 0$ sein müsse, die Korrektion in EB. $+ 0\rlap{.}''0174$ ausmache. Einen ähnlichen Wert leitet KAPTEYN aus der Vergleichung älterer und neuerer Pulkowaer und anderer Kataloge ab. Um ferner die aus den μ_δ Boss berechnete Apexdeklination $+ 35\rlap{.}°2$ in Übereinstimmung mit der aus den Radialbewegungen folgenden, $+ 25\rlap{.}°2$, zu bringen, erfordern die μ_δ die Korrektion $+ 0\rlap{.}''0136$, und ebenfalls ergibt sich mit dieser eine gute Übereinstimmung der Vertexdeklinationen des Sternstroms I, wenn man die Sterne nach der Größe der EB. gruppiert. KAPTEYN setzt schließlich die Korrektion $\varDelta = + 0\rlap{.}''0130 \cos \delta$.

Besäßen wir in genügender Anzahl EB. der schwächsten Sterne im Boss-System, so wäre eine Entscheidung über die Realität der Korrektion nicht schwierig, denn für die Sterne 12. Größe z. B. (mittl. EB. $= 0\rlap{.}''014$) würde sich ohne sie und mit ihr eine Differenz in der Apexdeklination von 93° ergeben. Vorläufig aber können wir über die EB. dieser Sterne gar nichts aussagen. Die Ergebnisse der Untersuchung von Herrn WIRTZ[5]) „Über die Bewegungen der Sterne 11. Größe" können nicht maßgebend sein, denn diese EB. sind ursprünglich relative, die erst mit Hilfe von aus einer angenommenen Sonnenbewegung abgeleiteten Mot. par. in absolute verwandelt sind. Die Fortsetzung dieser Untersuchung durch HÜGELER[6]) gibt diesen Umstand klar zu erkennen.

KAPTEYN hält die Bestimmung der systematischen Fehler der Deklinationen unseres Bezugssystems für eins der dringendsten Probleme der fundamentalen Astronomie. Ich stimme ihm hierin bei, muß aber die Forderung auch auf die Fehler der AR. ausdehnen. Die EB. der Fundamentalsterne bilden die Basis für die Bestimmung der EB. aller übrigen Sterne. Durch Meridiankreisbeobachtungen können wir dieses Ziel vorderhand nicht erreichen; deshalb schlägt KAPTEYN eine andere

---

[1]) Astr. Nachr. Bd. 212 u. 213. Die Diskussion der Münchner Beobachtungen durch Herrn KIENLE harrt noch der Drucklegung.

[2]) Res. der abs. Dekl.-Best. des Pulkowaer Kat. 1915 von J. BONSDORFF. Helsinki 1922.

[3]) Bull. of the astr. Instituts of the Netherlands Bd. 1, Nr. 14.

[4]) Popular Astronomy Bd. 30.   [5]) Astr. Nachr. Bd. 211.

[6]) Astr. Nachr. Bd. 216.

Methode vor, die uns sehr viel rascher zum Ziele führen soll. Wenn die mit den langbrennweitigen Instrumenten hergestellten, zur Bestimmung von Parallaxen dienenden Aufnahmen nach einigen Jahren wiederholt werden, so liefern sie uns die relative EB. der Hauptsterne. Durch Kombination zweier in AR. um 180° sich unterscheidenden Aufnahmen ergibt sich die mittlere Säkularparallaxe der Vergleichsterne; mit Hilfe dieser ihre parallaktische Bewegung und damit die Reduktion der relativen EB. in absolute. Voraussetzung ist Kenntnis der Apexkoordinaten.

Eine weitere Vervollkommnung des KAPTEYNschen Vorschlages hat VAN RHIJN[1]) gegeben und zugleich eine Anwendung auf die c-, O- und N-Sterne, bei denen die parallaktischen Bewegungen nur klein sind. Er findet als Korrektion der μ_δ Boss den sehr viel geringeren Wert $+ 0{.}''0064 \pm 0{.}''0013$, der mir aber trotz des geringen w. F. noch wenig verbürgt erscheint. Wenn VAN RHIJN[2]) ferner den Nachweis führt, daß der Einfluß selbst des KAPTEYNschen Wertes auf die mittleren Säkularparallaxen nur gering sei, so ist hierzu zu bemerken, daß er sich hierbei nur auf die Sterne bis zur 10. Größe beschränkt; bei schwächeren Sternen aber beginnt der Einfluß rapid zu steigen, bei Sternen 13. Größe wächst die Säkularparallaxe von $0{.}''005$ auf $0{.}''011$.

VAN DEN BOS[3]) stellt eine Vergleichung an zwischen Boss' EB. und mikrometrisch bestimmten von BURNHAM[4]), die aber nur relativ sind. Die in Zonen von 10° gemittelten Differenzen haben einen ausgesprochenen Gang. Als Korrektion der μ_δ Boss ergibt sich $+ 0{.}''011$, die der Autor jedoch für wenig sicher hält.

Bei der Wichtigkeit der Frage möchte ich auf die Darstellung der Beobachtungskataloge, besonders von AUWERS-BRADLEY, durch die FC. etwas näher eingehen.

Für die BRADLEYschen Deklinationen leitet AUWERS[5]) systematische Korrektionen Δz ab, die sich aus Fehlern der Teilung und der Unebenheit der Limbusfläche zusammensetzen; sie steigen bis über $1''$ an und verlaufen sehr diskontinuierlich. Die indirekten Methoden der Untersuchung können für die Realität der Resultate, deren Übereinstimmung auch sehr wenig befriedigend ist, keine sehr sichere Bürgschaft leisten. AUWERS selbst spricht sich in diesem Sinne aus (Bd. I, S. 190). An den Katalog (Bd. II) sind die unkorrigierten Δz (Bd. II, S. 252) angebracht.

Boss[6]), der die Art der Ableitung der Δz noch nicht kennen konnte, verwirft sie auf Grund von Widersprüchen zwischen den Δz (Bd. II, S. 252) und den Differenzen $\Delta \delta_\delta$ von BRADLEY gegen seinen ersten

---

[1]) Bull. of the astr. Instituts of the Netherlands Nr. 36.
[2]) Publ. Groningen Nr. 34, Kap. 4.
[3]) Bull. of the astr. Instituts of the Netherlands Nr. 26.
[4]) BURNHAM: Measures of proper motion stars.
[5]) AUWERS-BRADLEY Bd. 2, S. 252, und Bd. 1, S. 586.
[6]) Astr. Journal Bd. 23, S. 157.

Katalog[1]). Er begnügt sich damit, die Differenzen zwischen den unkorrigierten Deklinationen BRADLEYS und seinem Katalog festzustellen und diese als Korrektionen von BRADLEY zu benutzen; sie steigen bis zu 2″ an. Er behandelt also BRADLEY als Katalog II. Klasse, der somit zu den Bestimmungen der EB. keinen fundamentalen Beitrag liefert.

NEWCOMB[2]) bestimmt die Verbesserungen des soeben genannten Boss-Kataloges nach der Theorie der Planetenbewegungen, und indem er BRADLEY hiermit vergleicht, erhält er dessen absolute Korrektionen in der Äquatorzone, die, so nimmt er an, proportional mit der Poldistanz abnehmen. Für die Sterne nördlich von $+ 50°$ jedoch wird Δz nach AUWERS angebracht.

In der folgenden Tabelle gebe ich die numerischen Werte der Korrektionen, welche die FC.-Autoren an die Deklinationen BRADLEYS vor der weiteren Verwertung derselben bei der Herstellung der Normalsysteme angebracht haben; ich beschränke mich hier auf die Zone $+ 25°$ bis $- 15°$. Zugleich gebe ich die Reduktionen $\Delta \delta_\delta$ von AUWERS-BRADLEY auf die drei FC.

δ	Δz	$\Delta z'$	Δc Boss	Δc Newc.	$\Delta \delta_\delta$ Auw.	$\Delta \delta_\delta$ Boss	$\Delta \delta_\delta$ Newc.
$+ 25°$	$+ 0.''33$	$+ 0.''15$	$+ 1.''03$	$+ 0.''80$	$+ 1.''2$	$+ 1.''02$	$+ 1.''23$
20	$+ 8$	$- 24$	$+ 1.48$	$+ 0.86$	$+ 1.3$	$+ 1.59$	$+ 1.48$
15	$+ 64$	$+ 31$	$+ 1.26$	$+ 0.93$	$+ 1.3$	$+ 1.39$	$+ 0.94$
10	$- 20$	$- 53$	$+ 1.29$	$+ 0.99$	$+ 1.0$	$+ 1.42$	$+ 1.78$
5	$+ 45$	$+ 12$	$+ 0.72$	$+ 1.05$	$+ 1.0$	$+ 0.74$	$+ 1.03$
0	$+ 77$	$+ 44$	$+ 0.60$	$+ 1.11$	$+ 1.0$	$+ 0.75$	$+ 0.61$
$- 5$	$- 2$	$- 35$	$+ 1.24$	$+ 1.16$	$+ 0.9$	$+ 1.32$	$+ 1.32$
10	0	$- 33$	$+ 0.98$	$+ 1.21$	$+ 0.9$	$+ 1.09$	$+ 1.21$
15	$+ 18$	$- 15$	$+ 1.24$	$+ 1.27$	$+ 0.8$	$+ 1.16$	$+ 0.97$

Δz = Korr. der BRADLEY-Dekl. nach AUWERS Bd. 2, S. 252, sind an den Kat. Bd. 3 angebracht; $\Delta z'$ = definitive Korr. nach Bd. 1, S. 586.
Δc = Korr. der im Kat. Bd. 3 enthaltenen Dekl. nach Boss und NEWC.
$\Delta \delta_\delta$ = Reduktionen von Bd. 3 auf NFK, PGC und FC NEWC.

Hieraus geht hervor:
1. Der Verlauf der $\Delta \delta_\delta$ ist so gleichmäßig bei allen drei FC., auch wenn wir die Δz wieder in Abzug bringen, daß wir die BRADLEYschen Beobachtungen als in sich sehr homogen ansehen müssen; sie sind daher zur Ableitung von mittleren EB. durchaus geeignet. Gleiches gilt übrigens auch für $\Delta \delta_\alpha$ und ganz besonders für $\Delta \alpha_\alpha$ und $\Delta \alpha_\delta$.
2. Die systematischen Korrektionen der BRADLEYschen Deklinationen, wie sie von AUWERS und NEWCOMB abgeleitet sind, werden durch ihre FC. nicht bestätigt; die Übereinstimmung der ΔC und der $\Delta \delta_\delta$ bei Boss kann angesichts der Entstehung der ΔC nicht überraschen.

---

[1]) Catalogue of 627 principal Standard Stars.
[2]) Catalogue of fund Stars. Astr. Papers Bd. 8, Teil 2.

Wir müssen daher schließen, daß in den drei FC. den Bradley-Beobachtungen nicht die genügende Rechnung getragen ist, die sie verdienen.

3. Es muß auffallen, daß die $\varDelta\,\delta_\delta$ bei allen drei FC. von $+40°$ bis $-30°$ stark positiv und daß sie bei NFK. nahezu konstant sind; es liegt hiernach in dieser Zone das Bradley-System um rund $1''\!.3$ zu südlich.

4. Werfen wir diese Differenz auf die FC., so würde sie aussagen, daß wenn wir mit Kapteyn die FC. zur Zeit ihrer mittleren Epoche 1874 als richtig annehmen, ihre EB. zu korrigieren sind um $+0''\!.011$, in auffallender Übereinstimmung mit der oben diskutierten Korrektion von Kapteyn.

Andere Kataloge in gleicher Weise zu besprechen, muß ich mir an dieser Stelle versagen; Pulkowa habe ich bereits oben erwähnt. Aber auf folgendes muß ich aufmerksam machen: Betrachtet man die Reduktionen $\varDelta\,\delta_\delta$ der Kataloge auf PGC[1]) für $\delta=0°$, so ersieht man, daß diese von 1800 bis 1840 mit zwei Ausnahmen (Königsberg und St. Helena) negativ sind; sie würden somit eine negative Korrektion der EB. bedingen. Von 1840 bis 1890 ist der Vorzeichenwechsel befriedigend, und von 1890 an sind die $\varDelta\,\delta_\delta$ wieder durchweg negativ. 25 Kataloge geben hier im Mittel $\varDelta\,\delta_\delta=-0''\!.274$ und damit als Korrektion der EB. $+0''\!.011$, wie oben.

Daß hier etwas nicht in Ordnung ist, unterliegt keinem Zweifel. Meine Folgerung hieraus gebe ich am Schluß der Arbeit.

Wir wenden uns nunmehr zu den relativen Bestimmungen, wie sie mit Hilfe des Meridiankreises ausgeführt werden. Es ist eine bekannte Tatsache, daß selbst, wenn diese streng an einen FC. angeschlossen sind, sie dennoch gegen diesen wie auch unter sich mehr oder weniger große systematische Unterschiede aufweisen; es sind nachträglich Katalogreduktionen erforderlich. Würden jene einen nach AR., Deklination und Helligkeit streng systematischen und klar erkennbaren Verlauf haben, so gestaltete sich die Reduktion sehr einfach. Das ist aber keineswegs der Fall! Im allgemeinen ist die Anzahl der Vergleichspunkte, also der FC.-Sterne, nur gering; vor allem fehlen hierbei naturgemäß die schwächeren Sterne. Wie will man die H.-Gl. der relativen Kataloge, besonders der älteren, bestimmen? In welcher Reihenfolge soll man die systematischen Differenzen $\varDelta\,\alpha_\alpha,\varDelta\,\alpha_\delta,\varDelta\,\delta_\alpha\,\varDelta\,\delta_\delta$ und $\varDelta\,m$ untersuchen, und welchen Ansatz will man für sie wählen? Die Berechtigung von Fourier-Reihen od. dgl. ist keineswegs erwiesen, und graphische Ausgleichung ist nur ein Notbehelf. Welchen Einfluß konstante systematische Differenzen in Deklination, die in die EB. übergehen, auf die Apexdeklination haben, hat Herr Kobold[2]) dargelegt.

---

    [1]) PGC, App. III, und Roy, A. J.: Systematic corrections and weights of catalogs.

    [2]) Astr. Nachr. Bd. 203.

Die Reduktionstafeln von AUWERS und BOSS sind graphischen Kurvenzügen entnommen; sie nehmen daher einen sehr glatten Verlauf. Wirft man aber einen Blick hinter die Kulissen, so ändert sich das Bild. Man braucht sich nur einmal die Vergleichung der AG.-Zonen mit Pulkowa 1875 von AUWERS[1]) anzusehen, um zu verstehen, wie recht AUWERS hat, wenn er seine graphische Ausgleichung nur als rohe Interpolation bezeichnet.

Die Zuverlässigkeit der Reduktionen möge ein Beispiel beleuchten: Für einen Stern $\alpha = 8^h0^m0^s$, $\delta = +55°$, Gr. 9.0 (Cambridge-Harvard-AG.-Zone) gibt BOSS die Reduktion auf PGC $- 0^s.058$, AUWERS auf NFK $- 0^s.151$, dazu H.-Gl. des NFK $- 0^s.026$, insgesamt $- 0^s.177$. Die Differenz NFK—PGC ist jedoch nur $+ 0^s.002$[2]). Je nachdem man auf diesen oder jenen Katalog reduziert, in dem Resultat ergibt sich eine Differenz von $0^s.119$. Verglichen mit einem heutigen Katalog geht dieser Betrag in die EB. ein mit $0^s.119 : 50 = 0''.036$. Nach VAN RHIJN[3]) ist aber zu fordern für die EB. der Sterne 9. Größe der w. F. $\pm 0''.01$ (rund), und die mittlere EB. für diese Sterne beträgt $0''.03$. Ich könnte mit Leichtigkeit noch weitere Argumente für die Unzuverlässigkeit der Katalogreduktionen beibringen.

Es dürfte hier am Platze sein, die Frage aufzuwerfen, wie weit heute die AG.-Zonen zur Ableitung von EB. bereits geeignet sind. In einer Überschlagsrechnung, die VAN RHIJN[3]) anstellt, setzt er den w. F. einer Position dieser Zonen $= \pm 0''.4$ und damit den w. F. einer durch Verbindung mit heutigen Beobachtungen (Epochendifferenz 50 Jahre) abgeleiteten EB. $\pm 0''.010$, und wenn man noch alle älteren Kataloge hinzuzieht $\pm 0''.008$. So allgemein darf man m. E. die Frage nicht behandeln. In den siebziger Jahren sind nur zwei Zonen fertiggestellt, die Beobachtung etwa der Hälfte aller Zonen dehnte sich bis in die neunziger Jahre aus. Es ist somit die mittlere Epochendifferenz mit 50 Jahren zu groß angesetzt. Ähnlich steht es mit dem w. F. Die einzelnen Zonen sind von sehr ungleicher Genauigkeit; der w. F. eines Katalogorts schwankt zwischen $\pm 0''.33$ und $\pm 0''.70$, von dem sehr viel genaueren Katalog Berlin B abgesehen. Systematische Fehler, wie ich sie soeben besprochen habe, sind von VAN RHIJN gar nicht berücksichtigt. Wenn er nun aus seiner Rechnung bereits den Schluß zieht, daß zur Bestimmung der absoluten EB. der schwachen Sterne durch Anschluß dieser an AG.-Zonen-Sterne die für die EB. dieser Werte zu erlangende Genauigkeit nicht ausreicht, so erfährt dieser Schluß durch die obigen Einschränkungen eine wesentliche Stütze. Zur Bestimmung der absoluten EB. der schwachen Sterne durch mikrometrischen

---

[1]) Astr. Nachr. Bd. 161.

[2]) Boss, B.: Comparison of the Standard Catalogues of AUWERS and Boss for 1910. Astr. Journal Bd. 26.

[3]) Third report ... of the selected areas. Tafel 6 und S. 27.

Anschluß dieser an die AG.-Zonen-Sterne kann somit eine heutige Wiederholung der Beobachtung der AG.-Zonen keinen Beitrag liefern.

Damit sind wir bereits übergegangen zu der dritten Methode der Bestimmung der EB., der Methode des differentiellen mikrometrischen Anschlusses von Sternen unbekannter EB. an solche bekannter EB. Diese kommt hauptsächlich in Betracht für schwache Sterne, die mit dem Meridiankreis nicht mehr erreichbar sind. Es handelt sich heute hierfür wohl nur noch um die photographische Aufnahme. Selbstverständlich muß nun hier die Voraussetzung gemacht werden, daß die EB. der Vergleichssterne die erforderliche Genauigkeit besitzen. Diese trifft jedoch, wie wir gesehen haben, keineswegs zu. Beträgt tatsächlich der Fehler der EB. der FC.-Sterne in Deklination $0''013$, so erreicht er bereits die mittlere EB. der Sterne 11. Größe; es würden also nicht einmal die FC.-Sterne zu unserem Zweck ausreichen.

Noch mißlicher liegen die Verhältnisse, wenn auf der Platte überhaupt kein Stern mit bekannter EB. vorhanden ist.

Um diesen Übelständen zu begegnen, haben KAPTEYN und VAN RHIJN[1]) das folgende Verfahren vorgeschlagen: Die EB. der Plattensterne werden zunächst auf eine Gruppe von schwachen Sternen bezogen, die auch zur Bestimmung der Plattenkonstanten benutzt sind, deren mittlere relative EB. daher gleich Null ist und deren mittlere absolute Bewegung gleich ihrer mittleren parallaktischen Bewegung gesetzt wird. Diese letztere ist daher die Reduktion der relativen EB. der übrigen Sterne auf absolute EB. Wie die Autoren selbst sagen, hat dieses Verfahren seine Nachteile: Es setzt voraus, daß die Mot. par. der Vergleichssterne bekannt sind und daß ihre Peculiares sich aufheben. Es ist nicht fundamental und kann zur Ableitung der Sonnenbewegung nicht benutzt werden.

Da dieses Verfahren heute allgemein und vorbehaltlos zur Anwendung gelangt, sehe ich mich veranlaßt, hierauf etwas näher einzugehen, soweit der Raum es gestattet.

Die Bestimmung der Mot. par. setzt Kenntnis der Sonnenbewegung und der Säkularparallaxen voraus, also der Größen, die zu bestimmen unser Ziel ist. Wir bewegen uns also im Kreise. Freilich ergibt sich die Sonnenbewegung aus den Sternen mit bekannter EB. Aber genügt diese? In Publ. Groningen Nr. 29 führt die Diskussion zu den Resultaten:

1. In den bekannten Werten für die Geschwindigkeit der Sonne zeigt sich innerhalb der Parallaxen $0''0022$ bis $0''0243$ keine systematische Änderung, hingegen wohl für größere Parallaxen, die jedoch für die vorliegende Untersuchung belanglos ist.

2. Der Apex ist unabhängig von galaktischer Breite, EB. und Helligkeit der jeweils in die Untersuchung einbezogenen Sterne; es wird der Wert von WEERSMA[2]) $A = 17^{\mathrm{h}} 51^{\mathrm{m}}$, $D = + 31^\circ 4$ angenommen.

---

[1]) Publ. Groningen Nr. 28, S. 12, und Third report of the . . . selected areas, S. 28.   [2]) Publ. Groningen. 1908, Nr. 21.

Wenn man die für diese Resultate angeführten Argumente schon nicht als besonders beweiskräftig ansehen kann, so wird man sie auch auf Grund der neueren Forschungen vollends verwerfen müssen. Wir wissen heute: Jedes beliebig herausgegriffene Partialsystem führt zu einem anderen Apexwert; eine Vereinigung aller Werte zu einem Mittel ist nur möglich, wenn wir Richtung und Größe der Schwerpunkte der Partialsysteme kennen. Davon aber sind wir weit entfernt. Schon WEERSMA boten sich große Schwierigkeiten dar, die aus den verschiedenen Katalogen resp. Zonen sich ergebenden, stark voneinander abweichenden Apexdeklinationen miteinander zu vereinigen. Heute muß es als erwiesen gelten, daß die resultierenden Elemente der Sonnenbewegung abhängig sind von Spektraltypus und absoluter Größe, von der Größe der EB. und Radialgeschwindigkeit, von der galaktischen Breite, ja selbst von der AR. Ich verweise nur auf die Arbeiten von L. Boss[1]), DYSON und THACKERAY[2]), LUYTEN[3]), STRÖMBERG[4]) u. a.

Daß hier systematische Fehler im Spiele sind, unterliegt nach dem obigen keinem Zweifel; aber sie reichen nicht aus, es muß die Hypothese der Regellosigkeit der Sternbewegungen fallengelassen werden. Für die helleren Sterne, deren Bewegungen wir kennen, sind Vorzugsrichtungen erwiesen. Sollten sie nicht auch bei den schwachen und schwächsten Sternen vorhanden sein? Doch sicherlich! Mögen auch die Resultate COMSTOCKS[5]) noch wenig verbürgt erscheinen, so lassen doch neuere Arbeiten hierüber keinen Zweifel. Solche liegen vor vom Union Observatory[6]) (South Africa) und von SMART[7]). Nach diesen treten bei den Sternen der 12. Größe zwei Vorzugsrichtungen auf, die von denen der helleren Sterne merklich abweichen.

Damit erklären sich vollauf die großen Unterschiede in den Elementen der Sonnenbewegung.

Die für unser Problem besonders wichtige, in den größten Kreis Stern-Apex fallende Komponente v der EB. setzt sich hiernach zusammen aus der Mot. par., der Mot. pec., der Strombewegungskomponente, dem systematischen Fehler der EB. und dem aus der Unsicherheit des angenommenen Apex herrührenden Fehler. Daß die letzteren vier Bestandteile sich bei einer größeren Anzahl von Sternen aufheben, wird man füglich kaum annehmen können.

Und was heißt eine größere Anzahl von Sternen? Das ist ein Kapitel, über das sich manches sagen ließe, so eindeutig auch die Antwort ist. Die Statistik liefert nur zuverlässige Resultate, wenn ihr Massenerscheinungen zugrunde liegen; das wird nicht immer genügend beachtet. Ist die Bedingung erfüllt, dann kann man auch die Frage nach dem Ausschluß

---

[1]) Astr. Journal Bd. 28.   [2]) Monthly Notices Bd. 65 u. 79.
[3]) Lick Obs. Bull. Bd. 344.   [4]) Astroph. J. Bd. 56.
[5]) Astr. Journal Bd. 25 u. 28.   [6]) Circular of the Union Obs. Nr. 46.
[7]) Monthly Notices Bd. 84, H. 1 u. 3.

großer EB., die für unser Problem von Interesse ist, prinzipiell verneinen, wie KAPTEYN es tut, der aber bei der praktischen Durchführung seines Prinzips mehrfach auf Schwierigkeiten stößt, weil eben die Bedingung nicht erfüllt ist. Ebensowohl wie durch den Ausschluß Naturgesetze verdeckt werden können, wie KAPTEYN meint, ebenso gut kann auch, so behaupte ich, der umgekehrte Fall eintreten. Die Frage kann nicht prinzipiell, sondern sie muß von Fall zu Fall entschieden werden.

Wenden wir uns nunmehr zu den Säkularparallaxen! Die trigonometrischen Parallaxen, die in erster Linie von KAPTEYN [1]) und VAN RHIJN [2]) benutzt sind, erstrecken sich nur auf die nächste Umgebung der Sonne; zudem treffen beide Autoren noch eine Einschränkung, indem sie alle Sterne mit kleiner EB. ($< 0\rlap{.}''3$ resp. $0\rlap{.}''2$) ausschließen, dieser mit der Begründung, daß deren Parallaxen gegen die wahren, d. h. wohl gegen seine abgeleiteten, zu groß seien. Wenn er hierfür die Erklärung gibt, daß der Beobachter die positiven Werte begünstige, und um solche zu erzielen, die negativen Werte häufiger revidiere, so dürfte es ihm wohl schwer fallen, Belege hierfür zu erbringen. Im übrigen hat auch ein solcher Ausschluß nach Maßgabe der Größe der EB. seine Bedenken. Der Arbeit von HAAS[3]) entnehme ich, daß von 248 Sternen mit einer Parallaxe $> 0\rlap{.}''05$ 44 Sterne eine EB. $< 0\rlap{.}''3$ und 21 $< 0\rlap{.}''2$ haben. Das ist ein immerhin beträchtlicher Prozentsatz.

Aus der Diskussion der beiden genannten Autoren läßt sich ersehen, daß die trigonometrischen Parallaxen kein sehr zuverlässiges Material zur Lösung ihres Problems geliefert haben. Spektroskopische Parallaxen sind von VAN RHIJN nur zur Vergleichung mit seinen definitiven Werten herangezogen, mit dem Resultat, daß für die Riesen der späteren Spektralklassen die Bestimmung der absoluten Größen mit Hilfe der bislang benutzten Linien praktisch unmöglich sei, wogegen von SHAPLEY[4]) Widerspruch erhoben ist.

Bei dieser Sachlage müssen die EB. aushelfen. Soweit diese bekannt sind, also für die helleren Sterne, leiten KAPTEYN[5]) und seine Mitarbeiter aus ihnen für die mittleren Säkularparallaxen die Relation ab:

$$\log\left(\frac{\bar{h}}{\varrho}\right) = a + bm$$

Für die übrigen, schwächeren Sterne wird folgendes Verfahren eingeschlagen: Für zwei Gruppen von Sternen, 9. bis 10. und 12. Größe, werden aus den relativen Bewegungen, wie sie die Ausmessung der photographischen Platten ergibt, die relativen mittleren Säkularparallaxen bestimmt und damit die Änderung $\dfrac{d}{dm}\left(\dfrac{\bar{h}}{\varrho}\right)$ pro Größenklasse.

---

[1]) Publ. Groningen Nr. 8.   [2]) Publ. Groningen Nr. 34.
[3]) Veröffentl. Berlin-Babelsberg Bd. 3, Heft 3.
[4]) Harvard Bull. Bd. 788.   [5]) Publ. Groningen Nr. 29.

Aus der Differentiation der obigen Formel folgt:

$$\frac{d}{dm}\left(\frac{\overline{h}}{\varrho}\right) = \left(\frac{\overline{h}}{\varrho}\right) \cdot \frac{b}{\text{Mod}} \qquad \text{oder} \qquad \left(\frac{\overline{h}}{\varrho}\right) = -3.16\,\frac{d}{dm}\left(\frac{\overline{h}}{\varrho}\right)$$

wobei b für alle Größen als konstant $(= -0.137)$ angenommen wird. Aus den Beobachtnngen (l. c. Tafel 17) ergibt sich für die galaktische Zone $0°$ bis $\pm 20°$ und für $m = 10.3$: $\dfrac{d}{dm}\left(\dfrac{\overline{h}}{\varrho}\right) = +0.0056$, also $\left(\dfrac{\overline{h}}{\varrho}\right) = 0\rlap{.}''0177$. Hiernach würde also zu der obigen Formel für die schwächeren Sterne noch ein weiteres, vielleicht noch ein viertes Glied hinzutreten, das zu bestimmen aus dem vorliegenden Material nicht möglich ist. Deshalb beschränken sich wohl die Autoren unter Nichtbeachtung dieses Resultats darauf, ihre mittlere Säkularparallaxe nach der obigen Formel für alle Größenklassen bis 13.0 zu berechnen (Tafel 25); diese können somit nur als Extrapolationsprodukte angesehen werden, desgleichen auch die ihnen beigefügten w. F. Daß die extrapolierten mittleren Säkularparallaxen für Sterne 13. Größe bis auf $\pm 0\rlap{.}''0001$ verbürgt sein sollen, erscheint doch kaum glaubhaft, zumal sich schon gegen die Art der Ableitung der Konstanten a und b Bedenken erheben lassen, wovon ich jedoch an dieser Stelle absehen muß.

Van Rhijn gibt Publ. Groningen Nr. 34, Tafel 19 und 20, die mittleren Säkularparallaxen noch für die Sterne 14. Größe; auch diese beruhen nur auf Extrapolationen, die durch Beobachtungen nicht begründet sind.

Wir müssen hiernach schließen, daß auch das zweite Element, dessen wir zur Reduktion von relativen EB. auf absolute bedürfen, noch sehr unsicher bestimmt ist.

Nun liegen bereits Vergleichungen vor zwischen EB., die aus Meridiankreisbeobachtungen gewonnen sind, und photographisch bestimmten. In Publ. Groningen Nr. 28, Tafel 11, haben die Autoren eine solche gegeben, und zwar mit Boss PGC und einem Katalog von Gill. Nach Abzug einer unaufgeklärten konstanten Differenz zwischen Gill und Groningen im Betrage von $0\rlap{.}''038$ folgt aus den verbleibenden Resten der w. F. $\pm 0\rlap{.}''013$, und aus der Genauigkeit einer Position der drei Kataloge $\pm 0\rlap{.}''010$, nach der Ansicht der Autoren eine sehr gute Übereinstimmung. Ich will hierzu nur bemerken, daß:

$$\sqrt{0.013^2 - 0.010^2} = \pm 0\rlap{.}''008$$

ist, also gleich einer nicht zu unterschätzenden Größe, nämlich gleich der mittleren EB. der Sterne der 12. Größe.

Eine Vergleichung von Porter[1]) von 18 Sternen des PGC und 147, deren EB. aus allen einschlägigen Katalogen abgeleitet sind, mit photographisch bestimmten, offenbar relativen, gibt ein sehr unbefriedigen-

---

[1]) Astr. Journal Nr. 34.

des Resultat; die Differenzen selbst bei PGC-Sternen steigen mehrfach bis über $0''.1$. Die mittlere Differenz beträgt $0''.05$. Zweckmäßiger wäre es gewesen, wenn PORTER die EB. in jeder Koordinate gegeben hätte.

Zu einem gleich ungünstigen Resultat führt eine Vergleichung von SMART[1]). Eine Aufklärung darüber, welcher Methode die Fehler zur Last zu legen sind, wird voraussichtlich die nächste Zukunft bringen.

Zusammenfassung:

1. Positionen und EB. der FC. können den heutigen Ansprüchen nicht mehr genügen. Lediglich durch eine Bearbeitung des gesamten Materials von BRADLEY bis heute können die Fehler nicht beseitigt werden; neue fundamentale Beobachtungen, die von möglichst vielen Sternwarten nach einem einheitlichen Programm ausgeführt werden, sind dringend notwendig. Diese Arbeit sind wir der Nachwelt schuldig, denn ohne sie können sichere absolute EB. nie gewonnen werden.

2. Dieser Mangel der FC. gilt naturgemäß auch für die an sie angeschlossenen Kataloge; dazu treten die Unsicherheiten der Reduktionstafeln. Daher verdienen EB., die aus der Kombination solcher Kataloge abgeleitet werden, nur ein sehr geringes Vertrauen.

3. Photographische Aufnahmen zur Bestimmung relativer EB. sind durchaus erwünscht, denn nur dadurch wird die Grundlage geschaffen zur Ableitung von EB. der schwachen Sterne; sie können uns ferner Auskunft geben über die relative Bewegung gewisser Kategorien von Sternen gegen andere sowie über gröbere Fehler von einzelnen EB. der FC. Wieweit es möglich ist, mit ihrer Hilfe den Fehler $\varDelta\mu_\delta$ der EB. in Deklination der FC. nach dem Vorschlage KAPTEYNS zu bestimmen, ist eine offene Frage, die aber sehr der praktischen Nachprüfung zu empfehlen ist.

4. Die Reduktion relativer EB. auf absolute mit Hilfe solcher, die durch Meridiankreisbeobachtungen bestimmt sind, muß als unsicher bezeichnet werden. Dasselbe gilt für das von KAPTEYN und VAN RHIJN vorgeschlagene Verfahren, die Reduktion gleich der parallaktischen Bewegung der Vergleichssterne zu setzen.

Das Schlußresultat geht somit dahin: Unsere Kenntnisse der EB. sind noch sehr mangelhaft. Es ist ihnen als dem fundamentalen Element der Stellarstatistik in der nächsten Zukunft die größte Aufmerksamkeit zuzuwenden.

---

[1]) Monthly Notices Bd. 84, H. 3.

Die Sternleeren bei S Monocerotis.

Von **Max Wolf**, Heidelberg.

Mit 3 Tafeln und 2 Abbildungen.

1. Unsere Wintermilchstraße, wie sie sich über die Sternbilder Gemini, Monoceros, Canis maior usw. lagert, erscheint dem unmittelbaren Anblick recht verschieden von unserer Sommermilchstraße in Cygnus, Aquila, Scutum und Sagittarius. Hier ist sie ungemein abwechslungsreich, wie aus leuchtenden Wolken zusammengebaut und wieder in Fetzen zerrissen, während sie im Monocerosteil einen recht gleichmäßigen Eindruck erzeugt. Photographische Aufnahmen mit großem Bildfeld aus dem Winterteil erscheinen langweilig und strukturarm neben jenen aus dem Sommerteil. Das Auftreten so großer Kontraste von hellsten Wolken zu dunkelsten Leeren unmittelbar nebeneinander, wie in Cygnus und Scutum, fehlt hier; alles erscheint gleichmäßiger und weniger lichtstark. Die hauptsächlichsten Leeren liegen hier auffällig *neben* dem Milchstraßenzug in Taurus und Orion; man könnte fast meinen, die „dunkle Milchstraße" und ihre Nebel seien hier gegen die „helle Milchstraße" verschoben. Doch, das könnte es allein nicht ausmachen. Die Sternmassen sind wirklich hier weniger glänzend als im Winterarm. Besonders die Sterne der 9. bis 11. Größe scheinen hier weniger zahlreich zu sein als dort. Es ist merkwürdig, daß man den Grund dieses Unterschiedes im Aussehen noch nicht ganz sicher ermittelt hat, so augenfällig er ist. Die schwachen Sterne sind auch hier sehr zahlreich; z. B. in der im folgenden behandelten Gegend bei S Monocerotis sind sie zahlreicher als bei 52 Cygni, während allerdings die helleren in geringerer Anzahl vorhanden zu sein scheinen.

Die Tafel 1 gibt eine Tessaraufnahme von dem Winterarm der Milchstraße in Gemini und Monoceros vom 24. Januar 1909 bei 4 Stunden Belichtung. Das kleine Tessar hat 31 mm Öffnung und 145 mm Bildweite. Auf der Kopie sind die Sterne bis $12^{\mathrm{m}}0$ gut und bis fast $12^{\mathrm{m}}5$ noch als Spuren abgedruckt, so daß man annehmen kann, daß die Photographie eine Darstellung von der Verteilung aller Sterne bis zur 12. Größe gibt. Sie deckt eine Fläche von etwa $29 \times 36 = 1044$ Quadratgrad[1]).

Wenn auch die Milchstraße hier als fast gleichmäßiges Sternband von Norden (oben) aus Auriga durch Gemini herabkommt, so zeigt doch die genauere Betrachtung, daß auch hier in ihrer Mitte dunkle

---

[1]) Einem cm des Bildes entsprechen etwa $1^{\circ}5$ größten Kreises.

Kanäle ziehen, als ob sie die Zerschneidung in zwei Bänder anstrebten, ohne sie aber so gründlich zu erreichen, wie z. B. in Aquila-Cygnus und Cassiopeia. Erst weiter südlich und von der Orionseite her wird der Längsriß wieder stark. Hier in Monoceros ist er durch eine wenig auffällige Kette von vielverzweigten, fürs bloße Auge nicht wahrnehmbaren Kanälen vertreten. Etwas oberhalb (nördlich) der Mitte unseres Bildes finden wir sie rechts neben einem der hellsten Sterne der Aufnahme, γ Geminorum. Diesen und die anderen Hauptsterne des Bildes identi-

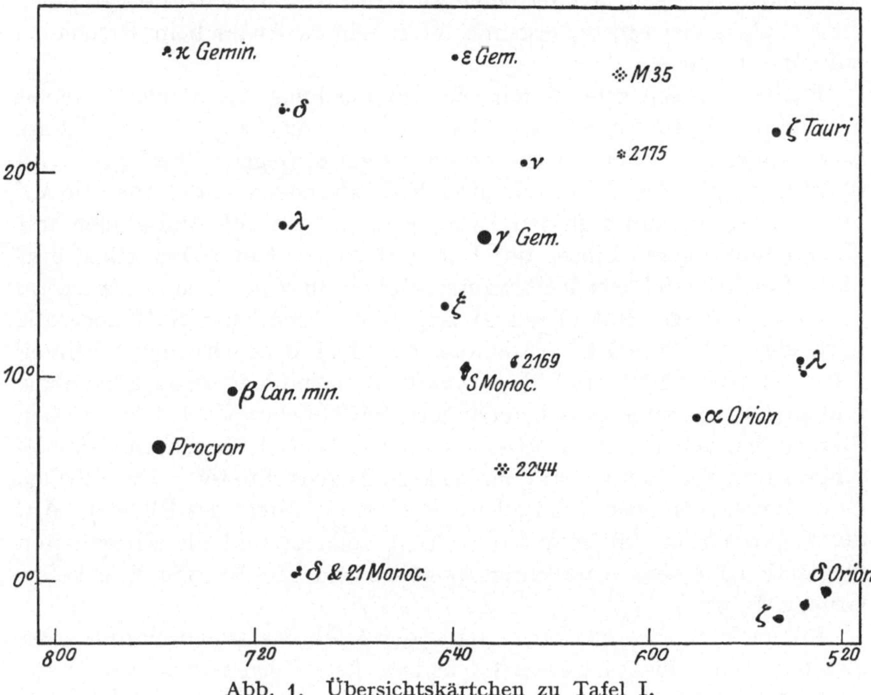

Abb. 1. Übersichtskärtchen zu Tafel I.

fiziert man aus dem hier abgedruckten Übersichtskärtchen (Abb. 1). Der Riß zieht in genau nordsüdlicher Richtung westlich von γ vorbei, wird nahe der Mitte unseres Bildes am merklichsten, gabelt sich in zwei Äste, deren westlicher mit der nebeligen Gruppe J₂ 2169 endet, während der östliche über die Nebelwolke von S Monocerotis (NGC 2264) sich verengend hinabreicht. Weiter südlich findet man ihn wieder erweitert und in zwei Arme gespalten. Der westliche Arm geht über den nebligen Sternhaufen NGC 2244 in den Orion, sich vielleicht mit den großen Dunkelwolken dort verschmelzend, während der südlich ziehende sich in ungefähr $6^{h} 50^{m} + 9°$ mit einem anderen sehr langen und dünnen Kanal zu einem kleinen „Kohlensack" vereinigt. Dieser lange, dünne, dunkle Arm kommt weit von oben aus Auriga, westlich neben λ Geminorum vorüberziehend.

Äußerst merkwürdig ist der vorhin genannte, am Sternhaufen 2244 vorüberziehende Arm. Es ist der nämliche, der, stellenweise nur wenige Minuten dick, in kontinuierlichem Zug α Orionis in größerem Abstand umzieht, so daß α Orionis wie mitten auf einem Schild aus schwächsten Sternen haftet. Er vereinigt sich dann mit jenem, der den „nebeligen Kreisfleck" um λ Orionis umgrenzt[1]). Dieser Fleck ist auf dem Bilde deutlich rechts am Rande zu erkennen.

Es ist ausgeschlossen, hier in Kürze eine Beschreibung der Kanäle und Nebel zu geben. Es sei bemerkt, daß man auf den Photographien die Kanäle am leichtesten erkennt, wenn man die Augen beim Beschauen nahezu schließt.

Recht lehrreich scheint mir die Untersuchung des „hellen" Nebels auf unserer Tessaraufnahme. Der mehrere Quadratgrad ausgedehnte zarte Schleier, der bei S Monocerotis die ganze Gegend überlagert, verdeckt hier mit seinem Licht die dunkle Straße so stark, daß man sie auf dem Tessarbild kaum finden kann, während sie auf Aufnahmen mit langbrennweitigeren Linsen durch ihre Sternarmut und Dunkelheit auffällt. Der helle Schleier liegt wahrscheinlich im Vordergrund der Leere.

Unser zweites Bild (Tafel 2) zeigt die Gegend um S Monocerotis nach einer $5^1/_4$stündigen Aufnahme mit dem Brucefernrohr (Öffnung 400, Brennweite 2030 mm). Die Einzelheiten sind hier besser erkennbar, und die Leeren treten zwischen den jetzt auffällig gewordenen schwachen Sternen deutlicher hervor, obwohl auch hier noch der leuchtende Nebelschleier ihre Verfolgung mit dem nackten Auge recht stört. Die nebelige Gruppe von S Monocerotis finden wir nahe der Mitte des Bildes[2]). Auf der Reproduktion (nicht auf der Originalplatte) sind die Einzelheiten ineinandergeflossen. Am rechten Rand des Bildes findet man die nebelige Gruppe J_2 2169.

Der helle Stern links oben, nahe der Ecke, ist ξ Geminorum. Der Abstand von S bis ξ ist etwa $3^1/_4°$. Der helle Stern rechts unten nahe der Ecke ist 13 Monocerotis.

Vergleicht man das Tessarbild mit dem Brucebild, so erkennt man große Unterschiede in der Auffälligkeit der Leeren; besonders in dem ganzen Gebiet nördlich von S. Kein Wunder, denn hier auf dem Brucebild treten die Sterne 10. bis 12. Größe weiter auseinander, so daß die Leere ihre Auffälligkeit dem Eindruck der schwächeren Sterne verdankt. Die Höhle um 2169 ist fast unkenntlich durch die leuchtende Masse, die sich ganz über sie gelegt hat. Ähnlich bei den nördlich von diesem Nebel befindlichen drei oder vier hellen Sternen. Deren südlichster (BD $+ 11°1204$; 7^m) liegt fast mitten in einer Leere, wo alle schwachen Sterne dezimiert sind. Man beachte auch das hübsche Dreieck aus schwachem Sternmaterial rechts oberhalb von S, das aus der Um-

<hr>

[1]) S.B. d. Heidelberg. Akademie d. Wiss. 1910, 3. Abh.
[2]) Vergl. auch d. Reflektorbild in A. N. 5303.

grenzung durch schmale Leeren entsteht. Hervorgehoben sei ferner die bekannte Regel in dem einseitigen Auftreten der leuchtendsten Nebelteile am Ende der tiefsten Leeren, welche auf die gerichtete Bewegung der verhüllenden Wolken oder des Vorganges, der sie erzeugt, hinweist.

Der knappe Raum verbietet ein Eingehen auf Einzelheiten. Es sollte hier nur eine Vorstellung von der Lage der zu untersuchenden Leere und der Großartigkeit des Milchstraßenbildes auch in diesem scheinbar armen Gebiet gemacht werden.

2. Ähnlich wie bei der Leere von 52 Cygni[1]) habe ich mir auch bei der Leere von S Monocerotis die Aufgabe gestellt, die Sternzahl außerhalb und in den Leeren nach verschiedenen Größenklassen getrennt zu ermitteln. Als geeignet zu diesem Zweck schienen mir die beiden Aufnahmen vom Brucefernrohr B 694 und 695, die ich am 18. Februar 1903 mit $5^h 18^m$ Belichtung erhalten hatte. Die Platten sind gut; speziell 694 zeigt die Sternscheibchen in den in Benutzung genommenen Teilen rund und recht scharf.

Auf den in Betracht kommenden Teil der Platte 694 wurde eine glasklare Kopie eines FUESSschen Millimeternetzes aufgepreßt, wobei dieses möglichst genau auf den Parallel von 1875.0 justiert wurde. Den Anblick dieser Anordnung gibt Tafel 3.

Die Umrisse der tiefsten Leeren hatte ich auf das Glas der Kopie der Meßplatte gemalt, so daß sie hier beim Abdruck als mattweiße Linien erkennbar werden. Das Netz ist von seiner Südostecke aus, als Nullpunkt des Zählung, nach oben (y) und nach rechts (x) nach Fünfmillimetern numeriert. Die Mitte der Originalaufnahme liegt in dem Fünfmillimeterquadrat $y = 9$, $x = 8$, etwas westlich von S Monocerotis, der durch drei Punkte in der Sternscheibe gekennzeichnet ist. Dieser Stern, eingebettet in den hellen Nebel NGC 2264, steht etwa in der Mitte der östlichen Leere nahe ihrer südlichen Verengung. Die größere, westliche Leere endet im Süden in dem hellen Nebel J_2 2169. Der durch zwei horizontallagernde Punkte gekennzeichnete Stern ist $+ 10°1159$. Die westliche Leere kommt in langem Zug von Norden herunter; der dort von ihr umschlossene Stern $+ 11°1204$ ist durch zwei aufrechtstehende Punkte markiert. Sie wendet sich unten in Ausläufern nach Osten und geht teilweise von Sternen überlagert in die östliche Leere über.

Meine erste Absicht war, beide Leeren zu zählen. Aus diesem Grunde habe ich im vergangenen Frühjahr sowohl die Gegend um S Monocerotis als auch diejenige um $10°1159$ durch Reflektoraufnahmen gleicher Zenitdistanz an die Harvardstandardgegend $9^h 0^m + 10°30'$ angeschlossen[2]). Leider läßt der Anschluß von $10°1159$ sehr viel zu wünschen übrig. Das Wetter erlaubte keine Wiederholung. Deshalb mußte ich

---

[1]) A. N. 5239.   [2]) Harvard Ann. Bd. 85, 1, S. 27.

darauf verzichten, aus der Aufnahme dieser 2—3 Grad von der Plattenmitte von B 694 abstehenden Gruppe die Abnahme der Sternzahl bezüglich der Größenklassen gegen die Feldränder hin zu ermitteln, und
habe mich deshalb bei der genauen Schätzung und Zählung auf die
Leere um S Monocerotis beschränkt, die so nahe der Plattenmitte liegt,
daß merkliche Fehler nicht entstehen können. Ich bin dabei nur soweit
aus der Mitte gegangen, als eine Abnahme der Sternzahl oder ein Fehler
in den Helligkeiten nicht nachweisbar war.

Es wurden drei verschiedene Abzählungen ausgeführt: a) eine rohe,
ohne Helligkeitsschätzung, bei der alle noch sichtbaren Sterne in jedem
Quadratmillimeter einiger quer durch die ganze Gegend ziehender
Streifen abgezählt wurden; das gezählte Feld deckte in 2375 Millimeterquadraten etwa 2 Quadratgrad. — b) eine Zählung über den größten
Teil des ganzen Feldes, aber nur bis zu den Sternen 11.5 Größe, wobei
die Helligkeit der Sterne so genau als möglich geschätzt wurde. Das
Feld betrug 12 000 Quadrate, die fast 10 Qnadratgrad deckten. —
c) eine genaue Zählung unter Schätzung aller Helligkeiten von ausgewählten Quadraten der sternreichen Gegenden nahe der Plattenmitte
und ebenso der sternärmsten Gegenden daselbst, in der Umgebung
von S Monocerotis. Das Feld deckte in 1430 Millimeterquadraten die
Fläche von etwas über einem Quadratgrad.

Die beiden Zählungen b und c wurden jeweils dreimal ausgeführt
und alle Abweichungen jedesmal nachgesehen und verbessert.

3. Zur Bestimmung der Helligkeiten ist, wie erwähnt, die nähere
Umgegend von S Monocerotis an die Standardgegend $9^h.0 + 10°.5$ angeschlossen worden. Die Reflektoraufnahmen vom 5. April 1924 von
je 40^m Belichtung mit auf 52 cm Durchmesser abgeblendetem Spiegel
sind um die Sternzeiten $9^h 21^m.7$ (Monoceros) und $11^h 38^m.1$ (Standard)
ausgeführt. Die mittleren Höhen waren $39°.2$ und $37°.7$, wonach ich
einen Extinktionsunterschied von $0^m.03$ anbrachte. Die in der Umgebung von S Monocerotis bestimmten Sterne habe ich rechtwinklig
ausgemessen. Ich gebe im folgenden ihre angenäherten rechtwinkligen
Koordinaten, bezogen auf den Stern Lpz II 3122 = BD $+ 9°.1333$, für
das m.Äquinox 1900.0. Dahinter stehen die aus der Standardgegend
aus Scheibendurchmessern abgeleiteten Helligkeiten (s. Tab. S. 317).

Diese Helligkeiten sind sicher wenig genau. Die Größen in. der
Standardgegend passen unter sich nach den Durchmessern auf der
Reflektoraufnahme recht schlecht; der m. F. einer Sternhelligkeit blieb
$\pm 0^m.25$ bei der Ausgleichung, obwohl die Durchmesser ganz geringe
Abweichungen hatten. Dazu kommt, daß die Zahl der Anschlußsterne
der Standardgegend (nur 20) zu gering ist. Weiter war die Zenitdistanz
(52—51°) zu groß, und die Bestimmung hätte an mehreren Abenden
wiederholt werden sollen. Das Wetter verhinderte es. — Für den vorliegenden Zweck dürfte trotzdem die Genauigkeit noch genügen.

x	y	Gr	x	y	Gr
+ 71″	+ 221″	7^{m}8	+ 919″	+ 294	13^{m}8
656	497	8.6	758	183	13.8 ?var.
+ 949	724	9.6	245	689	14.0
− 194	623	9.7	1030	617	14.0
+ 211	+ 206	10.1	+ 557	+ 663	14.0
+ 887	+ 487	10.7	+ 973	+ 659	14.2
968	480	10.9	361	183	14.3
200	545	12.2	+ 632	597	14.6
563	160	12.8	− 49	185	14.8
+ 461	+ 622	13.0	+ 532	+ 603	15.0
+ 445	+ 503	13.1	+ 286	129	15.0
190	103	13.1	+ 203	+ 462	15.5
764	240	13.2			
562	275	13.5			
+ 456	+ 154	13.7			

Die bestimmten Sterne wurden auf der Zählplatte (694) aufgesucht und aus ihren Durchmessern die mittleren Durchmesser für runde Größenklassen für diese Platte graphisch ermittelt. Diese Durchmesser, in Zehnteln der Strichkreuzarme des Stereokomparators, dienten als Normen für die Einordnung der gezählten Sterne, deren jeder unter das Strichkreuz gestellt, roh gemessen und in seine Größenklasse eingereiht wurde. Am unsichersten sind natürlich die Sterne 16. Größe abzugrenzen; schon aus dem Grunde, weil die Standardgegend nicht so weit reicht, also extrapoliert werden mußte. Es ist daher recht wahrscheinlich, daß ich zu schwache Sterne in die 16. Größenklasse hereingenommen habe.

Bei der Zählung wurde der Stereokomparator insofern gut ausgenutzt, als vor das linke Mikroskop die Platte B 695, vor das rechte die Platte B 694 (mit ihrem Netz) eingesetzt wurde, und so, bei der Zählung auf B 694, jeder Stern auf B 695 kontrolliert wurde. Ich glaube nicht, daß ein Stern zuviel gezählt worden ist; gegen das Übersehen eines Sternes schützte die dreimalige Wiederholung, die zu verschiedenen Zeiten vorgenommen worden ist, so gut als möglich.

Die Platte B 695 ist etwas schwächer als 694, gerade so viel, daß sie die Sterne 17. Größe von jenen der 16. Größe (meiner Skala) bequem auszuscheiden gestattete.

4. Die erste Zählung, die Übersichtszählung, die alle gut erkennbaren Sterne erfaßte, ging durch die folgenden Fünfmillimeterquadrate, deren Lage aus dem Bild (Tafel 3) zu entnehmen ist:

$$y = 25, \quad x = 1 \text{ bis } x = 20$$
$$y = 19, \quad x = 1 \;\;,,\;\; x = 20$$
$$y = 18, \quad x = 1 \;\;,,\;\; x = 5$$

$$y = 17, \ x = 1 \ \text{bis} \ x = \ 5$$
$$y = 16, \ x = 1 \ \text{,,} \ \ x = \ 5$$
$$y = 15, \ x = 1 \ \text{,,} \ \ x = \ 3$$
$$y = 14, \ x = 1 \ \text{,,} \ \ x = 20$$
$$y = \ 9, \ x = 1 \ \text{,,} \ \ x = 20$$

Das sind 95 Fünfmillimeterquadrate. Die Zählung ergab in diesem Feld:

     6064 Sterne in 44 völlig ungedunkelten Quadraten

     3756   ,,   ,, 41 angedunkelten Quadraten

     350   ,,   ,, 10 ausgesprochen sternleeren Quadraten

oder, auf Quadratgrade umgerechnet, Sterne auf 1 Quadratgrad:

     in Füllen: $B_m = 6865.02$, $\log B_m = 3.84$

     in Halbleeren: $B_m = 4563.28$, $\log B_m = 3.66$

     in Leeren: $B_m = 1743.42$, $\log B_m = 3.24$

Somit enthält die gezählte Fläche, bis zur Grenzgröße der Platten abgezählt, fast genau viermal soviel Sterne in den Füllen, als in den Leeren. Dabei sind aber tiefste Leeren und sternreichste Felder nicht streng genug geschieden, so daß das Verhältnis etwas zu niedrig herauskommt.

Die Grenzgröße ist natürlich unsicher. Ich schätze sie äußersten Falles als rund 17^m. Nach meiner Reihe von 52 Cygni[1]) käme der Logarithmus 3.84 auf die Helligkeit $16^m.8$ zu liegen. Aber nach Van Rhijn würde 3.84 etwa der Größe 18.6 entsprechen[2]). Das scheint mir nicht wahrscheinlich, wie der Vergleich mit länger belichteten Reflektoraufnahmen der Umgebung von S Monocerotis zeigt.

Entweder ist die Anzahl der schwachen Sterne viel größer, als gewöhnlich angenommen wird, oder aber, was recht leicht möglich ist, die Harvardskala paßt bei schwächeren Größen nicht mehr mit der Groninger zusammen.

Der Unterschied von $\log B_m = 3{,}24$ auf $\log B_m = 3.84$ bedeutet etwa 2 Größenklassen. *Der Dunkelnebel der Leeren würde demnach etwa 2 Größenklassen wegdunkeln.*

In den Halbleeren ist die abfangende Wirkung viel geringer. Sie entspricht im Durchschnitt dem Abfangen von etwas mehr als einer halben Größenklasse. Man hat also anzunehmen, daß die abfangende Wirkung durchaus nicht plötzlich einsetzt; die Sternabnahme erreicht nur allmählich ihren maximalen Betrag, wenn man von den Rändern gegen das Innere der Leeren fortschreitet. Im Sinne abfangender Wolken würde das besagen, daß die Wolken an den Rändern räumlich viel dünner, allmählich verlaufend, beschaffen sind, oder aber, daß die Dichte der Wolken von außen gegen ihre Mitten hin zunimmt.

---

[1]) A. N. 5239.   [2]) Groningen Publ. Bd. 27, S. 46.

5. Um möglichst große Sicherheit über das Verhalten der helleren Sterne zu erlangen, deren Zahl naturgemäß in so kleinen Bereichen für statistische Schlüsse viel zu gering ist, wurde ein großes Areal von rund 10 Quadratgraden speziell auf Sterne 6. bis 10. Größe durchgezählt. Diese Zählung ist dreimal zu verschiedenen Zeiten mit größtmöglicher Sorgfalt durchgeführt worden. Die abgezählte Fläche überdeckt folgende Fünfmillimeterquadrate:

$$y = 19, \quad x = 1 \text{ bis } x = 30$$
$$y = 18, \quad x = 1 \text{ ,, } x = 30$$

$$y = 4, \quad x = 1 \text{ bis } x = 30$$

Sie enthält den Hauptteil beider Leeren und nahezu eine gleich große Fläche von Füllen.

Die beiden Leeren fallen auf folgende Fünfmillimeterquadrate:

Östliche Leere:

$$y = 18, \quad x = {}^1/_2 7$$
$$y = 17, \quad x = {}^1/_2 6, 7, 8$$
$$y = 16, \quad x = {}^1/_2 6, 7, 8, {}^1/_2 9$$
$$y = 15, \quad x = 7, 8, {}^1/_2 9$$
$$y = 14, \quad x = 7, 8, {}^1/_2 9$$
$$y = 13, \quad x = 7, 8, {}^1/_2 9$$
$$y = 12, \quad x = {}^1/_2 6, 7, 8, {}^1/_2 9$$
$$y = 11, \quad x = {}^1/_2 6, 7, 8, {}^1/_2 9$$
$$y = 10, \quad x = 6, 7, 8, {}^1/_2 9$$
$$y = 9, \quad x = {}^1/_2 6, 7, 8, 9$$
$$y = 8, \quad x = 6, 7, 8, 9, {}^1/_2 10$$
$$y = 7, \quad x = {}^1/_2 5, 6, 7, 8, 9, 10$$
$$y = 6, \quad x = 5, 6, 7, 8, 9$$
$$y = 5, \quad x = 5, 6, 7, 8, 9$$
$$y = 4, \quad x = 5, 6, 7, 8, 9$$

das sind $51^1/_2$ Fünfmillimeterquadrate. Die westliche Leere umfaßt:

$$y = 19, \quad x = {}^1/_2 23, 24, 25, 26$$
$$y = 18, \quad x = {}^1/_2 25, 26$$
$$y = 17, \quad x = {}^1/_2 25$$
$$y = 16, \quad x = {}^1/_2 14, 15, {}^1/_2 16, 17-22, 24-26$$
$$y = 15, \quad x = {}^1/_2 14, 15, {}^1/_2 16, {}^1/_2 17, 19-25, {}^1/_2 26$$
$$y = 14, \quad x = {}^1/_2 15, 16, {}^1/_2 17, 18, {}^1/_4 19, {}^1/_4 20, 21-25$$
$$y = 13, \quad x = {}^1/_2 15, {}^1/_2 16, 18, 19, {}^1/_2 20, 21-25$$
$$y = 12, \quad x = {}^1/_2 15, {}^1/_2 16, {}^1/_2 21, 22, 23, {}^1/_2 24, {}^1/_2 25, {}^1/_2 26$$

$$y = 11, \quad x = \tfrac{1}{2}21,\ 22,\ \tfrac{1}{2}23,\ 26$$
$$y = 10, \quad x = 26,\ \tfrac{1}{2}27$$
$$y = \ 9, \quad x = \tfrac{1}{2}24,\ 25,\ \tfrac{1}{2}26$$
$$y = \ 8, \quad x = \tfrac{1}{2}24,\ \tfrac{1}{2}25$$

das sind 59 Fünfmillimeterquadrate.

Alle übrigen Quadrate fallen auf Sternfüllen.

Es wurden also gezählt $51\tfrac{1}{2} + 59 = 110\tfrac{1}{2}$ Leerenquadrate und $369\tfrac{1}{2}$ sternreiche Quadrate, zusammen 480 Quadrate. Jedes Fünfmillimeterquadrat hat 0,020075 Quadratgrad Fläche.

Die Zählung, nach Horizontalreihen bereits summiert, ist in folgender Tabelle zusammengestellt. Es bedeutet F = Fülle (sternreiche Gegend), L = Leere, OL = östliche Sternleere, WL = westliche Sternleere.

Es erhellt sofort aus der 4. Rubrik der Tabelle, daß die Sternzahl, wenn man die Sterne 6 bis 11^{m} zusammennimmt, in den *Leeren etwas größer* ist als in den Füllen. Das gleiche zeigt die 5. Rubrik für die Sterne $6^{m} - 9^{m}$; ebenso die drei nächsten Rubriken, wo die Summen für die Größen > 9, 9 und 10 aufgeführt sind. Bei der 11. Größe aber zeigt sich die Umkehr: Die Sterne 11^{m} sind in den Füllen zahlreicher als in den Leeren. Jedenfalls beginnt also die abfangende Wirkung der Leeren ungefähr bei den Sternen 11. Größe, ganz ähnlich wie bei den Leeren von 52 Cygni.

Es interessiert hier besonders die Feststellung, daß *die hellen Sterne in den Leeren zahlreicher sind* als in den Füllen. Sie ist von Bedeutung für das Verständnis der ganzen Erscheinung.

Sie gilt aber nicht für beide Leeren bzw. die ganze Fläche der Leeren. Behandelt man östliche und westliche Leere getrennt, wie es in den drei letzten Rubriken der Tabelle durchgeführt ist, dann erkennt man, daß die Anhäufung der hellen Sterne besonders in der Ostleere stattfindet. Wie man leicht erkennt, und was hier aus Platzmangel nicht weiter ausgeführt werden soll, findet die Anhäufung auch in der westlichen Leere statt, aber nur in ihrem südlichen Teil (analog wie in der Ostleere). Sie findet sich dort, wo die „leuchtende" Nebelmasse hervortritt. Auch der unmittelbare Anblick läßt erkennen, wie sich die aufleuchtenden Nebel je um eine Ansammlung hellerer Sterne drängen.

Die Erscheinung ist eigentlich nichts Neues. Sie ist uns bereits von anderen Sternleeren, bzw. Leerennebeln, bekannt, wie z. B. vom Orionnebel, Amerikanebel, dem Nebel bei 12 Monocerotis, bei π^{2} Cygni u. a. Aber bisher konnte sie meines Wissens nicht eindeutig durch Sternzahlen festgestellt werden, wie es jetzt hier möglich war.

Es scheint eben doch, als ob die Dunkelnebel nicht nur indifferent im Raum vor ferneren Sternen lagernde dunkle Wolken sind, sondern daß sie sich physisch an Sternansammlungen knüpfen (HERSCHEL). Die nächstliegende, allerdings nicht einfach zu lösende Aufgabe wäre

Anzahl der Sterne 6. bis 11. Größe in der Gegend von S Monocerotis.

y	F	L	OL	WL	F	L	F	L	F	L	F	L	F	L	F	L	OL	WL	OL	LW	OL	WL
	Anzahl der gezählten Quadrate				Anzahl der Sterne																	
					$6\text{--}11^{m}$		$6\text{--}9^{m}$		$>8^{m}\!5$		9^{m}		10^{m}		11^{m}		9^{m}		10^{m}		11^{m}	
19	26.5	3.5	0	3.5	30	2	1	0	0	0	1	0	5	0	24	2	0	0	0	0	0	2
18	28	2	0.5	1.5	34	3	6	0	1	0	5	0	6	1	22	2	0	0	1	0	0	2
17	24	6	2.5	3.5	32	8	2	1	1	0	1	1	6	0	24	7	1	0	0	0	1	6
16	17	13	3	10	20	13	1	0	0	0	1	0	7	5	12	8	0	0	2	3	1	7
15	17.5	12.5	2.5	10	21	8	2	0	0	0	2	0	5	1	14	7	0	0	0	1	2	5
14	18	12	2.5	9.5	23	7	5	0	0	0	5	0	5	3	13	4	0	0	0	3	1	3
13	19	11	2.5	8.5	17	11	0	3	0	0	0	3	3	3	14	5	3	0	0	3	1	4
12	22	8	3	5	29	12	1	2	0	2	1	0	10	2	18	8	0	0	1	1	1	7
11	24	6	3	3	33	6	3	3	0	0	3	3	7	1	23	2	1	2	0	1	2	0
10	25	5	3.5	1.5	31	5	4	1	0	0	4	1	5	1	22	3	1	0	1	0	1	2
9	24.5	5.5	3.5	2	32	13	2	7	0	3	2	5	6	2	24	4	4	1	2	0	1	3
8	24.5	5.5	4.5	1	26	14	2	7	0	0	2	7	9	2	15	5	6	1	1	1	4	1
7	24.5	5.5	5.5	0	22	12	1	2	1	1	0	1	1	3	20	7	1	0	3	0	7	0
6	25	5	5	0	29	11	6	4	0	1	6	3	8	4	15	3	3	0	4	0	3	0
5	25	5	5	0	31	10	2	0	0	0	2	0	8	5	21	5	0	0	5	0	5	0
4	25	5	5	0	36	7	5	1	0	0	5	1	6	4	25	2	1	0	4	0	2	0
Summe	369.5	110.5	51.5	59.0	446	142	43	31	3	7	40	25	97	37	306	74	21	4	24	13	32	42
Quadr.-Grd.	=7.42	=2.22	=1.03	=1.19																		
Sternzahl auf den Quadratgrad $\left\{\begin{array}{l} F: \\ L: \end{array}\right.$					60.1 / 64.0		5.8 / 14.0		0.4 / 3.2		5.4 / 11.3		13.1 / 16.7		41.3 / 33.4		20.3 / 3.4		23.2 / 11.0		31.0 / 35.5	

die, zu ermitteln, welchem Spektraltypus die in Betracht kommenden Sterne angehören, was für absolute Größen sie besitzen und ob sie besondere Temperaturverhältnisse verraten.

6. Die dritte Zählung umspannt ein viel kleineres Feld; dasselbe liegt aber fast ganz in der Mitte der Originalplatten. Es wurden die Sterne bis zur 16. Größe gezählt, während die (nach meiner Skala) der 17. Größe angehörigen Sterne möglichst „hartherzig" weggelassen wurden. Nur solche Quadrate wurden dafür ausgewählt, die einwandfrei ungetrübt schienen, und andererseits nur solche, die völlig sternarm waren. Ich glaube diese Forderungen durch die Auswahl folgender Quadrate erfüllt zu haben:

$$A. \ Füllen.$$

$$
\begin{aligned}
\text{Nr.} \quad 1-\ 4:\ &y = 15, \quad x = 1-4 \\
\text{,,} \quad 5-\ 8:\ &y = 14, \quad x = 1-3, \ {}^{1}/_{5}4, \ {}^{2}/_{5}4, \ {}^{2}/_{5}5 \\
\text{,,} \quad 9-12:\ &y = 13, \quad x = 1-4 \\
\text{,,} \quad 13-15:\ &y = 12, \quad x = 1-3 \\
\text{,,} \quad 16-18:\ &y = 11, \quad x = 1-3 \\
\text{,,} \quad 19-20:\ &y = 10, \quad x = 2,\ 4 \\
\text{,,} \quad 21-23:\ &y = \ 9, \quad x = 1-3 \\
\text{,,} \qquad 24\ :\ &y = \ 8, \quad x = 1 \\
\text{,,} \qquad 25\ :\ &y = \ 7, \quad x = 2
\end{aligned}
$$

Koordinaten (1875.0) der Mitte der Quadrate:

$$
\begin{aligned}
y = 15, \quad x = 1 :\ &\alpha = 6^{\mathrm{h}}37^{\mathrm{m}}9, \quad \delta = +10°51' \\
y = 15, \quad x = 4 :\ &\alpha = 6^{\mathrm{h}}36.2, \quad \delta = +10°51' \\
y = 11, \quad x = 3 :\ &\alpha = 6^{\mathrm{h}}36.8, \quad \delta = +10°17' \\
y = \ 7, \quad x = 2 :\ &\alpha = 6^{\mathrm{h}}37.3, \quad \delta = + \ 9°43'
\end{aligned}
$$

$$B. \ Leeren.$$

$$
\begin{aligned}
\text{Nr.} \quad 1,\ \ 2:\ &y = 16, \quad x = 7,\ 8 \\
\text{,,} \quad 3,\ \ 4:\ &y = 15, \quad x = 7,\ 8 \\
\text{,,} \quad 5,\ \ 6:\ &y = 14, \quad x = 7,\ 8 \\
\text{,,} \quad 7,\ \ 8:\ &y = 13, \quad x = 7,\ 8 \\
\text{,,} \quad 9,\ 10:\ &y = 12, \quad x = 7,\ 8 \\
\text{,,} \quad 11,\ 12:\ &y = 11, \quad x = 7,\ 8 \\
\text{,,} \quad 13,\ 14:\ &y = 10, \quad x = 7,\ {}^{4}/_{5}8 \\
\text{,,} \qquad 15:\ &y = \ 9, \quad x = 9 \\
\text{,,} \quad 16,\ 17:\ &y = \ 8, \quad x = {}^{3}/_{5}6,\ 9 \\
\text{,,} \quad 18-20:\ &y = \ 7, \quad x = {}^{4}/_{5}6,\ 8,\ 9 \\
\text{,,} \quad 21-24:\ &y = \ 6, \quad x = 5,\ 6,\ 8,\ 9 \\
\text{,,} \quad 25-28:\ &y = \ 5, \quad x = 5,\ 6,\ 7,\ 8 \\
\text{,,} \quad 29-32:\ &y = \ 4, \quad x = 5,\ 6,\ 8,\ 9 \\
\text{,,} \qquad 33\ :\ &y = \ 3, \quad x = 8
\end{aligned}
$$

Koordinaten:

$$y = 15, \quad x = 7 : \alpha = 6^h34^m.4, \quad \delta = +10°59'$$
$$y = 15, \quad x = 8 : \alpha = 6^h33.9, \quad \delta = +10°59'$$
$$y = 9, \quad x = 9 : \alpha = 6^h33.3, \quad \delta = +10°0'$$
$$y = 3, \quad x = 8 : \alpha = 6^h33.9, \quad \delta = +9°8'$$

Mehr Quadrate, die beiden Forderungen, zentraler Lage und Charakter, genügt hätten, ließen sich nicht finden.

In den folgenden beiden Tabellen sind die Resultate dieser Zählung zusammengestellt:

Sternzahl in Füllen:

Quadrat	9^m	10	11	12	13	14	15	16^m
1	0	2	0	5	4	10	25	54
2	0	0	1	3	5	13	16	80
3	1	0	0	1	3	8	15	78
4	0	0	0	0	3	4	24	64
5	0	0	1	1	2	8	21	71
6	0	1	0	0	2	9	20	81
7	1	1	0	0	12	7	20	86
8	0	1	1	2	7	10	38	72
9	0	0	1	0	6	14	29	76
10	0	1	1	4	6	11	19	75
11	0	0	0	2	5	12	26	73
12	0	0	1	4	6	5	21	75
13	0	0	0	2	4	10	22	74
14	0	0	1	1	4	9	25	55
15	0	0	2	3	5	8	16	74
16	0	0	0	0	3	8	20	77
17	0	2	1	1	4	6	19	72
18	0	1	0	3	2	5	16	86
19	0	0	1	0	5	14	17	58
20	0	1	0	1	7	13	17	71
21	0	0	1	2	7	12	26	87
22	0	0	1	2	11	12	21	70
23	0	0	1	6	9	16	19	54
24	0	0	1	2	9	13	21	71
25	0	0	0	2	5	11	28	69
Summe	2	10	15	47	136	248	541	1803
Auf 1 Qu.-Gr.	4	20	30	94	271	494	1078	3592
B_m	4	24	54	148	419	913	1991	5583

Die Gesamtzahl aller Sterne bis zur 16. Größe in den Sternfüllen ist 2802, was 5583.3 Sternen auf den Quadratgrad entspricht; dabei ist für die Größen 1—8 nur 0.4 zu ergänzen. Es wird also:

$$\log B_{16} = 3.747$$

Die Gesamtzahl der Sterne bis zur 16. Größe in den Leeren ist 554, wobei für die Größen 1—8 noch 3.2 Sterne pro Quadratgrad zu ergänzen

Sternzahl in Leeren:

Quadrat	9^m	10	11	12	13	14	15	16^m
1	0	1	0	0	1	2	4	12
2	0	0	1	0	2	1	4	4
3	0	0	0	1	1	2	5	9
4	0	0	1	2	0	2	1	8
5	0	0	0	1	3	1	4	6
6	0	0	1	1	1	1	2	6
7	0	0	0	2	1	1	1	1
8	2	0	0	0	2	1	0	4
9	0	1	0	0	1	1	0	5
10	0	0	0	2	1	2	8	11
11	1	1	0	2	2	0	2	1
12	0	2	0	1	2	2	4	13
13	0	0	0	3	2	3	3	4
14	1	1	0	0	0	1	1	6
15	0	0	0	0	0	1	2	4
16	0	1	0	0	0	2	5	5
17	1	0	1	0	1	3	1	7
18	0	0	0	0	1	1	3	12
19	0	2	1	2	1	1	5	11
20	0	0	2	0	2	2	6	10
21	1	0	1	0	1	3	5	10
22	1	0	0	1	3	3	3	15
23	0	1	3	2	0	3	5	10
24	0	1	0	0	3	3	0	9
25	0	0	1	2	2	2	5	12
26	0	0	1	1	0	1	3	5
27	0	1	3	0	2	1	2	4
28	0	1	0	2	0	4	3	15
29	0	0	2	0	2	2	4	10
30	0	1	0	1	1	2	5	11
31	0	0	1	1	3	3	3	14
32	0	0	1	2	1	0	3	9
33	0	0	0	2	1	1	2	13
Summe	7	14	20	31	43	58	104	276
auf 1 Q.-Grd.	11	22	31	48	67	90	161	427
B_m	14	36	67	115	182	272	433	860

wären, so daß die Anzahl aller Sterne bis zur 16. Größe in den Leeren
860.2 auf den Quadratgrad beträgt. Es wird also:

$$\log B'_{16} = 2.935$$

Der Quotient der Zunahme von Leeren zu Füllen ist daher:

$$6.5$$

für die Gegend bei S Monocerotis.

7. Um auch für die helleren Sterne eine möglichst einwandfreie Vergleichung der Sternzahlen zu erhalten, nehme ich für die Sterne von den
hellsten bis zur 11. Größe die Ziffern aus der obengegebenen Zählung

der Felder von 10 Quadratgraden zu Hilfe. Dabei nehme ich für die Sterne der 9. bis 11. Größe das Mittel aus jener Zählung und der eben besprochenen genauen Zählung, weil eben doch die genaue Zählung durch ihre günstigere Lage auf der Platte viel größeres Gewicht verdient. So ergibt sich die folgende Nebeneinanderstellung:

Füllen				m	Leeren			
$\log B_m$	$\log A_m$	B_m	A_m		A'_m	B'_m	$\log A'_m$	$\log B'_m$
3.75	3.56	5586.3	3592.4	16	427.0	856.8	2.63	2.93
3.30	3.03	1993.9	1077.9	15	160.9	429.8	2.21	2.63
2.96	2.69	915.9	494.1	14	89.7	268.9	1.95	2.13
2.63	2.43	421.8	271.0	13	66.5	179.2	1.82	2.25
2.18	1.97	150.8	93.6	12	48.0	112.7	1.68	2.05
1.76	1.55	57.2	35.6	11	32.1	64 7	1.51	1.81
1.33	1.22	21.6	16.5	10	19.2	32.6	1.28	1.51
0.71	0.67	5.1	4.7	9	11.1	13.4	1.04	1.13
—	9.60	—	0.4	>9	2.4	—	0.37	—

Vergleichung der Sternzahl in Füllen und Leeren.

Die Tabelle bestätigt die schon aus der Zählung der helleren Sterne gezogenen Schlüsse. Wir sehen auch hier, wie etwa von der 11. Größe

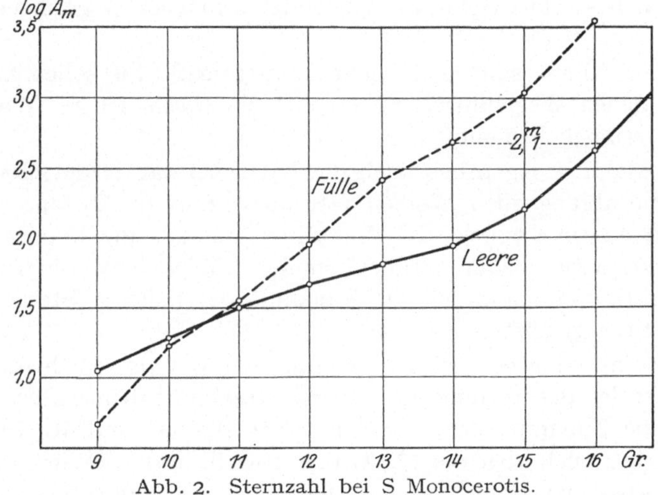

Abb. 2. Sternzahl bei S Monocerotis.

an die Abnahme der Sternzahl beginnt und bei der 14. bis 16. Größe auf etwas über 2 Größenklassen anwächst.

Wir sehen auch hier die schon oben besprochene merkwürdige Zunahme der Zahl der helleren Sterne *in* den Leeren.

Man erkennt die Verhältnisse am besten aus einer graphischen Darstellung des Logarithmus der Sternzahl als Funktion der Größenklasse. Auf dem Diagramm (Abb. 2) gibt die gestrichelte Kurve den Verlauf von $\log A_m$ für die Sternfüllen, die ausgezogene Kurve jenen für die Sternleeren $(\log A'_m)$:

Der Schnitt der beiden Kurven nahe der 11. Größe springt zuerst in die Augen. Ihre Parallelität bei den schwachen Sternen ist ziemlich sicher ausgeprägt. Bei der 14. bis 16. Größe beträgt die Abnahme der Sternzahl in den Leeren 2.1 Größenklassen.

Die Parallelität der Kurven bei den schwachen Größen spricht dafür, daß die Minderung der Sternzahl durch die lichtabfangende Wirkung einer dunklen Wolke verursacht wird. Andernfalls müßten sich die Kurven einander bei den schwächeren Sternen wohl wieder nähern. Wenigstens ist dies die einfachste Annahme.

Die Extinktion ist hier bei S Monocerotis viel größer als bei 52 Cygni. Sie beträgt hier fast 83% gegen 60% in jener Gegend. Die Extinktion nimmt aber hier auch langsamer zu; ihr Anwachsen verläuft über etwa 3 Größenklassen.

Die Dicke der abfangenden Wolke wäre also von diesem Betrage. Bei 52 Cygni ist es anders. Dort beträgt die Extinktion 1.0 Größen, bei einer Dicke von ebenfalls etwa 1.0 Größen, während wir hier bei S Monocerotis auf eine räumliche Ausdehnung von 3 Größen nur 2.1 Größen Extinktion finden. Die *Dichte* der Wolke von S Monocerotis wäre folglich — im Gegensatz zum ersten Eindruck — geringer als in Cygnus.

Solche Schlüsse sind noch recht unzeitgemäß. Sie sollen auch nur auf den neuen Weg hinweisen, der uns die *Dichte* solcher Wolken zu schätzen ermöglichen muß.

Interessant ist eigentlich auch die Tatsache, daß in der Monocerosgegend die abfangenden Wolken fast im nämlichen Abstand von uns beginnen, wie in Cygnus. In Monoceros scheinen die hypothetischen Massen ein wenig näher heranzukommen. Möglicherweise wird diese Abstandsbestimmung durch die Zunahme der helleren Sterne in den Leeren etwas gestört.

Ebenso auffällig wie bei der Zählung von 52 Cygni ist hier, wie wir bereits bei der Besprechung der Übersichtszählung bemerkten, die ungebührliche Zunahme der Sternzahl über die 15. und 16. (und, wie oben gezeigt, auch über die 17.) Größe, sowohl in den Füllen als in den Leeren. Aber man braucht sich deshalb noch nicht zu sorgen; denn die nächstliegende und wahrscheinlichste Erklärung ist die, daß die benutzte Größenskala unrichtig ist und gegen schwächere Größenklassen hin verschoben werden sollte.

Zusammenfassung: Es wird durch Helligkeitsschätzung und Zählung der Sterne bei der Sternleere von S Monocerotis gezeigt:

Die Minderung der Sternzahl hält sich über die schwachen Größen ziemlich konstant, was dafür spricht, daß die Leeren als scheinbar aufgefaßt werden könnten, vorgetäuscht durch ein im Wege liegendes lichtabfangendes Medium.

Die Extinktion beginnt vor den Sternen der 11. Größe einzusetzen. Ihr Höchstbetrag ist etwas über 2.1 Größenklassen.

Die Dichte der abfangenden Materie scheint in Monoceros, bei größerer Raumerstreckung, geringer zu sein als in Cygnus.

Die Zahl der Sterne 9. und 10. Größe ist in der Leere größer als außerhalb. Das beweist, daß ein physischer Zusammenhang zwischen der hypothetischen Wolke und einer Sterngruppe bestehen muß.

Problems of the O-Type Stars.

By **J. S. Plaskett**, Victoria B.C.

The O-type stars, although probably fewer in numbers than any other spectral class, are nevertheless exceedingly important in the accepted theory of stellar evolution and are the most interesting and at the same time the most enigmatic and puzzling of all the stars. The problems presented by the O-type stars are both numerous and difficult and though some advance has been made by a recent investigation of the writer, the results of which will be discussed in this monograph, the problems of their constitution and development still remain largely unsolved.

Under the title of the O-type stars two distinct classes, the emission and absorption O's are usually considered. The emission O-type or Wolf Rayet stars were the first discovered, and have attracted a great deal of attention and research without any very definite results. They are uniquely characterized among the stars by broad and diffuse emission bands due principally to the neutral atoms of hydrogen and helium and to the ionized atoms of helium, silicon, carbon, nitrogen and oxygen. No absorption and little continuous spectrum is present and the breadth and mixing of the bright bands makes accurate identification or measurement of radial velocity impossible so that they, with possibly also the novae, remain the most peculiar and enigmatic objects in the sky. The absorption O-type stars are characterized by generally broad but still fairly well measurable absorption lines of the same neutral and ionized atoms with occasional narrow emission of ionized helium λ 4686 and ionized nitrogen λ 4634, 4640 and were early grouped by Miss CANNON with the Wolf Rayet stars and recognized as being "earlier" in type than the B's. Although some similarities were previously recognized, it now seems to me desirable to include a third limited class of celestial objects, the planetary nebulae, with the O-type stars. It has been shown by WRIGHT and HUBBLE that their stellar nuclei are either absorption or emission O's and the similarity in distribution, mass and radial velocity with the absorption O's shown by my recent work fully justifies their inclusion with the two previous groups, each forming a sub-class of the O-type stars.

The most profitable method of discussing the problems of the O-type stars seems to be in briefly summarizing the known data about this most interesting group including the results obtained from my recent investigation as yet unpublished, and in referring to the problems in the order that they arise in this summary. Our knowledge about the O-type stars may be grouped and compared with the similar data of other spectral classes for this purpose preferably under the following heads — mass, motion, distance, distribution, luminosity, spectra, physical constitution and the relations between the three sub-classes of the group.

Mass. It has long been recognised from double star data that the average mass of the stars appears to be a function of the temperature, that the G-type are of about the solar mass, the F-type have two or three times, the A-type four or five, and the B-type about ten times the mass of the sun. LUDENDORFF's investigation of the mass ratios of spectroscopic binaries indicated that the masses of the O-type stars were probably considerably greater than the B's but as only two O-type binaries were available, the third he discussed, Boss 6142, being of B-type, this was somewhat uncertain. There are now available the mass ratios of 5 O-type binaries, the most remarkable being the double system 6°1309, of which the components have minimum masses of 75 and 63 times the sun. In the next most massive, 29 Can. Maj. the minimum mass of the bright component is about 40 times and the probable mass over 60 times the sun. The other three binaries have smaller mass ratios but there can be no reasonable doubt, from the evidence available, that the masses of the O's range from 10 to 80 times the sun with an average value of at least 40 times the solar mass.

This mass data refers only to the absorption O's as the velocities of the emission O's or Wolf Rayet stars are unmeasurable and we have no known means of determining their mass. It seems probable, however, as will be seen later from other considerations, that the emission are at least as massive as the absorption O's.

The investigations of CAMPBELL and MOORE have given us fairly definite ideas of the masses of the planetary nebulae or rather of their stellar nuclei, and, assuming a distance of 1000 light years, the masses of nine planetaries give values between 4 and 210 times with an average at least 50 times the sun. As it seems probable, as will be seen later, that the distance above given is considerably under-estimated, it seems certain that the masses of the planetaries are even higher than the absorption O's.

The abnormally high masses of these exceptional objects, the O-type stars, undoubtedly place them at the apex of the stellar evolutionary sequence and necessarily make them the hottest and brightest of the stars. EDDINGTON's investigations of the physical conditions in the interiors of the stars indicates that the radiation pressure outward in a star of 50 times the sun's mass is 85 per cent of the gravitational force inward and that this proportion rapidly increases with the mass. Such a massive body must hence be approaching a condition of instability and one of the most interesting of the problems of O-type stars is to investigate what effect this abnormal mass, and consequently high radiation pressure, has on the density and other physical conditions and whether it may be a contributory cause in inducing the emissive and planetary stages of these bodies.

Motion. The O-type stars are exceptional also in their motions as compared with the other spectral classes. The proper motions of the O-type stars are on the average only about one-hundredth of a second of arc per annum, much smaller than any other spectral class with the exception of the N-types which have proper motions of about the same order. In many cases the probable error of the determination is nearly as great as the measured proper motion, an indication that the true value is probably considerably lower than the measured. The proper motions of these bodies are hence almost immeasurably small and their distances must be correspondingly remote.

The residual radial velocities of the usual spectral types run about as follows, M variables = 35 km, S = 24, R = 21, N = 18, M = 17, K = 17, G = 15, F = 14, A = 11, B = 6.5. It would be expected, especially if any equipartition was present, that the very massive O's would have a lower velocity than the B's. The average radial velocity of 47 absorption O-types, all those within reach at Victoria, is, however, 27 km. per second and the residual radial velocity 25.5 km. The radial velocity of the emission O's is not measurable but the velocity of the planetaries as determined by CAMPBELL and MOORE is 30 km. An analysis of the space velocities of the absorption O-type stars based on radial velocities, proper motions and probable parallax gave a value of 53.5 km., practically exactly double, as it should be, of the radial velocities. The direction of the residual motion of the centroid is, however, nearly perpendicular to the galactic plane. This unusual and improbable result can probably be ascribed to errors in the very small proper motions of these stars and indeed the application of KAPTEYN's correction of $0''.013 \cos \delta$ to the proper motions in declination brings the residual motion nearly into the galactic plane. The discontinuity in the run of the residual velocities of the various spectral types between the O's and B's, which we have every reason to believe follow one another in the evolutionary sequence, is another outstanding problem of the O-type stars which will, however, be further discussed after a consideration of the distance and distribution.

Distance. While there are trigonometrical parallaxes for a few of the O-type stars, these values are so small and irregular and the method is admittedly so unreliable for such remote objects, that recourse must be had to statistical methods of obtaining the mean parallax. GYLLEN-BERG and HERTZSPRUNG by the parallactic motions of 17 O-type stars obtained a mean parallax of about 0″.0025, equivalent to a distance of 300 or 400 parsecs. This result appears to be vitiated by the inclusion of 5 southern stars of large parallactic motion and, if they are omitted, gives values in accord with the latest determination by the writer from the radial velocities and proper motions of 35 absorption O's. The proper motions are of very unequal weight and so the stars were divided into different groups and weighted, and the mean parallax determined from (a) the parallactic motions, (b) the peculiar motions and (c) a new method using a combination of both parallactic and peculiar motions. The mean parallax of 12 stars, mag. 2.9 to 5.2 is 0″.0013; of 14 stars, mag. 5.6 to 6.9 is 0″.0009 and of 9 stars, mag. 7.0 to 7.7 of poorly determined proper motions is 0″.0028. The mean parallax of 17 of the hotter "earlier" type O's, O 5, O 6, O 7, is 0″.0009 and of 18 "later" type, O 8, O 9, is 0″.0013. The mean parallax of the 35 O's is 0″.0011 equivalent to a distance of 900 parsecs, 3000 light years, and this mean distance would likely be slightly increased if due allowance were made for the less reliable proper motions.

The absorption O-type stars are hence nearly five times as far away from the sun on the average as the B-types and from seven to twelve or more times the distance of the other usual types. They are about three times the distance of the M and S variables and only approached in remoteness by the N's, deep red giant stars at the opposite extreme of stellar temperature and equally limited in number, which have about the same average distance as the O's. While no mean parallaxes have been obtained for the Wolf Rayet stars and the planetaries their similarity in constitution, distribution and motion seems to clearly point to the same order of distance as the absorption O's.

Distribution. The O-type stars both emission and absorption are closely confined to the plane of the galaxy, all except some southern Wolf Rayet stars in the Magellanic Clouds being within 18° of the galactic equator and the great majority within 8°. The distribution in galactic longitude is fairly uniform the only marked gap occurring at 0° to 20°, possibly explained by obscuring nebulosity. From considerations of the dispersion in absolute magnitude they appear to lie in an annular ring of mean radius about 900 parsecs, 3000 light years, with a probable breadth of 1000 parsecs and thickness of 300 parsecs.

The great distances of these stars as compared with the usual types seem to place them in a different region of the galactic system, four or five times as far removed as the B-types to which they most nearly conform in temperature, mass and other physical conditions. The reason for this discontinuity is not apparent and seems equally as inexplicable as the difference in the residual radial velocity previously discussed. A purely speculative hypothesis may, however, be tentatively advanced. If we imagine the galactic system to have been originally formed from an enormous spiral nebula in the manner so ably discussed by JEANS, it is possible that at a certain early epoch in the emission of the nebular condensations, some conditions of density or other factors in the nuclear peripheral ring may have produced condensations of greater than the average mass of 10^{34} grams, which may have attracted also a greater amount of neighbouring nebular matter, thus forming the very massive O-type stars. This may later have been followed by the normal conditions giving the B and other types. In the mean time, however, the O-types judging by analogy from VAN MAANEN's measured motions in severals spirals, will have moved out along the spiral arms with accelerating velocities, thus explaining the higher random velocities of the O's. The progression of velocity in the other spectral types which seem to occupy a different and rather limited region of space nearer the centre and are hence fairly well mixed, may reasonably be ascribed to equipartition. While this explanation of the discontinuity in velocity and distribution is only speculative, it at least serves to illustrate another problem of the O-typ stars.

Luminosity. The most reliable method of obtaining the absolute magnitude of the O-type stars is the usual process from the apparent magnitude and distance $M = m + 5 + 5 \log \pi$, but two auxiliary methods have also been used. Using the mean parallaxes previously given the mean absolute magnitude of 12 stars between 2.9 and 5.2 magnitude is $- 4.9$, of 16 stars between 5.6 and 6.9 mag. is $- 3.9$ and for 9 stars fainter than 7.0 but with uncertain proper motions is $- 0.45$. The mean absolute magnitude of the 17 O 5, O 6, O 7 stars is -4.1 and of the 18 O 8, O 9 stars is $- 3.4$, while the mean absolute magnitude of all 35 stars is $- 3.9$. Taking account of the uncertain proper motions of the fainter stars, the mean absolute magnitude of the absorption O-type stars has been conservatively placed as $- 4.0$ with a probable dispersion of about three magnitudes, definitely making this class much the brightest of the stars, only equalled by such exceptional objects as RIGEL and CANOPUS.

The second method is an appllication of EDDINGTON's theoretical treatment of giant stars based on considerations of the radiant energy, opacity, and average molecular weight of stars of known mass and gives an absolute visual magnitude of $- 1.9$ for an O-type star of temperature $20,000° K$ and of mass 10 times the sun and absolute magnitude $- 4\,7$ for star of mass 80. The average absolute magnitude is $- 3.7$, in fairly close agreement with the first method.

The third method depends on estimates of mass, density and surface brightness for these stars and can be simply calculated when these

factors have been determined. The mass has already been accepted as
10 to 80 times the sun. The density from data from eclipsing variables,
modified by considerations of the high radiation pressures in these
tremendously hot and massive bodies, probably lies between one-tenth
and one-hundredth the sun, with a preference for the lower value. The
surface brightness, assuming a temperature of 20,000° K, is calculated
from radiation laws as $- 3.5$ magnitudes compared with the sun.
RUSSELL's determination of the surface brightness of stars of different
types, with corrections applied from the interferometer measures of
stellar diameters, gives a value about $- 4.0$. The mean of these two
values $- 3.75$ mags. applied to masses from 10 to 80 times the sun gives
an average absolute magnitude of $- 3.2$ for density one-tenth and $- 4.9$
for density one-hundredth the sun, also in substantial agreement with
the adopted value of $- 4.0$.

The above values refer only to the absorption O's. The emission O's
if we assume, as seems probable from similarity of constitution, distribution and proper motions, the same order of distance are about three
magnitudes fainter while the nuclei of the planetary nebulae also assumed
at the same distance, are on the average nearly six magnitudes fainter
than the absorption O's. This decrease in total luminosity with so far
as can be judged, similar conditions of mass and temperature seems
only possible of explanation by the absorbing action of the emissive
matter in the Wolf Rayet stars and of the nebulous matter in the
planetaries. This curious condition will be again referred to but it
obviously forms another important and difficult problem of the O-type
stars.

Spectra. The classification of the absorption O-type spectra proposed by H. H. PLASKETT, which has been generally accepted, has been
adopted without change for the present investigation. The absorption
line O stars have been subdivided into 5 classes O 5 to O 9 which proceed
continuously with about the same gradation from the B's and form
a continuous sequence with the latter. Occasional narrow emission is
distinguished as usual by the suffix e to the type subdivision. The
distinguishing feature of these O-type spectra is the presence of lines
due to ionized helium and the ionized atoms of other non-metallic
elements and it may well be called a super-enhanced spectrum, representing the highest known stellar temperatures.

The emission O-type, Wolf Rayet spectrum is unique among stellar
spectra being distinguished by broad emission bands confused with and
overlapping one another, generally without either much continuous or
any absorption spectrum except for narrow stationary H and K which
probably take their origin outside the star. Very little progress has
been made in the elucidation of this mysterious spectrum and we still
know practically nothing of the physical cause of the broad emission.

However, some advance has been made by the discussion of recent measures of spectra, with different dispersions, of several of the Wolf Rayet stars. The majority of the emission bands have been identified, showing a similar constitution to the absorption O's. Using, as in the latter, the intensity ratios of several pairs of lines of ionized and neutral atoms, the spectra observed have been arranged in the order of excitation, provisionally divided into four sub-groups and shown to be generally similar in constitution and state of excitation to the absorption O's. Nothing, however, can be definitely said about the temperature when in emission, but it seems reasonably sure from the similarity in excitation that the underlying temperatures in the two classes are similar.

The planetary nebulae give of course an overlying emission nebular spectrum but their nuclei as shown by Wright and Hubble are either emission or absorption O-type stars, and their inclusion with the other two sub-classes seems justified.

Physical conditions. The probable mean density of the absorption O-types assumed above as between 0.1 and 0.01 times the sun, seems confirmed by the agreement of the absolute magnitudes thereby obtained with those directly computed but nothing can be said about the density of the Wolf Rayet stars or planetaries except that it is probably of the same order. The surface pressures as judged by the recent work of Fowler and Milne, Russell and others must be of the order of 10^{-4} atmospheres, entirely too low to explain the great breadth of the emission in the Wolf Rayet stars for which no probable cause has yet been adduced. The reversing layer temperatures of the absorption O's have been calculated from ionization theories by Saha, H. H. Plaskett and Fowler and Milne with rather widely differing values. For the different sub-types, O9 to O5, Saha gives $19\,000°$ to $24\,000°$, H. H. Plaskett $15\,000°$ to $22\,000°$ and Fowler and Milne $22\,000°$ to $30\,000°$. Another problem of the O-type stars would be to measure the spectral energy distribution in some of these stars and obtain by radiation laws an independent determination of the temperature.

Relations of three sub-classes. As has been indicated under the heading "spectra" the absorption and emission O-types are similar in constitution and both have been arranged in order of excitation by similar criteria, by the ratios of line intensities. It is impossible, however, to definitely say from these ratios whether the excitation of the emission stars is higher or lower than the absorption. Fortunately, however, there are four transition examples in which the broad emission is present in an apparently otherwise normal absorption star and it seems from these that the excitation of the emission is not higher than that of the absorption stars. The results given by the four available transition stars differ considerably among themselves. From their general character and the manner in which the absorption spectrum appears to be obscured and blotted out by emission it seems more likely that the WOLF RAYET stars are an accidental development in the normal course of evolution of the absorption O's than that this evolution is in the reverse order. It seems also that the emission may begin at any stage of the development, and is hence not a function of temperature. The transition examples further show that as the absorption lines become weaker the continuous spectrum also begins to weaken and both are approximately simultaneously superseded by the broad emission. It may be that this obscuration of the continuous spectrum is one of the causes of the smaller absolute magnitudes of the Wolf Rayet stars previously referred to. However, it is evident that the physical causes leading to the development of the one sub-class from the other and to the production of broad emission are among the principal problems of the O-type stars.

These problems are only added to when we consider the relations of the planetary nebulae to the other two sub-classes. The similarities in mass, velocity, distribution and spectrum form fairly convincing evidence of the intimate connection of the three classes but themanner and order in which they develope from one ano ther is still obscure.

It may perhaps be permissible to indulge in speculation about the development and the relations between the three subclasses of O-type stars, provided its tentative character is clearly kept in mind. The fundamental characteristic about all these bodies seems to be their exceptional mass, undoubtedly present in the absorption stars and planetary nuclei and probably present from their close relations with the others, in the Wolf Rayet. This characteristic may be the primary cause of the observed phenomena. Radiation pressure approaches equality with the gravitational force for stellar masses over 50 times the sun and conditions must be relatively unstable in such bodies near the surface.

It seems probable that this surface instability in the normally developing absorption O, in conjunction perhaps with some other disturbing cause, may easily give rise to the eruption of emissive material which gradually completely obscures the absorption and most of the continuous spectrum, in this way possibly accounting for their lower luminosity, and eventually resulting in a Wolf Rayet star. In similar manner the planetary nebulae, whose masses appear to be even greater, may be assumed to evolve from the most massive of the absorption and emission stars. Radiation pressure in such case may easily become overpowering, possibly assisted by the action of ionization, an important factor at such high temperatures. The interaction on the molecules at the surface layers of these repulsive forces, of the centrifugal forces of rotation and of the gravitational forces may lead, as recently discussed by Jeans, to the expanded and intricate but generally symmetrical forms of the planetary nebulae. The nucleus remains an absorption or emission O, with mass perhaps sufficiently reduced to be stable, while the surrounding matter emits the nebular spectrum by the excitation of the high temperature nucleus, offering no greater difficulties of explanation than such an object as the Orion Nebula. The low luminosity of the nuclear stars, at least six magnitudes fainter than the normal O's, and their relatively high intensity in the ultra-violet, require, for explanation by absorption, a peculiar selective effect of the surrounding nebulous matter.

The above speculations must not, however, be taken too seriously but may serve to point the way to possible lines of attack on the difficult problems of the constitution and relations of the O-type stars.

It seems worth while in conclusion to briefly discuss an auxiliary result of the recent investigation of the O-type stars. Narrow, sharp H and K lines of calcium and, wherever tested, sharp D lines of sodium are present in the spectrum of practically all the absorption O-type stars observed. Further the same lines always appear in the spectra of the Wolf Rayet stars, in which no stellar absorption lines occur, as soon as sufficient continuous or emission spectrum to provide the necessary differentiation is superposed.

The most striking characteristic of these sharp lines is, however, their stationary position showing that the ionized calcium and neutral sodium producing them is practically at rest in space relative to the local cluster, or in other words its velocity is practically always the reflex of the solar motion in that direction. Its residual radial velocity is hence zero, markedly different from the 25.5 km. per second of the O-type stars, while several individual examples show differences of 50 or 60 km. per second between the calcium velocities and the constant stellar velocities. The previous explanation, deduced from stationary calcium lines among the early B's, of a local calcium cloud surrounding the stars and moving with them through space must hence necessarily be abandoned, as the O-type stars are certainly moving in all directions with, in many cases, large velocities through the calcium vapour.

It seems necessary to suppose, therefore, that there are present throughout interstellar space widely, perhaps almost universally, distributed clouds of matter. These clouds must be exceedingly tenuous, else there would be appreciable general absorption, and must be practically at rest with respect to the stellar system. The stars are moving in all directions through these clouds but only the high temperature O and early B-type have sufficient exciting power to excite and ionize, just as HUBBLE showed in the nebulae, the neighbouring matter and produce the sharp stationary calcium and sodium lines observed.

Emission has not been seen, probably on account of the exceeding tenuity of the material and it has been further shown that the lines observed are the only ones of those likely to be excited, which can be readily observed by stellar spectrographs. It seems then that the presence of these tenuous stationary clouds throughout interstellar space has been firmly substantiated by this investigation of the O-type stars.

The problems of the O-type stars are, as have been clearly indicated in the preceding discussion, both numerous and difficult and while some progress has been recorded by recent work, it only appears to open out new and interesting avenues of research.

Die Durchmesser der Fixsterne.

Von **K. F. Bottlinger**, Berlin-Babelsberg.

Mit 1 Abbildung.

Über die Durchmesser der Fixsterne können wir mit wenigen Ausnahmen nur auf hypothetischem Wege etwas aussagen. Ja, eigentlich kennen wir nur den Durchmesser der Sonne direkt.

Bei den anderen Sternen haben wir prinzipiell zwei Möglichkeiten, die Durchmesser zu berechnen, 1. aus dem scheinbaren, dem Winkeldurchmesser, und der Parallaxe, und 2. bei den Bedeckungsveränderlichen aus der Lichtkurve und den Elementen der spektroskopischen Doppelsternbahn.

1. Die Bestimmung der Sterndurchmesser aus dem Winkeldurchmesser und der Parallaxe.

a) Mit Hilfe des MICHELSON-Interferometers ist es gelungen, bei bisher 4 Sternen die angulären Durchmesser direkt zu messen. Leider sind dies alles Sterne mit recht unsicherer oder so gut wie unbekannter Parallaxe. Die 4 Sterne sind in folgender kleinen Tabelle zusammengestellt. Unter ω, p, D sind der Winkeldurchmesser, die angenommene Parallaxe und der lineare Durchmesser, letzterer in der Einheit des Sonnendurchmessers, angegeben. Die weiteren Kolonnen werden weiter unten erklärt werden.

	Spektrum	ω	p	D	ω_c	c/T_i	c/T_c	T_i	T_e
α Orionis . .	cM2	0''.045	0''.010	480	0''.051	4.62	4.79	3100°	2980°
α Scorpii . .	cM2 + A	.040	.009	480	.045	4.65	(4.79)	3080	(2980)
α Tauri . . .	gK5	.021	.057	39	.035	3.83	4.47	3750	3200
α Bootis . .	gK0	.019	.108	19	.029	3.28	3.74	4350	3830

Die Werte für die linearen Durchmesser sind natürlich sehr unsicher, aber durchaus plausibel, zumal wir auch aus anderen Gründen wissen, daß Aldebaran und Arkturus Giganten, Beteigeuze und Antares aber Übergiganten sind.

b) Ein viel reichhaltigeres Material an Sterndurchmessern können wir aber erhalten, wenn wir die Sterne als schwarze Strahler ansehen. Sind h_* und $h_\odot$ die bolometrischen Helligkeiten, ϱ_* und $\varrho_\odot$ die linearen Durchmesser, T_* und $T_\odot$ die effektiven Temperaturen von Stern und

Sonne, und r die in astronomischen Einheiten gemessene Entfernung des Sternes, so ist:

$$\frac{h_*}{h_\odot} = \left(\frac{\varrho_*}{\varrho_\odot}\right)^2 \frac{1}{r^2} \left(\frac{c/T_\odot}{c/T_*}\right)^4$$

Logarithmisch schreibt sich diese Formel dann, wenn man $\varrho_\odot = 1$ setzt und statt r die Parallaxe p einführt:

$$\log \varrho_* = 0.2\, m'_\odot - 0.2\, m'_* - \log p + 2 \log c/T_* - 2 \log c/T_\odot + 5.311$$

m' bedeutet hier die bolometrische Helligkeit in Größenklassen. Unsere Helligkeitsmessungen ergeben gewöhnlich die visuellen oder photographischen Größen, und zwar auf das Zenith reduziert, d. h. mit der Extinktion der Erdatmosphäre bei senkrechtem Strahleneinfall behaftet. Die in der Formel benötigte Größe ist aber die auf den leeren Raum reduzierte Helligkeit. Es existieren verschiedene Tabellen, um die visuelle auf die bolometrische Größe zu reduzieren. Die neueste ist von BRILL gegeben[1]) und weicht nur wenig von den früheren ab. Ich habe noch die von EDDINGTON angewandte[2]) benützt. Für die Sonne ist die Korrektion $m_{\mathrm{bol}} - m_{\mathrm{vis}} = 0$ gesetzt. Bei höheren wie bei tieferen Temperaturen ist die bolometrische Helligkeit größer als die visuelle. Die Temperaturen der Sterne aus der spektralen Energieverteilung sind von verschiedenen Beobachtern gemessen worden, vor allem von WILSING[3]) im Gebiet der visuellen Wellenlängen und von ROSENBERG[4]) auf photographischem Wege. Beide Temperaturskalen gehen für die frühen wie die späten Typen sehr erheblich auseinander, während sie in der Mitte übereinstimmen. Eine Neureduktion von BRILL[5]), sowie die Ionisationstheorie von SAHA haben hier einige Klarheit geschaffen. Die Genauigkeit der Einzelbestimmungen der Sterntemperaturen ist ziemlich gering. Sehr viel größere innere Genauigkeit erreichen wir bei der Messung von Farbenindizes, die aber nur durch Vergleich mit den ebengenannten eigentlichen Temperaturbestimmungen Rückschlüsse auf die Temperatur gestatten. In einer Arbeit „Lichtelektrische Farbenindizes von 459 Sternen"[6]) habe ich von 104 Sternen, deren Farbenindizes lichtelektrisch gemessen waren und deren Parallaxen zugleich bekannt waren, die Durchmesser berechnet. Die Temperaturen hatte ich in der Weise aus

---

[1]) Veröffentlichungen der Universitätssternwarte Berlin-Babelsberg Bd. 5, H. 1, S. 17.

[2]) Zeitschr. f. Physik Bd. 7, S. 351.

[3]) Publikationen des Astrophysikalischen Observatoriums Potsdam Nr. 74 u. 76.

[4]) Abh. der Kaiserl. Leop. Carol. Deutschen Akademie der Naturforscher Bd. 51, Nr. 2: Photographische Untersuchung der Intensitätsverteilung in Sternspektren.

[5]) Spektralphotometrische Untersuchungen. Astr. Nachr. Nr. 5222—5223, 5234, 5254.

[6]) Veröffentlichungen Berlin-Babelsberg Bd. 3, H. 4.

den Farbenindizes gefunden, daß ich ein System suchte, in dem die Temperaturen der heißesten Sterne der Theorie von Saha angepaßt waren, während sie für die kühlsten mit den von Wilsing gemessenen übereinstimmten, die für α Orionis nahezu den gleichen Durchmesser ergaben, wie er von Pease mit dem Michelson-Interferometer gemessen worden war. Diese Temperaturen stimmen gut mit den neuerdings von Brill gefundenen überein. Mein Bestreben war dann, das Material noch weiter zu vermehren, indem ich von allen Sternen mit bekannter Parallaxe die Temperatur aus dem Spektrum entnahm. Dem Unterschied in Farbenindex und Temperatur zwischen Riesen und Zwergen gleichen Spektraltyps wurde natürlich Rechnung getragen. Besonders bei den Zwergen ist die Temperatur in hohem Grade eine eindeutige Funktion des Spektraltypus. Für die roten Zwerge vom Spektraltypus M wurden die gelegentlich gemessenen effektiven Wellenlängen mit denen der Riesensterne gleicher effektiver Wellenlänge verglichen und so die Temperaturskala hier extrapoliert, was natürlich erhebliche Unsicherheit mit sich bringt. Ferner gelang es mir durch Hinzuziehung von über 80 Riesensternen, deren absolute Helligkeit ich aus einem schwächeren Doppelsternbegleiter geschätzt hatte[1]), das Material bezüglich der Riesensterne erheblich zu vermehren. Ebenso wurden neben Gruppenparallaxen auch die hypothetischen Parallaxen der Cepheiden von Shapley in umfangreichem Maße angewandt.

Wird die Größe der Sonne zu $-26^{\mathrm{m}}\!.9$ und $c/T_{\odot}$ zu 2.20 angenommen, so wird die Formel für:

$$\log \varrho_* = 2.25 - 0.2\,m'_* - \log 1000\,p + 2\log c/T_*$$

(Die Formel für den Winkeldurchmesser ω, aus der die Parallaxe p herausfällt, wird entsprechend:

$$\log 1000\,\omega = 0.221 - 0.2\,m'_* + 2\log c/T_*)$$

Vergleicht man die vier interferometrisch gemessenen Sterndurchmesser mit den auf Grund des Farbenindex berechneten, so findet man, daß für α Orionis und α Scorpii die Werte ziemlich gut übereinstimmen, wogegen bei α Bootis und besonders bei α Tauri die Abweichungen groß sind, und zwar in dem Sinne, daß die interferometrischen Werte kleiner sind. In der Tabelle auf S. 338 ist unter ω_c der aus dem Farbenindex berechnete Winkeldurchmesser, ferner unter c/T_c und T_c die verwandten, unter c/T_i und T_i die aus dem *gemessenen* Winkeldurchmesser berechneten Temperaturwerte angegeben. Es sei bemerkt, daß die von Saha angegebenen Temperaturen noch höher liegen, für α Orionis bei 5000°.

Diese Abweichung zwischen gemessenem und errechnetem Durchmesser ist insofern auffallend, als eigentlich bei den extremen Typen

---

[1]) Atti della Pontificia Accademia delle Scienze Nuovi Lincei Anno LXXVII 1924.

und nicht bei den dazwischenliegenden die größten Abweichungen zu erwarten wären. Auch durch keine empirische lineare oder quadratische Formel wäre zu erreichen, daß c/T den drei Spektraltypen Sonne (G), Arkturus und Aldebaran (K), Beteigeuze (M) angepaßt würde, ohne bei den früheren Typen unmögliche Werte zu liefern. Wenn man nicht annehmen will, daß die Abweichung auf interferometrischen Messungsfehlern beruht, so kommt man zu dem Schlusse, daß das Strahlungsgesetz bei den späten Typen nicht mehr stimmt. Die Lösung wird vielleicht darin zu suchen sein, daß wir das c/T-System den K-Typen (α Tauri und α Bootis) anzupassen haben, da bei den M-Typen infolge des Auftretens der Titanoxydbanden die spektrale Energieverteilung sehr stark von der des schwarzen Körpers abweicht. Auch die Rückgängigkeit der lichtelektrischen Farbenindizes der späten M-Typen (Mb und Mc), worauf ich in meiner Farbenindexarbeit hinwies, beruht darauf, daß die Titanoxydabsorptionsbanden hauptsächlich in dem Wellenlängengebiet liegen, das bei den Gelbfiltermessungen wirksam ist. Es ist demnach nicht unwahrscheinlich, daß die auf Grund des korrektionsbedürftigen c/T-Systems errechneten Durchmesser für die M-Typen ziemlich richtig sind, da sich hier die Fehler im System und die Abweichung vom Strahlungsgesetz aufheben, wogegen die für die K-Sterne ermittelten Werte um beiläufig 35% zu groß sind.

2. Eine ganz andere Möglichkeit und frei von der Strahlungshypothese wie von Parallaxenmessungen zur Bestimmung der Fixsterndurchmesser bieten uns in einzelnen Fällen die Bedeckungsveränderlichen. Hier erhalten wir nämlich nicht anguläre, sondern wirkliche lineare Durchmesser.

Die Gestalt der Lichtkurve gibt uns die Größe der sich bedeckenden Körper in Teilen der Halbachse und liefert zugleich noch einige Bahnelemente, darunter die Neigung i. Wenn von solchen Sternen nun beide Komponenten im Spektrum meßbar sind, man also die spektroskopische Bahn, d. h. auch die Größe $(a_1 + a_2) \sin i$, berechnen kann, so ergeben sich aus der Lichtkurve und der spektroskopischen Bahn die absoluten Werte von $(a_1 + a_2)$ und damit die Dimensionen der Körper. Diese Durchmesser haben den Vorteil, daß sie in höherem Grade von Hypothesen frei sind als die aus dem Farbenindex berechneten. Sie beruhen erstens auf der Bedeckungshypothese. Wenn auch die aus der Lichtkurve abgeleiteten Werte in manchen Fällen mit beträchtlichen Fehlern behaftet sein mögen, so kann man doch in den meisten Fällen die Bedeckung als so gesichert betrachten, daß von einer Hypothese nicht mehr die Rede sein kann. Ferner beruhen die Durchmesser auf der Annahme, daß die Radialgeschwindigkeitsschwankungen nur oder nur im wesentlichen durch Bahnbewegung verursacht werden. Bei der Mehrzahl der in folgendem benutzten Sterne kann man das mit Bestimmtheit annehmen. Leider gibt es verhältnismäßig wenig Bedeckungsveränderliche, in deren Spektren beide Komponenten meßbar sind. Sehr verdienstvoll sind Untersuchungen von J. S. Plaskett in Victoria (Kanada) in dieser Hinsicht. Ich habe vor kurzem eine kleine Zusammenstellung über 17 solcher Doppelsternpaare veröffentlicht[1]), 12 Paare können davon nach meiner Ansicht als gesichert gelten, bei 5 Paaren ist die schwächere Komponente recht unsicher bestimmt, so daß im ganzen bisher 29 individuelle Sterndurchmesser zuverlässig angebbar sind.

3. In einem beigefügten Diagramm habe ich sämtliche von mir als diskutabel betrachteten Sterndurchmesser als Funktion des Spektraltypus eingetragen. Der besseren Darstellung wegen wurde als Ordinate nicht der Durchmesser ϱ, sondern die Größe $\log \varrho$ gewählt. Das Diagramm hat naturgemäß große Ähnlichkeit mit dem Russell-Diagramm, nur daß der Riesenast nicht wie dort horizontal liegt, sondern die Durchmesser von den frühen zu den späten Typen ansteigen. Beim Zwergast dagegen ist der Abfall geringer als im Russell-Diagramm. Es konnten so gegen 400 Durchmesser berechnet werden, und zwar:

a) 99 aus trigonometrischen und dynamischen Parallaxen. Der Wert von c/T stammt hier teilweise aus Farbenindexmessungen, teilweise wurde er nach dem Spektrum angenommen. Es wurden nur die als einigermaßen sicher geltenden Parallaxen verwertet[2]);

b) 39 aus Gruppenparallaxen (Taurus, Perseus, Ursa major, Plejaden);

c) 83 aus nach dem Spektrum eines nahen Begleiters geschätzten absoluten Helligkeiten;

d) 55 aus spektroskopischen Parallaxen;

e) 84 aus Cepheidenparallaxen nach Shapley[3]);

f) 2 Mittelwerte von 17 bzw. 24 c-Sternen, deren Farbenindizes gemessen und deren Säkularparallaxe aus den Eigenbewegungen bestimmt wurde[4]);

g) 29 aus Bedeckungsveränderlichen.

Ursprünglich hatte ich beabsichtigt, die spektroskopischen Parallaxen ganz aus dem Spiel zu lassen, weil einerseits für die Zwerge die trigonometrischen Parallaxen überlegen sind und andererseits für die Riesen die individuelle Trennung von normalen Exemplaren und Übergiganten nicht gelungen war. Neuerdings ist aber hierin ein großer Fortschritt zu verzeichnen. Unabhängig voneinander ist es den drei Sternwarten Mt. Wilson, Harvard und Victoria (Can.) gelungen, hier eine gute Trennung vorzunehmen. Die einzige bisher publizierte größere Liste

---

[1]) Atti della Pontificia Accademia delle Scienze Nuovi Lincei Anno LXXVII 1924.

[2]) Diejenigen Sterne, für welche bei Haas, Veröffentlichungen der Sternwarte Berlin-Babelsberg, das Gewicht der absoluten Größe G_m mindestens 4 ist.

[3]) Nach M. Güssow: Kritische Zusammenstellung sämtlicher Beobachtungsergebnisse der Veränderlichen vom δ Cepheitypus. Dissertation. Berlin 1924.

[4]) Meine Farbenindexarbeit, S. 41.

stammt von HARPER und YOUNG in Victoria und umfaßt ausschließlich
K- und M-Sterne[1]). Ich habe dieser Liste alle Sterne entnommen, deren

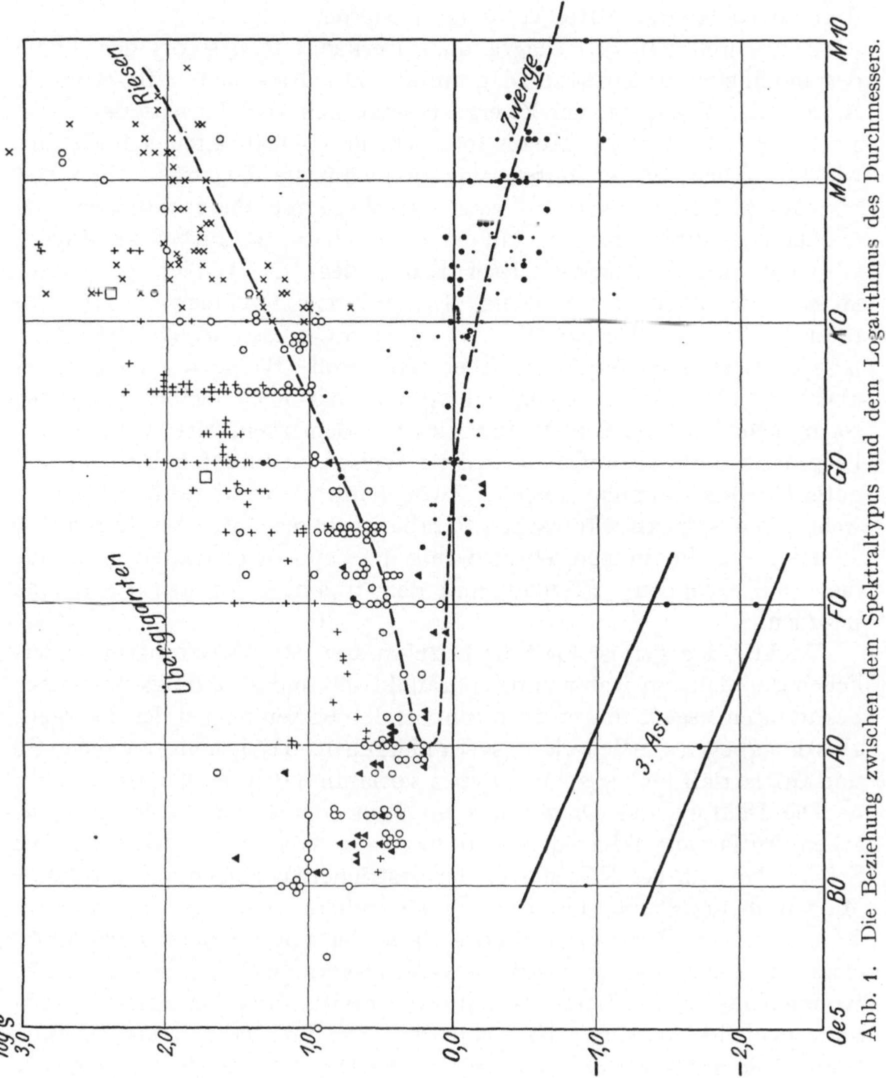

Abb. 1. Die Beziehung zwischen dem Spektraltypus und dem Logarithmus des Durchmessers.

Farbenindizes von mir gemessen worden waren, und konnte so 55 neue
Werte hinzufügen.

Im Diagramm bedeuten Dreiecke Bedeckungsveränderliche, Punkte
trigonometrische Parallaxen geringeren Gewichts, Vollkreise trigono-

---

[1]) Journ. of the Royal Astronomical Society of Canada Bd. 18, S. 49—54. 1924.

metrische oder dynamische Parallaxen größeren Gewichts, offene Kreise theoretische Parallaxen (Gruppen, Begleiter), × Kreuze spektroskopische Parallaxen, + Kreuze Cepheidenparallaxen. Die beiden Quadrate bedeuten die Mittelwerte für c-Sterne.

Betrachtet man das Diagramm, so erkennt man sofort den Typus des modifizierten RUSSELL-Diagramms. Der Riesenast trifft etwa bei A5 auf den Zwergast, der Zwergast setzt sich aber jenseits der Kreuzungsstelle bis über B0 hinaus fort. Die gestrichelte Linie soll die Entwicklungskurve eines Sternes von Sonnenmasse bedeuten. Von gG0 bis dK5 (G0-Riese bis K5-Zwerg) ist sie durch Beobachtungen hinreichend gestützt, darüber hinaus habe ich sie willkürlich verlängert. Oberhalb des Riesenastes dieser Kurve liegt das Gebiet der großen Massen, vor allem der c-Sterne. Die größten Durchmesser zeigen die roten c-Sterne 46 Aurigae (K2), μ Cephei, α Orionis und α Scorpii (M2). μ Cephei, bei dem alles für eine übermäßig große, Beteigeuze und Antares übertreffende absolute Helligkeit spricht, hat nach meiner Rechnung sogar mehr als 1200 Sonnendurchmesser. Bemerkenswert ist ferner am Diagramm, daß die auf verschiedene Weisen bestimmten Durchmesser gute Übereinstimmung zeigen. Zwar kann es nicht wundernehmen, wenn die aus spektroskopischen Parallaxen berechneten Werte mit den anderen übereinstimmen, ebensowenig die Cepheidenparallaxen, da alle diese auf Gruppen-, Säkular- und den trigonometrischen Parallaxen beruhen.

Wichtig dagegen ist, daß die Durchmesser der Bedeckungsveränderlichen durchaus im Rahmen der parallaktisch und strahlungstheoretisch bestimmten liegen, und zwar sowohl bei den Riesen als bei den Zwergen; allerdings liegen alle Bedeckungsveränderlichen im Diagramm zwischen B0 und G0, so daß für die späten Typen vorläufig noch die Kontrolle fehlt.

Die Deutung des Diagramms im Sinne der LOCKYER-EDDINGTONschen Auffassung[1]) hat demnach folgendermaßen zu geschehen. Die Sterne oberhalb des Riesenastes der gestrichelten Kurve haben größere Massen als die Sonne, und zwar um so größere, je weiter sie von dieser Linie abstehen. Über die wirkliche Masse der c-Sterne können wir kaum etwas aussagen, da von den Bedeckungsveränderlichen nur die eine Komponente von β Lyrae c-Charakter besitzt und bei diesem Stern noch manches ungeklärt ist. Von den zwei Sternen mit den größten bisher bekannten Massen kennen wir die Durchmesser leider in keiner Weise. B. D. + 6°1309, der von PLASKETT entdeckte Oe5-Stern, von dem jede spektroskopische Komponente mindestens 70 Sonnenmassen

---

[1]) Die neue Auffassung von EDDINGTON (Monthly Notices Bd. 84, S. 308ff.), wonach die absolute Helligkeit eines Sternes bloß Funktion seiner Masse sein soll, und der von den theoretischen Physikern scharf widersprochen wird, soll hier nicht berücksichtigt werden.

besitzt (es ist nämlich $m_1 \sin^3 i = m_2 \sin^3 i = 70\odot$), hat sich leider mit ziemlicher Sicherheit als Nichtbedeckungsveränderlicher herausgestellt. Immerhin scheint er dem verlängerten Zwergast anzugehören, also relativ hohe Dichte zu besitzen. Von v Sagittarii mit dem zusammengesetzten Spektrum cF2 + B wissen wir bisher nur, daß $m_1 \sin^3 i = 260\odot$ und $m_2 \sin^3 i = 56\odot$ ist. Der B-Stern hat hier die kleinere Masse. Vielleicht würde eine genaue photometrische Untersuchung dieses Sternes interessante Resultate zeitigen. Umgekehrt kennen wir von den c-Sternen, deren Durchmesser hier berechnet wurden, keine irgendwie sicheren Massen, es sci denn, daß man die theoretisch aus der Pulsationstheorie abgeleiteten Cepheidenmassen als richtig annimmt, die mir allerdings durchweg als etwas zu niedrig erscheinen.

Die Sterne auf dem überstehenden Ende des Zwergastes von A5 bis Oe5 sind offenbar Übergiganten am Ende ihrer Laufbahn als c-Sterne. Daß sie sich im Diagramm hier drängen, wogegen der Übergigantenraum nur verhältnismäßig dünn besät ist, beruht zum Teil wohl auf der zufälligen Auslese, da fast die ganze Perseusgruppe an dieser Stelle (B0 — B8) vertreten ist, zum Teil aber auch darauf, daß sie häufiger sind, wahrscheinlich weil die Verweilungsdauer in diesem Entwicklungszustand größer ist als auf dem Riesenast. Grundsätzlich muß man diese Sterne zu den Zwergen und nicht zu den Riesen rechnen.

Der Winkel zwischen Riesen und Zwergen bei den späten Typen ist nur sehr dünn besetzt. Vermutlich handelt es sich hier meistens um Sterne geringerer Masse im Riesenzustand. Bei den Zwergen findet sich ein Stern (Spektrum K2, $\log \varrho = -1.1$), dessen Durchmesser auffallend klein ist. Es existiert aber nur eine Parallaxenmessung, und der Spektraltypus ist vielleicht noch nicht genügend gesichert. Es ist der Schnellläufer C. Z. 5^h243 mit $8''$ Eigenbewegung.

Das sonderbarste aber sind die vier Sterne frühen bis mittleren Spektraltyps unterhalb der Hauptmasse der Sterne des Diagramms. Der kleinste (F0, $\log \varrho = -2.1$) ist gar von kleinerer Dimension als die Erde. Von zwei dieser Sterne (Siriusbegleiter und o_2 Eridani B) kennen wir die Masse und können so die mittlere Dichte ausrechnen, die sich zu dem 6000- bzw. 28 000 fachen der Sonnendichte ergibt. Der Mirabegleiter (B0, $\log \varrho = -0.9$) ist vielleicht noch unsicher. Die Parallaxenmessungen gehen bei diesem Stern sehr auseinander.

Über die Deutungsmöglichkeiten habe ich mich in meiner Farbenindexarbeit (S. 30—33) eingehend geäußert. Entweder sind die kleinen Dimensionen reell oder die Strahlung dieser Sterne kann kein Temperaturleuchten sein. Letzteres ist unwahrscheinlich, da sowohl die Energieverteilung wie die Linienausbildung in diesen Spektren nur wenig von denen der gewöhnlichen Sterne abweichen. Im übrigen hat EDDINGTON ebenfalls, und zwar vom atomtheoretischen Standpunkt aus[1]), die hohen Dichten als diskutabel bezeichnet. Ich habe daher diese Sterne, die offenbar pro Volumeinheit häufig sind und etwa 10% aller Sterne ausmachen mögen, provisorisch als den *dritten Ast* neben den Riesen und Zwergen bezeichnet. Jedenfalls haben wir den Sternen dieses Typus alle Aufmerksamkeit zuzuwenden[2]).

Wenn wir von diesen noch zweifelhaften Exemplaren absehen, sowie von den hier gar nicht besprochenen Sternen der Spektraltypen R und N (Kohlenoxydsterne), so deuten doch diese und auch andere Untersuchungen darauf hin, daß bei der großen Masse der Sterne das Strahlungsgesetz wenigstens mit guter Annäherung anwendbar ist. Allerdings werden bei den späten Typen, vor allem bei M, merkliche Korrektionen erforderlich sein. Auf jeden Fall können wir uns von den Dimensionen einer sehr erheblichen Zahl von Sternen schon heute ein klares Bild machen.

---

[1]) Monthly Notices Bd. 84, S. 308 ff.

[2]) Ich möchte vorschlagen, diese Sterne im Gegensatz zu den Riesen und Zwergen als Liliput-Sterne zu bezeichnen und in der Spektralklassifikation das Präfix l zu verwenden. o_2 Eridani wäre dann lA0, der Siriusbegleiter lF0.

Über Strahlungsgleichgewicht und Hellig-
keitsverteilung der Sonnenphotosphäre.

Von **R. Emden**, München.

In einer grundlegenden Arbeit: „Über das Gleichgewicht der Sonnenatmosphäre" hat SCHWARZSCHILD zum ersten Male den Begriff des
Strahlungsgleichgewichtes qualitativ und quantitativ erfaßt und damit
der Statistik der Gasmassen nebst ihren Folgeerscheinungen neue Wege
geöffnet. Für die Sonnenatmosphäre im Strahlungsgleichgewicht findet
er folgende Helligkeitsverteilung: Wird die Lage eines Elementes der
Photosphäre im Abstand r vom Zentrum angegeben durch einen $\sphericalangle i$,
$\sin i = \dfrac{r}{R}$, R der Radius der Photosphäre, so ergibt sich für die Strahlung, die von diesem Elemente in Richtung Erde ausgeht, der Betrag:

$$F(i) = \frac{A_0}{2}(1 + 2\cos i) \tag{1}$$

A_0 bedeutet die gesamte, aus diesem Elemente pro Flächeneinheit
austretende Strahlung, also die Menge, welche die Solarkonstante bestimmt. Daraus ergibt sich unmittelbar die Helligkeitsverteilung:

$$F(i) = F(0)\tfrac{1}{3}(1 + 2\cos i) \tag{2}$$

Die Übereinstimmung mit den vorliegenden Messungen ist, wie aus
der unten angegebenen Tabelle hervorgeht, vorzüglich. Vermutlich
getäuscht durch diese glänzende Bestätigung hat SCHWARZSCHILD aber
nicht beachtet, daß Gleichung (1) nicht richtig sein kann. Denn wird
$F(i)$ über die ganze Scheibe gemittelt, so muß sich der Wert A_0 ergeben;
statt dessen liefert Gleichung (1) den Wert $\tfrac{7}{6}A_0$. Wenn ich es unternehme, diesen kleinen Schönheitsfehler zu beseitigen, geschieht es nicht,
um eine brauchbarere Formel zu liefern, sondern um auf ein Verfahren
aufmerksam zu machen, das bei Behandlung von Strahlung, die weder
parallel, noch schwarz diffus ist, sich nützlich erweisen kann. Ich
skizziere erst die SCHWARZSCHILDsche Ableitung.

Angesichts des großen Radius der Sonne und der geringen Mächtigkeit
der für den Strahlungsprozeß in Betracht kommenden atmosphärischen
Massen genügt es, ein ebenes Problem anzusetzen. Die z-Achse werde
von oben nach unten gelegt. Die Atmosphäre wird nach *allen* Richtungen von Strahlung durchsetzt, die sich in Richtung $+z$ und $-z$

zu Energieströmen B und A zusammenfassen lassen. Diese werden beim Durchsetzen der Strecke dz um den Bruchteil $k\varrho\,dz = k\,dm$ geschwächt und nach dem KIRCHHOFFschen Gesetze um den Betrag $k\,dm\,E$ verstärkt; E das Emissionsvermögen eines schwarzen Strahlers von der Temperatur dieser Schicht. So ergeben sich die Beziehungen:

$$\frac{dB}{dm} = -kB + kE$$

$$\frac{dA}{dm} = \ \ kA - kE$$

mit der Folge:

$$\frac{d(B+A)}{dm} = -k(B-A)$$

$$\frac{d(B-A)}{dm} = -k(B+A) + 2kE$$

Im stationären Zustande, also bei Strahlungsgleichgewicht, müssen absorbierte und emittierte Strahlungsbeträge gleich sein:

$$k\,dm(B+A) = 2\,k\,dm\,E$$

Also folgt bei Strahlungsgleichgewicht:

$$B - A = \text{const} = 2\gamma$$

und daraus weiter, wie leicht ersichtlich:

$$B = -k\gamma m + B_0$$
$$A = -k\gamma m + A_0$$
$$E = -k\gamma m + E_0 \qquad E_0 = \frac{B_0 + A_0}{2}$$

Die von außen in die Atmosphäre eindringende Strahlung B_0 ist selbstverständlich gleich 0 zu setzen; A_0 ist durch die meßbare Solarkonstante bestimmt. So ergibt sich $\gamma = -\dfrac{A_0}{2}$, und wir erhalten die schließliche Lösung:

$$\left.\begin{aligned} B &= k\frac{A_0}{2}m \\[2mm] A &= k\frac{A_0}{2}m + A_0 = \frac{A_0}{2}(2 + km) \\[2mm] E &= k\frac{A_0}{2}m + \frac{A_0}{2} = \frac{A_0}{2}(1 + km) \end{aligned}\right\} \qquad (3)$$

m die über dem Niveau z liegende atmosphärische Masse pro Querschnittseinheit, A_0 die gesamte pro Oberflächeneinheit sekundlich auswärts gesandte Strahlung.

Wir betrachten nunmehr die in einer *bestimmten Richtung* fließende Energie *isoliert* und bezeichnen mit $F(i)$ speziell die Strahlung, die unter einem $\sphericalangle\, i$ zur Richtung $-z$ aufsteigt. Durchsetzt *parallele* Strahlung eine Schicht dm senkrecht, so werde sie um den Bruchteil $k_p\, dm$ geschwächt; k_p wohl unterschieden von dem bisher benutzten k für die vorhandene *diffuse* Strahlung. Die Strahlung $F(i)$ wird dann geschwächt um den Bruchteil $\dfrac{k_p\, dm}{\cos i}$, für sie gilt also entsprechend oben die Beziehung:

$$\cos i\,\frac{dF(i)}{dm} = k_p\,(E - F(i))$$

Damit ergibt sich für die unter dem $\sphericalangle\, i$ austretende Strahlung:

$$F(i) = \int\limits_0^\infty E \cdot e^{-k_p\, m\,\sec i}\, k_p\,\sec i\, dm \tag{4}$$

worin für E die ermittelte Lösung (3):

$$E = \frac{A_0}{2}\,(1 + k\,m)$$

zu setzen ist. So ergibt sich die schließliche Lösung:

$$F(i) = \frac{A_0}{2}\left(1 + \frac{k}{k_p}\cos i\right) \tag{5}$$

Nun nahm SCHWARZSCHILD kurzerhand die vorhandene Strahlung diffus wie schwarze Strahlung an, für welche er:

$$k = 2\,k_p \tag{6}$$

bestimmte, und erhielt so seine Lösung:

$$F(i) = \frac{A_0}{2}\,(1 + 2\cos i) \tag{7}$$

Mitteln wir aber $F(i)$ nach Gleichung (5) über die Photosphäre, so ergibt sich:

$$\frac{1}{\pi R^2}\int\limits_0^R F(i)\,2\pi r\,dr = A_0\int\limits_0^{\frac{\pi}{2}}\left(1 + \frac{k}{k_p}\cos i\right)\sin i\cos i\, di = A_0\left(\frac{1}{2} + \frac{k}{k_p}\cdot\frac{1}{3}\right)$$

und daraus für $k = 2\,k_p$ der falsche Wert $\tfrac{7}{6}A_0$. Eine richtige Energiebilanz erfordert, wie leicht ersichtlich, die Beziehung:

$$k = \tfrac{3}{2}k_p \tag{8}$$

Sie kann direkt abgeleitet werden, indem man an Stelle schwarzer Strahlung, den wirklichen Verhältnissen Rechnung tragend, eine Strahlung behandelt, die nach ganz bestimmtem Gesetze diffus ist. Auf solche Verhältnisse habe ich bereits in einer früheren Arbeit: „Über Strahlungsgleichgewicht und atmosphärische Strahlung" hingewiesen. Ich behandle den hier in Betracht kommenden Spezialfall.

In eine ebene Platté von der Mächtigkeit M und dem Absorptionskoeffizienten k_p dringe parallele Strahlung J unter einem $\sphericalangle\,i$ ein. Dann ergibt sich:

$$\text{Durchgelassene Strahlung} = J\,e^{-\frac{k_p\,M}{\cos i}}$$

$$\text{Absorbierte Strahlung} = J\,(1 - e^{-\frac{k_p\,M}{\cos i}})$$

Damit kann das Strahlungsgesetz einer ebenen Gasschicht abgeleitet werden.

Ist diese zwischen zwei schwarze Flächen gleicher Temperatur mit den Emissionsvermögen $E = \pi J$ eingeschlossen, so erhält sie pro Flächeneinheit in Richtung i eine Strahlung $2\pi J \sin i \cos i\, di$, die wie angegeben, absorbiert und folglich nach gleichem Gesetze wieder emittiert wird. Also ergibt sich das Strahlungsgesetz:

$$2\pi J\,(1 - e^{\frac{k_p\,M}{\cos i}})\sin i \cos i\, di \qquad (9)$$

Eine isotherme Atmosphäre von genügender Mächtigkeit emittiert folglich schwarze Strahlung. Eine Schicht von der geringen Mächtigkeit $\varDelta m$ hingegen absorbiert und emittiert nach dem Gesetze:

$$2\pi J \cdot k_p \varDelta m \sin i\, di \qquad (10)$$

Nach dem Halbraum wird folglich ausgestrahlt die Menge:

$$2\pi J \cdot k_p\, \varDelta m$$

Dies ist zugleich der Bruchteil der ganzen Zustrahlung πJ, der absorbiert wird. Daraus folgt:

$$\text{Absorptionsvermögen } k\varDelta m = \frac{\text{Absorbierte Strahlung}}{\text{Auffallende Strahlung}} = 2k_p \varDelta m$$

also für *schwarze* Strahlung der Schwarzschildsche Ansatz:

$$k = 2\,k_p$$

Um den Wert k in dem hier in Betracht kommenden Falle zu bestimmen, haben wir anzunehmen, daß über der absorbierenden Schicht $\varDelta m$ eine Atmosphäre im Strahlungsgleichgewicht liegt, in welcher folglich nach Gleichung (3) das Emissionsvermögen proportional der durchstrahlten Masse m zunimmt. Wir setzen also $E \sim \dfrac{A_0}{2}km = J_1\,k_p\,m$ und erhalten nach leichter Überlegung:

$$\text{Auffallende Strahlung} = 2\pi J_1 \int\limits_0^M k_p^2\, m\, dm \int\limits_0^{\frac{\pi}{2}} \sin i\, e^{-\frac{k_p m}{\cos i}}\, di$$

$$= \frac{2\pi}{3}\left[1 - e^{-\beta}(1 + \beta - \beta^2) - \beta^3 \int\limits_{\frac{\pi}{2}}^{\infty} \frac{e^{-x}}{x}\, dx\right]$$

$k_p M = \beta$ gesetzt.

$$\text{Absorbierte Strahlung} = k_p\,\varDelta m\,2\pi\,J_1\int\limits_0^M k_p^2\,m\,dm\int\limits_0^\beta \frac{\sin i}{\cos i}\,e^{-\frac{k_p m}{\cos i}}\,d i$$

$$= k_p\,\varDelta m \cdot \pi J_1\left[1 - e^{-\beta}(1+\beta) + \beta^2\int\limits_0^\infty \frac{e^{-x}}{x}\,dx\right]$$

$$\left.\begin{array}{l}\text{Absorptions-}\\ \text{vermögen}\end{array}\right\}\,k\,\varDelta m = \frac{\text{Absorbierte Strahlung}}{\text{Auffallende Strahlung}}$$

$$= \frac{3}{2}\,k_p\,\varDelta m\,\frac{1 - e^{-\beta}(1+\beta) + \beta^2\int\limits_0^\infty \dfrac{e^{-x}}{x}\,dx}{1 - e^{-\beta}(1+\beta-\beta^2) - \beta^3\int\limits_0^\infty \dfrac{e^{-x}}{x}\,dx}$$

Ist der Wert von $k_p M$ nicht allzu klein, so können die von β abhängigen Glieder $= 0$ gesetzt werden, und wir erhalten mit genügender Genauigkeit wieder:

$$k = \tfrac{3}{2}k_p \tag{8}$$

Die Erkenntnis der Bedeutung des Strahlungsausdruckes als tragendes Prinzip beim Aufbau von Gaskugeln nötigt, den geometrischen Verhältnissen der Strahlungsvorgänge gebührend Rechnung zu tragen. Da man es in der Regel mit Strahlung zu tun hat, die weder parallel, noch schwarz diffus ist, ergeben sich bekanntlich außerordentliche Schwierigkeiten. In vielen Fällen ist es aber mit einer für die praktische Anwendung weitaus hinreichender Genauigkeit möglich, wie eben gezeigt wurde, so zu verfahren, daß man mit Hilfe eines vorderhand unbekannten, den geometrischen Mischungsverhältnissen Rechnung tragenden Absorptionskoeffizienten k die Temperaturverteilung (Verteilung von E) ermittelt und diesen dann durch den experimentell ermittelbaren Absorptionskoeffizienten k_p für parallele Strahlung ausdrückt. Die Abweichung der Strahlung von Parallel oder Schwarz wird so auf den Wert des Absorptionskoeffizienten abgeschoben. Ist z. B. die Verteilung der Strahlungsintensität $J,(E)$, bei paralleler Schichtung gegeben durch die Beziehung:

$$J = \sum_1^q J_q (k_q m)^q \tag{10}$$

so ergibt sich wie oben

$$\left.\begin{array}{l}\text{Auffallende Strahlung} = 2\pi\sum\limits_1^q J_q\dfrac{q!}{q+2}\\[2mm] \text{Absorbierte Strahlung} = k_p\,\varDelta m\,,\;2\pi\sum\limits_1^q J_q\dfrac{q!}{q+1}\end{array}\right\} \tag{11}$$

Der Quotient liefert das gesuchte $k \Delta m$. Ist speziell:

$$J = J_q(k_p m) \tag{12}$$

gegeben, so kann q durch eine Polytropenklasse n ausgedrückt werden. Bei horizontaler Schichtung und konstantem Werte von g ist:

$$m \sim p \sim T^{n+1} \sim E^{\frac{n+1}{4}}$$

Daraus ergibt sich $q = \dfrac{4}{n+1}$, und folgt:

$$k = \frac{2n+6}{n+5} k_p \tag{13}$$

Wächst n von 0 bis ∞, so ist:

$$\tfrac{6}{5} k_p < k < 2 k_p \tag{14}$$

$n = 3$ entsprechend $q = 1$ ergibt wieder $k = \tfrac{3}{2} k_p$.

Geht man mit dem nunmehr ermittelten Werte $k = \tfrac{3}{2} k_p$ in Gleichung (5) ein, so ergibt sich:

$$F(i) = \frac{A_0}{2} (1 + \tfrac{3}{2} \cos i) \tag{15}$$

und die Helligkeitsverteilung

$$F(i) = F(0) \tfrac{2}{5} (1 + \tfrac{3}{2} \cos i) \tag{16}$$

Nimmt man an Stelle von Strahlungsgleichgewicht Aufbau nach einer Polytropen n an, so ist in Gleichung (4) zu setzen $E \sim m^{\frac{4}{m+1}}$; wodurch sich die Helligkeitsverteilung ergibt:

$$F(i) = F(o) \cos i^{\frac{4}{1+n}} \tag{17}$$

Für die in mannigfacher Hinsicht ausgezeichnete Polytrope $n = 3$ ergibt sich ein Kosinusgesetz der Helligkeitsverteilung; für $n = \infty$, Isothermie entsprechend, folgt (selbstverständlich!) konstante Helligkeit.

In der folgenden kleinen Tabelle sind wiedergegeben die Helligkeitsverteilungen nach der von SCHWARZSCHILD gegebenen Formel:

$$F(i) = F(0) \tfrac{1}{3} (1 + 2 \cos i) \tag{I}$$

welche, wie oben gezeigt, eine unrichtige Energiebilanz liefert; nach der daraufhin korrigierten Formel:

$$F(i) = F(0) \tfrac{2}{5} (1 + \tfrac{3}{2} \cos i) \tag{II}$$

und schließlich für polytropen Aufbau nach der Klasse n

$$F(i) = F(0) \cos i^{\frac{4}{n+1}} \tag{III}$$

speziell für $n = 3, 5$ und ∞. Zum Vergleiche sind die mittels Thermosäule und Bolometer gewonnenen Ergebnisse der Strahlungsverteilung beigefügt. (Nach MÜLLER: Photometrie der Gesteine, S. 323.)

$\dfrac{r}{R}$	Messung	Strahlungsgleichgewicht		Polytroper Aufbau $n=$		
		I	II	3	5	∞
0,0	1,00	1,00	1,00	1,00	1,00	Konstante Helligkeit
0,2	0,99	0,99	0,99	0,98	0,99	
0,4	0,97	0,95	0,95	0,92	0,94	
0,6	0,92	0,87	0,88	0,80	0,86	
0,7	0,87	0,81	0,83	0,71	0,80	
0,8	0,81	0,73	0,76	0,60	0,71	
0,9	0,70	0,63	0,66	0,44	0,58	
0,96	0,59	0,52	0,57	0,28	0,43	
0,98	0,49	0,47	0,52	0,20	0,34	
1,00	(0,40)	0,33	0,40	0,00	0,00	

Die Übereinstimmung ist bei Strahlungsgleichgewicht überraschend. Die scheinbar unzulässige Vereinfachung, den Strahlen aller Wellenlängen den gleichen Absorptionskoeffizienten zu geben, wird eben gemildert durch den Umstand, daß mit steigender Temperatur das Energiemaximum sich immer schärfer ausprägt. Die Polytropen stimmen, namentlich für kleine n, wesentlich schlechter, so auch für $n = 3$, trotzdem nach EDDINGTON das Innere der Gaskugeln nach dieser Polytropen gebaut ist. Aufbau im Strahlungsgleichgewicht und nach der Polytropen $n = 3$ sind wohl Lösungen derselben Differentialgleichung, aber durch verschiedene Integrationskonstanten ausgezeichnet. Im ersteren Falle ist nach Gleichung (3) E für die Oberfläche gleich $\frac{1}{2}E_{\mathrm{eff.}}$, im andern Falle $= 0$. In genügendem Abstande von $m = 0$ verschwindet der Einfluß der Oberflächenbedingung und verschmelzen die beiden Lösungen.

In einer Reihe von Arbeiten ist weiter der Einfluß der Diffusion auf die Helligkeitsverteilung behandelt worden. Da aber nach neueren Anschauungen Absorption und manche Vorgänge bei Diffusion durch die gleichen Grundvorstellungen geregelt werden, scheinen mir diese Untersuchungen einer eingehenden Revision bedürftig.

Über das Reizempfindungsgesetz und die Farbengleichung.

Von **E. Zinner**, München.

Die Helligkeitsmessungen der Sterne wurden bis vor kurzem im Vergleich zu den Ortsbestimmungen vernachlässigt. Dies zeigt sich am deutlichsten darin, daß wir wohl für die meisten helleren Sterne die Eigenbewegungen kennen, aber von einer Kenntnis des Lichtwechsels dieser Sterne, abgesehen von dem auffälligen Lichtwechsel einiger Sterne, noch keine Rede sein kann. Wohl liegen Versuche vor, aus der verschiedenen Helligkeit mancher Sterne in verschiedenen Helligkeitsverzeichnissen eine Veränderlichkeit dieser Sterne abzuleiten; jedoch ergab die darauf einsetzende Beobachtung in den meisten Fällen ihre Unveränderlichkeit und führte zu dem Schluß, daß diese Verschiedenheiten mehr auf Unterschiede der Verzeichnisse zurückzuführen seien. Ein einwandfreier Nachweis eines, wenn auch langsamen und geringen Lichtwechsels der meisten Sterne wird sich erst dann führen lassen, wenn die besseren Helligkeitsbestimmungen auf die ihnen eigentümlichen Fehler untersucht und miteinander verglichen worden sind. Solche Vergleichungen sind bereits geschenen, und zwar haben MÜLLER und KEMPF[1]) die Unterschiede verschiedener, fast gleichzeitiger Verzeichnisse gegen ihre Potsdamer Durchmusterung berechnet und fanden dabei Unterschiede, die von der Helligkeit und Farbe der Sterne abhängen. Sie haben aber diese Unterschiede nicht benützt, um aus den fremden und eigenen Helligkeiten neue genauere Werte abzuleiten, und auch nicht versucht, durch Heranziehung älterer Verzeichnisse langdauernde Veränderungen nachzuweisen Allerdings liegt ein solcher Versuch in der Überprüften Harvard-Durchmusterung[2]) vor, aus den verschiedenen auf den Harvard-Sternwarten gemachten Messungen Mittelwerte abzuleiten und mit den Werten aus anderen Verzeichnissen zu vergleichen. Da aber diese Mittelwerte selbst unsicher — auf die Eigentümlichkeiten der einzelnen Messungsreihen wurde keine Rücksicht

---

[1]) Photometrische Durchmusterung des nördlichen Himmels. Generalkatalog von G. MÜLLER und P. KEMPF. Publikationen des Astrophysikalischen Observatoriums zu Potsdam Bd. 17. 1907.

[2]) Revised Harvard Photometry. Annals of the Astronomical Observatory of Harvard College Bd. 50. 1908.

genommen[1]) — und die fremden Messungen nicht umgerechnet sind, so kommt dieses Verzeichnis für eingehende Forschungen nicht in Betracht. Es gibt also kein den ganzen Himmel umfassendes und die neuen Messungen berücksichtigendes Helligkeitsverzeichnis, noch liegen einigermaßen befriedigende Versuche vor, die älteren Messungen zum Nachweis langsamer Veränderungen zu verwenden. Aber angesichts der zahlreichen Bemühungen, für die hellen Sterne die absoluten Helligkeiten zu berechnen, ist es unbedingt notwendig, zuvor die scheinbare Helligkeit dieser Sterne und ihre Änderung so genau wie möglich abzuleiten. Da an eine Lösung dieser Aufgabe auf Grund der photographischen Aufnahmen zur Zeit nicht zu denken ist, so bleibt nichts anderes übrig, als die besten vorhandenen Beobachtungen am Fernrohre zu diesem Zwecke zu verwenden. Dabei wird man sich zuerst auf die helleren Sterne beschränken. Eine bloße Fortführung der von MÜLLER und KEMPF gemachten Vergleichungen würde nicht zum Ziele führen; vielmehr ist es wichtig, die Vergleichung möglichst auszudehnen und die Ursachen zu den aus den Vergleichungen sich ergebenden Unterschieden der verschiedenen Beobachtungsreihen festzustellen.

Zur Feststellung der Ursachen der Unterschiede kommt zuerst eine Prüfung der Beobachtungsverfahren in Betracht. Die ältesten Helligkeitsbeobachtungen bestanden in der Einteilung der Sterne in verschiedene Größenklassen, anfangend mit der 1. Größe für die hellsten Sterne und endigend mit der 6. Größe für die für das bloße Auge schwächsten Sterne, mit späterer Fortführung dieser Einteilung auf die nur im Fernrohr sichtbaren Sterne. Ein anderes Verfahren, das allerdings nur zur Kenntnis der gegenseitigen Helligkeiten der Sterne führte, verwandten WILHELM HERSCHEL und JOHN HERSCHEL. Sie reihten eine Anzahl von Sternen ihrer Helligkeit nach auf, wobei WILHELM HERSCHEL zur Bezeichnung der vorhandenen Helligkeitsunterschiede gewisse Kennzeichen benützte. Ungefähr zur selben Zeit begannen Versuche, die Helligkeiten der Sterne genauer zu bestimmen, indem man einen Stern mit einem künstlichen oder natürlichen Vergleichsstern verglich, beide gleich hell machte und aus dem Maße der Abschwächung des einen Sternes den Helligkeitsunterschied der beiden Sterne erschloß. Allerdings wird dabei nicht die Größe m des Sternes gemessen, sondern seine Lichtstärke J und die Größe berechnet nach der Formel:

$$\frac{J_1}{J_2} = 2.51^{m_2 - m_1} \qquad (1)$$

Dabei bedeutet die Zahl 2.51 die nach POGSONS Vorschlag für die Beziehung zwischen Lichtstärke und Größe maßgebende Zahl. Die Gleichung selbst, die eine Beziehung zwischen Reiz (= Lichtstärke)

---

[1]) Annals Bd. 50, S. 10. Nachträgliche Vergleichung der Beobachtungsreihen in Annals Bd. 64, S. 91—164. 1912.

und Empfindung (= Größe) darstellt, ist eine Folge des Fechner-Weberschen Reiz-Empfindungsgesetzes, wonach die Unterschieds-schwelle $\dfrac{dE}{dR} = \dfrac{1}{R}$ und also $E = \log R$ ist. Dieses Gesetz, dessen unbeschränkte Gültigkeit selbst von Fechner nicht behauptet worden ist — vielmehr nahm er für zu große bzw. zu kleine Reize eine Ungültigkeit des Gesetzes infolge der Schwächung bzw. der Eigenbeeinflussung der Sinne an —, wurde in der Sternforschung als uneingeschränkt gültig angesehen. Bei der Wichtigkeit des Gesetzes für die Helligkeitsmessungen ist es aber notwendig zu untersuchen, ob das Gesetz innerhalb des gewöhnlichen Messungsbereiches streng gültig ist oder nicht.

Die Nachprüfung des Fechnerschen Gesetzes kann im Versuchsraum oder am Sternenhimmel selbst geschehen. Das letztere Verfahren ist wegen der Anwendung des Gesetzes auf die Sternbeobachtungen vorzuziehen, selbst dann, wenn das unruhige Aussehen der Sterne eine so genaue Beobachtung wie im Versuchsraum ausschließt. Zur Nachprüfung ist es nötig, Helligkeitsschätzungen mit Helligkeitsmessungen zu vergleichen. Als Beobachtungsreihen kommen einerseits nur solche in Betracht, wo genaue Schätzungen nach einheitlichem Verfahren über ein möglichst großes Bereich ausgedehnt worden sind, und andererseits solche, wo die Lichtstärken der Sterne so sorgfältig wie möglich gemessen und mit Formel (1) in Größen umgewandelt worden sind. Die besten Schätzungsreihen sind die beiden großen von Argelander und Heis durchgeführten Größenschätzungen. Argelander hat in den Jahren 1838—42 alle dem bloßen Auge sichtbaren Sterne ihrem Ort und Größe nach in Karten eingezeichnet. Bei den Größenschätzungen richtete er sich nach den Größenklassen des Ptolemaios und bezeichnete die schwächsten als Sterne sechster Größe. Die Zahl seiner Sterne beträgt 3254 und ist im Sternverzeichnis seiner Neuen Uranometrie, Berlin 1843, mitgeteilt. Heis verglich in den Jahren 1845—72 die Helligkeiten der Sterne mit Argelanders Angaben, merkte die Abweichungen an und fügte viele von Argelander ausgelassene Sterne hinzu, besonders schwache, die er dank seiner scharfen Augen noch wahrnehmen konnte und als schwächste in der Klasse 6.7 vereinigte. Die Zahl der in seinem Buche: Neuer Himmels-Atlas, Köln 1872, mitgeteilten Sterne ist 5421. Die beste Messungsreihe ist die Potsdamer Durchmusterung Hierbei wurden die Lichtstärken der Sterne mit dem Zöllnerschen Photometer gemessen. Als Anhaltssterne dienten weiße oder wenig gefärbte Sterne. Es wurde immer das gleiche Beobachtungsverfahren angewendet, allerdings unter Benützung von verschiedenen Fernrohren mit Rücksicht auf die verschiedenen Sternhelligkeiten. Jedoch wurden die infolge der Verschiedenheit der Fernrohre herrührenden Unterschiede in Rechnung gezogen. Aus diesem Grunde erscheint die Potsdamer Durchmusterung als sehr geeignet zur Prüfung des Fechnerschen Gesetzes.

Allerdings hat sie den Nachteil gegenüber den Harvard-Verzeichnissen, daß sie nur die Sterne des nördlichen Himmels enthält. Aber dieser Nachteil fällt hier nicht sehr ins Gewicht, da die Größenverzeichnisse von Argelander und Heis auch nur die Sterne bis — 38° enthalten, wobei die Lichtauslöschung sich schon bei den Sternen von — 10° ab bemerkbar macht und die Größenangaben verfälscht. Dagegen kommt sie bei den nördlichen Sternen kaum in Betracht und braucht nicht berücksichtigt zu werden, wenn die Annahme, daß Argelander und Heis die Sterne nur in ihrer größten Höhe beobachteten, berechtigt ist.

Die Messungen der Potsdamer Durchmusterung sind allerdings 40 Jahre nach den Beobachtungen von Argelander und 20—40 Jahre nach den von Heis gemacht worden. Trotzdem können die Verzeichnisse unbesorgt miteinander verglichen werden, da etwa vorhandene Veränderungen aller Voraussetzung nach nur klein sind und die Ergebnisse nicht verfälschen können.

Bei der Vergleichung der Verzeichnisse von Argelander und Heis mit der Potsdamer Durchmusterung wurden folgende Gesichtspunkte berücksichtigt:

1. Es mußte dem Einfluß der Farbe auf die Helligkeitsschätzung Rechnung getragen werden. Daß ein solcher Einfluß vorhanden ist und im Sinne einer Dunklerschätzung der roten Gegenstände gegenüber gleich hellen weißen wirkt, ist seit Purkinje bekannt. Um in diesem Falle den Farbeneinfluß zu berücksichtigen, wurden die Farbenangaben der Potsdamer Durchmusterung benützt. Gemäß diesen Farbenangaben wurden die Sterne mit weiß, gelbweiß, weißgelb, gelb, rotgelb und gelbrot bezeichnet. Diese Farbeneinteilung geht für unsere Zwecke zu weit; deshalb wurden die Sterne nur in die Sammelgruppen weiß + gelbweiß, weißgelb und gelb + rotgelb + gelbrot geteilt. Dieses konnte um so unbesorgter geschehen, da zwischen Weiß und Gelbweiß einerseits und andererseits zwischen Rot, Rotgelb und Gelbrot kein wesentlicher Unterschied besteht, wie die Vergleichung der Potsdamer Farbenschätzungen mit denen von Osthoff oder mit den photographisch erhaltenen Farbenzahlen ergab.

2. Es mußte dem Einflusse des hellen Hintergrundes, nämlich der Milchstraße, Rechnung getragen werden. Es wurden daher die milchstraßennahen und milchstraßenfernen Sterne getrennt behandelt. Da Argelander und Heis ihre Sterne nach Sternbildern aufgeführt hatten, wurde zwischen milchstraßennahen und milchstraßenfernen Sternbildern unterschieden.

Unter Berücksichtigung dieser Gesichtspunkte wurden für jede Größenklasse und Unterklasse — Argelander und Heis hatten jede Größenklasse in zwei Unterklassen geteilt, z. B. 2.3 und 3.2, die im folgenden mit $2^1/_3$ und $2^2/_3$ bezeichnet werden sollen — die entsprechenden

Größen der Potsdamer Durchmusterung ausgeschrieben und gemittelt. Dabei wurden alle veränderlichen und Doppelsterne und die schwachen Sterne dicht neben hellen ausgelassen.

Die schwächsten milchstraßenfernen Sterne der Argelander-Größe 6 bzw. Heis-Größe $6^1/_3$ geben die Grenzgröße für das bloße Auge an. ARGELANDER konnte Sterne bis $7^m.30$ gelbweiß sehen und HEIS bis 8^m38 gelbweiß. Die Grenzgröße für HEIS ist erstaunlich und läßt zuerst eine Veränderlichkeit des Sternes vermuten; jedoch zeigten sich weder dieser Stern noch andere ähnliche schwache Sterne als veränderlich. Dagegen ist dabei in Betracht zu ziehen, daß HEIS ein so gutes Auge besaß, daß er alle Sterne punktförmig sah und Sterne wie δ und ε in der Leier trennen konnte, sowie bei Vergleichung seiner Sternkarten mit dem Himmel jede fremde Lichtquelle vermied und sich mit dem Sternenlicht begnügte. Kann also 8^m38 nur als die Grenzgröße eines scharfen Auges angesehen werden, so ist 7^m30 die eines gewöhnlichen Auges. Zum Vergleich möge erwähnt werden, daß die Grenzgröße für mein rechtes, etwas kurzsichtiges unbewaffnetes Auge 6^m7 gelbweiß und für HOFFMEISTER etwa 7^m2 ist.

Die Mittel für die Größenklassen und Unterklassen wurden benützt, um Kurven getrennt für die Farben und für die milchstraßennahen und -fernen Sterne zu zeichnen. Der Einfluß der Milchstraße macht sich darin bemerkbar, daß einerseits die schwächsten Sterne sich von dem hellen Hintergrund nicht mehr abheben und andererseits die Sterne bis zu 3. Größe in der Milchstraße schwächer geschätzt werden als außerhalb. Die Abschwächung beträgt bei

ARGELANDER

für weiß + gelbweiß: 0^m21, weißgelb: 0^m11, gelb + rotgelb + gelbrot: 0^m16

HEIS

0.19   0.17   0.14

also Beträge derselben Größenordnung. Die Sterne heller als 3^m scheinen durch das Licht der Milchstraße nicht beeinflußt zu werden. Infolge dieser Umstände wurden die Größenmittel der milchstraßennahen und -fernen Sterne unter Berücksichtigung der obigen Zahlen und der Anzahl der Sterne zu neuen Mitteln vereinigt, die in der Übersicht 1 unter F für jeden Beobachter und jede Farbe angegeben sind.

Diese neuen Mittel wurden zum Zeichnen der neuen Kurven benützt, aber nur für die Größenklassen 3—6. Da für die Sterne heller als 3^m die Mittel auf zu wenigen Sternen beruhen, mußte ein anderer Weg eingeschlagen werden, um den Kurvenverlauf zu sichern. Aus der Übersicht geht nämlich hervor, daß für denselben Beobachter von der 3. Größenklasse ab der Farbeneinfluß immer stärker wird und deshalb die denselben Größenklassen entsprechenden Mittel um so mehr auseinandergehen, je schwächer die Sterne sind. Man darf daraus den

Schluß ziehen, daß für hellere Sterne der Farbeneinfluß verschwindet. Das zeigt sich auch deutlich, wenn man für die Größenklassen 1—3 sämtliche Mittel desselben Beobachters einzeichnet. Die Lage der Mittel

Übersicht 1.

Vergleich der Größenschätzungen von ARGELANDER und HEIS mit den Größen der Potsdamer Durchmusterung.

Größe	Weiß + gelbweiß						Weißgelb			G+RG+GR		
	Zahl	F	K	K_1	K_2	Q	Zahl	F	K	Zahl	F	K
ARGELANDER:												
1	2	$0^m{.}55$	$0^m{.}56$	$1^m{.}06$	$0^m{.}79$	$0^m{.}55$	2	$0^m{.}34$	$0^m{.}56$	1	$1^m{.}15$	$0^m{.}56$
$1^1/_2$	2	1.43	1.35	65	1.54	1.57	1	—	1.35	—	—	1.35
$1^2/_3$	2	77	90	2.09	2.05	2.09	—	—	1.90	—	—	1.90
2	14	2.39	2.33	45	45	49	2	2 12	2.33	2	2.35	2.33
$2^1/_3$	5	61	68	77	77	83	2	84	68	6	62	68
$2^2/_3$	4	3.00	3.05	3.11	3.11	3.14	4	94	3.01	2	86	96
3	18	39	38	41	41	44	5	3.50	32	10	3.35	3.43
$3^1/_3$	28	85	74	75	76	72	4	66	62	9	55	51
$3^2/_3$	14	4.04	4.04	4.04	4.04	4.00	10	94	94	5	91	78
4	32	38	34	34	34	28	11	4.29	4.28	16	4.11	4.07
$4^1/_3$	37	69	64	64	64	56	11	56	54	18	40	37
$4^2/_3$	37	92	97	96	96	86	16	94	84	15	73	65
5	149	5.29	5.25	5.17	5.23	5.23	63	5.16	5.08	83	94	93
$5^1/_3$	43	53	52	48	47	48	19	35	34	21	5.18	5.18
$5^2/_3$	80	78	81	74	74	83	20	54	55	45	39	41
6	549	6.20	6.19	6.09	6.08	6.21	224	96	96	211	74	75
HEIS:												
1	2	0.55	0.56	1.06	0.79	0.56	2	0.34	0.56	1	1.15	0.56
$1^1/_3$	2	1.43	1.45	72	1.64	1.58	1	1.51	1.45	—	—	1.45
$1^2/_3$	2	77	95	2.13	2.10	2.10	—	—	95	—	—	95
2	15	2.41	2.35	47	47	50	2	2.12	2.35	2	2.35	2.35
$2^1/_3$	4	69	67	76	76	84	3	72	67	6	62	67
$2^2/_3$	5	3.08	3.11	3.16	3.17	3.15	3	3.09	3.09	2	86	3.02
3	14	35	43	45	46	45	5	52	42	10	3.28	33
$3^1/_3$	15	83	74	75	76	73	4	71	72	11	63	62
$3^2/_3$	11	92	4.04	4.04	4.04	4.01	7	95	4.01	6	96	90
4	38	4.31	34	34	34	29	24	4.28	27	19	4.14	4.15
$4^1/_3$	39	66	64	64	64	57	9	58	57	16	46	44
$4^2/_3$	37	85	92	91	91	87	20	79	83	24	67	71
5	108	5.22	5.21	5.19	5.19	5.17	42	5.12	5.11	63	95	98
$5^1/_3$	87	51	51	47	46	49	39	39	39	51	5.22	5.25
$5^2/_3$	157	80	82	75	75	84	65	67	70	89	50	53
6	478	6.23	6.24	6.15	6.13	6.22	191	6.13	6.14	164	84	87
$6^1/_3$	613	80	80	62	64	64	324	65	64	234	6.37	6.36

zeigt bis zur Größe $2^2/_3$ hin keinen Farbeneinfluß. Man ist daher berechtigt, die zur Darstellung der Gesamtmittel, gebildet aus den Mitteln für jede Farbe, dienende Kurve als für jede Farbe gültig anzusehen und sie zur Verlängerung der einzelnen Kurven für das Bereich der

helleren Sterne zu verwenden. Den Verlauf der somit erhaltenen Kurve
für jede Farbe zeigen die Zahlen K in der Übersicht. Die Kurven können
zur Untersuchung dienen, ob die sich auf das FECHNERsche Gesetz
stützende POGSONsche Gleichung gültig ist und wie stark sich der
Farbeneinfluß bemerkbar macht. Der Farbeneinfluß soll später besprochen werden. Deshalb ist zuerst die Kurve für die weißen und gelbweißen Sterne, wo die POGSONsche Formel unbeeinträchtigt vom Farbeneinfluß in Erscheinung tritt, zu behandeln.

Die Kurve verläuft für beide Beobachter zwischen der 3. und 5. Größe
geradlinig, krümmt sich aber an den Enden, und zwar für die größere
Helligkeit mehr als für die kleinere. In ARGELANDERS Kurve ist die
Krümmung für die kleinere Helligkeit nur schwach angedeutet, in der
Kurve von HEIS entsprechend seiner größeren Berücksichtigung schwächerer Sterne mehr ausgeprägt. Diese Kurven sind zum Vergleich mit
der POGSONschen Formel noch nicht verwendbar. Zuvor muß der Änderung der Pupille beim Beobachten Rechnung getragen werden. Die
Pupillenöffnung ändert sich bekanntlich selbsttätig bei verschiedenem
Lichteinfall. Bei hellem Licht verkleinert sie sich, während sie sich
bei schwachem öffnet. Das bewirkt, daß das für das Auge in Betracht
kommende Helligkeitsbereich ohne Schädigung oder Anstrengung des
Auges erweitert wird. Der Pupillendurchmesser ändert sich zwischen
3.3 und 8 mm. Dabei entspricht 3.3 sehr großen Helligkeiten, bis zu
1000 Meterkerzen gemäß LANS. Mit welcher Pupilländerung man
bei Sternenbeobachtungen rechnen muß, läßt sich aus Potsdamer
Beobachtungen ableiten. Es hatten sich nämlich zwischen den mit
den Fernrohren verschiedener Öffnung D und C_I angestellten Messungen
derselben Sterne Unterschiede ergeben, die von $+0^{m}.10$ bei $5^{m}.36$ bis
$+0^{m}.01$ bei $6^{m}.22$ abnahmen und darauf bis $0^{m}.08$ bei $6^{m}.88$ anstiegen.
Diese Unterschiede sind zweifellos die Wirkungen der Pupilländerung;
bei den Messungen mit den beiden Fernrohren wurden nicht dieselben
Anhaltssterne benützt, sondern das Mittel der Anhaltssterne bei D betrug
$6^{m}.8$, bei C_I aber nur $5^{m}.4$. Es muß also eine Pupilländerung wirksam
werden, wenn die Helligkeiten eines Sternes von z. B. $5^{m}.4$ das eine Mal
mit Hilfe schwacher Anhaltssterne von $6^{m}.8$ und das andere Mal mit Hilfe
gleichheller Anhaltssterne gemessen wird. Diese Wirkung zeigt sich in
den von MÜLLER und KEMPF abgeleiteten Unterschieden[1]) der Messungen mit den beiden Fernrohren. Die folgende Übersicht gibt in der
ersten Spalte die mittlere Helligkeit der einzelnen Gruppen, in der
zweiten und dritten Spalte den zugehörigen Unterschied berechnet und
durch eine Kurve ausgeglichen.

Über den Verlauf der Kurve K ist nicht viel bekannt, sie erreicht
bei $6^{m}.1$ anscheinend ihren kleinsten Wert. Außerhalb des Bereiches

---

[1]) Publikationen Bd. 17, S. IX.

Übersicht 2.

	F	K	K_1	K_2
5^m36	$+0^m10$	$+0^m10$	$+0^m07$	$+0^m08$
5.67	08	08	05	06
5.79	07	06	04	06
5.89	04	04	04	06
5.97	02	03	04	05
6.06	02	02	04	05
6.22	01	02	04	05
6.52	04	04	05	06
6.88	08	08	07	08

der Übersicht scheint sie nur langsam zu steigen; denn die 10 hellsten Sterne, zwischen 4^m87 und 5^m22, liefern das Mittel $+0^m10$ und die schwächsten Sterne, zwischen 7^m0 und 7^m2, noch nicht $+0^m09$. Es war also eine Formel zu finden, um für jeden Lichtstärkezuwachs die Pupillenänderung und damit den Gewinn oder Verlust an Helligkeit Δm zu finden Δm berechnet sich leicht aus $\Delta m = 5\,\mathrm{g}\left(\dfrac{r_0}{r}\right)$, da die Beziehung besteht $\dfrac{r^2}{r_0^2} = \dfrac{J_0}{J} = 2.51^{M-M_0}$, wo r und r_0 die Halbmesser der Pupille, J und J_0 das entsprechende ins Auge gelangende Licht, $M - M_0$ den Gewinn bzw. Verlust an Größenklassen darstellt. Für die Beziehung zwischen $\dfrac{r}{r_0}$ bzw. $r - r_0$ und den Potsdamer Größen m und m_0 wurden verschiedene Formeln aufgestellt. Gar nicht in Betracht kamen die Gleichungen:

$$r - r_0 = a(m - m_0)^n \quad \text{oder} \quad \frac{r}{r_0} = 1 + a(m - m_0)^2 - b(m - m_0)$$

oder $\dfrac{r}{r_0} = 1 + a(m - m_0)^2 - b(e^{m-m_0} - 1)$ oder $\left(\dfrac{r}{r_0}\right)^n = \dfrac{m}{m_0}$;

allerdings vermag diese letzte Gleichung in der Form $\left(\dfrac{r}{r_0}\right)^{4.1} = \dfrac{\lg J_0}{\lg J}$ die Beobachtungen von LANS befriedigend darzustellen, versagt aber bei den Sternhelligkeiten. Zu der Kurve der Übersicht 2 passende Gleichungen sind folgende:

$$\frac{r}{r_0} = 1 + \frac{(m - m_0)^2}{70} \tag{2}$$

und daraus:

$$\Delta m = 5\,\lg\left[1 + \frac{(m - m_0)^2}{70}\right] \tag{3}$$

oder:

$$\frac{r}{r_0} = e^{(m-m_0)^2\,[a - b\,(m - m_0)]} \tag{4}$$

und daraus:

$$\Delta m = (m - m_0)^2\,[0.04706 - 0.00820\,(m_0 - m)] \quad \text{für } m > m_0 \ \Bigg\}$$
$$\Delta m = -(m_0 - m)^2\,[0.04706 - 0.00820\,(m_0 - m)] \quad \text{für } m < m_0 \ \Bigg\} \tag{5}$$

Die Gleichungen (2) und (4) vermögen allerdings nicht die Beobachtungen von Lans darzustellen, was teils an der Unsicherheit dieser Beobachtungen, teils an dem Unterschied zwischen Tagessehen und Dämmerungssehen liegen kann. Mit den Gleichungen (3) und (5) erhält man die wegen Pupillenänderung berechneten Zahlen K_1 und K_2 der Übersichten 1 und 2. Dabei ist in Übersicht 1 als m_0 die Potsdamer Größe der 4. Größenklasse von Argelander und Heis, entsprechend der Mitte des Bereiches der bequemen Helligkeit, in Übersicht 2 aber die Größe 5.4 bzw 6.8 genommen. Wie die Übersicht 2 lehrt, stimmen die Zahlen K_1 besser zu K, jedoch genügen sie nicht, wie später gezeigt wird, für große $m - m_0$

Die unter Berücksichtigung der Pupillenänderung erhaltene Kurve ist weniger steil, die Krümmungen an ihren Enden sind weniger hervortretend, die geradlinige Strecke der Kurve ist verlängert. Für diesen Teil besteht die Beziehung:

$$\lg J - \lg J_0 = 0.4\,(m_0 - m) = 0.44\,(E_0 - E) \ \text{ bei } \ K_1 \atop \text{bzw.}\,0.446\,(E_0 - E) \quad \text{,, } \ K_2 \Big\} \ \text{für Argelander}$$

$$g J - \lg J_0 = 0,4\,(m_0 - m) = 0.464\,(E_0 - E) \quad \text{,, } \ K_1 \atop \text{bzw.}\,0.454\,(E_0 - E) \quad \text{,, } \ K_2 \Big\} \ \text{für Heiss,}$$

wo E_0 und E die den Potsdamer Größen m und m_0 entsprechenden Größenklassen sind. Offenbar ist hier die durch Pogsons Gleichung ausgedrückte Beziehung zwischen der Lichtstärke J und der Schätzungsgröße E nicht ganz erfüllt. Statt 2.51 müßte die Beziehungszahl 2.8 bzw. 2.9 stehen. Da außerdem die Größenschätzungen von Argelander und Heis wie alle anderen auf die Größenschätzungen des Ptolemaios zurückgehen, so muß, zum mindesten für alle Schätzungen mit bloßem Auge, die Beziehung bestehen: Ein Helligkeitsunterschied von einer Größenklasse entspricht einem Lichtstärkenverhältnis von 1 zu 2.8 bzw. 2.9, nicht wie bisher angenommen und bei den photometrischen Messungen verwirklicht, von 1 zu 2.5. Diese Beziehung gilt nur für den geraden Teil der Kurve, der übrigens bei der K_2-Kurve besser zum Ausdruck kommt. Dieser Teil der K_2-Kurve, der bei Argelander von $2^\mathrm{m}5 - 5^\mathrm{m}8$ und bei Heis von $2^\mathrm{m}7 - 5^\mathrm{m}8$ reicht, entspricht etwa dem Bereiche der bequemen Helligkeit, außerhalb dessen die schwachen Sterne, die das Auge anstrengen, und die hellen Sterne, die es blenden, liegen. Dieses Bereich ist für die verschiedenen Augen wohl etwas verschieden Innerhalb des Bereiches gilt das Fechnersche Gesetz streng, weshalb denn auch die Formel $\varDelta m = 0.4\,(M - M_0)$ zur Berechnung der Wirkung der Pupillenänderung berechtigt ist. Außerhalb des Bereiches gilt das Gesetz nicht mehr. Die Kurve, die bisher geradlinig war, krümmt sich, wie gesagt, an den beiden Enden — bei den geringeren Helligkeiten allerdings nur bei Heis —, und zwar ist der den hellen Sternen entsprechende Ast mehr gekrümmt als der andere.

Die beiden Kurvenäste lassen sich als Parabeläste darstellen, die im Bereich des geraden Stückes zusammenstoßen　Es ergaben sich für die nicht wegen Pupillenänderung verbesserten Kurven der milchstraßenfernen Sterne die Gleichungen:

$$a_1 (E - E_0) = 0.4 (m - m_0) - b_1 (m - m_0)^2 \left.\right\}$$
$$a^2 (E_0 - E) = 0.4 (m_0 - m) - b_2 (m_0 - m)^2 \left.\right\} \qquad (6)$$

und zwar für die Kurvenäste der kleineren und größeren Helligkeiten. E_0 und m_0 bedeuten die Schätzungs- bzw. Potsdamer Größe des Trennungspunktes der beiden Kurven.

Es liegt nahe, die Formel von PÜTTER[1]) für die Beziehung zwischen Reiz und Empfindung zur Darstellung der Kurven heranzuziehen. PÜTTER ging aus von einer Annahme über die Beziehung zwischen den Reizen und dem durch die Reize hervorgebrachten Stoffumsatz in den Sinnen. Er nahm dabei an, daß der Zustand der Erregung durch die jeweilige Anhäufung der Erregungsstoffe, des Endergebnisses des Stoffumsatzes, bestimmt ist. Für die Anhäufung y stellt er auf Grund gewisser vereinfachender Vorstellungen die Formel auf:

$$y = \frac{q_0 (1 + J)}{r [1 + q_0 (1 + J)]} \left\{ 100 + \frac{c \cdot r}{r - 1 - q_0 (1 + J)} e^{-[1 + q_0(1 + J)]t} + d \cdot e^{-rt} \right\} \quad (7)$$

wo J die Reizstärke, t die Zeit, r und q_0 gewisse Parameter, c und d Konstanten darstellen. Auf Grund dieser Gleichung berechnet PÜTTER die Nullschwelle, die dem kleinsten wahrnehmbaren Reize, also der Grenzgröße, entspricht, und die Unterschiedsschwelle, die dem Stufenwerte nach ARGELANDERS Auffassungen entspricht. Zur Berechnung der Unterschiedsschwelle verwendet er die Formel in vereinfachter Gestalt, indem nur der Fall $t = \infty$ betrachtet wird. Unter passender Wahl der q_0 und r lassen sich die Werte y und damit die Unterschiedsschwelle T gemäß der Formel $T = \dfrac{100\,a}{J_1} \cdot \dfrac{J_2 - J_1}{y_2 - y_1}$ finden. PÜTTER konnte Unterschiedsschwellen für Druckreize errechnen, die zu den von STRATTON beobachteten gut passen. Zur Darstellung der Lichtreize genügte die vereinfachte Gleichung (7) nicht, sondern sie mußte insofern erweitert werden, daß die Anhäufung y nicht nur als Funktion der Reizstärke, sondern auch als Funktion des jeweiligen Zustandes des reizbaren Sinnes dargestellt wird. Die Gleichung für y:

$$y = \frac{10 - 0.099\,J + \sqrt{(10 - 0.099\,J)^2 + 400\,J}}{2.02} \qquad (8)$$

gestattet Unterschiedsschwellen zu berechnen, die mit den von KÖNIG beobachteten vergleichbar sind. Obwohl die berechneten wie auch die

---

[1]) PÜTTER, AUGUST: Studien zur Theorie der Reizvorgänge. Pflügers Arch. f. d. ges. Physiol. Bd. 171, S. 201—261. 1918.

beobachteten Unterschiedsschwellen groß für sehr kleine und sehr
große Lichtreize sind, so ist die Übereinstimmung der Kurven der
beobachteten und der berechneten Schwellen nicht gut. Diese Abweichung suchte PÜTTER zu erklären einerseits durch den Einfluß der
Größe des leuchtenden Gegenstandes, andererseits durch das Vorhandensein verschiedener Schwellenwerte für die Stäbchen und Zapfen im
Auge. Dann stellte er auf Grund der Annahme, daß die Unterschiedsschwelle Exponentialfunktion der Reizstärke sei und die Gleichung
$\dfrac{dE}{dR} = f(e^R)$ bestehe, die Gleichung:

$$E = H\left(1 - e^{-\frac{R}{H}}\right) \tag{9}$$

auf, wo H den Grenzwert der Empfindung für unendliche Reizstärke
darstellt. Die Gleichung genügt dann den 3 Bedingungen, daß $\dfrac{dE}{dR}$
eine Exponentialfunktion von R ist, $E = 0$ für $R = 0$ und E endlich
für $R = \infty$ wird. Eine Prüfung der Formel (9) an Hand von Beobachtungen fand nicht statt. Obwohl PÜTTER seine Untersuchung
über die Unterschiedsschwelle nicht mit einer Formel, welche J als
Exponentialfunktion enthält, durchgeführt hat, also demgemäß zu seiner
Annahme über die Unterschiedsschwelle kein Grund vorliegt[1]) und
überdies die Formel (8) die Beobachtungen nicht völlig darstellt, ist
es doch erwünscht, seine Formel (9) auf die Kurven von ARGELANDER
und HEIS anzuwenden. Die Formel (9) liefert eine Kurve, deren mittlerer Teil angenähert geradlinig ist und deren Enden gekrümmt sind,
und zwar das den größeren Reizstärken entsprechende mehr. Die Kurve
ist also der Reizempfindungskurve ähnlich. Sie läßt sich aber nicht
zur Darstellung der Beobachtungen verwenden. Dagegen wird die
Kurve vergleichbar der Beobachtungskurve in der Form:

$$E = H\left(1 - e^{-\frac{R^p}{H}}\right), \qquad \text{wo} \qquad p = \frac{1}{2.1 \pm} \qquad \text{ist.} \tag{10}$$

Der Wert $p = \dfrac{1}{2.1}$ ergab die beste Übereinstimmung und wurde zur
Berechnung der Kurve nach der Formel:

$$\left.\begin{aligned}
Q &= 14.48 - 2.5\left[100\,\frac{\lg\dfrac{100}{13.5\cdot S - 12.4}}{\lg e}\right]^{2.1} \text{für ARGELANDER} \\[2ex]
&= 14.49 - 2.5\left[100\,\frac{\lg\dfrac{100}{13.5\cdot S - 12.4}}{\lg e}\right]^{2.1} \text{für HEIS}
\end{aligned}\right\} \tag{11}$$

---

[1]) Auf verschiedene Merkwürdigkeiten, besonders auf seine Fehlerberechnung,
kann hier nicht eingegangen werden; erwähnt werden soll nur das häufige Vorkommen von Druck- und von Rechenfehlern.

verwendet, wo S die Größen von ARGELANDER bzw. HEIS bedeuten. Der Vergleich von Q mit K_1 und K_2 in der Übersicht 1 zeigt die Berechtigung der durch K_2 dargestellten Annahme über die Pupillenänderung. K_1 gibt wohl eine bessere Darstellung der Potsdamer Werte der Übersicht 2 und paßt auch gut zu den Q-Werten der nicht größten Helligkeiten; aber bei den den größten Helligkeiten entsprechenden Größenklassen 1 und $1^1/_3$ weicht sie erheblich von Q ab und läßt damit erkennen, daß die Pupillenänderung nicht unbegrenzt wächst und selbst für große Lichtstärkenunterschiede nicht groß ist. Dies wird durch die Annahme K_2 mehr berücksichtigt. Allerdings ist auch hier bei der Größe 1 der Unterschied $K_2 - Q$ größer als gewöhnlich, was aber zum Teil auf die Unsicherheit des entsprechenden F-Wertes, der auf nur 5 Sternen beruht, zurückzuführen ist. In diesem Zusammenhang scheint es angebracht, eine Reizempfindungskurve zu betrachten, die nicht auf Größenschätzungen, sondern auf Stufenschätzungen beruht Im Frühjahr 1917 stellte ich zur Bestimmung meines Stufenwertes eine Reihe von Beobachtungen an, indem ich als geeigneten Sternhaufen das Haupthaar auswählte und von den im Feldstecher eben sichtbaren Sternen anfangend eine Stufenfolge bis zu den hellsten Sternen herstellte. Da es keine Sterne heller als 4. Größe in unmittelbarer Nachbarschaft des Haupthaares gibt, mußte von dem Grundsatze, nur benachbarte Sterne zu benützen, abgegangen werden. Die ganze Stufenfolge wurde in mondlosen Nächten zweimal mit dem linken Auge und einem Feldstecher von 28.5 mm Öffnung durchbeobachtet. Auf diese Weise gelang es, 79 Sterne zwischen der 1. und 9. Größe durch 148.6 Stufen zu verbinden. Wenn auch alle Sterne in der größten Höhe beobachtet worden waren, wurde doch die Lichtauslöschung berücksichtigt.

Werden nur die weißen und gelbweißen Sterne benützt, so lassen sich die Beobachtungen durch eine Kurve darstellen, deren gesicherter Teil in der folgenden Übersicht unter K wiedergegeben ist.

Stufe	K	K_2	Q	Stufe	K	K_2	Q	Stufe	K	K_2	Q
10	8^m27	8^m04	8^m13	60	5^m52	5^m49	5^m56	110	3^m75	3^m79	3^m75
20	7.66	7.47	7.45	70	5.09	5.09	5.19	120	3.35	3.42	3.36
30	7.07	6.94	6.88	80	4.77	4.77	4.83	130	2.81	2.94	2.91
40	6.54	6.44	6.40	90	4.48	4.48	4.48	140	2.11	2.29	2.33
50	6.00	5.94	5.96	100	4.14	4.16	4.12	150	1.05	1.28	1.24

K_2 ist berechnet unter Berücksichtigung der Pupillenänderung auf Grund der Gleichung (5), wobei $m_0 = 4.77$ gesetzt wurde. Die Kurve ähnelt sehr den Kurven von ARGELANDER und HEIS, zeigt allerdings die Krümmung bei den kleinen Helligkeiten deutlicher und läßt sich infolgedessen durch eine ähnliche Formel darstellen. Für sie gilt:

$$Q = 15.05 - 2.5 \left[100^{\dfrac{\lg \dfrac{100}{86.9 - 5.7\,S}}{\lg e}} \right]^{2.1} \tag{12}$$

Die Übereinstimmung zwischen K_2 und Q ist befriedigend, während die Anwendung der Formel (3) bei den Stufen 10 und 150 große Abweichungen bewirkt hätte.

Den Werten K_2 entspricht eine Pupillenänderung von 1 : 1.24, etwa von 5.6 mm für Stufe 150 bis 8.0 mm für Stufe 10, wobei der Stufe 80 eine Öffnung von 7.2 mm entsprechen würde.

Merkwürdig sind die Größen 14.48, 14.49 und 15.05, die dem kleinsten Reize entsprechen, und ihre starke Abweichung von der Grenzgröße, die für ARGELANDER bzw. HEIS 7.30 bzw. 8.38 und für mein linkes Auge bei Benützung des Feldstechers 9.4 betragen.

Bei allen Beobachtungen mit den Augen bewirkt die gelbe oder rote Farbe eines Sternes eine gewisse Abschwächung seiner Helligkeit gegenüber der eines weißen Sternes, wenn die Helligkeit beider Sterne gleichmäßig verringert wird. Es ist eine alte Erfahrung, daß farbige Sterne im kleinen Fernrohr schwächer gegenüber weißen Sternen erscheinen als im großen Fernrohr. Für farbige Sterne, die dem Auge sehr hell erscheinen, also oberhalb des Bereiches der bequemen Helligkeit liegen, besteht nach den früheren Auseinandersetzungen dieser scheinbare Helligkeitsverlust nicht. Der Grund dieses Verhaltens liegt in der Netzhaut. Diese besteht in dem gelben Fleck, der Stelle des deutlichsten Sehens, fast ausschließlich aus Zapfen, außerhalb davon aus Stäbchen. Die Zapfen sind farbenempfindlich, die Stäbchen nicht, dafür aber viel helligkeitsempfindlicher. Die Wirksamkeit der Zapfen ist an eine gewisse Helligkeit gebunden, wie beim Tagessehen und beim farbigen Dämmerungssehen. Bei verringerter Helligkeit arbeiten nur die Stäbchen; es handelt sich dann um das farblose Dämmerungssehen. In der Nacht kommt nur das Dämmerungssehen in Betracht, und zwar bei Beobachtung mit bloßem Auge bei Sternen bis zu $2^1/_2$. Größe das farbige und bei schwächeren Sternen das farblose Dämmerungssehen. Die Grenze zwischen beiden Arten von Dämmerungssehen entspricht der oberen Grenze des Bereiches der bequemen Helligkeit. Für gewöhnlich wird der Übergang vom Sehen mit den Zapfen zum Sehen mit den Stäbchen sich nicht bemerkbar machen. Gelegentlich ist dies der Fall, nämlich dann, wenn man farbige Sterne an dieser Grenze beobachtet. Die Sterne erscheinen zuerst farblos, bei längerem Hinsehen aber farbig und werden dabei zugleich etwas heller, um einige Stufen, entsprechend 0^m1 bis 0^m2. Für die als farbig gesehenen Sterne gibt es keine Helligkeitsabschwächung infolge der Farbe; dagegen setzt diese sogleich unterhalb der Grenze ein und nimmt immer mehr zu. Dies zeigt sich in den Schätzungen von ARGELANDER und HEIS. Bildet man die Unterschiede gelbweiß-weißgelb und gelbweiß-gelb, so lassen sich diese Unterschiede durch die Farbengleichungen:

$$F_1 = 0^m069\,(m - 3^m5), \qquad F_2 = 0^m115\,(m - 2^m6) \text{ für ARGELANDER und}$$
$$= 0.028\,(m - 2.1), \qquad\qquad = 0.080\,(m - 2.3) \quad \text{,, HEIS}$$

darstellen, wo F_1 und F_2 an die gelbweißen Werte anzubringen sind. Allerdings lassen sich durch diese Farbengleichungen nicht die beträchtlichen Unterschiede für die Grenzgrößen darstellen. Sie lauten

für ARGELANDER: $F_1 = 0\overset{m}{.}56$ statt 0.26, $F_2 = 1\overset{m}{.}24$ statt 0.54
„ HEIS: $= 0.39$ „   0.18, $= 0.90$ „   0.49

Zum Vergleich mögen noch hinzugefügt werden die Farbengleichungen, die sich aus den Stufenschätzungen WILHELM HERSCHELS mit bloßem Auge ergaben, wenn zur Umrechnung die Potsdamer Größen und Farben benützt werden:

$$F_1 = 0\overset{m}{.}051\,(m - 2\overset{m}{.}4), \qquad F_2 = 0\overset{m}{.}092\,(m - 2\overset{m}{.}4)$$

Man erhält also aus den 3 Beobachtungsreihen mit bloßem Auge Gleichungen, die einigermaßen vergleichbar sind. Es ergeben sich im Mittel für die Konstante von F_1: $0\overset{m}{.}049$ und von F_2: $0\overset{m}{.}096$. Diese Gleichungen gelten für 3 bis 4 Größenklassen, also für das Bereich der bequemen Helligkeit. Zu den Grenzgrößen hin scheinen, wie die Zahlen für ARGELANDER und HEIS beweisen, die F_1 und F_2 rascher zu wachsen, als ob die Farbengleichungen ein quadratisches Glied in bezug auf $m - m_0$ enthielten.

Die Wirksamkeit der Stäbchen, die in den Farbengleichungen zum Ausdruck kommt, entspricht ungefähr der Wirkung der photographischen Platte. In beiden Fällen übt der farbige Stern eine geringere Wirkung aus als der nichtfarbige. Man kann die ausgeprägte Stäbchenwirksamkeit, wie sie bei der Grenzgröße in Erscheinung tritt, mit der in den Farbenzahlen zutage tretenden Wirkung der photographischen Platte vergleichen. Zuvor ist es nötig, die den Farben der Potsdamer Durchmusterung entsprechenden Farbenzahlen und zwar gemäß der Reihenfolge, wo $A_0 = 0.00$ und $K_0 = 1.00$ ist, zu bestimmen. Aus der Vergleichung der Potsdamer Farben mit den Spektren des Draper-Verzeichnisses und der Umwandlung der Spektren in Farbenzahlen folgt für die gelbweißen Sterne im Mittel als Farbenzahl $+ 0\overset{m}{.}06$, für die weißgelben Sterne $+ 0\overset{m}{.}80$ und für die gelben Sterne $+ 1\overset{m}{.}10$. Daraus ergibt sich:

F_2: ARGELANDER $+ 0\overset{m}{.}56$, HEIS $+ 0\overset{m}{.}39$, Farbenzahl $+ 0\overset{m}{.}74$
F_2:   „ $+ 1.24$,   „ $+ 0.90$,   „ $+ 1.04$

Es sind also die F-Grenzwerte von ARGELANDER und HEIS den photographischen Farbenzahlen vergleichbar. So bedeutsam dies auch sein mag, so soll doch auf diese Übereinstimmung kein großer Wert gelegt werden, da die Bestimmung der Grenzgröße nicht sehr genau ist.

Bisher galt die Annahme, daß die Potsdamer Durchmusterung von systematischen Fehlern frei und deshalb als Ausgangsreihe für unsere Untersuchung geeignet sei. Es bleibt noch zu untersuchen, inwieweit die Potsdamer Größen durch die Berücksichtigung der festgestellten

Regeln verbessert werden können. Die Wirkung der Pupillenänderung besteht darin, daß ein hellerer Stern als der Anschlußstern etwas schwächer und ein schwächerer Stern etwas heller erscheinen und somit gemessen wird Dieser Nachteil läßt sich vermeiden, wenn man die Sterne, auch die Anhaltssterne, immer gleich hell einem unveränderlichen Vergleichstern macht, also immer bei gleicher Pupillenöffnung beobachtet. Dies läßt sich erreichen bei dem verbesserten Keilphotometer, jedoch nicht beim Photometer von ZÖLLNER. Einen gewissen Einfluß der Pupillenänderung haben MÜLLER und KEMPF, wie erwähnt, beim Vergleich der Messungen mit den Fernrohren D und C_I berücksichtigt. Sie haben für jedes der beiden Fernrohre Verbesserungen abgeleitet auf Grund der Übersicht 2 und der Annahme über den weiteren Verlauf der Kurve außerhalb des Bereiches $5^m.4 - 6^m.9$, einer Annahme, die durch diese Untersuchung bestätigt wird. Die Verbesserungen sind in abgekürzter Form in Übersicht 3 mitgeteilt, wo für jedes der beiden Fernrohre Spalte 1 die Potsdamer Größen, Spalte 2 die von MÜLLER und KEMPF abgeleiteten Verbesserungen und Spalte 3 die nach Formel (5) berechneten Verbesserungen K_2 enthält.

Übersicht 3.

Berücksichtigung der Pupillenänderung bei den Potsdamer Messungen.

	P. D.	M.+K.	K_2	P. D.	M.+K.	K_2
	$3^m.4$	$-0^m.09$	$-0^m.12$	$5^m.9$	$+0^m.00$	$+0^m.01$
bei C_I	3.9	8	8	6.4	3	4
	4.4	3	4	6.9	8	8
	4.9	0	1	7.4	9	12
	5.4	0	0			

	P. D.	M.+K.	K_2	P. D.	M.+K.	K_2
	$4^m.8$	$-0^m.11$	$-0^m.12$	$7^m.3$	$+0^m.00$	$+0^m.01$
bei D	5.3	10	8	7.8	6	4
	5.8	6	4	8.3	10	8
	6.3	0	1	8.8	11	12
	6.8	0	0			

Die Unterschiede zwischen M. + K. und K_2 sind nicht groß. K_2 hält die Mitte zwischen den beiden M. + K. - Reihen, deren Glieder MÜLLER und KEMPF zur Darstellung der Kurvenwerte K in Übersicht 2 verschieden groß annahmen. Mit den M. + K. - Verbesserungen sind natürlich nicht alle Wirkungen der Pupillenänderungen berücksichtigt, sondern nur eine mittlere Wirkung, wie sie dem Helligkeitsunterschiede zwischen einem Stern und dem Mittel der Anhaltssterne im Fernrohr D oder C_I entspricht. Tatsächlich konnte der Helligkeitsunterschied eines Sternes zu dem zugehörigen Anhaltsstern viel kleiner und damit auch die Wirkung der Pupillenänderung kleiner als durch die Übersicht 3 angegeben sein. Die durch die Nichtberücksichtigung dieses Verhältnisses im einzelnen entstehenden Fehler sind wohl nicht groß, wohl kaum mehr als $0^m\!.03$ für die Größe des Verzeichnisses und können auch so lange unberücksichtigt bleiben, als die in der Reihe der Anhaltssterne steckende Wirkung der Pupillenänderung noch nicht in Rechnung gezogen ist. Diese Wirkung muß sich in einer Verkleinerung des Größenunterschiedes zwischen dem größten und kleinsten Anhaltsstern zeigen. Der daraus entstehende Fehler ließe sich beseitigen, indem man mittels Gleichung (5) die Größe jedes Anhaltssternes verbesserte, ausgehend von $m_0 = 6.0$, das dem Mittel sämtlicher Anhaltsterne entspricht. — Außer mit den Fernrohren D und C_I haben MÜLLER und KEMPF auch mit C_{II} und C_{III} gemessen und die Messungen verglichen. Die mit C_I und C_{II} gemeinsam gemessenen Sterne lieferten keinen merklichen systematischen Unterschied. Dies war zu erwarten, da dieselben Vergleichsterne benützt wurden, also im Mittel keine Pupillenänderung in Betracht kam. Dagegen können in den Messungen mit C_{II} noch kleine, besondere Fehler infolge der Pupillenänderung vorhanden sein. Anders ist es mit den Fernrohren C_{II} und C_{III}. Hierbei wurden Anhaltsterne benützt, deren Reihen nicht vorher miteinander verglichen und ausgeglichen waren, sondern die endgültige Vergleichung wurde, nach einer vorläufigen, erst durch die Gegenüberstellung der beiden Messungsreihen durchgeführt. Deshalb kann diese Vergleichung nicht dazu dienen, eine mittlere Wirkung der Pupillenänderung abzuleiten. Eine besondere Wirkung der Pupillenänderung wird sicherlich bestehen. Wir sehen also, daß die einzelnen Messungen der Potsdamer Durchmusterung durch die Wirkung der Pupillenänderung verfälscht sind, aber infolge des geringen Messungsbereiches nur wenig, und daß die meisten Potsdamer Größen außer der von MÜLLER und KEMPF durchgeführten Verbesserung noch einer kleinen Verbesserung bedürfen.

Wie machen sich die Abweichungen des Reizempfindungsgesetzes vom FECHNERschen Gesetze bemerkbar? Die Formel POGSONS gilt für das Bereich der bequemen Helligkeit streng; allerdings ist die Beziehungszahl etwas anders, aber das kommt doch nur beim Vergleich mit den auf die Größenklassen des Ptolemaios gegründeten Schätzungsreihen in Betracht, nicht aber beim Vergleich mit den neueren photometrischen Sterngrößen. Außerhalb dieses Bereiches gilt die POGSONsche Formel nicht, sondern die Beziehung zwischen Reiz und Empfindung erfolgt so, als ob, ähnlich der Wirkung der Pupillenänderung, zu große Helligkeiten abgeschwächt und zu kleine vergrößert würden. Dadurch wird der Größenunterschied zweier Sterne kleiner gemessen als bei unbeschränkter Gültigkeit der POGSONschen Formel. Das Bereich der Gültigkeit beträgt 4 Größenklassen. Nun ist bei der Potsdamer Durchmusterung der Messungsbereich selten über 3^m ausgedehnt worden und, was noch wichtiger ist, der Unterschied zwischen Stern und Anhaltsstern beträgt fast nie mehr als 2^m, weshalb die Gültigkeit der Pogsonschen Formel für die Potsdamer Messungen von vornherein anzunehmen wäre. Bedenken kann nur der Umstand erwecken, daß das Bereich der bequemen Helligkeit bei Argelander und Heis dem Bereiche zwischen $+2^m$ und $+6^m$ entspricht und sich nicht deckt mit dem Messungsbereich der Potsdamer Beobachter, welche die Sterne bei scheinbar großer Helligkeit — entsprechen -1^m bis $+2^m$ für das bloße Auge — beobachteten. Da die Beobachter aber Sterne von der scheinbaren Helligkeit $2^m.1$ als etwas zu schwach bezeichneten, so ist die Annahme wohl gestattet, daß bei ihnen das Bereich der bequemen Helligkeit bei helleren Sternen als gewöhnlich lag. Diese Annahme wird unterstützt durch den Vergleich der Potsdamer Messungen mit denen Pritchards, die mit einem Keilphotometer gemacht und deshalb durch die Pupillenänderung und Abweichungen vom Fechnerschen Gesetz nicht beeinflußt sind. Würden die Potsdamer Messungen außerhalb des Geltungsbereiches von Pogsons Formel fallen, so müßten die Unterschiede der beiden Messungsreihen zu den hellen Sternen hin sehr wachsen. Tatsächlich ist in den Unterschieden nur ein kleiner Gang wahrnehmbar. Und dieser Gang, der mehr bei den Sternen mittlerer Helligkeit auftritt, läßt sich durch kleine Fehler in der Eichung des Keiles erklären. Man kann also mit Sicherheit annehmen, daß für die Potsdamer Messungen die Abweichung vom Fechnerschen Gesetz nicht oder höchstens ausnahmsweise in Betracht kommt.

Wie macht sich der Farbeneinfluß bemerkbar? Er äußert sich bei den farbig gesehenen Sternen gar nicht, erreicht dagegen bei den farbigen, als farblos gesehenen Sternen erhebliche Beträge. Bei den Potsdamer Messungen wurden die Sterne farbig gesehen. Es kommt daher ein Farbeneinfluß nicht in Betracht. Dies wurde auch durch die Untersuchung einiger farbigen Sterne bestätigt. Bei der Auswahl dieser Sterne wurde auf folgende Gesichtspunkte Rücksicht genommen: 1. ausgeprägte Farbe (gelb im Mittel), 2. lichtschwache Sterne ($6^m\pm$), 3. mindestens 4 Messungen mit 2 Paar nur weißen oder gelblichweißen Anhaltssternen, wobei jedes Paar der Anhaltsterne um $1^m.4$ im Mittel vom andern absteht. Die Vergleichung ergab für $D - C_I$ nur den durch die Pupillenänderung bedingten Betrag.

Wieweit wird die Vergleichung der Schätzungen von Argelander und Heis mit der Potsdamer Durchmusterung durch die letzten Feststellungen beeinflußt? Der Einfluß der Pupillenänderung auf die Potsdamer Größen ist, soweit nicht schon berücksichtigt, noch vorhanden, aber wohl nur klein. Die Abweichungen von dem Fechnerschen Gesetz kommen für die Potsdamer Messungen offenbar nicht in Betracht. Auch ist kein Farbeneinfluß vorhanden. Dagegen bleibt noch die Frage zu beantworten übrig, ob die Potsdamer Farbenangaben genau genug sind. Die Farben beruhen auf Schätzungen mit verschiedenen Fernrohren, wobei zuletzt alle Farben auf das Fernrohr D mit 13.5 cm Öffnung umgerechnet wurden. Vergleicht man diese Farbenangaben mit den Schätzungen von OSTHOFF oder KRÜGER oder mit den Farbenzahlen auf Grund der Spektren, so zeigen die Potsdamer Farben eine gewisse Unbestimmtheit; offenbar entspricht dieselbe Potsdamer Farbe einem ziemlich weiten Bereich von Farbenabstufungen. Auch zeigen sich gelegentlich große Abweichungen zwischen den Farben der Potsdamer Durchmusterung und denen anderer Beobachter, wobei der Fehler bei den Potsdamer Schätzungen zu liegen scheint. Es ist daher bei der Verwendung der Potsdamer Farben Vorsicht geboten, wenn auch ihrer Verwendung in so großen Gruppen wie weiß + weißgelb, gelbweiß und gelb + rotgelb + gelbrot kein Bedenken entgegensteht. Daher kann man mit Sicherheit annehmen, daß die Zahlenwerte F der Übersicht 1, wenn überhaupt, nur um wenige Einheiten sich ändern und die daraus abgeleiteten Folgerungen nach wie vor bestehen werden.

Die Reduktion von Fernrohrbeobachtungen wegen Kontrastfehlers.

Von **A. Kühl**, München.

Mit 1 Abbildung.

Nachdem durch einige vorhergehende Aufsätze[1]) die Nützlichkeit der Auffassung der Konturen optischer Bilder als Kontrasterscheinungen gezeigt wurde, sollen die folgenden Ausführungen dem Versuch gelten, die Ergebnisse jener Arbeiten auf eine für die praktische Nachprüfung und Anwendung geeignete, möglichst einfache Form zu bringen.

1. Das Kontrastgesetz. E. Mach[2]) und unabhängig von ihm H. v. Seeliger[3]) haben nachgewiesen, daß das menschliche Auge in ausgedehnten, *stetigen* Lichtverteilungen deutliche Licht- oder Schattengrenzen empfindet, wenn die partiellen zweiten Differentialquotienten der etwa nach den rechtwinkligen Koordinaten x und y variablen Lichtverteilungsfunktion $i\,(x\,y)$ gewisse Eigentümlichkeiten zeigen. Mach konnte auf Grund von Versuchen und theoretischen Überlegungen das empirisch befriedigende Gesetz aufstellen, daß die durch *Kontrast* gegen die Umgebung modifizierte Lichtempfindung einer Netzhautstelle einer Erregung e entspreche, die aus der tatsächlichen Intensität $i\,(xy)$ an dieser Stelle gewonnen wird durch die Operation:

worin bedeutet:

$$\left. \begin{aligned} e &= i - \frac{k}{2}\,\varDelta\,i \\[2ex] \varDelta\,i &= \frac{\partial^2 i}{\partial\,x^2} + \frac{\partial^2 i}{\partial\,y^2} \end{aligned} \right\} \tag{1}$$

und $\dfrac{k}{2}$ einen numerisch nicht näher angebbaren positiven Faktor von der Dimension einer Fläche.

---

[1]) Kühl, A.: a) Über Wesen und Veränderlichkeit der Konturen optischer Bilder. Vierteljahrsschr. d. Astr. Ges. Bd. 56, H. 3. Leipzig: Poeschel & Trepte 1921. — b) Die Konturen optischer Bilder als Kontrasterscheinungen. Im Manuskript in der Bibliothek d. Astrophys. Obs. Potsdam. — c) Anwendung der Kontrasttheorie auf das Fadenmikrometer. Zentralztg. f. Optik u. Mechanik Bd. 45, S. 27. 1924.

[2]) Mach, E.: Die physiologische Wirkung räumlich verteilter Lichtreize auf die Netzhaut. Wiener Sitzungsber. 1865—68 (4 Abhandlungen).

[3]) Seeliger, H.: Die Vergrößerung des Erdschattens bei Mondfinsternissen. Abh. d. Bayer. Akad. d. Wiss. 1896, München.

Nach der MACHschen Ableitung dieser Formel stellt $\dfrac{k}{2}\,\varDelta i$ die Abweichung der Intensität an der Stelle $x\,y$ von einem gewissen Mittelwert über die Intensität in der nächsten Umgebung dieser Stelle dar; das Gesetz besagt also: *Je nachdem die Abweichung der Intensität $i\,(x\,y)$ von dem Mittel ihrer Umgebung* (also das Vorzeichen von $\varDelta i$) *positiv oder negativ ist, wird diese Netzhautstelle dunkler oder heller gesehen als bei gleichmäßiger Beleuchtung der Nachbarstellen mit derselben Intensität $i\,(x\,y)$.*

2. Kontrastverlauf an Fernrohrbildern. Infolge der Lichtbeugung gibt es an optischen Bildern niemals Unstetigkeiten der Intensität, also im besonderen keine wirklichen scharfen Bildgrenzen, sondern nahe der geometrisch-optischen Bildgrenze einen allmählichen, nach bekannten Gesetzen verlaufenden Lichtabfall, der sich theoretisch erst merklich *außerhalb* der geometrischen Bildgrenze asymptotisch der Helligkeit des Untergrundes angleicht. Es liegt daher die Vermutung nahe, daß die vom Auge *gesehenen* scharfen Bildgrenzen durch *Kontrasterscheinungen* in der Nähe des geometrischen Bildrandes hervorgerufen werden. Es ist weiter anzunehmen, daß sich diese Kontrasterscheinungen durch das empirisch bewährte MACHsche *Kontrastgesetz* [Gl. (1)] darstellen lassen.

Berechnet man für die beugungstheoretisch bekannten Intensitätsverteilungen von Fernrohrbildern gleichförmig leuchtender Objekte die Funktion $\varDelta i$, die ich kurz als *Kontrastfunktion* bezeichne, so stellt sich heraus, daß sie in der Nähe des idealen Bildrandes stets die Form einer Wellenfläche hat mit einem *Wellental* innerhalb und einem *Wellenberg* außerhalb der *leuchtenden* Bildfläche. In Rücksicht auf die Wirkung des Vorzeichens von $\varDelta i$ folgt also, daß eben *innerhalb* des idealen Bildrandes des leuchtenden Bildes ein *heller* Kontraststreifen, eben *außerhalb* ein *dunkler* Kontraststreifen gesehen werden müßte. Die Nachprüfung mit *hinreichender* (vgl. Abschn. 4) Okularvergrößerung bestätigt diese Folgerung aufs deutlichste.

Aus der Deutlichkeit dieser Erscheinung ziehen wir die für numerische Auswertung nützliche Folgerung, daß der Faktor $\dfrac{k}{2}$ in Gl. (1) numerisch groß genug sei, um bei endlichen Werten von $\varDelta i$ in der Kontrastzone das Glied $\dfrac{k}{2}\,\varDelta i$ den (variablen) Wert i so weit überwiegen zu lassen, daß im *allgemeinen* die Untersuchung der Kontrastfunktion $\varDelta i$ allein hinreicht, um eine genügend angenäherte Lage der Kontrastextreme zu sichern. An den Stellen $\varDelta i = 0$ ist natürlich der Wert des ersten Gliedes in Gl. (1) zu beachten.

Identifiziert man hiernach genähert die Kontrastextreme mit den Extremstellen von $\varDelta i$ und beachtet, daß zwar an den Nullstellen von $\varDelta i$ die Intensität i maßgebend ist, daß aber wegen der Größe von $\dfrac{k}{2}$ auch die Nullstellen von e in ihrer unmittelbaren Nähe liegen, so zeigt der Vergleich der hauptsächlich für die Fernrohrmessung in Betracht kommenden Bildformen eine so weitgehende Übereinstimmung in der Lage der ausgezeichneten Kontraststellen zum idealen Bildrand, daß man für die Beobachtungspraxis unbedenklich folgende einheitliche Darstellung wählen kann:

Hat an *hellen* Bildscheiben, Bildstreifen oder Bilddecken der von der fraglichen idealen Randstelle gemessene *kleinste* Bilddurchmesser in *optischen Einheiten* (vgl. Abschn. 3) den Wert u, so haben längs ihm die ausgezeichneten Kontraststellen, die in Tab. 1 angegebene Entfernung vom geometrischen Bildrand. Die Entfernungen vom geometrisch-optischen Bildrand ins *Innere* des *hellen* Bildes sind durch ein Minuszeichen von den Entfernungen nach *außen* (Pluszeichen) unterschieden. Alle Entfernungen sind in *optischen Einheiten* angegeben.

In Abb. 1 ist die Tabelle graphisch dargestellt. Die Ordinatenachse gibt durch ihre Bezifferung oberhalb der Abszissenachse die „kleinsten" Objektdurchmesser in optischen Einheiten an und stellt gleichzeitig den Ort der Durchmessermittelpunkte dar. Die halben Objektdurchmesser sind als Abszissen zu den Ordinaten eingetragen, so daß der Ort der geometrischen Endpunkte der Objekthalbmesser (nämlich der Randpunkte der halben geometrisch-optischen Bilder) durch die Gerade OG dargestellt wird. Trägt man zu den Ordinatenhöhen nach Tab. 1 von der Geraden OG aus die Randabstände der ausgezeichneten Kontraststellen

Tabelle 1.

Lagen der ausgezeichneten Kontraststellen zum Bildrand von hellen Bildern.

Wahre Bildbreite o. E.	Innere Grenze des hellen Streifens $\Delta i = 0$	Maximum des Hellkontrastes $\Delta i = \mathrm{Min.}$	Grenze zw. beiden Kontraststreifen $\Delta i = 0$	Maximum des Dunkelkontrastes $\Delta i = \mathrm{Max.}$	Äußere Grenze des dunklen Streifens $\Delta i = 0$
0		0	$+1.4$	$+2.4$	$+4.3$
0.2		-0.1	$+1.3$	$+2.3$	$+4.1$
0.4		-0.2	$+1.2$	$+2.15$	$+4.0$
0.8		-0.4	$+0.95$	$+2.0$	$+3.9$
1.0		-0.5	$+0.9$	$+1.9$	$+3.8$
2.0		-1.0	$+0.5$	$+1.6$	$+3.7$
3.0		-1.5	$+0.25$	$+1.4$	$+3.6$
4.0		-2.0	$+0.1$	$+1.4$	$+3.6$
4.7		-2.35	$+0.05$	$+1.4$	$+3.6$
5.0	$(-2.5)^{1)}$	-1.8	0	$+1.4$	$+3.6$
6.0	$(-3\ \ )$	-1.4	0	$+1.4$	$+3.6$
8.2	-4.1	-1.4	0	$+1.4$	$+3.6$
9.0	-3.7	-1.4	0	$+1.4$	$+3.6$
10	-3.6	-1.4	0	$+1.4$	$+3.6$
∞	-3.6	-1.4	0	$+1.4$	$+3.6$

gestellt wird. Trägt man zu den Ordinatenhöhen nach Tab. 1 von der Geraden OG aus die Randabstände der ausgezeichneten Kontraststellen

---

[1] Für die eingeklammerten Stellen ist $\Delta i > 0$, aber relatives Minimum der Umgebung.

parallel zur Abszissenachse ein, so stellen die gestrichelten Kurven I, G, A die Orte der inneren, mittleren und äußeren Grenze der Kontraststreifen, die Doppelkurven H und D die Orte maximalen Hell- und Dunkelkontrastes dar. Bemerkenswert ist, daß die Kurven I, die an *beiden* Bildrändern entlanglaufenden Kurven $\Delta i = 0$, bei einem Bilddurchmesser von 8.2 o. E. in *eine* zusammenfließen, d. h. daß dann beide hellen Kontraststreifen unmittelbar aneinandergrenzen. Dieser Zustand bleibt bestehen bis zur Bildbreite 4.7 o. E., indessen wird in diesem Bereich in der Grenzlinie (Bildmitte) nicht mehr der Kontrastwert 0, sondern nur ein relatives Minimum erreicht; mit andern Worten die längs der Ordinatenachse zwischen 8.2 und 4.7 o. E. verlaufende I-Linie ist als Projektion der Sohle einer Talrinne anzusehen. Bei geringeren Bildbreiten als 4.7 o. E. hört die Grenze zwischen beiden Hellkontraststreifen auf; wir haben nur noch *ein* über der Bildmitte liegendes Kontrastmaximum. Gleichzeitig rücken die Kurven G, D, A nun mit abnehmender Bildbreite weiter vom idealen Bildrand ab. Die Abstände, die sie bei der Bildbreite Null haben, gelten für eine unendlich dünne *Lichtlinie* und für den ausdehnungslosen *Lichtpunkt*.

Will man die Darstellung von Abb. 1 gleichzeitig als Bild des Kontrastverlaufes an einer *Ecke* auffassen, so sind nahe der Spitze

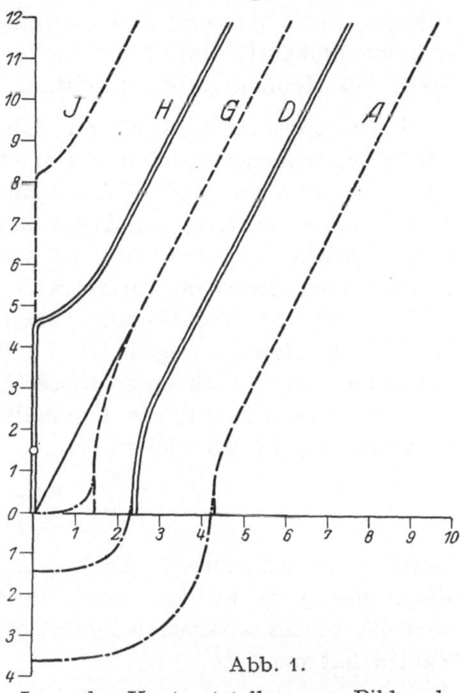

Abb. 1.

Lage der Kontraststellen zum Bildrand als Funktion des Bilddurchmessers.

------ J: Innere Grenze des Hellkontrastes.
====== H: Hellste Stelle des Hellkontrastes.
------ G: Grenze zw. Hell- u. Dunkelkontrastes.
o—G: Geometrischer Bildrand.
====== D: Dunkelste Stelle des Dunkelkontrastes.
------ A: Äußere Grenze des Dunkelkontrastes.
—·—·— : Änderung von $G\,D\,A$ an der Spitze einer Ecke.

die strichpunktierten Modifikationen der Kurven G, D, A und das Hellkontrastmaximum auf der Ordinatenachse bei der Bildbreite 1.5 o. E. (durch den kleinen Kreis angedeutet) zu beachten. Würde man sich für *diese* Auffassung den Kontrastverlauf als Gebirgsmodell plastisch vorstellen, so ergibt sich, daß bei größerer Bildbreite (größerer Entfernung von der Spitze einer Ecke) *jeder* Bildrand *innen* von einem *Wall* (Hellkontrast), *außen* von einem *Graben* (Dunkelkontrast) begleitet wird. Von der Bildbreite 8.2 o. E. an beginnen beide Randwälle miteinander zu verschmelzen, indem sie eine immer flacher werdende *Rinne* zwischen sich lassen, die bei der Bildbreite 4.7 o. E. aufhört, um von hier ab in *eine Gratlinie* überzugehen, die bei der Bildbreite 1.5 o. E. eine *Kuppe* aufweist. Während dieser Änderungen im Bildinnern entfernen sich die langsam *abflachenden* Gräben zeitweise ein wenig vom geometrisch-optischen Bildrand und fließen an der Spitze als ziemlich flache *Mulde* zusammen.

Abb. 1 geht in die Darstellung des Kontrastverlaufs für *schwarze* Bilder auf hellem Grunde über, wenn man die Vorzeichen der Kontrastextreme umkehrt, also H als Stelle des Dunkelkontrastes und D als Stelle des Hellkontrastes ansieht.

3. Die optische Einheit. Zur Übertragung der durch Tab. 1 und Abb. 1 angegebenen Maße in die am Bilde meßbaren Winkelbeträge ist daran zu erinnern, daß Objekte (und *Fernrohrbilder*), welche relativ zum Öffnungsverhältnis des Auges dieselben Winkelabmessungen haben, beugungstheoretisch (bis auf einen von der Öffnungsfläche des Systems abhängenden Proportionalitätsfaktor der Lichtstärke) zu den *gleichen* Lichtverteilungen Veranlassung geben. Deshalb gibt man für beugungstheoretische Betrachtung die Bildmaße vorteilhaft in diesen „relativen Einheiten" an, die ich kurz *optische Einheiten* nenne. Eine Abmessung u in optischen Einheiten (o. E.) stellt auf der Netzhaut daher folgende Abmessung ϱ in Millimetern dar:

$$u = \frac{\pi \cdot p}{\lambda \cdot f} \cdot \varrho \tag{2}$$

worin π die LUDOLPHsche Zahl, p den *Durchmesser* der Eintrittspupille des Auges, f die Luftbrennweite des Auges $= 17$ mm und λ die Luftwellenlänge des wirksamen Lichts, etwa 589 $\mu\mu$, bedeutet. Mit diesen Zahlen hat man:

$$\varrho = 3.2\,\mu\,\frac{p\,\mathrm{mm}}{u} \tag{2a}$$

d. h. z. B.: bei einer Pupillenöffnung von $p = 1$ mm entspricht einer optischen Einheit $u = 1$ o. E. auf der Netzhaut eine Strecke von $\varrho = 3.2\,\mu$. Da der Durchmesser eines Netzhautzäpfchens durchschnittlich 3.0 μ beträgt, so entspricht also für $p = 1$ mm eine o. E. ungefähr einem Zapfendurchmesser.

Die Abmessung ϱ mm auf der Netzhaut wird bekanntlich als scheinbare Größe α'' eines unendlich fernen Objektes (oder Fernrohrbildes) gesehen gemäß:

$$\alpha'' = \frac{\varrho}{f} \cdot 206\,265''$$

So ist nach Gl. (2):

$$\alpha'' = 38''\!.7\,\frac{u}{p\,\mathrm{mm}} \tag{3a}$$

Also bei 1 mm Pupillenöffnung wird eine o. E. durch einen Sehwinkel von rund 39 Bogensekunden veranschaulicht.

Nennt man diejenige Okularvergrößerung Γ eines Fernrohrs, welches seine Eintrittspupille (Objektivöffnung) $2R$ auf die Austrittspupille $p_0 = 5$ mm reduziert, die *Normalvergrößerung* Γ_0, wo also $\Gamma_0 = \dfrac{2R\,\text{mm}}{5}$, so ist bekanntlich:

$$\frac{\Gamma}{\Gamma_0} = \frac{p_0}{p}, \quad \text{also} \quad p = \frac{5\Gamma_0}{\Gamma}$$

Damit geht Gl. (2a) über in:

$$\rho = 0.64\,\mu \cdot u \cdot \frac{\Gamma}{\Gamma_0} \tag{2b}$$

d. h. die einer o. E. entsprechende Abmessung ρ auf der Netzhaut wächst proportional der Okularvergrößerung.

Führt man endlich noch die Mikrometerablesung β'' ein, welche der scheinbaren Bildgröße α'' entspricht, durch $\beta'' = \dfrac{\alpha''}{\Gamma}$, so geht Gl. (3a) über in:

$$\beta'' = 7''.7\,\frac{u}{\Gamma_0} \tag{3b}$$

d. h. die einer o. E. entsprechende Mikrometerablesung ist der Fernrohröffnung *umgekehrt* proportional.

4. Die hinreichende Okularvergrößerung. Bei Untersuchung der Frage, an welcher Stelle der in Abschn. 2 angegebenen Kontrastzone das Auge die Bildgrenze zu *sehen* pflegt, ergibt sich[1]) eine eigentlich nicht überraschende Schwierigkeit infolge der Zusammensetzung der Netzhaut aus *endlichen* Empfindungselementen, den Zäpfchen. Die Zusammenarbeit benachbarter Zapfen kann deswegen streng genommen natürlich nicht in der Abschätzung von zweiten Differentialquotienten (Δi) bestehen, sondern in Abschätzung von zweiten Differenzenquotienten. Solange nun mit größerer Pupillenöffnung beobachtet wird, werden diese Differenzenquotienten nach Gl. (2a) mit ziemlich großen Intervallen in optischen Einheiten $\left(du = \dfrac{d\varrho}{3.2\,\mu} \cdot p\ \text{mm}, \text{ wo } d\varrho = 3{,}0\,\mu \text{ ist}\right)$ abgestuft. Die Folge ist natürlich eine schwer übersehbare Verlagerung der ausgezeichneten Kontraststellen im Vergleich zu Tab. 1 und Abb. 1 — d. h. sowohl eine Erschwerung der Untersuchung als eine Herabminderung der Beobachtungsgenauigkeit, wie die Nachprüfung im einzelnen zeigt. Ein Blick auf Abb. 1 lehrt auch sofort, daß die Netzhaut die ausgezeichneten Kontraststellen nur dann nahe an ihrem theoretischen Ort zu sehen in der Lage ist, wenn das engste Intervall zwischen zwei solchen Stellen (nach Abb. 1 beträgt dies $u = 1.4$ o. E.) von *mindestens* 2 bis 3 Zapfendurchmessern ausgefüllt wird. Setzen wir daher in Gl. (2b)

---

[1]) Vgl. Anm. 1 c auf S. 372.

$\varrho = 2.5 \cdot 3\,\mu$ und $u = 1.4$ o. E., so ergibt sich zur Erfüllung dieser Forderung die Okularvergrößerung:

$$\frac{\Gamma}{\Gamma_0} \geqq \frac{7.5}{0.64 \cdot 1.4} = 8.35$$

in Worten: *Hinreichende Beobachtungsgenauigkeit kann nur bei mindestens 8—9facher Normalvergrößerung erwartet werden*, d. h. bei einer Austrittspupille des Fernrohrs von $p \leqq \dfrac{5}{8.35} = 0.60\,\text{mm}$.

5. Gesehene und gemessene Bildgrenzen. Wenn bei der Ausführung von Beobachtungen die eben abgeleitete Bedingung erfüllt war — und das ist bei fast allen astronomischen Messungen der Fall —, so gilt über die Lage einer gut definierten Bildgrenze folgende, soviel ich sehe *allgemeine*, Regel:

An alleinstehenden und kombinierten Bildern sieht das Auge eine gut definierte Grenze dort, wo eine zwischen beiden Randkontrasten liegende kontrastfreie Zone ($\Delta i = 0$) *an den Dunkelkontraststreifen angrenzt.*

Die Kontrastlage an kombinierten Bildern erhält man mit genügender Annäherung durch einfache Superposition der nach Tab. 1 und Abb. 1 erhaltenen Kontrastlagen der Einzelbilder. Die relative Stärke der Kontrastextreme ist der relativen Stärke der den Kontrast verursachenden Bildintensität proportional — bei *sehr* schmalen Bildern sind evtl. noch die Sonderbemerkungen über relative Kontraststärke in Abschn. 2 zu beachten.

Damit ist man einerseits in der Definition der physiologischen Bildgrenze der mühseligen Untersuchung des Intensitätsverlaufs im einzelnen enthoben, andererseits gewinnt man mit einem Schlage einen charakterisierenden und für numerische Auswertung hinreichenden Einblick in die bei verschiedenen Meßmethoden auftretenden Fehler.

Wenn man, um schematische Verhältnisse zu schaffen, bei irgendeiner Meßmethode unterscheidet: an der zu messenden *Fläche* die Intensitäten weiß oder schwarz, bei dem zur Messung dienenden *Meßbild* (Faden, Doppelbild od. dgl.) die Intensitäten weiß oder schwarz, und für den Untergrund eine zwischen beiden stehende Intensität als grau bezeichnet, so sind folgende 8 Fälle denkbar:

Intensität des Meßbildes:	Meßbild wird bewegt auf:	Meßbild wird bewegt in Richtung gegen Rand von:
a) weiß	grauem Grund	weißer Fläche
b) schwarz	grauem Grund	schwarzer Fläche
c) schwarz	weißer Fläche	grauem Grund
d) weiß	schwarzer Fläche	grauem Grund
e) schwarz	grauem Grund	weißer Fläche
f) weiß	grauem Grund	schwarzer Fläche
g) weiß	weißer Fläche	grauem Grund
h) schwarz	schwarzer Fläche	grauem Grund

Die Fälle a), b), e), f) würde man in üblicher Bezeichnungsweise als Messung durch Außenberührung, die Fälle c), d), g), h) als Messung durch Innenberührung zusammenfassen. Die Fälle e), f), g) h) ergeben an oder nahe der Berührungsstelle eine *einfache*, beiden Bildern gemeinsame Grenzkurve $\Delta i = 0$, die Fälle a), b), c), d) geben nahe des Berührungsmomentes eine *Verzweigung* der Grenzkurve $\Delta i = 0$, die das äußere Zeichen von verwickelten Phänomenen ist, die man in einem typischen Fall als *Tropfenbildung* bezeichnet.

In der Charakterisierung der Lage der Bildgrenze in verschiedenen Fällen beginnen wir mit der *gesehenen* Bildgrenze am gleichförmig hellen oder dunklen Einzelbild. Sie ist nach der oben angegebenen Regel durch die Stelle G in Abb. 1 gegeben, fällt also bei Bildern, deren kleinster Durchmesser größer ist als 5 o. E., mit der geometrisch-optischen Bildgrenze zusammen; bei schmäleren Bildern rückt sie mehr und mehr bis zum Betrage 1.4 o. E. vom geometrischen Bildrand nach *außen*. Sehr schmale Bilder erscheinen also in ihrem Durchmesser um den doppelten Betrag dieser Verschiebung verbreitert. Solche Verbreiterung an *hellen* Bildern wird als *positive*, an *dunklen* Bildern als *negative* Irradiation bezeichnet.

Die *gemessene* Bildgrenze ist im allgemeinen von der am Einzelbild *gesehenen* Grenze verschieden.

Denkt man sich in den Fällen e) und f) der Meßmöglichkeiten der leichteren Übersicht wegen das zu messende und das Meßbild von größerem Durchmesser als 5 o. E., so hat das helle Bild nach Abb. 1 außen einen „Dunkelgraben“, das dunkle Bild außen einen „Hellwall“ von derselben Höhe wie die Grabentiefe, weil die verursachenden relativen Intensitäten nahe der Berührung gleich sind. Kurz vor der geometrischen Randberührung, wenn die beiden Bilder den Abstand der Stellen G und A der Abb. 1 haben, heben sich hier beide Außenkontraste zu einer Strecke $\Delta i = 0$ auf. Längs dieser kontrastfreien Strecke von 3.6 o. E. Länge wird jetzt die tatsächliche Intensität i *empfunden*, wo vorher Dunkelkontrast herrschte; mit anderen Worten, der Rand des *hellen* Bildes scheint an der Berührungsstelle um den Betrag 3.6 o. E. gegen den Rand des schwarzen Bildes hingerückt. In den beiden genannten Fällen wird also an hellen und dunklen Bildern der Durchmesser um $2 \cdot 3.6$ o. E. zu *groß* gemessen.

In den Fällen g) und h) stellt das Meßbild — soweit es sich *innerhalb* der zu messenden Bildfläche befindet — einen nicht oder kaum mehr unterscheidbaren Teil dieser selbst dar, soweit es *außerhalb* ist, zeigt es an seinem Außenrand dieselben Dunkel- bzw. Hellkontraste wie das zu messende Bild — im Moment der Innenberührung haben Meßbild und zu messendes Bild an der Berührungsstelle eine Stelle ihres Außenkontrastes gemeinsam; es findet also eine geometrische Berührung ihrer unveränderten Grenzen G nach Abb. 1 statt; d. h. bei Bildbreiten über

5 o. E. ergeben diese beiden Fälle die *wahren* geometrischen Bilddurchmesser. Abgesehen ist hier natürlich von der rein *technischen* Frage, ob man bei dieser Methode das zu messende Bild genügend *ruhig* im Gesichtsfeld erhalten kann.

In den Fällen a) und b) heben sich bei geometrischer Berührung, wie die Anwendung von Abb. 1 leicht ergibt, die Grenzkontraste längs einer Geraden *senkrecht* zu den sich berührenden Bildlinien auf. Der Zwischenraum zwischen beiden Bildern beiderseits dieser Linie — wenn man sich die zu messende Fläche etwa von Kreisform denkt — stellt sich als zwei *dunkle* (bei weißen Bildern) oder *helle* (bei dunklen Bildern) Keile dar, die mit ihren Spitzen zusammenstoßen. Ihr Kontrastverlauf ist also durch Abb. 1 unter Einbeziehung der strichpunktierten Modifikationen der Kurven G, D, A direkt gegeben. Man erkennt sofort, daß die Bildgrenze (die Kurve G) sich von dem Berührungspunkt, wie oben erwähnt, zunächst längs eines kurzen Stücks senkrecht zu den geometrischen Konturen bewegt und sich dann in *zwei* Züge gabelt, die, in einem Abstand von $2 \cdot 1.4 = 2.8$ o. E. voneinander verlaufend, bei einer Keilbreite von 5 o. E. in die geometrischen Grenzen der beiden Bilder übergehen — d. h. also: im Moment der geometrisch-optischen Bildberührung scheinen dem Beobachter die Bilder noch um 2.8 o. E. auseinanderzustehen, wiewohl er schon eine deutliche Trübung (bei dunklen Bildern) oder eine deutliche Aufhellung (bei hellen Bildern), den berüchtigten „Tropfen", an der trennenden Brücke bemerkt.

Beim Fadenmikrometer u. dgl. kommen für diesen Fall Überlegungen in Betracht, wie sie weiter unten für die Randberührung bei einem Planetendurchgang angestellt sind. Führt der Beobachter beim *Doppelbildmikrometer* die Bilder nach dem eben erreichten Zustand seiner Meinung nach noch mehr zusammen, so überdeckt er sie tatsächlich und erhält, sofern man wieder an kreisförmige Objekte denkt, einen lanzettförmigen Überdeckungsraum, der sich dunkel bei dunklen Bildern, hell bei hellen Bildern zwischen die auseinanderweichenden Keilspitzen lagert und über seiner Mitte (s. den unteren Teil von Abb. 1) *einen* dunklen bzw. hellen Kontraststreifen hat, der bei einer Breite des lanzettförmigen Flächenstücks von 1.5 o. E. (s. den kleinen Kreis auf der Ordinatenachse von Abb. 1) an der Berührungsstelle besonders deutlich wird und daher gut eine gemeinsame Bildgrenze beider Einzelbilder vortäuschen kann.

Noch weiteres Überdecken führt bei einer Breite des lanzettförmigen Flächenstücks von 4.7 o. E. (s. Abb. 1) zur Spaltung des eben erwähnten Kontraststreifens und wird damit für den Beobachter subjektiv den Eindruck erwecken, als fange jetzt gerade die Überdeckung an, als sei also gerade jetzt eben die geometrische Bildberührung eingetreten.

Es ist a priori nicht zu sagen, wie sich ein Beobachter den hier vorliegenden Komplikationen gegenüber verhält — er kann z. B., die Breite der „trüben Brücke" schätzend, um dieses Stück die Bilder überdecken, oder er kann die Bilder einander nähern, bis er im Berührungspunkt die vermeintlich gemeinsame Grenze bemerkt, oder er kann subjektiv den Beginn der Überdeckung abwarten. Im zweiten Fall wird er die Bilddurchmesser um 1.5 o. E., im ersten um 2.8 o. E. und im dritten um 4.7 o. E. zu *klein* erhalten.

In den Fällen c) und d) nimmt das kurz vor der Berührung noch vorstehende Reststück einer etwa wieder kreisförmig gedachten Bildfläche je nach der Gestalt des Meßbildes so verschiedene Formentypen an, daß es zu weit führt, einen vollständigen Überblick zu geben. Daher ist vorgezogen, ein hierher gehöriges Beispiel im nächsten Abschnitt etwas eingehender zu behandeln.

6. Reduktionsbeispiele. *A. Beispiele zur Verlagerung der Bildgrenze bei schmalen Bildern* findet der Leser in den Irradiationsmessungen von AUBERT[1]). Bei einem Teil derselben ist indes zu beachten, daß die Beobachtungsbedingungen der Forderung hinreichender Okularvergrößerung nach Abschn. 4 nicht entsprechen. Dieser Umstand (z. B. Beobachtungen mit bloßem Auge bei voller Pupillenöffnung) verursacht auch bei Bildern von größerer Breite als 5 o. E. Irradiation durch Verschiebung der Kontraststellen je nach den Intervallen der *Differenzenquotienten*, und zwar ergibt sich dann an breiteren *hellen* Bildern Bild*verbreiterung*, an *dunklen* Bildern *Verschmälerung* durch Irradiation.

B. Beispiel zum Fall e) am Fadenmikrometer. G. STRUVE[2]), Neubabelsberg, stellte vom 4. IV. bis 9. V. 1921 Rektaszensionsmessungen der *Venus* vor und nach ihrer unteren Konjunktion an, indem er den Mikrometerfaden des TÖPFERschen Durchgangsinstrumentes dem erleuchteten Sichelrand nachführte. Für den in die Normalgleichungen als Unbekannte eingeführten Radius des Planeten fand er $9''.072$. Die Fernrohröffnung beträgt 190 mm, die angewandte Okularvergrößerung war $\Gamma = 300$, die Austrittspupille betrug daher $p = \dfrac{190}{300} = 0.63$ mm, d. h. die Forderung auf hinreichende Okularvergrößerung ist gerade erfüllt.

Die Venussichel hatte während der Beobachtungen eine durchschnittliche scheinbare Breite von $10''$, d. h. nach Gl. (3 b) eine Breite in optischen Einheiten von:

$$u = \frac{10''}{7''.7} \cdot \Gamma_0 = \text{ca } 50 \text{ o. E.}$$

---

[1]) HOFMANN: Die Lehre vom Raumsinn des Auges. S. 14 u. f. Berlin: Julius Springer 1920.

[2]) STRUVE: Rektaszensionsbeobachtungen des Planeten Venus usw. Astr. Nachr. Bd. 219, S. 281. 1923.

Nach Abschn. 5, Fall e), liegt eine Radienvergrößerung (Entfernung der Stelle A von G in Abb. 1 bei größerer Bildbreite) von 3.6 o. E. vor, also nach Gl. (3 b) von:

$$\beta = 7''.7 \cdot \frac{3.6}{38} = 0''.730$$

Da während der Beobachtungen das Mittel der Entfernung Erde — Venus 0.306 betrug, so ging von diesem Betrag bei der Reduktion des Radius auf Entfernungseinheit der Bruchteil $dr_♀ = \beta \cdot 0.306 = 0''.223$ ins Resultat ein. Der *wahre* Venusradius beträgt daher:

$$r_♀ = 9''.072 - 0''.223 = 8''.849$$

ein Wert, der geradezu überraschend genau mit dem besten Wert nach *Hartwig* ($r_♀ = 8''.833$) übereinstimmt.

C. *Beispiel für die Fälle a) und b) am Doppelbildmikrometer.*

1. HARTWIG[1]) hat Heliometermessungen der Venusdurchmesser von MAIN, KAISER und sich selbst unter Einführung eines unbekannten konstanten Meßfehlers für jedes Instrument ausgeglichen und fand für den Fehler folgende Werte:

Oxforder Heliometer (MAIN)	$- 1''.18$	$- 4.1$ o. E.
Straßburger Heliometer (HARTWIG) . . .	$- 1''.23$	$- 1.7$,.
Straßburger Heliometer (HARTWIG) . . .	$- 0''.93$	$- 3.1$,,
Airys Doppelbildmikrometer (KAISER) .	$- 0''.77$	$- 2.7$.,

Nimmt man jeweils als beugende Öffnung ein Ersatzobjektiv von der halben Öffnungsfläche der Heliometer (vgl. dazu das nächste Beispiel) an, so entsprechen die Fehler den in der dritten Spalte aufgeführten Werten in o. E., d. h. die Durchmesser sind tatsächlich um solche Beträge und in denselben Stufen zu klein erhalten, wie sie oben für diesen Fall angegeben wurden. Das HARTWIGsche Ergebnis für den Venusdurchmesser $2 r_♀ = 17''.666$ darf man infolge dieser Ausgleichung als den wegen Kontrastfehlers reduzierten *wahren* Wert ansprechen.

2. Es *scheint* nun, daß geübte Beobachter immer mehr zur Einstellung auf den Moment der *subjektiv* beginnenden Überdeckung, also Durchmesservergrößerung um 4.7 o. E. hinneigen, denn AUWERS[2]) leitete aus den Heliometerbeobachtungen des Venusdurchgangs von 1874 den Durchmesserwert $16''.820$ ab.

Für Abschätzung der Beugungswirkung der Objektivhälften auf jedes Einzelbild ersetzen wir jede durch eine *kreisförmige* Objektivöffnung von halber Fläche des ganzen Heliometerobjektivs. Da die benutzten Heliometer eine volle Öffnung von 76 mm besaßen, so wäre die Öffnung

---

[1]) HARTWIG: Untersuchung über die Durchmesser der Planeten Venus und Mars. Publ. d. Astron. Ges. XV. Leipzig 1879.

[2]) STRUVE, W.: Über den Einfluß der Diffraktion an Fernröhren auf Lichtscheiben. S. 85. St. Petersburg 1894.

des jede Hälfte vertretenden „Ersatzobjektivs" zu $38 \sqrt{2} = 53.74$ mm anzunehmen. Durchschnittlich wurde bei 150facher Okularvergrößerung beobachtet; die Ersatz-Austrittspupille für jede Objektivhälfte maß daher $p = \dfrac{53.74}{150} = 0.358$ mm, war also hinreichend klein. Aus der zur „Bildberührung" beider Einzelbilder vorgenommenen Überdeckung um 4.7 o. E. würde nach Gl. (3 b) (für das Ersatzobjektiv ist $I'_0 = \dfrac{53.74}{5} = 10.75$) ein Mikrometerbetrag von $\beta = 7''.7 \cdot \dfrac{4.7}{10.75} = 3''.367$ folgen. Hiervon mußte bei der Reduktion auf Entfernungseinheit (Erde—Venus in unterer Konjunktion $= 0.28$) der Betrag: $2\,dr_{\varphi} = 0.28 \cdot 3''.367 = 0''.943$ ins Resultat eingehen. Der wahre Durchmesser beträgt daher $16''.820 + 0''.943 = 17''.763$, während der HARTWIGsche Wert $17''.666$ und der STRUVEsche $17''.698$ war. Die Mißstimmigkeit wird also hiermit fast ganz beseitigt.

3. Ein interessantes hierher gehöriges Beispiel besonderer Art ist das folgende: Nach AUWERS[1]) wurde bei Gelegenheit des Venusdurchgangs von 1874 in Bahia blanca und Punta Arenas eine Durchmessermessung an der schmalen Venussichel eben außerhalb der Sonnenscheibe ausgeführt; es ergab sich der Wert $17''.31$. Zweifelsohne *erwartet* der Beobachter in solchem Fall die Berührung der sehr feinen Sichelspitzen daran erkennen zu können, daß die nebeneinander gelegten Bilder *eine* gemeinsame Spitze zeigen. Das ist nach Abb. 1 aber nur dann möglich, wenn die Spitzen so *überdeckt* werden, daß die Hellkontrast-Maxima bei der Sichelbreite 1.5 o. E. zusammenfallen, woraus eine Durchmesser*verringerung* von gleichem Betrag resultiert. Da die Instrumente dieselben waren wie oben und sich auch die Entfernung Erde—Venus nicht geändert hatte, mußte ins Resultat eingehen der Betrag: $2\,dr_{\varphi} = \dfrac{7''.7 \cdot 1.5 \cdot 0.28}{10.75} = 0''.301$.

Der wahre Durchmesserwert beträgt daher $17''.31 + 0''.301 = 17''.61$, d. h. es wird wieder nahezu der HARTWIGsche Wert erreicht. Wenn man bedenkt, welche Meßschwierigkeiten so nahe der Sonnenscheibe vorliegen, so muß dies Ergebnis ganz besonders befriedigen.

D. Beispiel für Fall c) beim Planetendurchgang. Bekanntlich lassen sich bei Beobachtungen von Planetendurchgängen aus den Intensitätsverhältnissen allein keine Anhaltspunkte für die beobachteten Momente der berüchtigten „Tropfenbildung" entnehmen; um so mehr muß der Versuch reizen, aus der Kontrasttheorie diese Lücke zu füllen.

Verfolgt man die zur zweiten inneren Berührung schreitende dunkle Planetenscheibe in ihrer Annäherung an den Sonnenrand, so wird bald die schmalste Stelle des Lichtbandes zwischen beiden Rändern die Breite 4.7 o. E. erreichen. Von diesem Moment an hört, nach Abb. 1 hier beginnend, das Lichtband auf als „Streifen" zu erscheinen; es nimmt den

<hr>

[1]) AUWERS: Die Venusdurchgänge von 1874 und 1882. Berlin 1898 V, 724.

Charakter eines ,,Linienelementes'' an (*ein* Hellkontrast-Maximum über der Bildmitte), das nun nach beiden Seiten an Länge langsam wächst, während seine *subjektive* Bildbreite (vgl. Kurve *G* in Abb. 1 unterhalb der Ordinate 4.7 o. E.) *konstant* bleibt und das tatsächliche Abnehmen seiner Intensität durch vermehrte Kontrastwirkung zunächst fast ganz ausgeglichen wird. So kann der wirkliche Beginn der Randerscheinungen nicht gut an dem *Helligkeitsabfall* beobachtet werden, wohl aber scheint mir, daß der Entstehungsmoment des Lichtlinienelementes als erste Phase ziemlich gut angebbar wäre, wenn ein Beobachter darauf geschult wird.

Bei Fortsetzung der Annäherung des Planetenrandes an den Sonnenrand wird natürlich auch eine ,,Trübung'' der Lichtlinie subjektiv merkbar werden, aber ich sehe keine Möglichkeit, den ,,Beginn der Trübung'' an einen bestimmten Zustand der Annäherung zu knüpfen.

Je mehr aber nun die eigentliche geometrische Randberührung herannaht, um so mehr erhält die Nähe der Berührungsstelle die Form zweier mit der Spitze zusammenstoßender Lichtkeile. Man wird daher die Erscheinung gut erfassen, wenn man auch hier wieder die in Abb. 1 eingetragenen Modifikationen der Kontraststreifen ins Auge faßt. Denkt man sich die Randentfernung auf 1.5 o. E. herabgesunken, so herrscht an der Berührungsstelle *ein* relativer Maximalkontrast — weitere Randannäherung zertrennt dies Maximum in *zwei*, deren jedes zu einer Stelle, an der die beiden Lichtkeile momentan 1.5 o. E. breit sind, seitlich abwandert. Hierbei wird die Trübung zwischen ihnen an der Berührungsstelle fast plötzlich auffallend kräftig, so daß hier ein zweites gutes Beobachtungsmoment erscheint.

Im Augenblick der wirklichen geometrischen Randberührung berühren sich die geometrischen Keilspitzen. Die Intensität an der Berührungsstelle ist sehr gering, die hier zusammenfallenden Nullkontrastkurven *G* legen zwischen die beiden, bei den Keilbreiten 1.5 o. E. liegenden Hellkontrast-Maxima (O in Abb. 1), welche subjektiv jetzt fast plötzlich als die eigentlichen Keil*spitzen* erscheinen, scheinbar ebenso plötzlich eine auffällige schwarze Brücke. Dieser Moment als Beginn der ,,Tropfenbildung'' gibt daher eine dritte Beobachtungsnotiz für die Erscheinung. Der weitere Verlauf bietet für das Vorliegende kein Interesse.

Struve[1]) zitiert nach Auwers Modellbeobachtungen obiger Erscheinung mit einem Fernrohr von 117 mm Öffnung bei 171 facher Vergrößerung. Da die Fernrohr-Austrittspupille hierbei $p = \dfrac{117}{171} = 0.68$ mm betrug, so darf nach Abschn. 4 noch keine sonderlich genaue Erfassung der Erscheinung erwartet werden. Es fand sich:

---

[1]) Vgl. Anm. 2 auf S. 382.

a) schwache Trübung   Randabstand $-0''.77 = -2.34$ o. E.
b) sehr merkliche Trübung   „ $-0''.52 = +1.58$ „ „
c) sehr starke Trübung   „ $-0''.17 = -0.52$ „ „
d) Tropfenbildung   „ $-0''.06 = -0.18$ „ „
e) Tropfen sehr dick   „ $+0''.07 = +0.21$ „ „

Im Vergleich dazu folgt aus der Kontrastbetrachtung:

o) Entstehung des „Linienelements"   Randabstand -4.7 o. E.
b) Sehr merkliche Trübung   „ -1.5 „ „
d) Tropfenbildung   „ 0.0 „ „

Der Moment o) ist nicht beobachtet worden. Die Momente a), c), e) können nicht identifiziert werden; dagegen passen die Momente b) und d) sehr gut zu den betreffenden Kontrastmomenten. Da für das benutzte Instrument eine optische Einheit $= 0''.392$ ist, so hätte man allein durch Kontrastreduktion der Momente b) und d) die Berührung nur noch um $\dfrac{0.08 + 0.18}{2}$ o. E. $= 0''.043$ zu früh registriert, also immerhin ein sehr brauchbares Resultat erhalten. Dabei ist, wie bemerkt, die Vermutung zulässig, daß eine stärkere Okularvergrößerung von $\Gamma \geqq 200$ die Genauigkeit noch hätte etwas erhöhen können.

Die Anwendung der einfachen Regel des Abschn. 5 in Verbindung mit den in Tab. 1 und Abb. 1 enthaltenen numerischen Angaben wird also tatsächlich in weitem Maße der hier interessierenden Erscheinungsgruppe gerecht. Zur festeren Begründung der abgeleiteten numerischen Ergebnisse und zur Ausmerzung der über die Beobachtungen mit „Tropfenbildung" noch bestehenden Unsicherheit wären Messungen von einer größeren Anzahl verschiedener Beobachter höchst erwünscht. Sollten diese keine *allgemeingültige* Neigung für ein bestimmtes *subjektives* Berührungsmoment ergeben, so denke ich doch, daß es möglich wäre, durch Schulung der Beobachter für *eins* der fraglichen Momente in Zukunft alle Kontrastfehler aus Präzisionsmessungen zu eliminieren.

Über die Abhängigkeit der photographisch effektiven Wellenlängen vom chromatischen Korrektionszustand des Objektivs.

Von **Östen Bergstrand**, Upsala.

Mit 1 Abbildung.

Die Methode der effektiven Wellenlängen hat sich als genaue Methode zur Bestimmung der Farben der Sterne gut bewährt. Besonders in den Fällen, wo es sich um Untersuchungen in größerem Umfange von schwächeren Sternen handelt, ist diese Methode imstande gute Dienste zu leisten. Die erfolgreiche Anwendung derselben setzt aber voraus, daß das benutzte Instrument zu diesem Zweck besonders geeignet ist und daß eine eingehende Untersuchung gewisser für die Methode charakteristischer Fehlerquellen vorausgegangen ist. Die wichtigste dieser Fehlerquellen ist die Abhängigkeit der effektiven Wellenlängen von den Bildstärken der Sterne auf der Platte. In gewissen Fällen hat es sich auch erwiesen, daß die Fokussierung eine hervortretende Rolle spielen kann.

Wenn man von mehr zufälligen Umständen, wie z. B. der Einwirkung der Zenitdistanz und des während der Exponierung herrschenden Luftzustandes, absieht, hat man in jedem einzelnen Falle drei Hauptfaktoren als für die effektive Wellenlänge eines während einer gewissen Zeit exponierten Sterns bestimmend zu betrachten: in erster Linie natürlicherweise die Intensitätsverteilung innerhalb des photographischen Gebietes des Spektrums des Sterns, ferner die Empfindlichkeit der benutzten Platte für verschiedene zu demselben Gebiete gehörenden Wellenlängen und schließlich der chromatische Korrektionszustand des Objektivs. Diese drei Faktoren setzen sich miteinander und mit der Exponierungsdauer in einer mehr oder minder komplizierten Weise zusammen.

Im Falle einer vollkommenen Achromasie, z. B. bei Benutzung eines Reflektors, kommt der dritte der genannten Hauptfaktoren in Wegfall; man hat nur mit den beiden ersten zu rechnen, und die ganze Sache wird wesentlich vereinfacht. Es scheint also a priori zweckmäßiger zu sein, bei den hier in Frage stehenden Untersuchungen Reflektoren zu verwenden, um so mehr als diese eine größere Amplitude für die effektive Wellenlänge geben als die Refraktoren. Leider ist das brauchbare Feld der Reflektoren in der Regel ziemlich klein, und für größere Durchmusterungsarbeiten müssen daher die photographischen Refraktoren mit ihrem beträchtlich größeren Felde in Betracht kommen.

Ich will zunächst den Fall vollkommener Achromasie (Reflektoren) mit einigen Worten erwähnen. Bei gegebenem Stern und gegebener Plattensorte kommt ein gewisses wirksames Spektralgebiet mit einer gewissen Intensitätsverteilung in Betracht. Zum Erzeugen der schwächsten meßbaren Bilder wirkt nur diejenige Wellenlänge, welcher die maximale Intensität entspricht. Je kräftiger die Bilder werden, desto mehr von den übrigen zum genannten Spektralgebiet gehörenden Wellenlängen gelangen zur Wirkung. Wenn nun das Maximum nicht mit dem Schwerpunkt der Intensitäten zusammenfällt, wird mit zunehmender Bildstärke eine Verschiebung der resultierenden Wellenlänge stattfinden. Die Abhängigkeit der effektiven Wellenlänge von der Bildstärke kann also in der Weise charakterisiert werden, daß die schwächsten Bilder Werte der effektiven Wellenlänge geben, die etwa mit der Wellenlänge des Intensitätsmaximums übereinstimmen, die stärkeren Bilder hingegen Werte, die sich der Wellenlänge des Intensitätsschwerpunkts allmählich nähern.

Aus meinen mit dem großen Meudoner Reflektor gemachten Untersuchungen[1]) ergab sich, daß für die gelblichen und die rötlichen Sterne der Einfluß der Bildstärke im Durchschnitt ziemlich klein war, daß für die weißen Sterne aber eine ausgeprägte Zunahme der effektiven Wellenlänge mit der Bildstärke vorhanden war. Zu etwa demselben Ergebnis kam später M. WOLF[2]) durch seine Untersuchungen am großen Reflektor der Heidelberger Sternwarte. Die Erscheinung wird wahrscheinlich dadurch erklärt, daß das Intensitätsmaximum für die farbigen Sterne im allgemeinen nahe mit der mittleren Wellenlänge des photographischen Spektralgebiets zusammenfällt, für die weißen Sterne hingegen exzentrisch nach der violetten Seite zu liegt. Doch scheinen die einzelnen Sterne, auch wenn sie demselben Typus angehören, sich untereinander etwas verschiedenartig zu verhalten.

Wenn wir nun zum Falle der unvollständigen Achromasie (Refraktoren) übergehen, haben wir auch den letzten der drei obengenannten Hauptfaktoren in Betracht zu ziehen. Die Sache wird dann komplizierter und die einzelnen Instrumente können sich in ganz verschiedener Weise verhalten. ROSENBERG hat nachgewiesen[3]), daß hier in gewissen Fällen eine sehr auffällige Abhängigkeit der effektiven Wellenlängen

---

[1]) Recherches sur les couleurs des étoiles fixes. Nova Acta R. Soc. Scient. Upsal., Ser. IV, Bd. 2, Nr. 2, S. 23 u. f. 1909.

[2]) Vierteljahrsschr. d. Astr. Ges. Jg. 55, S. 100. 1920.

[3]) Astr. Nachr. 1921, Nr. 5109.

von der Fokussierung auftreten kann. Wie bereits von LINDBLAD[1]) andeutungsweise hervorgehoben worden ist, ist diese Erscheinung mit größter Wahrscheinlichkeit auf die Gestalt der Kurve für das sekundäre Spektrum, d. h. auf die Art der chromatischen Korrektion des Objektivs, zurückzuführen.

Durch die unvollständige Achromasie des Objektivs wird einerseits das oben erwähnte, für den Reflektor geltende, photographische Spektralgebiet vermindert, andererseits wird die Intensitätsverteilung innerhalb desselben wesentlich verändert. Der Einfluß der Fokussierung beruht natürlich darauf, daß infolge der ungleichen Abstände der

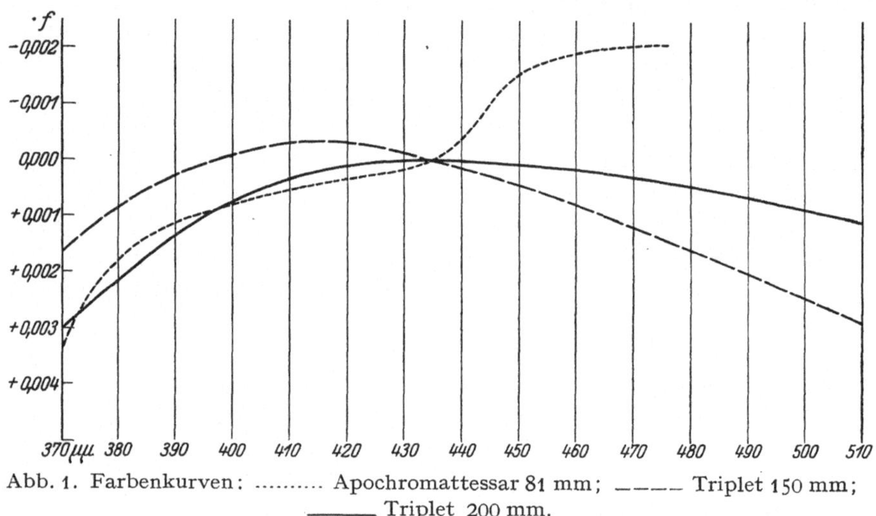

Abb. 1. Farbenkurven: ·········· Apochromattessar 81 mm; — — — — Triplet 150 mm; ———— Triplet 200 mm.

Brennpunkte der verschiedenen Wellenlängen vom Objektiv die Intensitätsverteilung nicht dieselbe für alle Fokaleinstellungen wird.

Die Wirkungen der eben erwähnten Umstände will ich hier mit einigen Beispielen aus der betreffs dieser Frage bisher nur spärlich vorliegenden Praxis beleuchten. ROSENBERG hat[2]) bezüglich des Verhaltens eines der Tübinger Sternwarte gehörigen Apochromattessars von ZEISS (Öffnung 81 mm, Brennweite 842 mm) einige sehr interessante Resultate mitgeteilt. Ich[3]) selbst habe über einige Versuche berichtet, die ich teils mit dem Objektiv I des ZEISS-HEYDEschen Doppelastrographen der Upsalaer Sternwarte (Öffnung 150 mm, Brennweite 1485 mm), teils mit dem Objektiv des neuen ZEISSschen Astrographen derselben Sternwarte (Öffnung 200 mm, Brennweite 995 mm) angestellt habe. Diese zwei Objektive sind beide vom Triplet-Typus.

---

[1]) LINDBLAD, B.: Die photographisch-effektive Wellenlänge als Farbenäquivalent der Sterne. Arkiv f. mat., astr., fys. Bd. 13, Nr. 26, S. 5. 1918.

[2]) ROSENBERG, H.: l. c.   [3]) Astr. Nachr. 1924, Nr. 5280.

Der ungefähre Verlauf der Farbenkurven der drei genannten Objektive geht aus der hier mitgeteilten Figur hervor. Die Farbenkurve für das Apochromattessar verdanke ich einer freundlichst gegebenen brieflichen Mitteilung von ROSENBERG, die entsprechende Kurve für das 150 mm-Triplet ist von LINDBLAD nach der l. c., S. 4, beschriebenen Methode bestimmt, und die Kurve für das 200 mm-Triplet endlich ist von Y. ÖHMAN mittels Aufnahmen mit Objektivprisma bestimmt.

ROSENBERG fand, daß für das Apochromattessar die Fokussierung einen bedeutenden Einfluß auf die effektiven Wellenlängen ausübt. Mit zunehmender Fokaleinstellung nimmt die effektive Wellenlänge sowohl für die weißen als für die farbigen Sterne sehr stark ab. Gleichzeitig geht eine wesentliche Veränderung in der Art der Abhängigkeit der effektiven Wellenlänge von der Bildstärke vor sich: bei den intrafokalen Bildern nimmt die effektive Wellenlänge mit wachsender Bildstärke ab, bei den extrafokalen trifft gerade das Gegenteil ein. Die erwähnten Erscheinungen finden offenbar ihre Erklärung in dem einseitig schiefen Verlauf der Farbenkurve innerhalb des größten Teils des photographischen Spektralgebietes. Diese Eigentümlichkeit in der chromatischen Korrektion des in Frage stehenden Objektivs ist so stark ausgeprägt, daß sowohl für weiße als für gelbe Sterne das Intensitätsmaximum (etwa die für die schwächsten Bilder gültige effektive Wellenlänge) schon bei einer Variation von 1 mm in der Fokaleinstellung sich um $20-25\ \mu\mu$ verschiebt.

Das 150 mm-Triplet bildet in bezug auf den Verlauf der Farbenkurve einigermaßen das Gegenteil zum Apochromattessar. Der Scheitel der Farbenkurve liegt hier ziemlich weit nach der violetten Seite zu (etwa bei $\lambda\ 415\ \mu\mu$). Infolgedessen verläuft die Kurve in der Umgebung der mittleren photographischen Wellenlänge etwas schief; die Schiefe hat aber hier die entgegengesetzte Richtung von der beim Apochromattessar vorherrschenden. In voller Übereinstimmung hiermit fand ich, daß die effektive Wellenlänge in diesem Falle mit zunehmender Fokaleinstellung wächst. Dieser Zuwachs ist für die weißen Sterne mäßig (etwa $5-6\ \mu\mu$ für eine Änderung von 1 mm in der Fokaleinstellung), für die farbigen Sterne entschieden größer (etwa $10-15\ \mu\mu$). Dieses letzterwähnte Verhalten erklärt sich in natürlicher Weise durch den Umstand, daß die Farbenkurve im kurzwelligen Teil des photographischen Gebiets mehr symmetrisch verläuft als im langwelligen.

Das in der hier in Frage stehenden Hinsicht günstigste Verhalten zeigt das 200 mm-Triplet. Wie ich l. c. dargelegt habe, läßt sich hier (innerhalb der Grenzen für diejenigen Fokaleinstellungen, die überhaupt brauchbare Bilder geben) kein ausgeprägter Gang in den effektiven Wellenlängen mit variierender Fokussierung spüren. Dieses Ergebnis gilt für die Sterne aller Typen. Wie aus der Abbildung ersichtlich ist, nimmt die Farbenkurve dieses Objektivs gewissermaßen eine Zwischenstellung zwischen den für die beiden anderen Objektive gültigen ein, und sie verläuft ziemlich symmetrisch in der Umgebung der mittleren photographischen Wellenlänge (etwa 435 $\mu\mu$). Diese Gestalt der Farbenkurve scheint also insofern die vorteilhafteste zu sein, als der Einfluß der Fokussierung auf die effektiven Wellenlängen möglichst gering wird. Indessen gibt dieses Objektiv eine verhältnismäßig kleine Amplitude für die effektive Wellenlänge, insbesondere was die Sterne von den früheren Spektralklassen (B bis F) betrifft. Dieser Nachteil steht wahrscheinlich mit der ziemlich großen chromatischen Abweichung im ültravioletten Spektralgebiet im Zusammenhang. Die Firma ZEISS hat ähnliche Triplete berechnet, für welche die Farbenkurve etwa bei λ 405 $\mu\mu$ eine Wendung nach oben macht, so daß die Korrektion der kürzesten Wellenlängen besser wird. Es wäre recht interessant, ein solches Objektiv in bezug auf seine Verwendbarkeit zur Bestimmung effektiver Wellenlängen zu prüfen. Leider dürfte das brauchbare Feld dieser Objektive wahrscheinlich ziemlich klein sein.

Die obigen Bemerkungen zusammenfassend dürfte man sagen können, daß die Art der Achromasie des Objektivs eine wichtige Rolle bei der Bestimmung effektiver Wellenlängen spielt. Die Kurve für das sekundäre Spektrum soll nicht nur möglichst flach innerhalb des photographischen Spektralgebiets sein; sie soll außerdem möglichst symmetrisch um die mittlere photographische Wellenlänge verlaufen. Daß diese Forderung an ein zur Bestimmung effektiver Wellenlängen gut geeignetes Objektiv zu stellen ist, war ja theoretisch vorauszusehen. Mit der obigen kleinen Auseinandersetzung habe ich nur hervorheben wollen, daß die bisher vorliegende praktische Erfahrung direkte Beweise für die Richtigkeit dieser Auffassung liefert.

Zwölf Jahre lichtelektrischer Photometrie auf der Berliner Sternwarte.

Von **P. Guthnick**, Neubabelsberg.

Der Gedanke, die lichtelektrischen Erscheinungen für die Photometrie der Gestirne nutzbar zu machen, wurde mir erstmals im Jahre 1899 durch den mir befreundeten Physiker H. KREUSLER nahegelegt, der sich in jener Zeit des Photoeffektes des Platins zur Photometrie des ultravioletten Lichtes bediente. Ich erwog damals, angeregt durch frühere Versuche von MINCHIN, die Möglichkeit der Benutzung der Widerstandsänderungen des Selens zu einem Astrophotometer, wurde aber von KREUSLER auf den weit aussichtsreicheren Photoeffekt der Metalle aufmerksam gemacht. Die Angelegenheit blieb aber dann ruhen bis zum Jahre 1911, hauptsächlich wegen des damals noch sehr unentwickelten Zustandes der lichtelektrischen Technik. Um diese Zeit, abgesehen von gelegentlichen Versuchen in den Jahren 1908, 1909 und 1910, hatte ich eine photometrische Durchmusterung der helleren Sterne der 2. bis 7. Größe mit spektralen Besonderheiten oder sonstigen Verdachtsmomenten auf noch unendeckt gebliebene Veränderliche mit kleinen Helligkeitsschwankungen begonnen. Ich bediente mich dabei eines von TOEPFER gebauten ZOELLNERschen Photometers in Verbindung mit einem 16 cm-Refraktor. Diese Durchmusterung, über deren Ergebnisse einige kleine vorläufige Mitteilungen in den Astronomischen Nachrichten erschienen sind[1]), ließ erkennen, daß hier ein weites vielversprechendes Forschungsgebiet vorhanden sei, daß aber seine Erschließung ein genaueres Meßprinzip erforderte, als es in dem ZOELLNERschen Photometer gegeben war. Da inzwischen durch die bekannten Arbeiten von ELSTER und GEITEL die lichtelektrische Technik große Fortschritte gemacht hatte, so lag es nahe, nunmehr einen Versuch mit ihr zu machen. Im Sommer 1912 konnte ich, dank der freundschaftlichen Aushilfe der Physiker POHL und v. WARTENBERG mit den nötigen Apparaten, auf der alten Sternwarte zu Berlin die ersten Versuche an hellen Sternen mit einer ganz primitiven selbstgebauten Apparatur ausführen, die so günstig ausfielen, daß mir der damalige Direktor der Sternwarte, H. STRUVE, die Mittel zur Fort-

---

[1]) A. N. 4516, 4519, 4570.

führung der Versuche mit einer neuen, dem besonderen Zweck angepaßten Apparatur zur Verfügung stellte. Ich setzte mich nun mit den
Herren Elster und Geitel in Verbindung, und mit ihrer Unterstützung
kam das erste praktisch brauchbare (von Günther und Tegetmeyer
gebaute) lichtelektrische Sternphotometer der Berliner Sternwarte
zustande, das im Sommer 1913 in Betrieb genommen wurde. Die beiden
späteren Apparate, von denen der letzte Anfang 1924 vollendet war,
unterscheiden sich nicht wesentlich von dem ersten, dessen Prinzip
durchaus beibehalten worden ist. Insbesondere haben sich bewährt
die Art der Aufhängung des Elektrometers, die bewegliche Verbindung
der Zellenanode mit dem Elektrometerfaden und die durch ihre Einfachheit und Wirksamkeit ausgezeichnete Methode der Aufladungszeiten. Das letzte Photometer, das vorläufig zum Gebrauch am 125 cm-
Reflektor der Sternwarte bestimmt ist, unterscheidet sich von den beiden
älteren dadurch, daß bei ihm vier Photozellen in beliebiger Abwechslung benutzbar sind. Dieselben sind beständig betriebsfertig, so daß
ohne weiteres von einer Zelle zur anderen übergegangen werden kann.
Haben die Zellen stark verschiedene selektive Empfindlichkeit, was
durch die geeignete Wahl der Alkalimetalle erreicht wird, so bietet die
neue Anordnung, besonders wenn sie noch mit einem geeigneten System
von Farbenfiltern verbunden wird, ein wertvolles Mittel zur Bestimmung
von Farbenäquivalenten. Sie besitzt wesentliche Vorzüge vor dem
älteren Einzellenapparat, dessen Verwendung zur Bestimmung von
Farbenäquivalenten nur unter erheblichem Lichtverlust möglich ist.

Im folgenden soll eine gedrängte Übersicht über die hier mit der
lichtelektrischen Methode bisher erzielten Ergebnisse dargeboten
werden nebst einigen Betrachtungen über deren Bedeutung.

Gleich das erste Objekt, dessen Beobachtung zur Prüfung der Leistungsfähigkeit der Methode aufgenommen wurde, der merkwürdige,
von Frost aufgefundene spektroskopische Doppelstern mit nur $4^1/_2$stündiger Periode, β Cephei, erwies sich als ein Haupttreffer. Es ergaben
sich kleine Helligkeitsschwankungen von weniger als $0^m_.1$ Umfang,
die mit der Periode der Radialgeschwindigkeitsschwankungen verlaufen[1]). Der Charakter der Helligkeitsschwankungen war unerwartet.
Auf Grund der bisherigen Erfahrungen hatte ich wegen des sehr frühen
Spektrums des Sternes (B 1) Algol- oder β Lyrae-Typus vermutet,
in Wirklichkeit ergab sich jedoch ein Lichtwechsel, der dem der kurzperiodischen δ Cephei-Sterne nahezustehen scheint, aber sich von
diesem durch die viel beträchtlichere Veränderlichkeit der Lichtkurve
unterscheidet. Es wurde nun sogleich Umschau nach ähnlichen Fällen
unter den damals bekannten spektroskopischen Doppelsternen gehalten,
leider kam der einzige analoge Fall, β Canis majoris, wegen seiner süd-

---

[1]) A. N. 4701, 4903. Veröff. Berlin-Bab. I, 1; II, 3.

lichen Deklination nicht in Betracht. Erst 1918 wurde ich auf den ganz ähnlichen Fall 12 Lacertae (Spektrum B2, Periode $0^d.193$) aufmerksam, dessen spektroskopische Elemente 1915 an mir damals nicht zugänglicher Stelle von YOUNG mitgeteilt worden waren. Bemerkenswert ist die fast völlige Gleichheit der Perioden beider Sterne, ihres Spektrums und des Charakters ihres Lichtwechsels[1]). Trotz der starken und schnellen Veränderlichkeit der Lichtkurven sind die Perioden bisher vollkommen konstant geblieben. Inzwischen waren mir und meinem Mitarbeiter PRAGER (seit Anfang 1914) bei mehreren anderen Sternen Helligkeitsschwankungen aufgefallen, deren wahrscheinliche Verwandtschaft mit den beiden genannten Fällen uns zunächst noch verborgen blieb, die aber in der Folge mehr und mehr offenbar wurde. Ich führe namentlich mit ihren Spektren an: γ Bootis (F0n), β Ursae majoris (A0), g Ursae majoris (A5n), α Lyrae (A0), γ Lyrae (A0p, Si 4128, 4131!), α Cygni (cA2c), δ Cygni (A0n), γ Ursae minoris (A2n), α Persei (cF5), α Aurigae (gG0)[2]). Sterne mit sehr schnellen Radialgeschwindigkeitsschwankungen nach Art von β Cephei, β Canis majoris oder 12 Lacertae faßt man neuerdings unter dem Namen β Canis majoris-Typus zusammen. HENROTEAU hat alle bis 1923 bekannt gewordenen Sterne dieser Art zusammengestellt[3]). Außer den drei genannten Sternen führe ich aus dieser Liste an: δ Ceti (B2), 20 Tauri (B5), η Aurigae (B3n), g Ursae majoris (A5n), η Lyrae (B3), τ Cygni (Fn), γ Ursae minoris (A2n), δ Aquilae (F0n), und als noch zweifelhaft: γ Bootis (F0n) und β Ursae majoris (A0). Nicht von allen diesen Sternen ist die Periodizität der Radialgeschwindigkeitsschwankungen bereits nachgewiesen, vielmehr scheinen auch aperiodische Fälle vorzukommen. Die festgestellten Perioden liegen sämtlich unter $0^d.3$, die bisher kürzeste ist die von γ Ursae minoris mit $0^d.108$. Von den Sternen der HENROTEAUschen Zusammenstellung sind außer den bereits vorher angeführten noch unter lichtelektrischer Beobachtung gewesen: 20 Tauri, η Aurigae, τ Cygni und η Lyrae, der letzte nur wenige Male und nur in der allerersten Zeit. Spektroskopische Perioden sind abgeleitet für δ Aquilae ($0^d.157$), γ Ursae minoris ($0^d.108$), β Cephei ($0^d.190$), 12 Lacertae ($0^d.193$), g Ursae majoris ($0^d.155$), τ Cygni ($0^d.143$), β Canis majoris ($0^d.257$), 20 Tauri ($0^d.08$?), σ Scorpii ($0^d.247$), ν Eridani ($0^d.237$). Die Radialgeschwindigkeit von δ Ceti scheint unregelmäßig veränderlich zu sein[4]). Die Geschwindigkeitskurven sind fast in allen Fällen sehr stark und schnell veränderlich. Die lichtelektrischen Messungen bieten bezüglich der Helligkeitsschwankungen ein ganz ähnliches Bild dar. Für γ Bootis, β Ursae majoris,

---

[1]) A. N. 4983, 5156, Jubiläumsnummer; B. Z. II, 38.

[2]) Veröff. Berlin Bab. I, 1; II, 3. A. N. 4823, 4852, 4870, 4903—4904, 5074, Jubiläumsnummer, 5156. Festschrift Elster u. Geitel.

[3]) Publ. Ottawa V, 362; VIII, 77.

[4]) Publ. Ottawa V, 427.

γ Lyrae und α Aurigae konnten Perioden abgeleitet werden, nämlich bzw. $0\overset{d}{.}290$, $0\overset{d}{.}312$, $0\overset{d}{.}142$ und $0\overset{d}{.}072$. Außer dieser sehr kurzen Schwankung wurde bei α Aurigae einmal eine etwas langsamere Schwankung von wahrscheinlich $0\overset{d}{.}16$ Dauer beobachtet. Die Helligkeitsschwankungen von β Cephei, 12 Lacertae und γ Bootis sind nach den bisherigen Beobachtungen ununterbrochen vorhanden. Die Periode von γ Bootis war trotz der starken Veränderlichkeit der Lichtkurve 1914—23 ebenfalls sehr konstant. Dagegen war der schnelle Lichtwechsel von β Ursae majoris, γ Lyrae und α Aurigae nur zeitweise vorhanden. Der letzte Stern zeigte außerdem beständige langsame Helligkeitsschwankungen, die jedoch nicht mit der Periode der 104tägigen spektroskopischen Bahnbewegung in Beziehung zu stehen schienen. Merkwürdig ist, daß seine mittlere Helligkeit zur Zeit der kurzen Schwankungen merklich erhöht war. Bei β Cephei, 12 Lacertae und β Ursae majoris ist Übereinstimmung der spektroskopischen und photometrischen Perioden nachgewiesen, bei γ Bootis machen die wenigen bisher vorliegenden Radialgeschwindigkeiten die Übereinstimmung wahrscheinlich. Die zeitweise Unterbrechung des Lichtwechsels bei einigen der vorgenannten Sterne hat ihr Gegenstück in den Radialgeschwindigkeiten von β Canis majoris, die ebenfalls zeitweise frei von kurzen Schwankungen sind[1]). Merkwürdig sind auch die Vertauschungen der Maxima mit den Minima in den Geschwindigkeitskurven einiger Sterne (z. B. ν Eridani)[2]), für die es wahrscheinlich auch photometrische Analoga gibt. Nach einer brieflichen Mitteilung von Frost unterliegen auch die Spektra von α und γ Lyrae kleinen schnellen Veränderungen; eine Reihe von Radialgeschwindigkeitsbestimmungen von γ Lyrae am 24. April 1916 ergab eine Welle von $2\overset{h}{.}6$ Dauer und 15 km Amplitude. Auch die Helligkeitsschwankungen folgen nicht immer der Periode von $0\overset{d}{.}142$, sondern waren häufig kürzer oder länger und ohne erkennbare Periodizität; sie erinnern darin an die Schwankungen der Radialgeschwindigkeit von δ Ceti. Die Helligkeitsschwankungen von α Lyrae, α und δ Cygni sind einander und denen von γ Lyrae ganz ähnlich, nur daß bisher keinerlei Periode in ihnen zu erkennen war. Die Helligkeit von α Cygni unterliegt ähnlich wie die von α Aurigae auch nicht unbeträchtlichen langsamen Schwankungen. Die Schwankungen von α Persei waren etwas langsamer und flacher als die der vorgenannten Sterne; auch in der Radialgeschwindigkeit dieses Sternes könnte man schnelle Schwankungen vermuten[3]). Von γ Ursae minoris liegen mehrere längere Reihen von Helligkeitsmessungen bereits aus dem Sommer 1914 vor, da die schnelle Veränderlichkeit damals bei Gelegenheit der Beobachtung

---

[1]) Lick Obs. Bull. IX, 155.

[2]) Publ. Ottawa V, 59.

[3]) A. N. 4300, 4599. Lick Obs. Bull. I, 23; VII, 99.

elektrischen Messungen zeigten 1915 und 1916 recht regelmäßige Hellig-keitsschwankungen von $0^m_.04$ Umfang, die einer Periode von $0^d_.952$ folgten. Die Lichtkurve war eine merkwürdige Mischung einer aus-geprägten Algolkurve und einer Kurve mit mehreren Wellen. Vielleicht gehört dieser Stern ebenfalls zur β Canis majoris-Gruppe.

Von sonstigen neu aufgefundenen Veränderlichen seien noch α Dra-conis und γ Cygni etwas näher betrachtet[1]). Das Spektrum des er-steren Sternes ist A0p (Si 4128, 4131!), das des letzteren cF8. Ersterer ist ein spektroskopischer Doppelstern mit 51 Tagen Periode. Beide Sterne wurden einmal in einem langsam verlaufenden, flachen Minimum beobachtet, das bei α Draconis 50 Tage dauerte und $0^m_.08$ tief war, bei γ Cygni mindestens 10 Tage dauerte und $0^m_.05$ bis $0^m_.06$ tief war. Zu anderen Zeiten war die Helligkeit konstant. Aus diesen einzigen Beobachtungen läßt sich noch kein zuverlässiges Bild von der Art des Lichtwechsels machen, der an Verdunkelungen nach Art der von ε Aurigae erinnert.

Noch manche andere Sterne sind im Laufe der Jahre schwach ver-änderlich gefunden worden; die meisten bedürfen noch weiterer Er-forschung, einige bieten kein besonderes Interesse. Sie seien nur mit gelegentlichen kurzen Bemerkungen aufgezählt: α Orionis (außer der bekannten langsamen Schwankung wurde einmal eine kurze von $11^1/_2$ tägiger Periode beobachtet), α Ursae majoris ($6^d_.1$ Periode), ξ Ge-minorum, ξ und c Persei, δ Cassiopejae (Bedeckungsveränderlicher), λ und ω Orionis, Boss 1607 (Bedeckungsveränderlicher), ζ Cephei, α Canis minoris (langsame Schwankung), 8 Lacertae, σ Cygni, 20 Canum venaticorum, HR 8226, 13 Cephei, o Andromedae[2]). PLASKETTS Massen-stern $+ 6° 1309$ und o Herculis wurden konstant gefunden[3]). Besondere Studien betrafen und betreffen noch die Bedeckungsveränderlichen β Lyrae und u Herculis, die Nova Aquilae 3, die δ Cephei-Sterne, ζ Ge-minorum, δ Cephei und Polaris[4]). Die Ergebnisse dieser Untersuchungen harren zum Teil noch der Veröffentlichung. Die wichtigsten sind: die schon erwähnte Veränderlichkeit der Lichtkurve von β Lyrae, eine ebensolche, jedoch geringere bei u Herculis, sehr langsame und geringe der Lichtkurven von δ Cephei und ζ Geminorum, stärkere der Lichtkurve von Polaris, Bestimmung der Farbenkurve dieser Ver-änderlichen.

Ein interessantes, bereits 1913 in Angriff genommenes Arbeitsfeld

---

[1]) Veröff. Berlin-Bab. I, 1; II, 3.

[2]) Veröff. Berlin-Bab. I, 1; II, 3. A. N. 4852, 4891, 4903—4904, 5074, 5076, 5088, 5156, 5263; B. Z. I, 21; II, 13, 25.

[3]) B. Z. IV, 23. Jahresbericht Berlin-Bab. V. J. S. **58**, 83.

[4]) Veröff. Berlin-Bab. II, 3; III, 4. Berl. Ber. 1917, XII. A. N. 4980, 5037/38, 5061, 5098, Jubiläumsnummer, 5196; B. Z. I, 9, 11; II, 10, 13. Naturwiss. 1918, 593, 713.

Vielleicht gehören auch manche noch ungeklärte ältere Fälle hierhin, wie Miller Barrs Veränderliche RU Cassiopejae und PD 2201, d Serpentis, λ Geminorum, ζ Lyrae, o Herculis, δ Ursae majoris, SY Ursae majoris, Y Aquilae PD 695 u. a.[1]). Die genauere Untersuchung wird wahrscheinlich bei manchem der vorgenannten Sterne die Vermutung sehr schneller Helligkeitsschwankungen nicht bestätigen, ich habe aber geglaubt, diese Möglichkeit der Aufklärung dieser Fälle andeuten zu sollen.

Eine weitere Gruppe von Sternen, für die die Möglichkeit vorliegt, daß sie den vorstehend betrachteten Sternen nahestehen, von denen ich jedoch zunächst annehmen möchte, daß sie etwas Besonderes vorstellen, umfaßt Objekte wie γ Orionis, PD 207, π Cassiopejae, β Lyrae u. a.[2]). Der mit der bekannten spektroskopischen Periode verlaufende periodische Lichtwechsel von β Lyrae und π Cassiopejae soll außer Betracht bleiben. Die hier zu erörternden kleinen Helligkeitsschwankungen, die man vorläufig als aperiodische bezeichnen muß, bestehen darin, daß die normale Helligkeit sich sprungweise um einige Hundertstel einer Größenklasse nach oben oder unten ändert, in den Zwischenzeiten aber, die von unbestimmter Dauer sind, konstant bleibt. Die Sprünge vollziehen sich innerhalb weniger Tage oder auch Stunden, die Zwischenzeiten sind in der Regel viel länger. Diese Veränderungen bewirken bei β Lyrae und π Cassiopejae, daß der instantane Zug der Lichtkurve zuweilen oberhalb, zuweilen unterhalb des mittleren Zuges, diesem parallel, verläuft. Es erscheint nicht ausgeschlossen, daß diese Sterne in ausgedehnten absorbierenden Medien sich befinden, die ihr Licht zuzeiten mehr, zu anderen Zeiten weniger verdunkeln. Eine zweite Möglichkeit ist die zuerst angedeutete.

Sieht man zunächst von der letzten Gruppe von Veränderlichen ab, so kann man, wie dies auch bereits in früheren Mitteilungen[3]) mehrfach geschehen ist, als Direktive für die Forschung die Arbeitshypothese annehmen, daß trotz der Verschiedenheit im einzelnen die Sterne von der Art wie β Canis majoris, β Cephei, α und γ Lyrae, δ Ceti usw. eine im wesentlichen einheitliche Gruppe bilden, und daß diese Gruppe den δ Cephei-Veränderlichen und ihren Verwandten nahesteht. Es ist ferner auch früher schon die Vermutung ausgesprochen worden, daß diese Veränderlichen Doppelsternsysteme in statu nascendi seien[4]). Vielleicht ist es besser zu sagen, daß die besprochenen Erscheinungen (einschließlich der δ Cephei- und verwandten Symptome) Anzeichen

---

[1]) G. u. L. der Veränderlichen. A. N. 4516, 4519, 4818. Ap. J. IX, 546. Katalog vermuteter Veränderlichen von Gore und in Harvard Annalen XIV. Usw.

[2]) Veröff. Berlin-Bab. I, 1; II, 3. A. N. 4903—4904, 5156; B. Z. I, 21; II, 11. Berl. Ber. 1917, XII.

[3]) Veröff. Berlin-Bab. II, 3, S. 130.

[4]) Naturwiss. 1920, 211. Henroteau: Journal Canada R. A. S. XVI, 58.

für eine bevorstehende bzw. vor nicht langer Zeit eingetretene Spaltung in ein Doppelsternsystem seien. Sollten indessen die weiteren Forschungen zeigen, daß diese Erscheinungen sehr viel allgemeiner verbreitet sind, als es jetzt den Anschein hat, so wird man eher an Vorgänge zu denken haben, die mit der normalen Entwicklung des einzelnen Sternes verbunden sind.

Ich kehre nach diesen Abschweifungen zu den unmittelbaren Ergebnissen der lichtelektrischen Messungen zurück. Eine dritte Gruppe von Sternen, die sich als veränderlich erwiesen, zeigen Helligkeitsschwankungen, die periodisch und deutlich mit den periodischen Schwankungen ihrer Radialgeschwindigkeit verknüpft sind, ohne daß es sich um Bedeckungs-, normalen δ Cephei- oder β Canis majoris-Typus zu handeln scheint. Die ersten Fälle dieser Art, die hier gefunden wurden, sind o Persei, φ Persei und α Geminorum; später kamen hinzu ζ_1 Ursae majoris, Boss 46 und manche andere[1]). Es sollen nur die charakteristischsten Fälle kurz besprochen werden. Bei α Geminorum wurde 1914 ein mit der 2.9tägigen spektroskopischen Periode der Komponente α_1 verlaufender Lichtwechsel beobachtet, der aus drei nahe äquidistanten Maxima und Minima bestand. Die wahre Amplitude des Lichtwechsels muß mehr als $0^{m}\!.2$ betragen haben. Spätere gelegentliche Nachprüfungen des Sternes ergaben keinen oder nur sehr geringen Lichtwechsel mit mehr als drei Wellen, so daß bei diesem Stern auch kurze Schwankungen nach Art der von α Lyrae usw. in Frage kommen. Bei o Persei ergab sich ein mit der spektroskopischen Periode von $4^{d}\!.42$ verlaufender Lichtwechsel, der zeitweilig aus mindestens sieben, hier aber nicht mehr äquidistanten Maxima und Minima bestand. Die Lichtkurve ist stark veränderlich. Auch bei diesem Stern kommt Verwandtschaft mit der Gruppe α Lyrae usw. in Betracht. Fast noch merkwürdiger ergab sich der Lichtwechsel von φ Persei mit der 126tägigen spektroskopischen Periode, der aus sieben nahe gleichlangen Wellen besteht, von denen die eine oder andere gelegentlich ausfällt und durch Zeiten konstanter Helligkeit ersetzt wird. Der periodische Lichtwechsel von π Cassiopejae bestand 1920 aus zwei Maxima und Minima. Sein Umfang betrug nur $0^{m}\!.04$. Ganz ähnlich ist der Lichtwechsel von Boss 46, nur erreicht sein Umfang fast $0^{m}\!.2$. Die Form der Lichtkurve, insbesondere die Lage der Maxima und Minima, ist eigentümlichen langsamen Veränderungen unterworfen. In keinem dieser Fälle kommen Bedeckungen oder ellipsoidischer Lichtwechsel in Betracht. Die Verhältnisse bei diesen Sternen sind aus der folgenden Zusammenstellung (s. S. 398) zu ersehen.

Außer bei α Geminorum sind in den Spektren aller dieser Sterne beide Komponenten sichtbar.

---

[1]) Veröff. Berlin-Bab. I, 1; II, 3. A. N. 4885, 4903—4904, 5061, 5156; B. Z. I, 8, 21; II, 11. Atti Acc. Pont. 1923, VI.

Stern	Spektrum	Spektroskopische Periode	Amplitude (K_1)	Umfang des Lichtwechsels
α_1 Geminorum . . .	A0	$2^{\mathrm{d}}928$	32 km	$0^{\mathrm{m}}2+$
φ Persei	Bpevr	126.6	27 ,,	0.1
o Persei	B1k	4.419	112 ,,	0.07+
ζ_1 Ursae majoris . .	A2p	20.536	71 ,,	0.05±
π Cassiopejae . . .	A5	1.964	118 ,,	0.04
Boss 46	B3k	3.522	217 ,.	0.19

Nicht ohne weiteres mit diesen Sternen zusammenstellen möchte ich den merkwürdigen Stern α Canum venaticorum[1]). Sein Spektrum zeigt bekanntlich verwickelte Veränderungen der Lage, des Aussehens und der Intensität der Linien, die einer $5\frac{1}{2}$tägigen Periode folgen. Linienverdoppelungen scheinen die Duplizität des Sternes zu sichern. Die Helligkeit schwankt mit der $5\frac{1}{2}$ tägigen Periode um durchschnittlich $0^{\mathrm{m}}05$. Die Lichtkurve hat in der Regel ein Maximum und Minimum, zuweilen treten aber auch kleine sekundäre Wellen auf, überhaupt ist die Form der Lichtkurve ziemlich stark veränderlich. Die Beobachtungen von 1914, 1915, 1916, 1920 bis Ende Mai zeigten stets diesen Lichtwechsel, der sich mit einer Periode von $5.^{\mathrm{d}}470$ und einer konstanten Ausgangsepoche durchaus befriedigend darstellen läßt. Die Periode ist die der spektralen Erscheinungen. Der Umfang der Helligkeitsschwankungen betrug im Mai 1920 nur noch $0^{\mathrm{m}}03$. Im Juni und Juli 1920 wurde der Lichtwechsel unmerklich. Im folgenden Jahre war er wieder vorhanden mit einem Umfange von $0^{\mathrm{m}}04$. Maximum und Minimum lagen wieder bei den früheren Phasen (die gegenteilige Angabe in der Jubiläumsnummer der A. N. beruht auf einem Versehen), aber die Messungen zeigen eine ungewöhnlich starke Streuung um den mittleren Zug der Lichtkurve und deuten damit eine starke Unruhe auf dem Stern an. Für das Jahr 1922 liegen leider nur vier Messungen vor, nach denen das Maximum und das Minimum auf ihre alten Phasen gefallen und die Amplitude wieder nahe $0^{\mathrm{m}}05$ gewesen zu sein scheint. Im Jahre 1923 endlich ergaben die Messungen wieder eine glatte Lichtkurve von $0^{\mathrm{m}}05$ Amplitude, indessen war im Gegensatz zu den früheren Jahren das Maximum sehr spitz, das Minimum sehr flach. Trotz der Unterbrechung des Lichtwechsels im Sommer 1920 werden auch die neuen Maxima und Minima durch die alten Elemente gut dargestellt.

Ein ebenfalls etwas außenstehender Fall ist ε Ursae majoris[2]). Das Spektrum ist A0p. Der Stern ist ein bekannter langperiodischer, spektroskopischer Doppelstern. Ludendorff vermutete außer den langsamen Schwankungen der Radialgeschwindigkeit noch schnelle. Die licht-

---

[1]) Veröff. Berlin-Bab I, 1; II, 3. Dort auch die spektroskopische Literatur. Ferner A. N. 4903—4904, Jubiläumsnummer; B. Z. II, 25, 27.

[2]) Festschrift Elster u. Geitel. Veröff. Berlin-Bab. II, 3. A. N. 4903—4904.

elektrischen Messungen zeigten 1915 und 1916 recht regelmäßige Helligkeitsschwankungen von $0^m.04$ Umfang, die einer Periode von $0^d.952$ folgten. Die Lichtkurve war eine merkwürdige Mischung einer ausgeprägten Algolkurve und einer Kurve mit mehreren Wellen. Vielleicht gehört dieser Stern ebenfalls zur β Canis majoris-Gruppe.

Von sonstigen neu aufgefundenen Veränderlichen seien noch α Draconis und γ Cygni etwas näher betrachtet[1]). Das Spektrum des ersteren Sternes ist A0p (Si 4128, 4131!), das des letzteren cF8. Ersterer ist ein spektroskopischer Doppelstern mit 51 Tagen Periode. Beide Sterne wurden einmal in einem langsam verlaufenden, flachen Minimum beobachtet, das bei α Draconis 50 Tage dauerte und $0^m.08$ tief war, bei γ Cygni mindestens 10 Tage dauerte und $0^m.05$ bis $0^m.06$ tief war. Zu anderen Zeiten war die Helligkeit konstant. Aus diesen einzigen Beobachtungen läßt sich noch kein zuverlässiges Bild von der Art des Lichtwechsels machen, der an Verdunkelungen nach Art der von ε Aurigae erinnert.

Noch manche andere Sterne sind im Laufe der Jahre schwach veränderlich gefunden worden; die meisten bedürfen noch weiterer Erforschung, einige bieten kein besonderes Interesse. Sie seien nur mit gelegentlichen kurzen Bemerkungen aufgezählt: α Orionis (außer der bekannten langsamen Schwankung wurde einmal eine kurze von $11^1/_2$ tägiger Periode beobachtet), α Ursae majoris ($6^d.1$ Periode), ξ Geminorum, ξ und c Persei, δ Cassiopejae (Bedeckungsveränderlicher), λ und ω Orionis, Boss 1607 (Bedeckungsveränderlicher), ζ Cephei, α Canis minoris (langsame Schwankung), 8 Lacertae, σ Cygni, 20 Canum venaticorum, HR 8226, 13 Cephei, o Andromedae[2]). PLASKETTS Massenstern $+ 6° 1309$ und o Herculis wurden konstant gefunden[3]). Besondere Studien betrafen und betreffen noch die Bedeckungsveränderlichen β Lyrae und u Herculis, die Nova Aquilae 3, die δ Cephei-Sterne, ζ Geminorum, δ Cephei und Polaris[4]). Die Ergebnisse dieser Untersuchungen harren zum Teil noch der Veröffentlichung. Die wichtigsten sind: die schon erwähnte Veränderlichkeit der Lichtkurve von β Lyrae, eine ebensolche, jedoch geringere bei u Herculis, sehr langsame und geringe der Lichtkurven von δ Cephei und ζ Geminorum, stärkere der Lichtkurve von Polaris, Bestimmung der Farbenkurve dieser Veränderlichen.

Ein interessantes, bereits 1913 in Angriff genommenes Arbeitsfeld

---

1) Veröff. Berlin-Bab. I, 1; II, 3.

2) Veröff. Berlin-Bab. I, 1; II, 3. A. N. 4852, 4891, 4903—4904, 5074, 5076, 5088, 5156, 5263; B. Z. I, 21; II, 13, 25.

3) B. Z. IV, 23. Jahresbericht Berlin-Bab. V. J. S. **58**, 83.

4) Veröff. Berlin-Bab. II, 3; III, 4. Berl. Ber. 1917, XII. A. N. 4980, 5037/38, 5061, 5098, Jubiläumsnummer, 5196; B. Z. I, 9, 11; II, 10, 13. Naturwiss. 1918, 593, 713.

für die lichtelektrische Methode ist die Bestimmung der Farbenäquivalente der Sterne[1]). Anfangs geschah die Bestimmung in der Weise, daß die zu untersuchenden Sterne nacheinander mit zwei Zellen verschiedener selektiver Empfindlichkeit (z. B. mit einer Na-Zelle und mit einer Rb-Zelle) durchgemessen wurden. Setzt man dann z. B. für das Spektrum A0 die Na- und die Rb-Helligkeit gleich, so erhält man Farbenäquivalente, die den photographisch-visuellen Farbenindizes entsprechen, aber eine kleinere Amplitude haben. Später wurde der Gebrauch von Gelbfiltern und schließlich von Blau- und Gelbfiltern mit einer einzigen Zelle eingeführt, wobei die Helligkeit der Sterne mit und ohne Filter bzw. mit Gelbfilter und mit Blaufilter gemessen und die erhaltenen Unterschiede als Farbenindizes angesehen wurden. Mittels der letzten Methode hat insbesondere BOTTLINGER seine Untersuchung über die Farbenindizes von 459 Sternen ausgeführt. Ein Versuch, in drei verschiedenen Spektralgebieten zu messen und so zwei unabhängige Farbenindizes (zur Bestimmung der absoluten Helligkeit) zu erhalten, scheiterte an der zu geringen Verschiedenheit der mit einer einzigen Zelle und nicht zu stark schwächenden Farbenfiltern erhaltenen Spektralgebiete. Für diese Aufgabe ist der eingangs erwähnte Vierzellenapparat gebaut worden, der vorläufig am 125 cm-Reflektor verwendet wird und mit dem voraussichtlich hinreichend verschiedene Spektralgebiete zugänglich werden.

Die lichtelektrische Einrichtung ist endlich dazu benutzt worden, die von LANGLEY und ABBOT beobachteten kurzen unregelmäßigen Schwankungen der Sonnenstrahlung zu prüfen[2]). Ein bereits 1914 unternommener Versuch, die Sonnenhelligkeit mittels Messungen der Helligkeit der großen Planeten zu prüfen, führte einerseits zunächst bei Mars zur Entdeckung von Helligkeitsschwankungen, die teils mit der Rotationsperiode verlaufen, teils anscheinend mit meteorologischen Vorgängen auf dem Planeten zusammenhängen, und machte anderseits auf Grund der Messungen von Saturn die Existenz schneller Schwankungen der Sonnenstrahlung zweifelhaft. Da der Versuch, mit einem Spezialapparat unmittelbare absolute Helligkeitsmessungen der Sonne auszuführen, an der Ungunst des mitteleuropäischen Klimas scheiterte und auch Anschlüsse des Mondes an helle Sterne sich als nicht ausführbar erwiesen, so wurden die Messungen an den großen Planeten fortgesetzt. Mars erwies sich für die Prüfung der Sonnenhelligkeit als ganz unbrauchbar, Jupiter ist nur mit großer Vorsicht zu benutzen, da er in der Opposition 1920 Helligkeitsschwankungen mit der Rotationsperiode zeigte, die zur Zeit ihrer größten Ausbildung $0^{m}\!.14$ betrugen.

---

[1]) Veröff. Berlin-Bab. II, 3; III, 4. A. N. 4763, 5037—5038, 5088; B. Z. I, 9.
[2]) Veröff. Berlin-Bab. I, 1; II, 3. A. N. 4903—4904, 4934, 4938, 4976, 5067: B. Z. II, 22. Naturwiss. 1918, 133.

Dagegen ist Saturn weit besser geeignet, wenigstens wenn man von der unmittelbaren Umgebung der Opposition absieht, wo der SEELIGERsche Ringeffekt merklich wird. Das Ergebnis aller Messungsreihen läßt sich dahin zusammenfassen, daß es nicht gelungen ist, mit Sicherheit Schwankungen der Sonnenhelligkeit nachzuweisen, die im photographischen Spektralgebiet 1% des Mittelwertes übersteigen. Bezüglich des Jupiter ergab sich nebenbei eine Bestimmung des bis dahin für unmerklich gehaltenen Phasenkoeffizienten, der 1917—1918 den erheblichen Wert von $0^m\!.015$ erreichte, jedoch in den verschiedenen Oppositionen etwas verschieden ist. Die Messungen des Saturn in der Umgebung der Opposition lieferten eine glänzende Bestätigung der SEELIGERschen Ringtheorie. Der Verlauf des Ringeffektes entspricht völlig der Voraussage der Theorie und läßt erkennen, daß die Dichte des Ringes sehr viel geringer ist, als man vorher angenommen hatte. Diese Feststellung ist vielleicht das schönste der bisherigen Ergebnisse der lichtelektrischen Photometrie. Zur weiteren Prüfung der Sonnenhelligkeit sind am 125 cm-Reflektor Helligkeitsmessungen. an kleinen Planeten geplant, von denen mehrere an diesem Instrument meßbar sein werden.

Zum Schluß möchte ich noch kurz über die bisherigen Erfahrungen bei lichtelektrischen Messungen am 125 cm-Reflektor berichten, da sie für jeden von Interesse sind, der sich mit solchen Messungen zu beschäftigen beabsichtigt. Das lichtelektrische Photometer ist bisher nur im Cassegrain-Fokus des Reflektors benutzt worden. Die Optik des Cassegrain-Systems besteht wie bei den Reflektoren auf dem Mt. Wilson außer dem Hauptspiegel und dem Konvexspiegel aus einem dicht über dem Hauptspiegel angebrachten ebenen, um 45° gegen die optische Achse geneigten Hilfsspiegel, der die vom Konvexspiegel kommenden Strahlen unten seitwärts aus dem Tubus hinaus reflektiert. Dieser Hilfsspiegel reflektiert aber unvermeidlich auch eine beträchtliche Menge direkten Himmelslichtes, das, neben dem Konvexspiegel in den Tubus einfallend, ihn trifft. Infolgedessen hat man im Cassegrain-Fokus trotz des kleinen Öffnungsverhältnisses von 1 : 20 stets mit beträchtlichem Himmelsgrundeffekt zu rechnen, dessen Berücksichtigung eine sehr lästige Beigabe ist. Dieser Effekt ist im Cassegrain-Fokus wahrscheinlich sogar größer als im Newton-Fokus, obgleich für diesen das Öffnungsverhältnis 1 : 7 beträgt. Zu diesem Nachteil kommt dann noch die Schwerfälligkeit des großen Reflektors, der ihn zu lichtelektrischen Messungen verhältnismäßig wenig geeignet macht. Aus diesen Gründen scheint mir der frühere Plan für ein lichtelektrisches Spezialinstrument[1]) heute noch mehr als damals der zweckmäßigste zu sein. Dieser Plan besteht im wesentlichen darin, daß ein Refraktor von mög-

---

[1]) Veröff. Berlin-Bab. II, 3, S. 141.

lichst großer Öffnung und möglichst großem Öffnungsverhältnis (etwa
1 : 3 oder 1 : 4) gewählt wird. Das Objektiv braucht nicht achromatisch
zu sein. Mittels zweier rechtwinkliger Prismen wird das Licht durch die
Deklinationsachse und die Stundenachse bis zu deren unterem Ende geleitet, wo der lichtelektrische Apparat fest aufgestellt wird. Das erste
Prisma, das im Tubus gegenüber der Deklinationsachse sich befindet, muß
dem Brennpunkt des Objektives so nahe sein, daß es keine ungebührliche Größe zu haben braucht. Mit diesem Prisma, und nötigenfalls
auch mit dem zweiten Prisma in der Kreuzung von Deklinations- und
Stundenachse, sind Korrektionslinsen zur Verlängerung der Brennweite
und zur notdürftigen Kompensation der sphärischen und chromatischen
Fehler des Objektives zu verbinden. Diese Anordnung ermöglicht
vollkommenen Schutz des lichtelektrischen Apparates gegen äußere
Einflüsse und weitgehendste Bequemlichkeiten für den Beobachter,
dessen Leistungsfähigkeit dadurch erheblich gesteigert werden würde.
Die Kosten einer solchen Spezialeinrichtung würden verhältnismäßig
gering sein.

Ionisation und Atomtheorie.

Von **G. Schnauder** †, Potsdam.

Der Aufsatz lag in erster Ausarbeitung vor, als sein Verfasser plötzlich starb. Er wurde nach Ergänzung der Formeln so abgedruckt, wie er sich im Nachlaß vorfand.   K.

Um den Begriff der Ionisation und der damit zusammenhängenden Größen so klar zu erfassen, wie es nötig ist, um den gegenwärtigen Stand der Forschung und die augenblicklichen Schwierigkeiten voll zu überblicken, müssen wir zwar eine allgemeine Kenntnis der heutigen Atomtheorie voraussetzen, wollen aber doch in einer ganz knappen Übersicht die für uns wesentlichen Züge dieser Theorie herausarbeiten.

Da wir später die SAHAsche Theorie nach einer Methode entwickeln werden, die sich von speziellen Atomvorstellungen möglichst frei hält und sich im wesentlichen auf statistisch-energetische Überlegungen stützt, so werden wir diesen Gesichtspunkt hervorzuheben versuchen und auf in diesem Sinne Nebensächliches verzichten.

Ein Atom besteht aus einem *Kerne*, d. h. einem positiv elektrisch geladenen Zentrum, sowie aus negativen *Elektronen*. Beide Arten von Teilchen besitzen gegen- und untereinander potentielle Energie, die sich äußert in dem Bestreben der Kerne, eine der Zahl ihrer positiven Einheiten [*Kernladungszahl* Z[1])] gleiche Anzahl von Elektronen möglichst fest bei sich zu vereinen, d. h. sich zu *neutralisieren* und zu *stabilisieren*.

Einen Z-Kern, der Z Elektronen bei sich hat, nennen wir ein *neutrales Atom*; fehlen ihm aber noch m Elektronen, so nennen wir das Ganze einen *m-fach ionisierten Z-Trümmer* und bezeichnen die Zahl m als seine *Ionisierungszahl*[2]). Atome und Trümmer werden wir häufig bequem unter der Bezeichnung *Teilchen* oder *freie Teilchen* zusammenfassen.

Wie man sich auch modellmäßig den Bau eines Teilchens und die dazu führenden Möglichkeiten denken mag, stets wird man die Voraussetzung zugrunde legen, daß das einzelne Teilchen von sich aus einem Zustande möglichst hoher Stabilität, d. h. möglichst kleiner potentieller Energie seiner Bestandteile, zustrebe, und daß ein solcher Zustand in jedem Falle eindeutig existiere. Die chemische Konstanz der Atome

---

[1]) Bekanntlich variiert nach unserer heutigen Kenntnis Z zwischen 1 und 92, also jedenfalls bis zu einer endlichen Grenze.

[2]) Es ist also $0 \leqq m \leqq Z$.

und die Konstanz ihrer Spektren legt bekanntlich diese Grundannahme sehr nahe. In diesem Zustande nennen wir ein Teilchen *unerregt*.

Die Erfahrungen der Atomphysik lehren, daß die verschiedenen Elektronen eines unerregten Teilchens durchaus nicht gleich zu bewerten sind. Es bedarf nämlich ganz verschiedener Energiebeträge, je nachdem man das eine oder das andere Elektron aus dem Verbande des Teilchens zu entfernen trachtet. Innerhalb desselben erscheinen also die Elektronen energetisch gruppiert und nach den neuesten Erfahrungen hat es fast den Anschein, als ob jedem Elektron im Verbande sein bestimmter, von allen anderen verschiedener Energiebetrag zur Losreißung zukommt, was nicht ausschließt, daß diese *Energieniveaus* sich in ausgeprägten und z. T. sehr engen Gruppen anordnen.

Wir verstehen nun unter *der Ionisationsenergie* eines bestimmten Elektrons eines bestimmten Teilchens diejenige Energiemenge, die nötig ist, um aus dem ruhenden, unerregten Teilchen *das* Elektron loszureißen und im Unendlichen zur Ruhe zu bringen. Zur Beschreibung dieser Energie bedarf es also eines Symboles mit mindestens 3 Parametern, nämlich: 1. die Kernladung Z, 2. die Ionisationszahl m, 3. eine Art Nummer des Elektrons im Teilchenverbande. Nach allen Erfahrungen ist letztere eine Funktion einer nicht genau angebbaren Zahl (mindestens 3) von ganzzahligen Parametern (Quantenzahlen).

Da gegenwärtig gar kein Grund vorliegt zu der Annahme, daß die hiernach möglichen Fälle von Ionisation nicht sämtlich tatsächlich vorkommen können, so ist schon hier klar, daß eine strenge Behandlung des Ionisierungsvorganges jedenfalls sehr verwickelt ist. Entscheidend kommt indessen hinzu, daß man neben unerregten auch noch erregte Teilchen zu unterscheiden hat. Durch quantenmäßige Aufnahme von Energiebeträgen, die im Verhältnis zur Ionisationsenergie klein sind, können die Elektronen eines Teilchens *erregt*, d. h. auf höhere *Energieniveaus* gehoben werden, ohne den Verband zu verlassen. Da sie auf diesen Niveaus eine Zeitlang *(stationär)* zu beharren vermögen, bis sie unter Energieabgabe bzw. -aufnahme in andere übergehen, so ist es möglich, daß Elektronen aus angeregten Zuständen heraus ionisiert werden, wozu dann ein Energiebetrag hinreicht, der kleiner ist als die oben definierte Ionisationsenergie. Macht die hierdurch bedingte Verwickelung der Verhältnisse die Aufgabe bei einer beliebigen Mischung von Teilchen scheinbar unlösbar, so sind nun einige Umstände aufzuführen, durch die die Sachlage in vielen Fällen einer einfachen Analyse zugänglich gemacht wird.

a) Eine erste allermeist gemachte Annahme ist die, daß *in einem Elementarprozesse der Ionisation nur ein einziges Elektron* abgetrennt bzw. angelagert werden solle. Die Berechtigung dieser Annahme ist nicht unbedingt sicher. Es ist für Helium nachgewiesen, daß in einem Vorgange zugleich zwei Elektronen abgerissen werden können. Hierzu war allerdings eine ganz besonders starke Anregung erforderlich; wenn man aber bedenkt, daß gerade das Helium dasjenige Element ist, das in normaler Weise bei weitem am schwierigsten zu ionisieren ist, daß es sein erstes Elektron erst bei einer Erregungsstärke hergibt, bei der alle anderen Elemente sicher das erste, meistens längst das zweite, und z. T. schon das dritte Elektron haben hergeben müssen, so wird man die Berechtigung des ausgedrückten Zweifels anerkennen. Eine Prüfung der Sachlage dürfte indessen sehr schwierig sein und unsere Annahme ist daher vorläufig als unkontrollierbare Arbeitshypothese zu betrachten.

b) Weit weniger bedenklich ist eine zweite wichtige und stark ver-einfachende Annahme. Erfahrungsgemäß wachsen bei festem Z die Ionisationsenergien mit m ziemlich stark (z. B. bei den Erdalkalien zwischen $m = 0$ und $m = 1$ auf fast das Doppelte). Da nun die für uns durch Temperatur und Druck gegebenen Anregungsbedingungen nach allen Erfahrungen in einem bestimmten Raumgebiete nur sehr wenig um einen Mittelwert streuen, so liegt es sehr nahe, in erster Näherung immer nur zwei *benachbarte* Atomzustände als miteinander in Reaktion befindlich zu behandeln.

c) Drittens endlich zeigen die Arbeiten von BOHR, daß es für jedes Teilchen eine *höchstquantige Gruppe* von bis zu 4 Elektronen gibt — sog. Leucht- oder Valenzelektronen —, die energetisch nahezu gleichwertig sind. Ihre Anzahl bezeichnen wir als *Valenz* σ. Indem man nun die geringen Unterschiede zwischen diesen Elektronen zunächst beiseite läßt, kommt man in die Lage, für das ganze Teilchen eine *einheit-liche* Ionisationsenergie $\chi(Z, m)$ einzuführen als diejenige Arbeit, die nötig ist, um dem Teilchen ein erstes Valenzelektron zu entreißen und dieses im Unendlichen zur Ruhe zu bringen. Die Annahme von σ *völlig* gleichberechtigten Elektronen bedingt allerdings, daß man nun dem Teilchen eine σ-fache *Ionisationswahrscheinlichkeit* zusprechen muß und damit gewinnt der Begriff der Valenz für die Theorie der Ionisation eine unmittelbare Bedeutung.

1. Ionisation und Statistik (erster Teil). Wir gehen nun dazu über, die entwickelten Vorstellungen in einer zunächst sehr elementaren, aber ziemlich allgemeinen Betrachtung anzuwenden. Da wir uns in der SAHAschen Theorie ausschließlich mit den Verhältnissen in Sternatmosphären zu beschäftigen haben, so seien diese zunächst ganz kurz allgemein charakterisiert. In dem Entwicklungszustande keines einzigen Sternes hat man bis jetzt eine dauernde und als stetig zu betrachtende Entwicklung bemerken können. Diese ist also normalerweise so langsam, daß man unbedenklich für eine erste Annäherung jeden Stern als im Zustande des Gleichgewichtes betrachten kann. Zum mindesten gilt das hinsichtlich der im Sterne enthaltenen trägen Atome und Elektronen, während man bei der Strahlung, die ja de facto dauernd in den Raum hinaus verlorengeht, vorsichtiger von einem *stationären* Zustande sprechen sollte.

Man pflegt den Gleichgewichtszustand nach zwei Gesichtspunkten hin zu betrachten, nämlich einmal als „mechanisches", insoweit es in einem Ausgleiche des infolge der Schwere eintretenden Gasdruckes mit dem Impuls der Strahlung besteht und uns an dieser Stelle nicht weiter interessiert, zweitens aber als „chemisches", insoweit die Konzentrationen der verschiedenen Teilchenarten in ihrem Ausgleiche betroffen werden. Bezüglich dieses *Reaktionsgleichgewichtes* läßt sich nach der Weise der Chemiker eine sehr allgemeine einfache Betrachtung anstellen, die wir nun vorführen.

Die Materie in der Sternhülle wollen wir uns völlig dissoziiert, d. h. aus lauter einatomigen Molekülen bestehend denken und wollen nur die Ionisation betrachten. Wir nehmen weiter an, die Hülle sei im Mittel unelektrisch, so daß gerade so viele Elektronen vorhanden sind, als zur vollen Sättigung der Kerne nötig sind. Wir denken uns einen Posten der Materie dem Sterne entnommen und auf den *Grundzustand* völliger Sättigung aller Kerne gebracht; es mögen dann M Mole sein. Der Prozentsatz der Atome mit einem Z-Kerne sei $\varkappa(Z)$, welche Größen wir *Grundkonzentrationen* nennen. Es ist also:

$$\sum_{1}^{92} {}_{Z}\varkappa(Z) = 1 \tag{1}$$

Nun denken wir uns die Materie in den Stern zurückgebracht und werden dabei beobachten, daß von den Atomen mehr oder weniger Elektronen abgespalten werden. Es sei dann $x(Z, m)$ derjenige Prozentsatz der Z-Kerne, der jetzt die Ionisationszahl m aufweist. Er heißt der *Konzentrationsgrad der m-fachen Ionisation* der Z-Atome und hat die Bedeutung eines Mittelwertes (vgl. später). Setzen wir die *Avogadrosche Hypothese* als gütig voraus, dann ist infolge der Ionisation die ursprüngliche Molzahl M gewachsen auf:

$$M' = \sum_{1}^{92} {}_{Z}[M(Z,0) + M(E) \sum_{1}^{Z} {}_{m} M(Z,m)] \tag{2}$$

worin bedeuten:

$$M(Z,0) = M \cdot \varkappa(Z) \left[1 - \sum_{1}^{Z} {}_{m} x(Z,m)\right]$$

die Molzahl der noch neutralen Z-Atome,

$$M(Z,m) = M \cdot \varkappa(Z) \, x(Z,m)$$

die Molzahl der (Z, m)-Teilchen,

$$M(E) = M \cdot \varkappa(Z) \sum_{1}^{Z} {}_{m} m \cdot x(Z,m)$$

die Molzahl aller überhaupt frei gewordenen Elektronen.

$$\left.\vphantom{\begin{array}{c}1\\2\\3\\4\\5\\6\end{array}}\right\} \tag{3}$$

Durch Division dieser Zahlen durch M' ergeben sich unmittelbar die *Gleichgewichtskonzentrationen* bzw. $k(Z, 0)$, $k(Z, m)$, $k(E)$ der verschiedenen nun vorhandenen Teilchenarten.

Die absolute Häufigkeit der pro Sekunde und Mol eintretenden Spaltungsprozesse von der Art $T(Z, m) \to T(Z, m + 1) + E$, wo T ein Teilchen und E ein freies Elektron bedeutet, ist als Wahrscheinlichkeit proportional der Zahl der vorhandenen ungespaltenen $T(Z, m)$, d. h. proportional zu $k(Z, m)$. Dagegen ist diejenige der aus einem $T(Z, m + 1)$ und einem E ein $T(Z, m)$ zurückbildenden Prozesse proportional dem Produkte $k(Z, m + 1)\, k(E)$. Im Gleichgewichte ist die Zahl der zerfallenden und der neugebildeten $T(Z, m)$ gleich und es ergibt sich somit für jede Teilchenart eine Gleichung der Art:

$$k(Z, m + 1) \cdot k(E) : k(Z, m) = K(Z, m, T, P, \ldots) \quad \left.\begin{array}{l} Z = 1, 2, 3, \ldots 92 \ldots \\ m = 0, 1, 2, \ldots Z \end{array}\right\} (4)$$

Dabei ist die Richtigkeit dieser Überlegung wesentlich bedingt durch die oben gemachten Annahmen a) und b). Die rechte Seite von (4) enthält die Konzentrationen nicht mehr, sondern nur die Größen Z und m sowie eine Anzahl von Parametern T, $P \ldots$, von denen der allgemeine Zustand der Materie sonst abhängt. Wichtig ist vor allem, daß auf der rechten Seite keine Daten eingehen, die auf die anderen noch vorhandenen Teilchen speziell Bezug haben. Setzen wir nun für die k ihre Werte nach (2) und (3) ein, so ergibt sich:

$$\frac{x(Z, m + 1)}{x(Z, m)} \sum_{Z}^{92} \left(\frac{x(Z)}{N} \sum_{m}^{Z} m \cdot x(Z, m) \right) = K(Z, m, T, P \ldots) \quad (5)$$

wo:
$$N = 1 + \sum_{Z}^{92} \left(x(Z) \sum_{m}^{Z} m \cdot x(Z, m) \right)$$

Für ganz spezielle Fälle haben SAHA und besonders RUSSELL diese Formeln schon abgeleitet; man sieht, daß sie auch im allgemeinsten Falle sehr einfach bleiben.

Wir betrachten den Fall eines überall nur einfach ionisierten Gasgemisches, wie ihn RUSSELL behandelt hat. Es sind also alle $m = 1$, so daß wir statt $x(Z, m)$ einfach $x(Z)$ schreiben können, außerdem wird $x(Z, 0) = 1 - x(Z)$ und (5) geht damit über in:

$$\frac{x(Z)}{1 - x(Z)} \sum_{Z}^{92} \frac{x(Z)}{N}\, x(Z) = K(Z, T, P, \ldots)$$
$$N = 1 + \sum x(Z)\, x(Z) \qquad (6)$$

Beachtet man nun (1), so findet man ohne weiteres:

$$\frac{x(Z)}{1 - x(Z)} \frac{\overline{x}}{1 + \overline{x}} = K(T, P, \ldots)$$

wo:
$$\overline{x} = \frac{\sum x(Z)\, x(Z)}{\sum x(Z)} \qquad (7)$$

der nach den Grundkonzentrationen gebildete *mittlere Ionisationsgrad* ist.

2. Entwicklung der Grundformel der Sahaschen Theorie nach der Methode von Darwin-Fowler. (Ionisation und Statistik, zweiter Teil.) Wir gehen nun an die Ableitung der Fundamentalformel für die Theorie von Saha, d. i. die Ermittelung der im vorigen noch unbestimmt gelassenen Funktion $K(Z, m, T, P, \ldots)$. Saha selber hat für sie keine Ableitung gegeben, sondern hat sie im wesentlichen fertig den Arbeiten von Nernst, Eggert, Tetrode und Stern entnommen. Der von diesen Forschern eingenommene Standpunkt ist im ganzen nicht einheitlich, denn soweit die Überlegungen von Nernst in Frage stehen, kommt die klassische Thermodynamik zur Anwendung, während die Bestimmung der wesentlichen Formelkonstanten durch Tetrode und Stern auf quantentheoretischer Grundlage erfolgte. Gegenüber dieser für eine Gesamtdarstellung nicht eben vorteilhaften Duplizität sind wir neuerdings durch die überaus wertvollen Arbeiten von C. G. Darwin und R. H. Fowler in die glückliche Lage versetzt worden, völlig konsequent einen rein quantenstatistischen Standpunkt vertreten zu können, soweit ein solcher überhaupt haltbar sein kann, d. i. bis an die Grenze, wo notwendig spezielle atomistische Vorstellungen eingeführt werden müssen. Die Methode ist so außerordentlich klar und so erstaunlich anwendungsfähig, daß wir nicht umhin können, sie, wenn auch knapp, in ihren Grundzügen hier darzulegen, wobei wir natürlich alles auf unseren speziellen Zweck zuschneiden werden.

Unser im vorstehenden bereits klar umrissenes Ziel ist die Bestimmung der Ionisationsgrade unserer Gasmischung in einem bestimmten Zustande und wir sahen, daß damit die Ermittelung der Anzahlen der Teilchen in den verschiedenen Zuständen identisch ist. Genau genommen hatten wir oben nur einen ganz bestimmten Zustand ins Auge gefaßt. Nach der Vorstellungsweise der Physik durchläuft indessen ein *System* von Teilchen fortwährend andere und andere Zustände, und beobachtbar ist nur ein gewisser *mittlerer Zustand*. Da wir die materiellen Teilchen nicht zu numerieren vermögen und da es bei der Energie gar nur auf den Betrag ankommen soll, den ein Teilchen gerade trägt, so muß man sich notwendig damit begnügen, einen solchen mittleren (Makro-) Zustand als hinreichend charakterisiert anzusehen, wenn man die *mittleren Anzahlen* derjenigen Teilchen anzugeben vermag, die sich in den einzelnen *Elementargebieten,* d. h. in den einzelnen möglichen atomaren Zuständen befinden. Hierbei hängt wieder alles davon ab, wieviele Parameter man erfahrungsgemäß zur Beschreibung eines solchen atomaren Zustandes in Betracht zu ziehen hat.

Da nun die Zahl der zur Verfügung stehenden Kerne und Elektronen sowie der verfügbare Energiebetrag endlich sind, da ferner für jedes Teilchen nur eine endliche Anzahl (höchstens Z) von Ionisationszahlen und auch nur eine endliche Zahl von Energieniveaus in Frage kommen, so gibt es, soweit wir bislang unsere Betrachtungen überhaupt gespannt haben, jedenfalls nur eine endliche Anzahl von Realisationsmöglichkeiten der Mikrozustände unseres Systems. Wir bezeichnen diese Zahl stets mit C. Da ferner alle $T(Z, m)$ und alle Elektronen vertauschbar sind, so gibt es für einen bestimmten Makrozustand $\mathfrak{Z}$ ebenfalls nur eine bestimmte endliche Zahl von Realisierungsmöglichkeiten, die wir mit $c(\mathfrak{Z})$ bezeichnen wollen. Der *Mittelwert* $\overline{\mathfrak{Z}}$ von $\mathfrak{Z}$ ist dann gegeben durch eine Beziehung:

$$C \cdot \overline{\mathfrak{Z}} = \sum c(\mathfrak{Z}) \cdot \mathfrak{Z} \tag{A}$$

wobei die Summe über alle *zugelassenen* Zustände $\mathfrak{Z}$ zu erstrecken ist. Diese Art der Mittelbildung setzt voraus, daß alle Realisierungsmöglichkeiten als gleichberechtigt zu betrachten sind und *a priori* liegt gar keine Veranlassung vor, dies nicht anzunehmen. Nur auf Grund spezieller, der hier betriebenen Statistik fremder Vorstellungen könnte man überhaupt auf eine andere Möglichkeit geführt werden.

DARWIN-FOWLER zeigen nun, wie man sowohl C wie auch die Summe rechter Hand in (A) auch in den scheinbar kompliziertesten Fällen durch Anwendung einiger Kunstgriffe sehr leicht gewinnen kann und ferner, daß ein so gewonnenes $\overline{\mathfrak{Z}}$ in allen Fällen eine „vernünftige" Eigenschaft — normal property im Sinne von JEANS — ist, indem ihre Schwankung $\sqrt{(\mathfrak{Z} - \overline{\mathfrak{Z}})^2}$ sehr klein gegen $\overline{\mathfrak{Z}}$ selber wird. Die Betrachtungen ergeben dann weiter, daß für alle in Frage stehenden Mittelwerte die Größe eines bestimmten Parameters ϑ maßgebend ist, der die Verteilung der Energie über die verschiedenen $T(Z)$-Arten charakterisiert und in einen einfachen Zusammenhang mit dem üblichen *Temperaturbegriffe* gebracht wird. Letztere wird also ohne Zuhilfenahme der Begriffe *Entropie* oder *Zustandswahrscheinlichkeit* eingeführt. Gerade deshalb aber führt das neue Verfahren fast ohne alle Schwierigkeiten zu allen den modernen Ergebnissen der Quantentheorie und ist sicher zu einer großen Zukunft befähigt.

Um die Gedankengänge recht klar hervortreten zu lassen, wählen wir zunächst einen besonders einfachen Fall. Gegeben sei eine feste Anzahl M von Teilchen mit beliebigen Kernen Z in deren resp. Anzahlen $M(Z)$, so daß $M = \sum\limits_{1}^{92} z\, M(Z)$ ist. Wir sehen von Ionisation einmal völlig ab, lassen aber zu, daß jedes $T(Z)$ durch Energieaufnahme aus einem gewissen unerregten Zustande auf höhere Energieniveaus gelangen könne. Ein Energieniveau sei mit $\varepsilon(Z, a)$ bezeichnet, wo $a = 0, 1, 2, \ldots \alpha(Z)$ sein mag[1]). Der Fall $a = 0$ bedeute den un-

---

[1]) Es steht experimentell fest, daß nach Aufnahme einer ganz bestimmten Energiemenge Ionisation eintritt; diese wird als Grenzfall der Anregung betrachtet.

erregten Zustand, von dem aus wir die $\varepsilon(Z, a)$ rechnen wollen. Es sei zunächst vorausgesetzt, daß für irgendein Z nicht alle $\varepsilon(Z, a)$ kleine, ganzzahlige Multipla eines für die $T(Z)$ charakteristischen „$\varepsilon(Z)$-Quantums" seien, obwohl ganze, wenn nur endliche Gruppen dieser Art zugelassen sind. Wohl aber dürfen sämtliche $\varepsilon(Z, a)$ überhaupt sehr hochzahlige Multipla eines universalen „Urquantums" sein, doch kann die Energie auch als Kontinuum gedacht werden. Es soll damit, wie wir schon andeuteten, zum Ausdruck kommen, daß es für die Energie nur auf die Beträge ankommt, nicht aber darauf, welche. der ja möglicherweise vorhandenen „Individuen der Energie" ein Teilchen bei sich hat.

Wir setzen weiter voraus, daß die zur Verteilung verfügbare, gewissermaßen „freie" Energie F in jedem Augenblicke bis auf einen verschwindend kleinen Bruchteil auch wirklich verteilt sei. Da wir andererseits die Möglichkeit quantenhaften Energieaustausches im Auge behalten müssen, so besagt dies, daß die Zeit, während deren ein Energiebetrag von einem zu einem anderen Teilchen „unterwegs" ist, verschwindend klein sei im Vergleich zu seiner *Verweilzeit* auf einem Teilchen.

An eine formelmäßige Darstellung der *Frequentierung* der verschiedenen Niveaus durch die Elektronen wollen wir im Augenblicke nicht denken, wollen aber auch nicht etwa annehmen, daß sie rein nach Zufall erfolge. Für alle Fälle sichern wir uns, indem wir für jedes $\varepsilon(Z, a)$ ein *apriorisches Gewicht* $p(Z, a)$ ansetzen, im Sinne einer nicht-statistisch gegebenen Wahrscheinlichkeit seiner Benutzung. Wir werden später sehen, in welcher Weise solche Gewichte bestimmt werden können.

Ist nun $M(Z, a)$ die Zahl der Teilchen im Zustande Z und a, so haben wir nach vorstehendem zunächst einmal folgende Gleichungen:

$$M = \sum_{Z}^{92} \sum_{a}^{\alpha(Z)} M(Z, a) \tag{1}$$

$$F = \sum_{Z}^{92} F(Z) = \sum_{Z}^{92} \sum_{a}^{\alpha(Z)} M(Z, a)\, \varepsilon(Z, a) \tag{2}$$

Es sind dies feste Bedingungsgleichungen, an die jeder Mikrozustand gebunden ist.

Die Energieniveaus streben also einer bestimmten endlichen Grenze zu und es sind dann grundsätzlich zwei Fälle möglich:

a) Entweder gibt es kleinste Energieelemente, dann muß notwendig die Zahl der Niveaus, d. h. $\alpha(Z)$ endlich sein, da sonst eine Konvergenz der Niveaus unmöglich wäre;

b) oder die Energie ist ein Kontinuum, — und dann *kann* die Zahl $\alpha(Z)$ unendlich sein, braucht es aber nicht.

Wir werden uns hier aus formalen Gründen auf den ersten Standpunkt stellen. Eine *eindeutige* Stellungnahme zu der Frage scheint uns heute überhaupt nicht möglich sein zu.

Wir greifen zunächst ein bestimmtes Z heraus, betrachten die Gleichung (1) und fragen uns, wie viele Möglichkeiten bestehen, sie zu realisieren, *wenn die $M(Z, a)$ fest vorgegeben* sind. Wir nennen einen solchen Komplex der $M(Z, a)$, der mit (1) und (2) verträglich ist, einen Zustand $\mathfrak{Z}$ (besser *Makrozustand*) unserer Gesamtheit der $T(Z)$. Eine einfache Überlegung[1]) zeigt nun, daß unter diesen Zustand

$$c'(\mathfrak{Z}) = \frac{M(Z)!}{\prod\limits_{a}^{\alpha(Z)} M(Z, a)!} \tag{3'}$$

Möglichkeiten fallen, wenn nämlich die einzelnen Niveaus als gleichberechtigt angesehen werden. Man hätte diesen also das „*Zustandsgewicht*" $c'(\mathfrak{Z})$ zu erteilen, wenn man irgendeine Größe über alle Zustände mitteln wollte. Werden dagegen den Niveaus apriorische Gewichte $p(Z, a)$ zuerteilt, so wird das Zustandsgewicht:

$$c(\mathfrak{Z}) = M(Z)! \prod\limits_{a}^{\alpha(Z)} \frac{p(Z, a)^{M(Z, a)}}{M(Z, a)!} \tag{3}$$

denn daß $M(Z, a)$ Teilchen das Niveau $\varepsilon(Z, a)$ *zugleich* einnehmen, ist ein Ereignis, dem nur mehr das Gewicht $p(Z, a)^{M(Z, a)}$ zukommt.

Wenn wir nun dazu übergehen, *alle* Zustände zu umfassen, so handelt es sich um zweierlei: Erstens brauchen wir eine zweckmäßige Formel, die alle möglichen Zustände und keine anderen in sich begreift, und die gestattet, das *Gesamtgewicht aller Zustände* $C = \sum c(\mathfrak{Z})$ festzustellen. Zweitens muß dabei besondere Rücksicht auf die Bedingungsgleichung (2) genommen werden. Diese Aufgabe lösen DARWIN-FOWLER durch folgende Überlegung: Bildet man mit einem komplexen Parameter u die Reihe:

$$f(Z, u) = \sum\limits_{a}^{\alpha(Z)} p(Z, a)\, u^{\varepsilon(Z, a)} \tag{4}$$

---

[1]) Man denke sich alle $T(Z)$ in eine Reihe gelegt; diese enthält $M(Z)$ Glieder. Man bilde alle Permutationen dieser Reihe und schreibe sie untereinander; es sind das bekanntlich $M(Z)!$. Man teile das ganze quadratische Gebiet in Kolonnen gemäß den vorgegebenen $M(Z, a)$, indem man etwa im Sinne wachsender a einteilt. Man betrachte die Glieder der ersten Kolonne und streiche darin alle Zeilen aus, welche dieselben Glieder enthalten, nur in anderer Reihenfolge. Da die $M(Z, a)$ Glieder der ersten Kolonne $M(Z, a)!$ Permutationen zulassen, so wird sich die erste Kolonne auf den $M(Z, a)^{\text{ten}}$ Teil reduzieren. Unter den so reduzierten Zeilen betrachte man nun die zweite Kolonne und verfahre wie vorher. Man kommt wieder zu einer Reduktion auf den $M(Z, a)^{\text{ten}}$ Teil usw. Das Ergebnis des ganzen zu Ende geführten Reduktionsprozesses wird sein, daß man nur

$$\frac{M(Z)!}{M(Z, a)!}$$

verschiedene Zeilen übrigbehält.

und erhebt sie in die Potenz $M(Z)$, so kann man auf sie den Multi-
nominalsatz anwenden und erhält, wenn man (3) betrachtet:

$$\begin{aligned}
f(Z,u)^{M(Z)} &= \sum_{\Sigma M(Z,a)=M(Z)} \frac{M(Z)!}{\prod\limits_{a}^{\alpha(Z)} M(Z,a)!} \cdot \prod\limits_{a}^{\alpha(Z)} p(Z,a)^{M(Z\,a)}\, u^{\varepsilon(Z,a)} \\
&= \sum c(\mathfrak{Z})\, u^{F(Z)}
\end{aligned} \right\} \tag{5}$$

Nun läßt sich aber die Reihe (5) rechter Hand auch als einfache
Potenzreihe schreiben nach wachsenden Potenzen von u, und es bedeutet nichts Wesentliches, wenn wir überall mit unendlichen statt
mit endlichen Reihen arbeiten, denn wir betrachten alsbald nur
mehr den Koeffizienten des einen Gliedes mit der Potenz $F(Z)$, denn
alle anderen interessieren ja nicht[1]). Ist nun $f(Z,u)$ eine *analytische*
Funktion, so läßt sich der Cauchysche Satz anwenden und man hat
unmittelbar:

$$C = \frac{1}{2\pi i} \oint \frac{[f(Z,u)]^{M(Z)}\,du}{u^{F+1}} \tag{6}$$

wobei das Integral zu erstrecken ist längs einer beliebigen geschlossenen,
durch das komplexe Gebiet den Punkt $u=0$ einmal umlaufenden
Kurve. Über die tatsächliche Ausrechnung von C werden wir später
ausführlich sprechen. Wesentlich erscheint an dieser ganzen Dar-
stellungsweise noch, daß von der Stirlingschen Formel an keiner Stelle
Gebrauch gemacht wird. Dafür tritt hier die Diskussion einer analytischen Funktion in den Vordergrund. Diese nennen Darwin-Fowler
„partition function", welchen Namen wir beibehalten wollen als *Partitionsfunktion*. Sie ist mit der von Planck benutzten *Zustandssumme*
aufs engste verwandt.

Wir wenden uns nun der Ermittelung zweier ganz besonders wich-
tiger *Mittelwerte* zu, nämlich der mittleren Energie $\overline{F(Z)}$ der Z-Teilchen
und der mittleren Anzahl $\overline{M(Z,a)}$ der auf dem $\varepsilon(Z,a)$-Niveau befindlichen Teilchen. Dabei ist einfach nach (A) zu verfahren.

a) Nach (2) ist:

$$C\overline{F(Z)} = \sum_{M(Z)} M(Z)! \prod\limits_{a}^{\alpha(Z)} \frac{p(Z,a)^{M(Z,a)}}{M(Z,a)!} \cdot \sum\limits_{a}^{\alpha(Z)} M(Z,a)\cdot\varepsilon(Z,a) \tag{7}$$

Betrachten wir aber (5), so sehen wir alsbald, daß:

$$u\frac{d}{dz}[f(Z,u)]^{M(Z)} =$$

$$\sum_{M(Z)} \left\{ M(Z)! \prod\limits_{a}^{\alpha(Z)} \frac{p(Z,a)^{M(Z,a)}}{M(Z,a)!} \sum\limits_{a}^{\alpha(Z)} M(Z,a)\cdot\varepsilon(Z,a)\cdot u^{\sum\limits_{a}^{\alpha(Z)} M(Z,a)\cdot\varepsilon(Z,a)} \right\} \tag{8}$$

---

[1]) Es entfällt damit die prinzipielle Wichtigkeit der oben gemachten Anmerkungen über die Kontinuität oder Diskontinuität der Energie für die vorliegende Methode.

so daß also der Koeffizient von $u^{\Sigma\cdots}$ in (8) gerade der Ausdruck unter der inneren Summe von (7) ist. Indem man genau wie vorher ansetzt, findet man aus CAUCHYS Satz:

$$\begin{aligned} C\overline{F(Z)} &= \frac{1}{2\pi i}\oint \frac{u\dfrac{d}{du}[f(Z,u)]^{M(Z)}\,du}{u^{F+1}} \\ &= \frac{1}{2\pi i}\oint M(Z)\,u\frac{d}{du}\lg f(Z,u)\cdot\frac{[f(Z,u)]^{M(Z)}}{u^{F+1}}\,du \end{aligned} \qquad (9)$$

b) Geht man mit $M(Z,a)$ in (A) ein, so bekommt man:

$$C\overline{M(Z,\mathfrak{a})} = \sum M(Z,\mathfrak{a})\,M(Z)!\prod_{a}^{\alpha(Z)}{}'\frac{p(Z,a)^{M(Z,a)}}{M(Z,a)!} \qquad (10)$$

wobei bedeuten soll Π', daß für $a=\mathfrak{a}$ im Nenner $[M(Z,a)-1]!$ und im Zähler $p(Z,a)^{[M(Z,a)-1]}$ zu schreiben ist. Die Zahl der Glieder unter der Summe bleibt unverändert, aber an die Stelle von (1) und (2) treten nun die Gleichungen:

$$M(Z) = \sum_{a}^{\alpha(Z)}[M(Z,a)] - 1 \qquad (1')$$

und:

$$F(Z) = \sum_{a}^{\alpha(Z)}[M(Z,a)\cdot\varepsilon(Z,a)] - \varepsilon(Z,\mathfrak{a}) \qquad (2')$$

und die Anwendung des Multinomialsatzes erfolgt dieses Mal auf den Ausdruck $[f(Z,u)]^{M(Z)-1}$. Man findet, daß der Koeffizient von $u^{F(Z)}$ wird:

$$\frac{1}{2\pi i}\oint\frac{[f(Z,u)]^{M(Z)-1}\,du}{u^{F-\varepsilon(Z,\mathfrak{a})+1}}$$

so daß man hat:

$$C\overline{M(Z,\mathfrak{a})} = \frac{p(Z,\mathfrak{a})\,M(Z)}{2\pi i}\oint\frac{u^{\varepsilon(Z,\mathfrak{a})}}{f(Z,u)}\cdot\frac{[f(Z,u)]^{M(Z)}\,du}{u^{F+1}} \qquad (11)$$

Der nächste Schritt, den wir tun, ist eine Erweiterung unserer Aufgabestellung im Sinne einer *Zusammenfassung aller möglichen Z*. Es gelten dann die Gleichungen (1) und (2) unverändert, die (3) bestehen ebenfalls weiter, jedoch nicht mehr im Sinne von Zustandsgewichten; denn da unsere Gesamtheit jetzt alle Z umfaßt, so ist jeder Zustand eines Z mit allen möglichen Zuständen aller anderen Z zu kombinieren, und daraus ergibt sich, daß das Gewicht eines Zustandes der Gesamtheit nunmehr ist:

$$c(\mathfrak{Z}) = \prod_{z}^{92}\left\{M(Z)!\prod_{a}^{\alpha(Z)}\frac{p(Z,a)^{M(Z,a)}}{M(Z,a)!}\right\} \qquad (12)$$

Wenn man nun fragt, ob dieser Ausdruck sich vielleicht auch zu dem Multinominalsatze in Beziehung setzen läßt, so sieht man leicht ein, daß dies in der Tat möglich ist; wie schon die einfache Entstehung von (12) aus (3) lehrt, ist nämlich:

$$[f(Z, u)]^{M(Z)} = \sum_{(1)} (12)\, u^{(2)} \tag{13}$$

welche abgekürzte Schreibweise wohl verständlich ist. Es bedeuten die Klammern die entsprechend bezeichneten Formeln. Bei der so überaus durchsichtigen Struktur des ganzen Verfahrens ist es klar, daß man in (13) alle möglichen Zustände und nur diese umfaßt, wenn man die Glieder mit $u^{(2)} = u^F$ zusammennimmt, wobei es wieder gleichgültig ist, ob man sich die $f(Z, u)$ als endliche oder unendliche Reihen denkt. Sobald alle $f(Z, u)$ analytische Funktionen sind, ist wieder der Cauchysche Satz der Helfer, der aus der in der Form einer nach steigenden Potenzen von u geschrieben gedachten Formel (13) den Koeffizienten von u^F herausgreift und damit als Gesamtgewicht aller nach (1) und (2) erlaubten Zustände liefert:

$$C = \frac{1}{2\pi i} \oint \frac{\prod\limits_{z=1}^{92} [f(Z, u)]^{M(Z)}}{u^{F+1}}\, du \tag{14}$$

und ebenso sieht man in den oben durchgerechneten Fällen a) und b) leicht, daß:

$$C \overline{F(\zeta)} = \frac{1}{2\pi i} \oint M(\zeta)\, u\, \frac{d}{du} [\lg f(\zeta, u)] \cdot \frac{\prod\limits_{z=1}^{92} [f(Z, u)]^{M(Z)}}{u^{F+1}}\, du \tag{15}$$

$$C M(\zeta, \mathfrak{a}) = \frac{p(\zeta, \mathfrak{a})\, M(\zeta)}{2\pi i} \oint \frac{u^{\varepsilon(\zeta, \mathfrak{a})}}{f(\zeta, u)} \cdot \frac{\prod\limits_{z=1}^{92} [f(Z, u)]^{M(Z)}}{u^{F+1}}\, du \tag{16}$$

Wir bemerken noch, daß es ganz allgemein möglich ist, das Verfahren noch auf beliebig viele andere Verteilungsformen der Energie zu erweitern. Z. B. haben Darwin-Fowler gezeigt, wie man sinngemäß die kinetische Energie und die Energie des Strahlungsfeldes mit berücksichtigen kann, und sind bei dieser Gelegenheit auf überaus einfache Ableitungen der Maxwellschen Verteilung und des Planckschen Strahlungsgesetzes gekommen. Für jede neue Erweiterung ist nur die Einführung der richtigen Partitionsfunktion nötig, und alles Weitere wickelt sich dann in demselben Schema ab, das wir soeben kennen gelernt haben. Die hierdurch mögliche Allgemeinheit der Entwicklungen scheint bei weitem größer als alle auf andere Weise bisher in Betracht gezogenen Fälle.

Es handelt sich nun nur noch um die *tatsächliche Auswertung der Integrale* (6), (9), (11) bzw. (14), (15), (16). Die für diese schließlich gewählte Schreibweise hat ihren guten Grund; man sieht nämlich, daß sie alle die Form haben:

$$I = \frac{1}{2\pi i} \oint G(u)\,[\Phi(u)]^F \frac{du}{u} \tag{17}$$

wobei zu setzen ist in:

$$\Phi(u) = \frac{f(Z,u)^{\frac{M(Z)}{F}}}{u}, \qquad G(u) = \begin{cases} 1 & (6) \\[2mm] M(Z)\,u\,\dfrac{d}{du}\,[\lg f(Z,u)] & (9) \\[2mm] \dfrac{u^{\varepsilon(Z,\mathfrak{a})}}{f(Z,u)} & (16) \end{cases}$$

und in:

$$\Phi(u) = \frac{\prod\limits_{\alpha\;1}^{92}[f(Z,u)]^{\frac{M(Z)}{F}}}{u}, \qquad G(u) = \begin{cases} 1 & (14) \\[2mm] M(\zeta)\,u\,\dfrac{d}{du}\,[\lg f(\zeta,u)] & (15) \\[2mm] \dfrac{u^{\varepsilon(\zeta,\mathfrak{a})}}{f(\zeta,u)} \cdot p(\zeta,\mathfrak{a})\,M(\zeta) & (16) \end{cases}$$

Die Behandlung der Integrale (17) stützt sich auf einen allgemeinen Satz, den DARWIN-FOWLER ableiten: Wenn $\Phi(u)$ eine analytische Funktion ist, deren aufsteigende Potenzentwicklung mit irgendeiner negativen Potenz beginnt und die lauter positive, reelle Koeffizienten hat, dann gibt es auf der reellen u-Achse eine und nur eine Stelle $u = \vartheta$, an der $\Phi(u)$ einen Sattelpunkt hat, und zwar findet in der reellen u-Richtung ein Minimum statt, aber auf dem ϑ-Kreise um den Nullpunkt ein Maximum. Dieses ist das einzige auf diesem Kreise, sobald nicht alle Potenzen, die in der Reihe vorkommen, von der Form sind $\alpha + \beta\,r$ ($r = 0, 1, 2, 3, \ldots$); und gerade diesen Fall hatten wir oben ausgeschlossen. Nun ist aber F in allen praktischen Fällen eine ungeheuer große Zahl und daher entstammt der Wert des Integrales fast ausschließlich aus der unmittelbaren Umgebung der Stelle $u = \vartheta$ beiderseits der reellen Achse. Infolgedessen setzten auch DARWIN-FOWLER $u = e^{ix}$ und entwickeln den Integranden nach Potenzen des Winkels x, wobei dann wegen des durch die Größe von F erzeugten starken Abfalles die Reihe schon hinter dem ersten Gliede abgebrochen werden darf. Auch über $G(u)$ wird dabei die Annahme gemacht, daß es eine analytische Funktion sei. Führt man die einfache Rechnung aus, so erhält man für I den allgemeinen Ausdruck:

$$I = \frac{1}{2\pi i} \oint G(\vartheta)\,[\Phi(\vartheta)]^F\, e^{-\frac12 F x^2 \vartheta \left(\frac{\Phi''}{\Phi}\right)} \pm \cdots i\,dx \tag{18}$$

Man zieht die konstanten Teile heraus und hat ein einfaches Wahrscheinlichkeitsintegral. Es ist nun weiter ganz unbedenklich, die Integrationsgrenzen beiderseits gegen Unendlich auszudehnen und dann erhält man ohne weiteres:

$$I = \frac{G(\vartheta)\,[\Phi(\vartheta)]^F}{2F\left(\dfrac{\Phi''}{\Phi}\right)\vartheta} \tag{19}$$

Diese in Anbetracht der Kompliziertheit der Problemstellung überraschende Lösung wird nun noch weit einfacher, wenn man die Anwendungen macht; denn da in den Integralen (14), (15), (16) $\Phi(u)$ stets dasselbe ist, so nimmt die Gleichung (A) einfach die Form an:

$$\overline{\mathfrak{z}} = G(\vartheta) \tag{20}$$

d. h. aus den Formeln (15) und (16) wird, wenn wir wieder Z und a statt ζ und $\mathfrak{a}$ schreiben:

$$\overline{F(Z)} = M(Z)\,\vartheta\,\frac{d}{d\vartheta}\,[\lg f(Z,\vartheta)] \tag{21}$$

$$\overline{M(Z,a)} = \frac{p(Z,a)\,M(Z)\,\vartheta^{\varepsilon(Z,a)}}{f(Z,\vartheta)} \tag{22}$$

Nun bleibt noch der Parameter ϑ zu ermitteln, der durch die Bedingung festgelegt ist, daß $\Phi(\vartheta) = \mathrm{Min.}$ sein soll. Durch logarithmische Differentiation von $\Phi(u)$ und Nullsetzen finden wir aber unmittelbar:

$$0 = -\frac{1}{\vartheta} + \sum_{1}^{92}{}_{Z}\,\frac{M(Z)}{F}\,\frac{d}{d\vartheta}\,[lg f(Z,\vartheta)] \tag{23}$$

oder auch:

$$F = \sum_{1}^{92}{}_{Z}\,M(Z)\,\frac{d}{d\vartheta}\,[\lg f(Z,\vartheta)] \tag{23 a}$$

Summiert man (21) über alle Z, so ergibt sich, wie es sein muß und gewissermaßen als Kontrolle, wieder die Gleichung (23 a).

Die im dritten Abschnitte angestellten Betrachtungen ermöglichen es uns, die soeben entwickelte Methode auf einen Fall des *Ionisationsgleichgewichtes* anzuwenden. Wesentlich ist für uns dabei, daß in den Formeln (5)—(7), nur benachbarte Ionisationszustände auftreten. Wir wollen hier den allgemeinen Fall eines höchstens einfach ionisierten, sonst aber beliebigen Gasgemisches behandeln. Die Gründe, weshalb wir den allgemeinsten Fall beliebiger Ionisation nicht in Angriff nehmen können, werden später klar werden.

Wir bezeichnen stets neutrale Atome durch das Argument 0, Einfachtrümmer durch 1 und Elektronen wie bisher mit E. Wir unterscheiden die Teilchen weiter nach Kernladung Z, Anregungszustand a

und kinetischer Energie $\varkappa$. Was die letztere angeht, so wollen wir zunächst annehmen, daß sie in diskreten Stufen auftrete und erst später die Kontinuität durch einen Grenzübergang herstellen.

Der Gang der Entwicklung wird nun der sein, daß wir zuerst einen bestimmten Ionisationszustand, gekennzeichnet durch feste prozentuale Anteile der verschiedenen Teilchen, und Elektronen betrachten und dann erst dazu übergehen, den Fall einer Reaktion und die Mittelbildung über alle möglichen Zustände zu erörtern.

Die Betrachtung eines bestimmten Zustandes schließt sich unmittelbar an das Vorhergegangene an. Alle *ab ovo* vorhandenen Anzahlen — in unserem Falle also diejenigen der Einfachtrümmer und der Elektronen — bezeichnen wir mit $m(\ldots)$, alle *augenblicklich* vorhandenen dagegen mit $M(\ldots)$. M schlechthin sei die Gesamtzahl aller Kerne und Elektronen zusammen, d. h. die Maximalzahl der bei einfacher Ionisation überhaupt denkbaren freien Teilchen. Da nun jedes Atom nur aus einem Trümmer und einem Elektron besteht, so gelten folgende festen Bedingungsgleichungen:

$$M = m(E) + \sum_{Z}^{92} m(Z,1) \quad \text{und} \quad m(E) = M(E) + \sum_{Z}^{92} M(Z,0)$$
$$M(Z,0) + M(Z,1) = m(Z,1); \quad Z = 1,2, \ldots 92$$

Ein bestimmter Makrozustand des Gemisches ist gegeben durch die Angabe aller $M(Z, m, a, \varkappa)$, worin: $Z = 1, 2, \ldots, 92$; $m = 0, 1$; $a = 0, 1, \ldots, a(Z)$; $\varkappa = 0, 1, \ldots\ldots\ldots$

Hier können $a(Z)$ und $\varkappa$ zwar sehr große Zahlen werden, bleiben aber jedenfalls kleiner als die Gesamtenergie des Systems.

Für die *Zählung der Energie* benötigen wir einen *Grundzustand* des Gemisches, den wir wie folgt festlegen: Es seien keine ionisierten Teilchen vorhanden, sondern nur neutrale Atome und evtl. freie Elektronen; alle diese Partikeln seien unerregt und unbewegt. In diesem Zustande schreiben wir dem Systeme die Energie Null zu. Führen wird dem Gemische nun eine Energie E zu, so wird diese sich über die verschiedenen Verteilungsmöglichkeiten verbreiten, und zwar ziehen wir folgende Posten in Betracht:

1. Die Anregungsenergie der $T(Z, 0)$ und $T(Z, 1)$ . . $= B(\ldots)$
2. die kinetische Energie der $T(Z, 0)$ und $T(Z, 1)$ und der E . . . . $= K(\ldots)$
3. die als (potentielle) Ionisationsenergie in den $T(Z, 0)$ gebundene Energie . . . . $= I(\ldots)$

Die Bezeichnungen B, K, I (ohne Argument) sollen immer den entsprechend untergebrachten Gesamtbetrag bedeuten. Außerdem sei $B + K = F$.

Nach den oben vorbereitend geführten Betrachtungen kann gar
kein Zweifel mehr sein, wie wir nun die Zahl der Realisierungsmöglichkeiten eines nach der Numerierung der Teilchen bestimmt heraus-
gegriffenen Zustandes zu ermitteln haben. Die Anregungszustände
sind überhaupt unverändert zu behandeln, daneben treten nur jetzt
die völlig analogen Stufen der kinetischen Energie, so daß wir nach
(14) haben:

$$c'' = \frac{1}{2\pi i} \oint k(E)^{M(E)} \prod_{z=1}^{92} \prod_{j=0}^{1} [k(Z,j)\, b(Z,j)]^{M(Z,j)} \frac{du}{u^{F+1}} \qquad (25)$$

wo $b(\ldots)$ und $k(\ldots)$ die entsprechenden Partitionsfunktionen bedeuten.

Daß wir c'' schreiben, soll daran erinnern, daß wir hier noch nicht
die Zustandswahrscheinlichkeit eines durch die $M(\ldots)$ charakterisierten Zustandes vor uns haben, denn in c'' ist noch gar nicht berücksichtigt, daß ein und derselbe Zustand auf mannigfache Weise entstehen kann. An dieser Stelle nun unterscheidet sich die gegenwärtige
Aufgabe von dem Früheren. Folgendes ist klar: Vertauschen wir die
Z-Trümmer (jedes Z für sich) und die Elektronen unter sich, auch
so weit sie in den Atomen verbunden sind, so ändert sich der Zustand
nicht. Solcher Vertauschungsmöglichkeiten gibt es aber alles in allem:

$$m(E)! \prod_{z=1}^{92} m(Z,1)!$$

Diese führen aber durchaus nicht alle zu wesentlich verschiedenen
Zuständen, worunter wir solche verstehen, die nicht aus identischen
Atomen bestehen. So oft man nämlich die neutralen Atome, die noch
freien Trümmer und die noch freien Elektronen je unter sich vertauschen
kann, entstehen keine wesensverschiedenen Zustände, und solcher Vertauschungen gibt es:

$$\prod_{z=1}^{92} M(E)!\, M(Z,0)!\, M(Z,1)!$$

Die Zahl der Vertauschungen, die zu zwar mikroskopisch verschiedenen,
aber makroskopisch gleichen Zuständen führen, ist also nur:

$$q = \frac{m(E)!}{M(E)!} \prod_{z=1}^{92} \frac{m(Z,1)!}{M(Z,0)!\, M(Z,1)!} \qquad (26)$$

Ehe wir darangehen, auf Grund dieser Formel das Zustandsgewicht
abzuleiten, ist es nötig, eine Bemerkung von grundsätzlicher Bedeutung
einzuschalten. Bei unseren bisherigen Betrachtungen hatten wir die
Trümmer und Atome immer insofern als gleichberechtigt behandelt,
als wir ihnen von vornherein die gleiche Wahrscheinlichkeit zuschrieben,
Elektronen zu verlieren und wieder einzufangen, was nicht unbedingt

der Fall zu sein braucht und im übrigen nur auf Grund nichtstatistischer Überlegungen zu entscheiden ist. In den Arbeiten von FOWLER wird nun diesem Punkte eine gewisse Bedeutung zuerkannt, und es ist nötig, hierauf genauer einzugehen.

Den Begriff der *Valenz* σ hatten wir oben schon erläutert als eine näherungsweise Aussage dahin, daß ein Atom σ energetisch nahezu gleichwertige Elektronen auf der höchsten Quantenstufe besitzt. An der Hand von EHRENFEST und TERKAL für Moleküle angestellter Überlegungen nimmt nun FOWLER an, die Valenzelektronen seien symmetrisch um das Atom gelagert, wodurch die EHRENFESTschen Formeln unmittelbar anwendbar werden. Es scheint uns indessen diese Art der Begründung unzulässig, denn während es nach allen neueren Erfahrungen durchaus erlaubt ist, ein Molckül als eine symmetrische Anordnung von Atomen zu betrachten, deren Konfiguration einigermaßen „starr", d. h. um gewisse Symmetriepunkte schwingungsfähig ist, sind im Gegensatze dazu die Atome selber höchst bewegliche Gebilde zu nennen. Es erscheint in hohem Maße zweifelhaft, ob es überhaupt möglich ist, ein Atom so aufzufassen, daß seine Valenzelektronen um gewisse symmetrisch gelagerte Librationspunkte zu schwanken scheinen, und nur dann wäre eine glatte Übernahme der EHRENFESTschen Formeln möglich. Die Einführung der Valenz auf Grund einfacher Symmetriebetrachtungen scheint uns also nicht zulässig.

Wohl aber glauben wir, daß eine andere Art der Betrachtung gestattet ist, die zu einem ähnlichen Ergebnisse führt, und in welcher der Charakter der Valenz als einer *apriorischen Wahrscheinlichkeit* noch schärfer hervortritt. Hat nämlich ein Teilchen σ Valenzelektronen bei sich, so ist die Wahrscheinlichkeit dafür, daß es innerhalb einer Zeiteinheit von einer zur Ionisation führenden Störung betroffen wird, etwa σ-mal so groß, als wenn es unter sonst gleichen Umständen nur ein einziges Elektron in höchstquantiger Bahn hätte, denn es wird sich im Zeitmittel σ-mal sooft ereignen, daß eines der Valenzelektronen sich gerade an der für die Ionisation besonders in Betracht kommenden äußeren Sphäre des Atomes befindet. Andererseits scheint es nach den vor allem von EDDINGTON angestellten Überlegungen keineswegs so, als ob die Wahrscheinlichkeit, ein Elektron wieder einzufangen, überhaupt wesentlich von der Valenz abhängen könnte, denn danach wird ein Elektron nur in allernächster Nähe des Kernes gebunden, in einer Entfernung von ihm, die 100—1000mal kleiner ist als die äußere Sphäre des Atoms, wo also nach den Erfahrungen der Röntgenspektroskopie von einem wesentlichen Einflusse der Valenzelektronen gar keine Rede mehr sein kann. Auf Grund dieser beiden Überlegungen werden wir aber sagen dürfen, daß die Existenzwahrscheinlichkeit eines mit σ Valenzelektronen ausgestatteten Teilchens nur $1/\sigma$ ist. Wir werden daher einem Gasgemische, in dem von den einzelnen

Teilchenarten $M(Z, j)$ $(j = 0,1)$ Individuen vorhanden sind, ein Gewicht zulegen:

$$\frac{1}{\prod\limits_{z}^{92}{}_{1}\, \sigma^{M(Z,0)}} \tag{27}$$

Hierbei ist zu beachten, daß es genügt, für die neutralen Atome Gewichte anzusetzen, denn bei der Ionisation ergibt sich mit Notwendigkeit aus jedem Atome ein und nur ein Einfachtrümmer, so daß die Einführung eines besonderen Gewichtes für die Trümmer nicht erlaubt ist.

Auf Grund der bisher angestellten Überlegungen können wir also nach Formel (26)—(27) für das Zustandsgewicht eines lediglich durch die $M(Z, j)$ gekennzeichneten Zustandes ansetzen:

$$c = \frac{m(E)!}{M(E)!} \prod\limits_{z}^{92}{}_{1}\, \frac{m(Z, 1)!}{M(Z, 0)!\, M(Z, 1)!\, \sigma^{M(Z,0)}} \cdot c'' \tag{28}$$

Um nun hieraus das Gesamtgewicht aller überhaupt möglichen Zustände zu erhalten, haben wir (28) über diese zu summieren und sind dabei gebunden für die $M(\ldots)$ an die Gleichungen (24) und für die Energie an die weitere Bedingung:

$$E = F + \sum\limits_{z}^{92}{}_{1}\, M(Z, 0)\, X(Z) \tag{29}$$

wo $X(Z)$ die Ionisationsenergie eines Z-Atomes ist in dem früher definierten Sinne. Die Anwendung des Cauchyschen Satzes erfolgt einfach unter Erweiterung von (25) bei der Summation mit:

$$u^{\sum\limits_{1}^{92} Z\, M(Z,0)\, X(Z)}$$

so daß folgt:

$$\left.\begin{aligned}
C = {}&\frac{1}{2\pi i} \oint \frac{m(E)!}{M(E)!} \sum \frac{m(Z, 1)!\, u^{M(Z,0)\, X(Z)}}{M(Z, 0)!\, M(Z, 1)!\, \sigma(Z)^{M(Z,0)}} \\
&\cdot k(E)^{M(E)} \prod\limits_{z}^{92}{}_{1} \prod\limits_{j}^{1}{}_{0} \, [k(Z, j)\, b(Z, j)]^{M(Z,j)}\, \frac{du}{u^{E+1}},
\end{aligned}\right\} \tag{30}$$

was man mit Rücksicht auf (24) leicht umformt in:

$$\left.\begin{aligned}
C = {}&\frac{1}{2\pi i} \oint k(E)^{m(E)} \prod\limits_{z}^{92}{}_{1} \, [k(Z, 1)\, b(Z, 1)]^{m(Z,1)}\, \frac{du}{u^{E+1}} \\
&\frac{m(E)!}{M(E)!} \sum \frac{m(Z, 1)!}{M(Z, 0)!\, M(Z, 1)!} \prod\limits_{z}^{92}{}_{1} \left\{ \frac{k(Z, 0)\, b(Z, 0)\, u^{X(Z)}}{k(Z, 1)\, b(Z, 1)\, \sigma(Z)\, k(E)} \right\}^{M(Z,0)}
\end{aligned}\right\} \tag{31}$$

Wir bezeichnen zur Abkürzung in Zukunft:

$$H(Z) = \frac{k(Z,0)\,b(Z,0)\,u^{X(Z)}}{k(Z,1)\,b(Z,1)\,\sigma(Z)\,k(E)} \tag{32}$$

$S(\ldots) = $ die ganze in (31) auftretende Summe.

Man erkennt ohne weiteres, daß bei der Bildung von $C\overline{M(Z,0)}$ genau dieselben Formeln auftreten, nur steht in der $S(\ldots)$ entsprechenden Summe S' auf dem Bruchstriche noch ein Faktor $M(Z,0)$ gemäß unserer alten Formel (A). Danach aber sieht man ebenso leicht, daß diese Summe S' den Wert hat:

$$S' = H(Z)\,\frac{d}{d\,H(Z)}\,S(\ldots)$$

Die Auswertung der Integrale entspricht im großen und ganzen dem, was wir schon früher sahen, doch treten einige kleine Verwicklungen dadurch ein, daß sich auf die Formel (31) der Binomialsatz nicht anwenden läßt.

Photographic Determinations of Stellar Parallaxes.

By **Frank Schlesinger,** New Haven.

In many astronomical problems that depend for their solution upon precise measurements, the chief difficulty to be overcome is not in attaining the requisite accuracy in the measurements themselves, but in planning the observations in such a way as to disentangle phenomena that run nearly or quite the same course. This is particularly true of phenomena that are annually periodic, for then the difficulty may become quite insuperable because of the great number of effects having this period. To cite one example out of many, it is almost impossible to determine a system of absolute positions of stars that shall be free from periodic errors; the size, the shape and position of the instrument, refraction, aberration, the quality of the seeing and therefore the amount of personal equation, and many other effects directly concerned, all vary with the season and all must be accurately evaluated or eliminated before we can hope to attain thoroughly satisfactory right ascensions and declinations.

The trigonometric determination of stellar parallaxes is one of the problems that fall into this troublesome class, and the whole history of this subject may be said to be concerned with attempts to separate parallax from other effects, for it is nearly a century since instrumental methods reached the degree of precision necessary to deal with at least the larger stellar parallaxes. The limitations of space prevent me from recounting this history; in what follows I have set down what seems to me the best procedure for the photographic determination of parallax, as indicated by our experience up to the present time.

1. Method of reduction. We suppose that we have at hand a series of photographs, showing near their centers the images of the star whose parallax we are seeking to determine, the *parallax* star. These photographs cover an interval of several years and have been taken at such seasons as to exhibit as nearly as may be the maximum shifts on account of parallax. The plates also show the images of several or many neighbouring stars, distributed over the plate, from which we select for measurement a number of *comparison* stars.

Before we can compare one plate with another, and so derive the shifts due to parallax and proper motion, it is necessary to clear the measures of certain other causes of difference. These may be classified into two groups, the first of which comprises *orientation, scale,* and *zero* corrections; the second, *refraction, aberration, precession,* and *nutation.* Those in the first group are to be determined from the measures themselves, and, under these circumstances, as is well known, we need pay no attention to those in the second group. The reason for this is that the correction of scale (a) of the co-ordinate X is of the form $a \cdot X$; that for orientation (b) is proportionate to the other co-ordinate Y, and takes the form $b \cdot Y$; while the zero correction (c) is applied as a constant to all the measures. The sum of the three is $a \cdot X + b \cdot Y + c$. It is obvious that this is also the form assumed by any correction whatever that we may take to be linear over the narrow area covered by a plate; consequently, as TURNER first pointed out, any such correction is automatically applied when we determine the scale, etc., from the measures themselves. The more rigorous expressions for the corrections in the second group contain terms in X^2, Y^2 and $X \cdot Y$, not to mention those of still higher order; but all of these are inappreciable on ordinary parallax plates, the distance from the parallax star to any comparison star being in general less than half a degree.

In reducing all the plates to the same scale, etc., it makes no practical difference which of them is assumed to be the standard. Having then chosen one of the plates for this purpose, or the mean of several, the usual method is to deduce a, b, and c from a least-squares solution. The observation equations, of which there will be one for each of the n comparison stars, take this form (we confine our attention to the case in which co-ordinates in only one direction have been measured):

$$a \cdot X_1 + b \cdot Y_1 + c + (X_1' - X_1) = v_1 \left.\right\}$$
$$a \cdot X_2 + b \cdot Y_2 + c + (X_2' - X_2) = v_2 \left.\right\} \tag{1}$$
$$\text{etc.}$$

Here X_1, X_2 refer to the standard plate and X_1', X_2' to the plate that we are reducing. The analytical expressions will be simplified if we suppose that a certain constant has been subtracted from the co-ordinates on the standard plate, so as to make $[X]$ and $[Y]$ each equal to zero. The normal equations are then as follows:

$$[X^2]a + [X \cdot Y]b \qquad\qquad = - [X(X' - X)] \left.\right\}$$
$$[X \cdot Y]a + [Y^2]b \qquad\qquad = - [Y(X' - X)] \left.\right\} \tag{2}$$
$$+ n \cdot c = - [X' - X] = - [X'] \left.\right\}$$

Whence

$$
\left.
\begin{aligned}
a &= \frac{[Y(X'-X)][X\cdot Y] - [X(X'-X)][Y^2]}{[X^2][Y^2]-[X\cdot Y]^2} \\
&= 1 - \frac{[X\cdot X'][Y^2] - [X'\cdot Y][X\cdot Y]}{[X^2][Y^2]-[X\cdot Y]^2} \\
b &= \frac{[X(X'-X)][X\cdot Y] - [Y(X'-X)][X^2]}{[X^2][Y^2]-[X\cdot Y]^2} \\
&= \frac{[X\cdot X'][X\cdot Y] - [X'\cdot Y][X^2]}{[X^2][Y^2]-[X\cdot Y]^2} \\
c &= -\frac{1}{n}[X']
\end{aligned}
\right\} \quad (3)
$$

Finally we compute for the parallax star:

$$
X'_\pi + a\cdot X_\pi + b\cdot Y_\pi + c \tag{4}
$$

This quantity, which we shall designate by m', may be called the solution of this plate. A comparison of the values of m', m'' from various plates enables us to determine the parallax.

The labor of computing the plate-constants by this method, for so large a number of plates as are included in an ordinary series, is very considerable. The following method is much shorter and leads to precisely the same results.

The plate-constants are of no interest in themselves in the present connection. Their computation may be altogether avoided, and the numerical work much abridged, by expressing m' directly in terms of the observed quantities X'_1, X'_2, etc. To effect this we substitute in (4) the values of a, b, and c as given in equations (3), and then collect the terms in X'_1, those in X'_2, etc:

$$
\left.
\begin{aligned}
m' = X'_\pi + X_\pi & \\
- X'_1 &\left\{ \frac{X_1(X_\pi[Y^2]) - Y_1(X_\pi[XY]) + Y_2(Y_\pi[X^2]) - X_1(Y_\pi[XY])}{[X^2][Y^2]-[XY]^2} + \frac{1}{n} \right\} \\
- X'_2 &\left\{ \frac{X_2(X_\pi[Y^2]) - Y_2(X_\pi[XY]) + Y_2(Y_\pi[X^2]) - X_2(Y_\pi[XY])}{[X^2][Y^2]-[XY]^2} + \frac{1}{n} \right\}
\end{aligned}
\right\} \quad (5)
$$

ect.

In this notation the comparison stars are distinguished from each other by subscripts. The co-ordinates on the standard plate are distinguished from those upon others by the omission of the primes. It will be noticed that the quantities in parentheses are the same for all the comparison stars, and those in the curved brackets the same for all the plates. These coefficients in curved brackets are called *dependences* and are designated by D_1, D_2, etc. In comparing the values of m from different plates we are at liberty to add a constant to them all, and we may accordingly omit the term X_π that appears outside the brackets in equation (5). This leaves the very simple expression:

$$m' = X'_\pi - [D \cdot X'] \tag{6}$$

Similarly for any other plate:

$$m'' = X''_\pi - [D \cdot X'']$$

If these dependences have been accurately computed they will fulfill the three conditions:

$$[D] = 1, \quad [DX] = X_\pi, \quad [DY] = Y_\pi \tag{7}$$

The use of dependences for the reduction of parallax plates has two important advantages; in the first place there is a great saving of time in making the reductions. It takes only about half an hour to compute the dependences, and this need be done only once for each region. To reduce a plate we have then only to multiply (by means of CRELLE's Tables or a multiplying machine) each X by its dependence and subtract the sum of these products from the corresponding X of the parallax star. If there are several exposures on a plate it takes very little longer to reduce these exposures separately by means of dependences than to reduce the mean of the exposures, and the former has the advantage of affording a valuable check on the measurements through the accordance of the solutions for the separate exposures.

2. Number of comparison stars. Whatever the method of solution employed, the dependences should always be computed, for in addition to saving much time in the reductions, they are invaluable for an intelligent discussion and solution of many questions of procedure. We shall take up the more important of these. First, how many comparison stars should be employed? As the solution of each plate has the form:

$$m = X_\pi - D_1 X_1 - D_2 X_2 - \cdots D_n X_n$$

The square of the probable error of this solution is:

$$\varepsilon_m^2 = \varepsilon_\pi^2 + D_1^2 \varepsilon_1^2 + D_2^2 \varepsilon_2^2 + \cdots D_n^2 \varepsilon_n^2$$

where ε_π, ε_1, etc., are the probable errors of measurement for the parallax star and the various comparison stars. Assuming these to be all equal to ε:

$$\varepsilon_m = \varepsilon \sqrt{1 + [D^2]} \tag{8}$$

If the stars are well distributed on the plate, that is if their mean X and their mean Y are respectively equal to X_π and Y_π then each dependence is equal to $\dfrac{1}{n}$ and equation (8) becomes:

$$\varepsilon_m = \varepsilon \sqrt{1 + \frac{1}{n}} \tag{9}$$

The values of this factor for various values of n are as follows:

$$n \sqrt{1 + \frac{1}{n}} \qquad\qquad n \sqrt{1 + \frac{1}{n}}$$

n		n	
3	1.15	8	1.06
4	1.12	9	1.05
5	1.10	10	1.05
6	1.08	20	1.02
7	1.07	40	1.01

We see from this that the accidental error diminishes very slowly with the number of comparison stars, the probable error with four stars being only ten percent greater than with forty, and the gain for eight stars as compared with four being only five percent. Even these small gains are not wholly realized in practice, because of the presence of other inevitable sources of error that do not depend upon the number of comparison stars. If the parallax star is measured twice on each plate, as is customary, these figures would require some modification; but it is well known that such duplicate measurement does not increase very much the weight of the position of the parallax star. Making all possible allowance for the effect of such duplicate measurement, the gain of eight comparison stars over four comes out about seven percent in the probable error of the solution of a plate.

In addition to accidental error there is another point of view that should be taken into account in deciding upon the most advantageous number of comparison stars. The photographs can yield only relative parallaxes; to obtain the absolute values we must add, not (as usually stated) the average of the parallaxes of the comparison stars, but a quantity that may be called their dependence mean:

$$D_1 \pi_1 + D_2 \pi_2 + \ldots D_n \pi_n$$

where π_1, π_2, etc., are the parallaxes of the several comparison stars. In general these parallaxes are unknown and the best we can do is to assume that each of them has a parallax equal to the average that applies to stars of that class of brightness. In any particular case this assumption may be appreciably in error, and it is obvious that this error is likely to be less as the number of comparison stars increases. In work of this kind with modern telescopes the average magnitude of the comparison stars is about 10 or fainter and their average parallax is $''003$, or less. On the other hand, in the best current work the average probable error of a parallax derived from fifteen to twenty plates is about $''009$. In how many cases will the error due to the uncertainty of the dependence mean of the parallaxes of the comparison stars equal the probable error of the parallax? A consideration of the distribution of the parallaxes of faint stars leads to the conclusion that such cases would be rare even if only one comparison star were employed, and is extremely rare if we use four. Or to put it another way, if we compute the amount by which it is necessary to increase the probable error of the parallax of the parallax star, to allow for the uncertainty in the parallaxes of the comparison stars, this correction will be found to be insensible in every case that has arisen in practice. We see therefore that both from this point of view and from that of accidental error of measurement, the gain in accuracy that comes from the use of more than four or five comparison stars does not justify the additional labor involved.

How is the accidental error affected by the distribution of the comparison stars? The most favorable case is the one we have already considered, the position of the parallax star coinciding with the mean of the positions of the comparison stars, and all the dependences being equal. In any other case the probable error of the solution of a plate is, from equation (8), $\varepsilon \sqrt{1 + [D^2]}$. The values of $\sqrt{1 + [D^2]}$ for various typical sets of four dependences are as follows, the sum of the four dependences in each set adding up of course to unity:

A	B	C	D	E	F	G
0.25	0.2	0.2	0.1	0.1	0.1	0.0
0.25	0.2	0.2	0.3	0.2	0.1	0.33
0.25	0.3	0.2	0.3	0.3	0.4	0.33
0.25	0.3	0.4	0.3	0.4	0.4	0.34
$\sqrt{1 + [D^2]} =$ 1.118	1.122	1.131	1.131	1.140	1.158	1.155

This table shows that the dependences may be very unequal without much loss in accuracy, or in other words the parallax star need not be near the mean position of the comparison stars. In Case F two of the comparison stars are about four times as distant as the other two from the parallax star; even in this extreme case the accidental error is no larger than for Case G which corresponds to *three* ideally distributed comparison stars.

Thus far we have discussed the relative error for different numbers and different situations of comparison stars and we see that both considerations are unimportant, a conclusion that is completely supported by a study of published results. There are, however, other than these purely geometrical conditions that need consideration in the selection of comparison stars, and these are referred to in what follows.

3. Guiding Error. Telescopes of the great or considerable focal lengths that are employed for parallax determinations cannot be well guided by means of any of the slow motions, whether mechanical or electric, with which modern instruments are usually provided. The double-slide plate-carrier, first employed by COMMON, is a great improvement in this respect and enables the observer to follow the apparent wanderings of the stars in the field by moving little more than the plate itself. But even this device does not bring about perfection, and some wandering of the images on the sensitive film is inevitable. These wanderings have been proved to be sensibly the same for all the stars in such narrow fields as we are here concerned with, but they are not registered to an equal extent. Every excursion of a very bright star from its mean position leaves its impression in the developed image; but a short excursion of a faint star cannot overcome the so-called inertia of the film and so remains unrecorded. The resulting error is aggravated by the fact, familiar to all observers, that stars show a tendency to depart always in the same direction from their mean positions during the whole course of an exposure, and hence images will show a progressive displacement corresponding to their intensity. This is undoubtedly the largest source of accidental error in parallax and other astrometric investigations. Guiding error that attains $0''\!.1$ for each magnitude is not uncommon. It is therefore imperative that the image of the parallax star should match those of the comparison stars. As the parallax star is almost always a comparatively bright object it is necessary to reduce its brightness by some artifice. This is now universally done by means of a rotary sector whose opening can be adjusted to admit any fraction of the light of the parallax star between about 0.002 and 0.5, thus permitting any reduction in apparent magnitude between 0.7 and about 7.0. For greater reductions diffraction makes its appearance and elongates the image of the parallax star, but between the limits specified the resulting images can in no way be distinguished from those of correspondingly faint stars; and much experience at several observatories shows that the measures on these reduced images are as accurate as on those of faint stars. As comparison stars of the tenth magnitude or fainter are employed a reduction of seven magnitudes does not suffice for bright stars, and so for them some supplementary device is necessary. Two have been successfully employed: (1) the plate has been desensitized by a known amount in the small area upon which the image of the parallax star falls, and then the rotating sector employed to effect a further reduction; and (2) a second rotating sector has been employed, geared to the first so as to occult all but one in ten, say, of the passages of the first opening across the beam of the parallax star. With such devices the image even of Sirius (magnitude, -1.6) has been reduced to equality with those of faint comparison stars.

It has been found necessary, in order to eliminate guiding error with sufficient thoroughness, that this reduction in apparent magnitude be made with considerable accuracy, that is to about one quarter of a magnitude. This can readily be done if all the comparison stars are equally bright; but if not, and if we assume that the guiding error varies directly with the diameter ($2r$) of the image (as must be approximately true if no great range in diameter is involved), then the diameter of the image of the parallax star must be reduced to equality with the sum $2D_1r_1 + 2D_2r_2 + \ldots + 2D_nr_n$. From this point of view we see that it is desirable that the comparison stars should all be of the same brightness, and I regard this condition as being much more important than that their number should be large or that their geometrical distribution should be ideal.

For the first plate of a particular parallax star the opening of the sector can be computed in advance to give the required reduction; but as it is not feasible to select the comparison stars until at least this first place has been examined, the computed reduction is usually not close enough. For the second and succeeding plates the careful observer will modify the sector opening until a good match is secured.

Many double stars show well separated and measurable images on parallax plates, but unless the two components are nearly equal in brightness such doubles yield results appreciably inferior to those for single stars. The reason is that neither component is accurately matched with the comparison stars but an effort is made to reduce the *mean* of their images. It is better to abandon such attempts and to disregard entirely one or the other of the components; or to secure two sets of plates or images, for one of which the brighter component, and for the other the fainter component, are accurately reduced.

4. Atmospheric Dispersion. As air refracts and disperses light the image of a star in every telescope is a vertical spectrum with the blue end uppermost. This effect is by no means small; the angular separation between the G-line at 4308 Å and the D-lines at 5893 Å is as follows for various zenith distances:

Zenith distance	Dispersion
0°	0''.0
15°	0.2
30	0.5
45	0.8
60	1.4
75	3.0

As first pointed out by RAMBAUT, the photographic position of a star depends upon where the maximum intensity of its continuous spectrum falls, as well as upon the curve of color sensitiveness of the plates employed and the color-curve of the objective. If ζ is the zenith distance and β the refraction constant for yellow light, then the refraction is, near enough for the present purpose:

$$\beta \, \mathrm{tang} \, \zeta$$

For light of any other color the refraction can be represented by:

$$(\beta + \varDelta\beta)\,\mathrm{tang}\,\zeta$$

$\varDelta\beta$ being practically constant for a given telescope and given plates. The relative displacement is therefore:

$$\varDelta\beta \cdot \mathrm{tang}\,\zeta$$

The following table gives these displacements in right ascension and declination, for various declinations and small hour angles, computed for latitude $45°$ and for $\varDelta\beta = 0\rlap{.}{''}1$; this latter is a somewhat extreme value but one that applies for some telescopes and plates and for stars of the M type as compared with those of the A type of spectrum.

Displacement in Right Ascension

	$\delta=-20°$	$\delta=-10°$	$\delta=0°$	$\delta=+10°$	$\delta=+20°$	$\delta=+30°$	$\delta=+40°$	$\delta=+50°$	$\delta=+60°$	$\delta=+70°$
$t=0°$	$\rlap{.}{''}000$	$\rlap{.}{''}000$	$\rlap{.}{''}000$	$\rlap{.}{''}000$	$\rlap{.}{''}000$	$\rlap{.}{''}000$	$\rlap{.}{''}000$	$\rlap{.}{''}000$	$\rlap{.}{''}000$	$\rlap{.}{''}000$
$3°$	.009	.007	.005	.005	.004	.004	.004	.004	.004	.004
$6°$	.018	.013	.010	.009	.008	.008	.007	.007	.008	.008
$9°$	.027	.020	.016	.014	.012	.011	.011	.011	.011	.012
$12°$	.036	.026	.021	.018	.016	.015	.015	.015	.015	.016
$15°$	.046	.033	.027	.023	.021	.019	.019	.019	.019	.020

Displacement in Declination

	$\delta=-20°$	$\delta=-10°$	$\delta=0°$	$\delta=+10°$	$\delta=+20°$	$\delta=+30°$	$\delta=+40°$	$\delta=+50°$	$\delta=+60°$	$\delta=+70°$
$t=0°$	$\rlap{.}{''}214$	$\rlap{.}{''}143$	$\rlap{.}{''}100$	$\rlap{.}{''}070$	$\rlap{.}{''}047$	$\rlap{.}{''}027$	$\rlap{.}{''}009$	$\rlap{.}{''}009$	$\rlap{.}{''}027$	$\rlap{.}{''}047$
$3°$	.215	.143	.100	.070	.047	.027	.010	.009	.027	.046
$6°$	.215	.144	.100	.070	.047	.027	.010	.009	.027	.046
$9°$	.216	.145	.101	.071	.047	.027	.010	.009	.026	.046
$12°$	.221	.147	.102	.072	.048	.028	.010	.008	.026	.045
$15°$	.224	.148	.104	.073	.049	.028	.010	.008	.026	.045

It will be seen that the displacements in declination are large, but they do not vary sensibly with the hour angle and so are practically the same for any given field. While the displacements in right ascension are in general much smaller they vary nearly in direct proportion to the hour angle and change their signs as the object goes through the meridian. Thus even for objects that culminate near the zenith a systematic error amounting to $\rlap{.}{''}04$ may be incurred if the observations are made at hour angles up to $15°$ on both sides of the meridian. Several possible expedients for avoiding such errors suggest themselves. One would be to derive the parallax from measures in declination. Unfortunately this would always involve a considerable increase in accidental error, as even in the most favorable case the range of parallactic shift in declination cannot in practice amount to much more than half that in right ascension, and for most objects the reduction would be much greater. A second expedient would be to employ comparison stars of the same spectral type as that of the parallax star; but these types are almost always unknown for such faint objects, and even if this were not so, there would be few cases in which such a choice would be possible, especially if the parallax star were of somewhat unusual type, like B or M. A third expedient would be to introduce an additional unknown in our solutions, depending upon the difference of spectrum between the parallax star and the comparison stars, but this would again greatly increase the accidental error; for, in order that this expedient should fulfill its purpose and enable the observer to secure the maximum shift due to parallax, the evening plates would be taken at western hour angles, and the morning at eastern. Under these circumstances the introduction of an unknown whose coefficient is a function of the hour angle would greatly reduce the weight of the parallax, because the two sets of coefficients would always change signs together. A fourth expedient is to use a nearly monochromatic filter, so that no matter what the spectrum of the star, it is only light of a certain wavelength that affects the plate; this would have the obvious disadvantage of unduly lengthening the exposure. A fifth and final expedient is to restrict the observations as near as may be to the meridian, and this is now the universal practice in spite of the disadvantage of a considerable (but not serious) increase in accidental error; with this restriction the parallax factors will in practice average about ± 0.80, while without it they would average ± 0.95; and the accidental error is increased in this same ratio.

With visual refractors as contrasted with reflectors and photographic refractors, atmospheric dispersion is easily eliminated, or in other words, it is not necessary to confine the observations to as small hour angles. There are two reasons for this; first, "yellow sensitive" plates are employed, and in this part of the spectrum, atmospheric dispersion being prismatic rather than normal, a greater difference in effective wave-length is necessary to produce a required displacement as compared with the blue region. Again, as such plates are sensitive to blue as well as to yellow light, a yellow filter is usually employed with visual refractors, and this has the effect of partially introducing the fourth expedient mentioned above; this combination having the effect of using a relatively small extent of spectrum and thus making the place of the star at any hour angle more nearly independent of the spectral type. This constitutes an advantage of the visual refractor over both other types, but in my opinion it is far outweighed by the disadvantage of lengthened exposures; with equal apertures and focal lenghts and with the best yellow sensitive plates that are forthcoming, the exposures average not less than five times those with the reflector or the photographic refractor.

It is not necessary that the observations be confined strictly to the meridian, it is just as good to confine them to any one hour angle.

Still better, to give the observer as much freedom as possible, it will be sufficient if the last plates in a series are taken so that the mean hour angles for the morning observations shall be equal to those for the evening observations, or more precisely (but still not quite rigorously), so that the following condition is fulfilled:

$$[p \cdot P \cdot t] = 0$$

where p is the weight; P, the parallax factor, and t, the hour angle.

Even when there is no difference in spectrum involved, atmospheric dispersion may still affect the relative positions of stars through inequality in brightness, a subtle fact first noted by Bergstrand. The reason for this lies in the character of the curve of color sensitiveness of ordinary photographic plates. This curve has an effective maximum near 4400 Å; the sensitiveness falls off somewhat abruptly on the side of greater wave-lengths, but more gradually on the side of lesser. Consequently, while for short exposures or for faint objects it is only the light at 4400 Å that affects the plates, for longer exposures and for brighter objects more and more of the spectrum comes into play, and the image expands. But on account of the shape of the color-curve of the plate, a limit is reached on the side of longer wave-lengths sooner than for the shorter, and as the latter are uppermost on account of atmospheric dispersion the image on the plate creeps toward the zenith as the exposure is prolonged, or is higher for a bright object than for a faint. This effect is in general somewhat smaller than that due to differences in spectrum, but is still very important. Here is an additional and a strong reason for selecting comparison stars of the same magnitude, and equating the parallax star closely with them; and for confining the observations to the same hour angle.

5. Distortion of the Film. In the early history of photographic astrometry no effect was feared more than the movement or displacement of the sensitive film that occurs between exposure and drying after development. Later investigations, carried out under a great variety of conditions, have amply proved that these fears are groundless. Measurements upon thousands of star images indicate that with some simple precautions in the manner of drying the plates, the probable value of this distortion, in a sense analogous to probable error, is about .0007 mm, whereas the total probable error of measurement (including many other sources of inaccuracy) is, under the most favorable circumstances, about .0015 mm. The total annihilation of the distortion would therefore decrease the latter quantity to .00133 mm. But even this slight decrease is to be expected only in exceptional cases. The probable error of measurement of a single image on a parallax plate secured with a long-focus telescope is usually about .0025 mm, which would be reduced to about .0024 mm if the distortion were wholly absent.

The process which seems to be successful in obviating large distortions is simply to dry the plates on edge, being careful to keep them out of the sunlight and away from any source of artificial heat; and the use of an electric fan or any similar device for hastening the drying must be avoided. If the observer wishes for any reason to hasten the process, it appears that this can be done without harmful effects by soaking the developed plate in alcohol for a minute or two and then drying on edge as before.

Two important methods owe their origin to this fear of distortion, and both are known now to be needless, at least in the present connection. Thus the *réseau* had its beginnings in this way, and while for certain classes of work (notably for the Astrographic Catalogue) other advantages may justify the continuance of its use, it really adds nothing at all to the accuracy of parallax measurements, for which it was formerly employed. Similarly KAPTEYN's method, in which undeveloped plates are to be preserved in darkness for many months in order to permit exposures at various seasons to be impressed close together on the same plate, has also now been abandoned for parallax measurements, though here again certain special investigations of a different character may profit through its use.

6. Distribution of the observations with regard to time. It has not always been sufficiently recognized how much an otherwise excellent series of parallax plates may lose in value because of a faulty distribution. The following example well illustrates this point; it is taken from recently published work and is not as extreme as might be cited from other actual examples. The data concerning the plates are as follows:

Parallax Factor	Epoch years	Weight	Parallax Factor	Epoch years	Weight
+ .87	− .82	0.5	− .54	+ .44	0.6
+ .84	− .81	0.7	− .65	+ .46	0.5
+ .80	− .81	0.9	− .74	+ .48	0.7
+ .62	− .76	0.5	− .77	+ .49	0.7
+ .94	+ .26	0.7	− .79	+ .49	0.6
+ .78	+ .20	0.5	− .80	+ .50	0.7

The epochs of the plates are referred to the mean date of all the plates.

This series is now to be solved for parallax (π), introducing a constant (c) as a second unknown, and the proper motion (μ) as a third; the left-hand sides of the normal equations are:

$$+ 7.60\,c - 0.06\,\mu + 0.36\,\pi = \ldots$$
$$- 0.06 \quad\; + 2.58 \quad\; + 2.79 \quad\;\; = \ldots$$
$$+ 0.36 \quad\; + 2.79 \quad\; + 4.57 \quad\;\; = \ldots$$

If this solution is carried out it will be found that the weight of the parallax comes out only 1.53; had the distribution of the observations been better, this weight would have been much more, and with an ideal distribution would have been three times as great; in other words, the observer has in effect lost two thirds of his material because of faulty distribution. The point is made even more impressive by the following computation: suppose that the observer waits another season and secures a single good plate at the epoch 1.26, just a year later than the fifth plate; combining this with the fifth and the twelfth plate the weight of the parallax from these three alone comes out 1.41, practically equal to that derived from the twelve plates.

The trouble with such a series is obviously the entanglement of the proper motion with the parallax, due to the large cross-term between them; or, to go back one step, it is due to the fact that all the plates with positive parallax factors were secured before any of the negative plates, instead of alternating to some extent with them.

It is not at all essential that a parallax series should contain an equal number of morning and evening plates; or otherwise expressed, that the number of positive parallax factors should equal the number of negative. If there are twelve plates in the series, then other things being equal, the relative weights of the derived parallax corresponding to various combinations are as follows:

$$
\begin{array}{lll}
9 \text{ positive,} & 3 \text{ negative} & \ldots \ldots \quad 0.75 \\
8 \quad ,, & 4 \quad ,, & \ldots \ldots \quad 0.89 \\
7 \quad ,, & 5 \quad ,, & \ldots \ldots \quad 0.97 \\
6 \quad ,, & 6 \quad ,, & \ldots \ldots \quad 1.00 \\
5 \quad ,, & 7 \quad ,, & \ldots \ldots \quad 0.97 \\
4 \quad ,, & 8 \quad ,, & \ldots \ldots \quad 0.89 \\
3 \quad ,, & 9 \quad ,, & \ldots \ldots \quad 0.75 \\
\end{array}
$$

There is thus no serious loss in weight even if the evening plates are twice as numerous as the morning or *vice versa*.

This circumstance makes it far easier than it otherwise would be to secure a good distribution. Other things being again equal, a series that begins and ends in corresponding seasons (that is, begins and ends with positive parallax factors, or begins and ends with negative factors) is much better than one in which the opposite is true. A nearly ideal set is one that contains an equal number of plates, say four, in each of five consecutive seasons, and thus covers a little more than two years. It is about as good if the same total number of plates are secured in four seasons, corresponding to the above five consecutive ones, but with the middle one omitted. In either of these cases, and in others that can readily be formulated, the cross-terms are small and the weight of the parallax is not much less than the quantity $[p \cdot P^2]$.

7. The Results. Photography is so effective a tool for the determination of parallaxes that a single photographic telescope of long focus, situated where the conditions are average, can in a year supply the material for about 150 determinations with a true probable error of about $".010$. In 1914 or thereabouts several large telescopes began to devote a good deal of their time to this problem, with the result that during the past ten years our knowledge of this kind has been vastly improved both as regards quantity and quality:

The writer has compiled and partially distributed a catalogue giving all the determinations of trigonometric parallaxes (as well as some other kinds) that were available either through publication or letter up to the beginning of 1924. This catalogue contains 1682 stars, for many of which two or more determinations are available. Almost all this material is derived from photographs; indeed, not less than 95 percent of the total weight of all the trigonometric determinations is derived from photographs secured in very recent years, and this percentage will doubtless increase from year to year.

In compiling this catalogue much attention was devoted to computing the true reliability of the various series. It was found, as might have been predicted, that the comparison of the various series with each other yields somewhat larger probable errors than are derived and published from the comparison of the several observations in each series. But the increase is not very great, usually being included between 10 percent and 30 percent.

A detailed study of all the results reveals two classes of systematic error, the first being a constant error that applies to all the determinations of any one observatory but which of course varies from observatory to observatory; the other is a periodic error, depending upon the right ascension. It is not difficult to see how the first can arise. For one thing it may be due in part to failure to observe at the same average hour angle for both morning and evening plates. Or it may be due to the observer's treatment of parallaxes that come out negative. Such results are certainly in error, and for them the computer is more likely than for other results to reject a discordant plate, to remeasure a discordant plate, or to secure additional plates. It hardly needs to be pointed out that any of these expedients, while they may lead to more precise results for the particular stars involved, surely introduce a systematic tendency in the output as a whole. Such a policy can only be justified from the latter point of view if the computer would treat in precisely the same way an equal number of derived parallaxes known to be in *excess* of the truth; but unfortunately there is no way in which we can place our fingers with certainty upon such cases. In the writer's opinion, an important part of the constant differences would disappear if no plates were ever rejected or remeasured and if no star were ever put back upon the observing programme for the purpose of securing additional plates. In several of the most recent series the freedom from such constant error is very gratifying, amounting to $''.003$ or less. In a few series it is considerably larger and causes other than those we have mentioned probably operate. It is worth while to remark that failure to observe at the same hour angle, and such treatment of negative results as we have described, both lead to positive errors; and this agrees with the facts, since in almost every case the error has this sign.

The origin of the periodic error is more obscure. Only a few of the published series are sufficiently rich to enable us to evaluate this kind of error, but in all cases where the investigation is possible a small error of this kind seems to be present, with a semi-amplitude of the order of $''.003$ or $''.004$. In a few years, when the material will have perhaps twice as much weight as it has now, a more instructive investigation can be carried out and very possibly this may indicate how such errors can be avoided. For the present we must content ourselves with its empirical elimination. Perhaps its origin will be found in the change of the optical properties of objectives and mirrors arising from changes of temperature. Such changes have been proved to exist for several telescopes and probably apply to all. It must be remembered that some stars are observed on winter mornings and on summer evenings; while others are observed on summer mornings and winter evenings; and thus whatever optical change may occur will enter with its full effect in the parallaxes and will give rise to a periodic error of the general character that has been found to exist. It is, however, difficult to see how this error can be appreciable in the differential measurement of stars of the same magnitude separated by such short arcs, averaging less than $20'$. Another puzzling circumstance is that the curves showing the run of this periodic error at different observatories appear to bear very little resemblance to each other.

Of the 1682 stars in this catalogue of trigonometric parallaxes, only 205 (12 percent) lie south of the equator, and only 100 (6 percent) lie south of $-10°$ declination. The southern skies are therefore as yet virgin soil in this department of astronomy. Two large photographic refractors are projected to repair this deficiency, one of which will probably be in operation in 1925.

The solution of a series of parallax measurements yields incidentally a determination of the proper motion. This is not quite the same quantity that appears in our star catalogues since the latter is absolute, while that from the plates is relative in precisely the same sense that the parallax is relative. The difference between the two may be regarded, aside from accidental errors, as the mean (or dependence mean) of the absolute proper motions of the comparison stars in each field. That these proper motions can be used to derive valuable information concerning the stream motions of faint stars was pointed out by the writer in 1917. KAPTEYN, just before his death in 1922, reiterated the importance of such results, but from a different point of view, namely the determination of systematic errors in proper motions. Several observatories have already carried out investigations along these lines, with the net result that the systematic error feared by KAPTEYN is probably not present, or at least has a much smaller value than he supposed. All of these investigations refer to stars north of the equator or not very far south, and it is of the greatest importance that similar observations be made in the near future upon more southerly stars.

The Magellanic Clouds.

By **Harlow Shapley**, Cambridge, Mass.

With the recent determination at the Harvard Observatory of standardized sequences of apparent magnitudes in the Small Magellanic Cloud, we have reached a new phase in the study of stellar organization. Having found, as one product of the photometric work, the distance of this star cloud, we are now in a position to determine for several types of stars and nebulae the absolute luminosities which were heretofore uncertainly known. We shall also be able to measure the maximum luminosity reached by stars in different evolutionary stages. Until now this has been accomplished only for stars in globular clusters, which have been considered, by some students of the subject, to be highly specialized regions. There can be no doubt, however, but that the Magellanic Clouds are comparable with star clouds of the Galaxy in practically every respect. In the following discussion a summary is given of our present knowledge of these objects.

Both of the Magellanic Clouds are visible to the naked eye. The center of the smaller is in the constellation Tucana. It is only fifteen degrees from the south pole. The galactic latitude and longitude are — 44° and 268°, respectively. The Large Cloud is in Doradus, in declination — 69°, with galactic coordinates — 33° and 247°. Photometric observations have been made almost exclusively at the Boyden Station of the Harvard Observatory, at Arequipa, Peru, and the spectroscopic observations have been made at the same place and at the station of the Lick Observatory at Santiago, Chile.

Long exposures with short focus instruments have shown that the extent of the clouds is greater than appears to the unaided eye. The principal feature seen visually in each cloud is an elongated patch of stars and nebulae with irregular extensions. Appropriate photographs show, however, that both clouds are nearly circular in outline. The average angular diameter, from measures in four position angles, is $3°.6$ for the Small Cloud, and $7°.2$ for the Large Cloud[1]).

The 24-inch Bruce refractor and other telescopes at the Boyden Station have been used throughout an interval of thirtyfive years in accumulating photographs of the two Magellanic Clouds. From a study

---

[1]) Harvard Bull. 796, 1923.

of the material available in 1906 Miss LEAVITT announced the discovery of 1777 variable stars (H. A. **60**, 87). In addition to these variables, many stars with spectra of peculiar type have been found on the Harvard plates. Miss CANNON has recently given the positions and provisional apparent magnitudes of all stars in the Large Magellanic Cloud known to have spectra of Class O and of the peculiar P Cygni type[1]).

Star clusters of the condensed, open, and nebulous types are present in both clouds, and also a considerable number of gaseous nebulae. The radial velocities of the nebulae, as given by their bright lines, have been made the subject of a special study by Dr. R. E. WILSON, while at the southern station of the Lick Observatory. Since practically all the known gaseous nebulae are closely confined to the Milky Way except those in the Magellanic Clouds, we can safely assume that the radial velocities revealed by their spectra indicate the velocities of the clouds themselves. The velocity of only one object in the Small Magellanic Cloud, N.G.C. 1644, was determined. Its value is $+$ 168 km/sec. In the Large Cloud, however, the velocities of seventeen nebulae were measured. The values range from $+$ 251 to $+$ 309, with an average of $+$ 276 km/sec[2]).

HERTZSPRUNG has considered the possibility that the Magellanic Clouds are moving in parallel paths and that the distribution of observed radial velocities throughout the Large Cloud is appropriately related to the distance from the convergent point of the motion of the two clouds[3]). Nothing is yet known of the proper motions of the stars of the Magellanic Clouds. Considering the great distance, appreciable apparent motion across the line of sight cannot be expected in the relatively short interval throughout which observations have been made.

The most important product so far of the study of the Magellanic Clouds is the period-luminosity curve. Some twenty years ago Miss LEAVITT determined the periods and provisional magnitude limits for twenty-five of the brighter variables. It appeared at once that the magnitude and period were related. Stars of short period were found to be fainter. Throughout the interval studied, the apparent magnitude was nearly a linear function of the logarithm of the period. This relation has since been extended to stars of shorter period in the Small Magellanic Cloud and to variable stars in globular clusters. Moreover, the curve has been transformed from apparent brightness to absolute luminosity. This important change could be made by taking three steps: 1. Assuming the generality of the phenomenon of Cepheid variation throughout the whole stellar system. 2. Determining the parallax and absolute magnitude of typical bright galactic Cepheids of known period as a means of establishing the zero point of the period-luminosity curve. 3. Noting that

---

[1]) Harvard Bull. 801. 1924.

[2]) Lick Observatory Publications **13**, 188. 1917.

[3]) Monthly Notices **80**, 782. 1920.

in clusters and the Magellanic Clouds relative absolute magnitudes can be substituted at once for relative apparent magnitudes[1]).

With reasonable assumptions, therefore, it appears that the actual luminosity of a Cepheid variable is known when the period of variation has been determined. And with luminosity known, the distance is immediately computed as soon as the apparent magnitude on a standardized photometric system has been measured.

With the aid of the provisional system of magnitudes given by Miss LEAVITT, the distance of the Small Magellanic Cloud has been estimated by HERTZSPRUNG, KAPTEYN, and the present writer. The revision of the magnitude system[2]) and the determination of the periods of nearly a hundred other variable stars, has permitted a new measurement of the distance of the Small Cloud. The revised determination is

$$D = 31{,}000 \text{ parsecs}$$

which corresponds to a parallax of $0''{.}000032$ and a distance of 102,000 light years. The estimated probable error is fifteen per cent. The uncertainty arises from errors in the mean parallax of galactic Cepheids of determined proper motions, from the remaining errors in the apparent magnitude, and from the uncertainty in the adopted correction for the average color of Cepheid variables.

The distance now adopted is three times that first computed by HERTZSPRUNG, a discrepancy due mainly to the provisional nature of the original magnitude system.

With the distance known, we compute the average linear diameter of the Small Cloud to be 6,500 light years, or ten times the distance from the Earth to the Orion Nebula. We are now also able to estimate the actual brightness of some of its giant stars. Although we can tell very little of the stars comparable with our Sun, since they would be of apparent magnitude twenty-three, we are able to study the distribution of the stars with absolute magnitudes greater than zero. The visual magnitudes of some of the brightest objects in the system are found to be near the tenth magnitude. The corresponding absolute brightness is therefore greater than -7, which makes them considerably the brightest stellar objects on record.

A number of the brightest stars in the Small Cloud are of late spectral class. Assuming that the usual relation of surface luminosity and spectrum applies to these objects as well as to analogous stars near the Sun, we compute that their linear diameters are of the order of 10^9 kilometers. This value is of the same order of magnitude as the major axis of the orbit of Jupiter, and indicates that a large number of the brightest stars in the Magellanic Clouds exceed Antares and Betelgeux in linear diameter.

---

[1]) Harvard Circular 255. 1924.   [2]) Harvard Circular 255. 1924.

Little is yet known concerning the variable stars in the Large Magellanic Cloud. The distance has been estimated, however, at 35,000 parsecs with the aid of measurements of the angular diameters of several faint globular clusters that seem to be members of the system[1]). In the near future a determination based on the periods and revised magnitudes for some of the variable stars will be deduced.

A still more remote object, which appears to resemble the Magellanic Clouds, is the faint nebulosity N.G.C. 6822, discovered by BARNARD and analyzed at Mount Wilson, Cordoba, and Harvard[2]). Probably it is a million light years distant. The Small Cloud at that distance would resemble it closely in angular dimensions and in discernible stellar contents.

The future work on the Magellanic Clouds will include more detailed investigations of the light curves and periods of the variable stars, a revision of the period-luminosity curve, a consideration of the general luminosity law for giant stars, a study of the distribution and luminosity of the scores of faint nebulae that have not yet been announced, and the measurement of the brightest stars in the numerous involved clusters. The study of the general luminosity law is now under way at Harvard; it should throw light on the nature and evolution of galactic star clouds. It is well to remember that probably both Magellanic Clouds have been in the galactic plane and indistinguishable from other galactic clouds at some time since the Paleozoic Era.

---

[1]) Harvard Bull. 775. 1922.   [2]) Harvard Bull. 796. 1923.

On the Reflection of Light in a Close Binary System.

By **Joel Stebbins**, Madison[1]).

Among the eclipsing variable stars which have been studied in recent years there are a number which in addition to the eclipse variation show a marked increase of light as the fainter component moves from inferior to superior conjunction. The physical interpretation of this so-called "radiation effect" is obviously the increased brightness of the faint companion on the side toward the primary, for in the systems where this effect is conspicuous the least distance between the surfaces of the two components does not much exceed the diameter of the main body. This increase of intensity on one side of the companion may be attributed in part to reflection, but because of the energy of radiation received at a small distance from the central star, any body would be heated up to incandescence. In the case of ALGOL it may be shown that the surface of the companion toward the main star must receive per unit area more radiation than is emitted from the surface of the sun, and since a condition of equilibrium has been reached, the companion must be incandescent on one side, though probably it is hot enough on its own account to be self-luminous independent of the primary.

If we knew the law of reflection from the surface of such a companion, it would be possible to compute for a system of given dimensions what fraction of the total light-variation between minima is due to reflection by the second body, and subtracting this effect from the total variation there would remain the simple heating effect combined with the variation due to ellipsoidal figure of the bodies. As even for the planets of the solar system the observed light-variations with phase are not in good agreement with any theoretical law, the best we can hope for in the case of a stellar system is some rough indication of how the light is reflected.

---

[1]) The following investigation was carried on more than ten years ago, while I was spending a sabbatical year at Munich, and I was much indebted to Geheimrat SEELIGER for various suggestions. I am now glad to join with other students of SEELIGER in honor of his anniversary, which should be an occasion for congratulations from astronomers of every land.

The two laws of diffuse reflection which have been seriously advanced for application to celestial bodies are LAMBERT's law and the LOMMEL-SEELIGER law. The latter we may call, following SEELIGER, the absorption law. These give for the amount of light dq which incident on a surface element ds at the angle i is reflected at emanation angle ε:

$$\left.\begin{array}{ll} \text{LAMBERT's law,} & dq_1 = \dfrac{J\,\mu_1\,ds}{\pi}\cdot\cos i \cos \varepsilon \\[3mm] \text{Absorption law,} & dq_2 = \dfrac{J\,\mu_2\,ds}{\pi}\cdot\dfrac{\cos i \cos \varepsilon}{\cos i + \cos \varepsilon} \end{array}\right\} \tag{1}$$

where J is the intensity of the light incident on ds, μ_1 the albedo according to LAMBERT's definition, and μ_2 the albedo according to SEELIGER's definition[1]).

The amount of light reflected by the companion of ALGOL has been computed for the position at superior conjunction, and the formulae in the previous paper[2]) give the albedo on LAMBERT's law for any eclipsing variable where the light-curve shows the radiation effect. It is proposed now to discuss for ALGOL the variation due to reflection at all phases of the companion, first according to LAMBERT's law and then according to the absorption law. Since in ALGOL the bright body gives more than ten times the light of the companion, the effect of the fainter body upon the primary may be neglected. For components of nearly equal intensity the interchange of radiation must produce two brighter regions where the surfaces are nearest, especially when the tidal attraction makes the periods of revolution and rotation the same. The revolution of such a twin system, even when observed in the plane of the orbit, would give only a minute radiation effect, what is gained in the full phase of one component being lost in the new phase of the other.

As preparation for the discussion of the illumination of a dark companion close to its primary we have to determine the illumination of a surface element ds by a luminous sphere part of whose apparent disk is below the horizon of ds. We assume that the sphere radiates according to the cosine law so that the disk is of apparent uniform intensity. To take into account any darkening at the limb would make the problem impossibly complicated at the very outset.

As viewed from ds let the center of the sphere of angular radius δ be at altitude h, or zenith distance z. If J is the light intensity at unit distance from a unit surface of the sphere, then the light received by ds at incidence angle i from an element $d\sigma$ of the luminous surface is:

$$d\,(d\,L) = J\,ds\,d\sigma\cos i$$

[1]) MUELLER: Photometrie der Gestirne pp. 40, 57, 63, usw.

[2]) Astrophysical Journ. **33**, 397. 1911.

Let v be the apparent angular distance of $d\sigma$ from the center of the disk and Θ the angle about the center, measured from the vertical, then

$$\cos i = \cos z \cos v + \sin z \sin v \cos \Theta$$

$$d\sigma = \sin v \, d\Theta \, dv$$

When all of the sphere is above the horizon the light received is:

$$dL = 2J \, ds \, \cos z \int_0^\pi d\Theta \int_0^\delta \sin v \cos v \, dv$$

the integral of the second term from $\cos i$ vanishing because of the symmetry[1]). Therefore:

$$dL = J\pi \, ds \, \cos z \sin^2 \delta \tag{2}$$

With part of the disk below the horizon we have, if:

$$\Theta_0 = \pi - \operatorname{arc\,cos} (\operatorname{ctn} z \operatorname{ctn} v)$$

$$dL = 2J \, ds \, \cos z \int_0^\pi d\Theta \int_0^h \sin v \cos v \, dv + 2J \, ds \, \cos z \int_0^{\Theta_0} d\Theta \int_h^\delta \sin v \cos v \, dv +$$

$$+ 2J \, ds \, \sin z \int_0^{\Theta_0} \cos \Theta \, d\Theta \int_h^\delta \sin^2 v \, dv$$

$$dL = J\pi \, ds \, \cos z \sin^2 h + 2J \, ds \, \cos z \int_h^\delta [\pi - \operatorname{arc\,cos}(\operatorname{ctn} z \operatorname{ctn} v)] \sin v \cos v \, dv +$$

$$+ 2J \, ds \, \sin z \int_h^\delta \left(\sqrt{1 - \operatorname{ctn}^2 z \operatorname{ctn}^2 v}\right) \sin^2 v \, dv$$

Separating the integrals of the second term we put:

$$dL = I + II_a + II_b + III$$

and find:

$$I + II_a = J\pi \, ds \, \cos z \sin^2 \delta$$

$$II_b = -2J \, ds \, \cos z \int_h^\delta \operatorname{arc\,cos} (\operatorname{ctn} z \operatorname{ctn} v) \sin v \cos v \, dv$$

Putting $x = \operatorname{ctn} z \operatorname{ctn} v$, this becomes:

$$II_b = -J \, ds \, \cos z \operatorname{ctn}^2 z \left[\frac{\operatorname{arc\,cos} x}{x^2 + \operatorname{ctn}^2 z} - \sin z \tan z \operatorname{arc\,cos} \frac{x \csc z}{\sqrt{x^2 + \operatorname{ctn}^2 z}} \right]_1^{\operatorname{ctn} z \operatorname{ctn} \delta}$$

$$= -J \, ds \, \cos z \left\{ \sin^2 \delta \operatorname{arc\,cos} (\operatorname{ctn} z \operatorname{ctn} \delta) - \cos z \operatorname{arc\,cos} (\csc z \cos \delta) \right\}$$

For the last term of dL we put $x = \cos v$ and find:

$$III = -J \, ds \, \sin z \left[x\sqrt{1 - x^2 \csc^2 z} + \sin z \operatorname{arc\,sin} (x \csc z) \right]_{\sin z}^{\cos \delta}$$

$$= -J \, ds \, \sin z \left\{ \cos \delta \sqrt{1 - \csc^2 z \cos^2 \delta} - \sin z \operatorname{arc\,cos} (\csc z \cos \delta) \right\}$$

---

[1]) Mueller, p. 36.

Collecting the terms we have:

$$dL = J\,ds\,\{\pi\cos z \sin^2\delta - \cos z \sin^2\delta \arccos (\operatorname{ctn} z \operatorname{ctn}\delta)$$
$$+ \arccos (\cos z\cos\delta) - \sin z \cos\delta \sqrt{1 - \csc^2 z \cos^2\delta}\} \tag{3}$$

The terms in the $\{\}$ will afterwards be referred to as $f(\varphi)$, and may be called the "effective" apparent area of the bright disk as seen from ds.

We now consider a spherical companion in a binary system illuminated by the primary. Let:

$ds =$ the element of the reflecting surface of the companion,

$\varphi =$ the angle between the normal of ds and the line joining the centers of the spheres,

$\varepsilon =$ the angle between the normal of ds and the direction to the earth; this is the angle of emanation from ds,

$\alpha =$ the angle at the center of the companion between the line to the earth and the line to the center of the primary,

$\psi =$ the azimuth angle about the line of centers,

$\varkappa =$ the radius of the companion, the radius of the bright body being unity,

$a =$ the distance between centers, in the circular orbit, the radius of the bright body being unity,

$\gamma = \varkappa/a$,

$\Gamma = \sqrt{1 + \gamma^2 - 2\gamma\cos\varphi}$, the distance of ds from the center of the bright body being $a\,\Gamma$,

$f(\varphi) =$ the effective apparent area of the bright body as seen from ds; this is defined in (3),

$J =$ the light emitted by unit surface of the bright body,

$L_1 = \pi J =$ the total light emitted by the bright body in any given direction,

$dL =$ the total light received by ds,

$L_2 =$ the total light reflected by the companion in the direction of the earth,

$\mu_1 =$ the albedo of the companion on LAMBERT's law,

$\mu_2 =$ the albedo on the absorption law,

$S_1 = \tfrac{2}{3}\gamma^2 + \tfrac{1}{2}\gamma^3$, for LAMBERT's law, see (7),

$S_2 =$ the corresponding expression for the absorption law, see (10) and (11),

$h =$ the apparent brightness of ds on absorptions law, see (12).

1. Lambert's Law. The light received by ds is:

$$dL = J\,ds\,f(\varphi)$$

and on LAMBERT's law the light reflected by ds in the direction ε is:

$$dL_2 = \frac{\mu_1}{\pi}\cdot dL\cos\varepsilon = \frac{J\mu_1\,ds}{\pi}\cdot f(\varphi)\cos\varepsilon$$

Therefore:

$$L_2 = \int \frac{J\mu_1}{\pi} f(\varphi) \cos \varepsilon \, ds \qquad (4)$$

Because of the boundary conditions this integration for varying phase becomes complicated and needlessly tedious, and the solution may be approximated by dividing the companion into zones of equal illumination.

The area $\varDelta s$ of a zone-segment of width ψ and length $2l$ with middle at angle φ is:

$$\varDelta s = \varkappa^2 \, d\psi \, [\cos(\varphi - l) - \cos(\varphi + l)] = 2\varkappa^2 \sin\varphi \sin l \, d\psi$$

Since $\cos \varepsilon = \cos \alpha \cos \varphi + \sin \alpha \sin \varphi \cos \psi$, the total light sent toward the earth by such a zone is:

$$\varDelta L_2 = \frac{4 J \mu_1 \varkappa^2}{\pi} \int\limits_0^{\text{arc cos}(-\text{ctn}\,\alpha\,\text{ctn}\,\varphi)} \sin\varphi \sin l \, f(\varphi) \, (\cos\alpha\cos\varphi + \sin\alpha\sin\varphi\cos\psi)\, d\psi$$

or:

$$\varDelta L_2 = \frac{4 J \mu_1 \varkappa^2}{\pi} \cdot \sin l \sin\varphi \, f(\varphi) \left\{ \cos\alpha\cos\varphi \,\text{arc cos}(-\text{ctn}\,\alpha\,\text{ctn}\,\varphi) + \right.$$
$$\left. + \sin\alpha\sin\varphi \sqrt{1 - \text{ctn}^2\alpha\,\text{ctn}^2\varphi} \right\} \qquad (5)$$

When $\varphi \gtreqless \dfrac{\pi}{2} - \alpha$, the second term vanishes and arc cos $(-\text{ctn}\,\alpha\,\text{ctn}\,\varphi)$ $= \pi$; when $\varphi \gtreqless \alpha - \dfrac{\pi}{2}$, then $\varDelta L_2 = 0$.

To find $f(\varphi)$ from (3) we substitute:

$$\sin z = \frac{\sin\varphi}{\Gamma}, \qquad\qquad \sin\delta = \frac{1}{a\Gamma},$$

$$\cos z = \frac{\cos\varphi - \gamma}{\Gamma}, \qquad\qquad \cos\delta = \frac{\sqrt{a^2\Gamma^2 - 1}}{a\Gamma},$$

$$\text{ctn}\,z = \frac{\cos\varphi - \gamma}{\sin\varphi}, \qquad\qquad \text{ctn}\,\delta = \sqrt{a^2\Gamma^2 - 1},$$

and derive:

$$f(\varphi) = \frac{1}{a^2}\left\{ \frac{\pi(\cos\varphi - \gamma)}{\Gamma^3} - \frac{\cos\varphi - \gamma}{\Gamma^3}\,\text{arc cos}\left(\frac{\cos\varphi - \gamma}{\sin\varphi} \cdot \sqrt{a^2\Gamma^2 - 1}\right) + \right.$$
$$\left. + a^2\,\text{arc cos}\left(\frac{\sqrt{a^2\Gamma^2 - 1}}{a\sin\varphi}\right) - \frac{\sqrt{(a^2\Gamma^2 - 1)(a^2\sin^2\varphi - a^2\Gamma^2 + 1)}}{\Gamma^2} \right\} \qquad (6)$$

With different values of α we can compute from (5) and (6) the light from each zone, and then take the sum for all of the zones. This was done for ALGOL with the elements found in the first work with the selenium photometer[1]; $\varkappa = 1.14$, $a = 4.77$, $\gamma = 0.239$. The results of the computation are in Table I, the tabulated quantity being $\varDelta L_2/\pi J_1 \mu_1$.

The zones were taken of width $2\,l = 20°$, with middle points at $\varphi = 10°$, $30°$, and $50°$, and then of width $2\,l = 10°$ with middle points at $\varphi = 65°$, $75°$, and $85°$.

Table I.

α	$0°$	$30°$	$60°$	$90°$	$120°$	$150°$
$\varphi = 10°$	0.01128	0.00975	0.00564	0.00063	0.00000	0.00000
30	.02089	.01809	.01044	.00384	.00000	.00000
50	.01216	.01053	.00751	.00461	.00143	.00000
65	.00177	.00158	.00153	.00121	.00064	.00005
75	.00032	.00034	.00037	.00038	.00021	.00007
85	.00001	.00002	.00004	.00003	.00001	.00000
Sum	0.0464	0.0403	0.0255	0.0107	0.0023	0.0001
Relative Δ mag.	$0^{\mathrm{M}}.0500$	$0^{\mathrm{M}}.0434$	$0^{\mathrm{M}}.0275$	$0^{\mathrm{M}}.0115$	$0^{\mathrm{M}}.0025$	$0^{\mathrm{M}}.0001$
Eq. (8)	0.0500	0.0433	0.0275	0.0117	0.0025	0.0001
Difference . .	0	-1	0	$+2$	0	0

For the simple case of $\alpha = 0$ there was found in the previous paper[1]) the expression for LAMBERT's albedo μ_1:

$$S_1 = \tfrac{2}{3}\,\gamma^2 + \tfrac{1}{2}\,\gamma^3 \left.\vphantom{\frac{L_2}{L_1 S_1}}\right\}$$
$$\mu_1 = \frac{L_2}{L_1 S_1} \qquad (7)$$

For convenience S_1 is given in Table II.
The sum in the first column of Table I is:

$$\frac{\sum \Delta L_2}{\pi\,J\,\mu_1} = \frac{L_2}{\mu_1\,L_1} = 0.0464$$

and this being another approximation for S_1, may be compared with $S_1 = 0.0449$ for $\gamma = 0.239$ in Table II. Since for ALGOL $L_2/L_1 = 0.049$, either value of S_1 gives $\mu_1 = 1.1$, showing that reflection alone will not account for the radiation effect in ALGOL.

The sum in Table I for each place α being proportional to the light reflected according to LAMBERT's law, we may consider the relative change in magnitude as proportional to this sum. The variation in magnitude of the system may be represented by the empirical relation:

$$\Delta \text{ mag.} = 0^{\mathrm{M}}.0117 + 0^{\mathrm{M}}.0250 \cos\alpha + 0^{\mathrm{M}}.0133 \cos^2\alpha \qquad (8)$$

showing that to a close approximation any reflection on LAMBERT's law would impress upon the light-curve an effect represented by a $\cos^2\alpha$-term the sign of which is opposite to the effect due to ellipsoidal figure. Therefore if in 1910 the selenium light-curve of ALGOL showed a radiation effect of 0.05 magnitude, there would be on LAMBERT's law an ellipsoidal effect corresponding to $0^{\mathrm{M}}.0133$ in (8), which would give RUSSELL's

---

[1]) Astrophysical Journal **33**, 397, 1911.

constant $z = 0.024$. In the photo-electric light-curve of 1920[1]) there were terms in both $\cos\theta$ and $\cos^2\theta$ $(\theta = \alpha + \pi)$. Since the derived dimensions of the system were about the same in 1910 and 1920, we may take the LAMBERT reflection results as of the same form but of reduced amount in 1920.

$$\left.\begin{array}{ll} \text{1920, observed } \Delta \text{ mag.} = \text{const.} - 0\overset{M}{.}0164\cos\theta - 0\overset{M}{.}0148\cos^2\theta \\ \text{LAMBERT,} \qquad \Delta \text{ mag.} = \text{const.} - 0.0164\cos\theta + 0.0087\cos^2\theta \\ \text{Difference} \qquad\qquad\qquad\qquad\qquad\qquad\qquad\quad 0.0235\cos^2\theta \end{array}\right\} \quad (9)$$

On LAMBERT's law we should allow for an ellipsoidal effect of $0\overset{M}{.}0235\cos^2\theta$, or $z = 0.043$ as against $z = 0.027$ which was derived without regard to the additional term of the reflection.

2. Absorption Law. On the basis of the absorption law of diffuse reflection we may derive the light from a companion body at different phases in the same manner as for LAMBERT's law, but the integrations become very complicated. As a first simplification we substitute for the sphere of light πJ a point source of intensity πJ; that is, a unit surface at unit distance from the point source receives a total light πJ. The integration over the surface of the companion includes the area visible from the point source, or from $\varphi = 0$ to $\varphi = \text{arc}\cos\gamma$.

Considering first the case of superior conjunction, $\alpha = 0$, the light received by an element ds is:

$$dL = \frac{\pi J\, ds \cos z}{a^2\, \Gamma^2}$$

and on the absorption law the amount reflected toward the earth, $\varepsilon = \varphi$, is:

$$dL_2 = \frac{\mu_2\, dL}{\pi} \cdot \frac{\cos\varphi}{\cos z + \cos\varphi} = \frac{J\mu_2\, ds}{a^2\, \Gamma^2} \cdot \frac{\cos z \cos\varphi}{\cos z + \cos\varphi}$$

Substituting as before, $ds = \varkappa^2 \sin\varphi\, d\psi\, d\varphi$, $\cos z = (\cos\varphi - \gamma)/\Gamma$:

$$L_2 = 2\pi J \mu_2 \gamma^2 \int\limits_0^{\text{arc}\cos\gamma} \frac{\sin\varphi \cos\varphi\, (\cos\varphi - \gamma)\, d\varphi}{\Gamma^2\, (\cos\varphi - \gamma + \Gamma\cos\varphi)}$$

or since $\cos\varphi = \dfrac{1 + \gamma^2 - \Gamma^2}{2\gamma}$, $\sin\varphi\, d\varphi = \dfrac{\Gamma\, d\Gamma}{\gamma}$, $L_1 = \pi J$, the above reduces to:

$$L_2 = \mu_2 L_1 \int\limits_{1-\gamma}^{\sqrt{1-\gamma^2}} \frac{(1 + \gamma^2 - \Gamma^2)(1 - \gamma^2 - \Gamma^2)\, d\Gamma}{\Gamma[1 - \gamma^2 - \Gamma^2 + \Gamma(1 + \gamma^2 - \Gamma^2)]}$$

---

[1]) Astrophysical Journ. **53**, 114. 1921.

Integrating[1]:

$$L_2 = \mu_2 L_1 \left[-\Gamma + (1+\gamma^2)\log\Gamma + \frac{\gamma^4}{4-\gamma^2}\log(1-\Gamma) + \right.$$

$$\left. + \frac{2\gamma}{2+\gamma}\log(1+\gamma+\Gamma) - \frac{2\gamma}{2-\gamma}\log(1-\gamma+\Gamma) \right]_{1-\gamma}^{\sqrt{1-\gamma^2}}$$

$$L_2 = \mu_2 \Gamma_1 \left\{ 1-\gamma-\sqrt{1-\gamma^2} + (1+\gamma^2)\log\sqrt{\frac{1+\gamma}{1-\gamma}} + \frac{\gamma^4}{4-\gamma^2}\log\frac{1-\sqrt{1-\gamma^2}}{\gamma} \right.$$

$$\left. + \frac{2\gamma}{2+\gamma}\log\frac{1+\gamma+\sqrt{1-\gamma^2}}{2} - \frac{2\gamma}{2-\gamma}\log\frac{1-\gamma+\sqrt{1-\gamma^2}}{2(1-\gamma)} \right\} \qquad (10)$$

which we write:

$$L_2 = \mu_2 L_1 S_2$$

and therefore:

$$\mu_2 = \frac{L_2}{L_1 S_2} \qquad (11)$$

for the albedo on SEELIGER's definition. The value of S_2 for different values of γ has been computed and is in Table II.

Table II.

γ	S_1	S_2	γ	S_1	S_2
0.10	0.0072_{15}	0.0053_{11}	0.20	0.0307_{33}	0.0221_{24}
.11	$.0087_{18}$	$.0064_{13}$	.21	$.0340_{36}$	$.0245_{25}$
.12	$.0105_{19}$	$.0077_{14}$	.22	$.0376_{38}$	$.0270_{26}$
.13	$.0124_{20}$	$.0091_{15}$	.23	$.0414_{39}$	$.0296_{28}$
.14	$.0144_{23}$	$.0106_{16}$	.24	$.0453_{42}$	$.0324_{29}$
.15	$.0167_{24}$	$.0122_{17}$	.25	$.0495_{44}$	$.0353_{30}$
.16	$.0191_{26}$	$.0139_{19}$	.26	$.0539_{45}$	$.0383_{32}$
.17	$.0217_{28}$	$.0158_{19}$	.27	$.0584_{48}$	$.0415_{33}$
.18	$.0245_{30}$	$.0177_{21}$	.28	$.0632_{51}$	$.0448_{35}$
.19	$.0275_{32}$	$.0198_{23}$	.29	$.0683_{52}$	$.0483_{36}$
.20	$.0307$	$.0221$	.30	$.0735$	$.0519$

For the 1910 elements of ALGOL $\gamma = 0.239$, $S_2 = 0.0321$, $L_2/L_1 = 0.049$, and:

$$\mu_2 = 1.5$$

which, as for LAMBERT's law, shows that the light from the companion is greater than can be accounted for by reflection. It will be noted from the relative values of S_1 and S_2 that for γ between 0.10 and 0.30 the computed value of μ_2 will be roughly forty per cent larger than μ_1.

For the partial phases of the companion we proceed as follows: The light reflected toward the earth by the surface element ds is given by:

$$dL_2 = \frac{J\mu_2 ds}{a^2 \Gamma'^2} \cdot \frac{\cos z \cos \varepsilon}{\cos z + \cos \varepsilon}$$

---

[1] This integration was given to me by Professor J. B. SHAW.

while the apparent brightness h of the element in the direction ε is:

$$h = \frac{d L_2}{d s \cos \varepsilon} = \frac{J \mu_2}{a^2 \Gamma^2} \cdot \frac{\cos z}{\cos z + \cos \varepsilon} \tag{12}$$

We divide the apparent disk of the companion into 5 rings of equal area, and each ring into 18 equal segments. The middle points of the segments of each ring have the emanation angles:

$$\varepsilon = 18^\circ\!.44, \quad 33^\circ\!.21, \quad 45^\circ\!.00, \quad 56^\circ\!.79 \quad \text{and} \quad 71^\circ\!.56$$

The values of φ and z are computed from the relations:

$$\cos \varphi = \cos \alpha \cos \varepsilon + \sin \alpha \sin \varepsilon \cos \zeta$$

$$\cos z = \frac{\cos \varphi - \gamma}{\Gamma}$$

where ζ is the angle about the center of the apparent disk; α is taken in steps of 30° from 0° to 150°, while for each α there are 18 values of ζ from 0° to 330°.

Adding the light from as many of the 90 segments as are illuminated, the total light from the companion at any phase is:

$$L_2 = \frac{\pi \varkappa^2}{90} \sum h = \frac{\pi \varkappa^2}{90} \cdot J \mu_2 \sum \frac{h}{J \mu_2}$$

or as before, $L_1 = \pi J$, and:

$$\frac{L_2}{L_1} = \frac{\mu_2 \gamma^2}{90} \sum \frac{\cos \varphi - \gamma}{\Gamma^2 (\cos \varphi - \gamma + \Gamma \cos \varepsilon)} \tag{13}$$

where φ and Γ are computed for the middle of each segment. This computation was carried out by Mr. J. D. BOND with the results in Table III, the tabulated quantity being $L_2/\mu_2 L_1$.

Table III.

α	$0°$	$30°$	$60°$	$90°$	$120°$	$150°$
Ring 1	0.00928	0.00748	0.00300	0.00014	0.00000	0.00000
2	.00816	.00670	.00343	.00099	.00000	.00000
3	.00697	.00564	.00382	.00182	.00002	.00000
4	.00556	.00474	.00425	.00273	.00069	.00000
5	.00257	.00391	.00472	.00408	.00222	.00007
Sum	0.0325	0.0285	0.0192	0.0098	0.0029	0.0001
Relative Δ mag.	$0^{\text{M}}\!.0500$	$0^{\text{M}}\!.0438$	$0^{\text{M}}\!.0295$	$0^{\text{M}}\!.0151$	$0^{\text{M}}\!.0045$	$0^{\text{M}}\!.0001$
Eq. (14) . . .	0.0500	0.0440	0.0295	0.0144	0.0045	0.0008
Difference . .	0	$+2$	0	-7	0	$+7$

As in the case of LAMBERT's law the reflection by the companion of ALGOL as given in Table III is approximated by a relation similar to (8):

Absorption, Δ mag. $= 0^{\text{M}}\!.0144 + 0^{\text{M}}\!.0250 \cos \alpha + 0^{\text{M}}\!.0106 \cos^2 \alpha$ (14)
The $\cos^2 \alpha$-coefficient is some twenty per cent smaller than with

LAMBERT's law, and there is a corresponding correction to the ellipsoidal constant z, but the relation (8) or (14) expresses within a thousandth of a magnitude the computed variation due to diffuse reflection.

3. Application to Other Stars. We can now use (7) and (11) to compute the albedoes in other stars, and besides Algol we select λ Tauri[1]), and two of the stars studied by DUGAN, viz. Z Draconis[2]) and RT Persei[3]).

Table IV.

Method	Algol	Algol	λ Tauri	Z Draconis	RT Persei
	Selenium	Photo-Electric	Photo-Electric	Visual	Visual
γ	0.239	0.244	0.224	0.270	0.283
L_2/L_1	0.049	0.032	0.105	0.045	0.026
μ_1	1.1	0.7	2.7	0.8	0.4
μ_2	1.5	1.0	3.8	1.1	0.6

The elements of λ Tauri are somewhat indeterminate because of the small range of variation, and also because of the known third body in the system, so the high values of the computed albedo are uncertain. Z Draconis seems to be much like Algol, but for RT Persei the radiation effect is only 0.02 magnitude, and can be ascribed wholly to reflection without assuming an unusually high albedo. In fact, in any system with γ about equal to 0.25, a radiation effect of one or two hundredths of a magnitude may be looked upon as due wholly to reflection.

Summary. For diffuse reflection from the companion of a binary system, according to LAMBERT's law and the absorption law, expressions for the albedo have been derived from the full phase in (7) and (11). It has been shown that reflection in the partial phases will mask to some extent the effect due to ellipsoidal figure, the results for Algol being in Tables I and III, and in (8), (9), and (14). A comparison of the computed albedoes for several stars is given in Table IV.

Finally, since the radiation effect usually amounts to only a few hundredths of a magnitude, it should not be forgotten that in the present state of photometry there is no prospect of securing anything but a rough indication of how light may be reflected in a stellar system.

---

[1]) Astrophysical Journ. **51**, 217. 1920.
[2]) Astrophysical Journ. **39**, 418. 1914.
[3]) Astrophysical Journ. **39**, 426. 1914.

Das Problem der Veränderlichkeit der Sonnenstrahlung.

Von **Walter E. Bernheimer**, Wien.

Mit 7 Abbildungen.

Die bedeutenden Fortschritte der Astrophysik in den letzten Jahrzehnten haben unsere Erkenntnisse über den inneren Aufbau und die Entwicklung der Fixsterne und die mannigfaltigen Probleme der Stellarastronomie gewaltig gefördert. Durch die Erfolge gesichert, erweiterte sich das Forschungsgebiet und die einst nur spekulativen Erörterungen über die Größe des Sternsystems, die Natur der Sternhaufen und Nebel bewegen sich heute auf gesunder, durch ein reiches Beobachtungsmaterial gestützten Grundlage.

Diese kraftvolle Erweiterung des Gesichtskreises unserer Wissenschaft mag es mit sich gebracht haben, daß die Probleme unserer engsten Welt mehr in den Hintergrund getreten sind. So brachten die letzten Jahre wenig Fortschritte in der Physik der Planeten, abgesehen etwa von den schönen Arbeiten von COBLENTZ[1]) und MENZEL[2]) über die Planetentemperaturen, und viel Dunkel ist hier noch aufzuhellen.

Aber auch die Sonnenphysik beschäftigt heute — im Gegensatz zur Bedeutung und Vielseitigkeit der zu lösenden Probleme — neben den Theoretikern EMDEN, LINDBLAD, LUNDBLAD, MILNE, WILSING u. a. im wesentlichen nur die Schulen von HALE und ABBOT. Greifen wir als das wichtigste das Problem der Sonnenstrahlung und ihrer Veränderlichkeit heraus, so ist es einleuchtend, daß diese Frage in doppelter Weise von größter Bedeutung ist. Ihre Klarstellung gäbe uns einerseits gewichtige Aufschlüsse über die Vorgänge in unserer Atmosphäre, und das Zutreffen einer von F. NANSEN, H. CLAYTON u. a. vermuteten Zwangsbeziehung zwischen wechselnder Sonnenstrahlung und Luftdruckschwankungen der Erdhülle würde auch in prognostischer Hinsicht neue Wege zeigen. So ist es begreiflich, daß die Meteorologie heute dem Problem der Veränderlichkeit der Sonnenstrahlung die intensivste Beachtung schenkt und gespannt einer Lösung entgegensieht.

Andererseits ist aber die Lösung des Problems der Veränderlichkeit der Sonnenstrahlung für die Astrophysik von mindestens ebenso weittragender Bedeutung. Ganz abgesehen davon, daß es ein dringendes

---

[1]) LICK Obs. Bull. VIII, Nr. 266.   [2]) Astroph. J. Bd. 58, S. 65.

Gebot ist, unsere Kenntnis über den nächsten Fixstern energisch zu fördern, wird eine Klarstellung der Frage der Schwankungen der Sonnenstrahlung dem Problem der veränderlichen Sterne neue, gewichtige Impulse verleihen. Gelingt es, die Realität der Schwankungen nachzuweisen, so hätten wir in der Sonne einen variablen Zwergstern vor uns, an dem wir zugleich leicht die Veränderungen der Flecken und Fackelbildungen, der Kontrastschwankungen Mitte—Rand studieren, mit den gleichzeitigen Änderungen der Gesamtstrahlung vergleichen könnten und wären so in der Lage, in den Mechanismus der Lichtschwankungen anderer Variablen, deren Vorgänge an der Oberfläche sich der Beobachtung entziehen, einzudringen.

Das Problem der Veränderlichkeit der Sonnenstrahlung, wie es ja erst heute zu fassen ist, liegt meines Erachtens nicht darin, auf die möglichen Ursachen dieser Schwankungen einzugehen, verschiedene Theorien aufzustellen und sie sodann auf ihre Stichhaltigkeit zu prüfen. Es liegt vielmehr darin, die bisherigen Beobachtungen solcher Schwankungen zu untersuchen und vorerst einmal festzustellen, ob diese Schwankungen reell sind und durch Veränderungen auf der Sonne selbst hervorgerufen werden. Erst dann kann das Problem weiter gefaßt werden, so wie es erst ein Cepheidenproblem gibt, seitdem die Helligkeitsschwankungen dieser Veränderlichen einwandsfrei klargestellt wurden.

1. Zur Geschichte der Messung der Sonnenstrahlung — Mittelwert der Solarkonstante. Nach vereinzelten Versuchen, mit Pyrheliometern Messungen der Sonnenstrahlung aufzuführen, ist 1902 von LANGLEY in *Washington* nach seiner verbesserten Methode zum erstenmal die Sonnenstrahlungmessung zu einem systematisch durchgeführten Beobachtungsprogramm geworden, das später auf dem *Mt. Wilson* (1905 bis 1920) von ABBOT fortgesetzt wurde. Neben einigen gelegentlichen Stationen, wie *Bassour* (Algier), *Arequipa* (Peru), *Mt. Whitney, Hump Mountain*, ist es dann vor allem 1918—1920 die Station *Calama* (Chile), die seit August 1920 auf einen Berg nahe Calama, *Montezuma* genannt, verlegt wurde, wo die glänzendsten klimatischen Verhältnisse herrschen. Die Beobachtungen des Mt. Wilson wurden seit Herbst 1920 ebenfalls in ein wesentlich günstigeres, mit Montezuma vergleichbares, Gebiet verlegt, nämlich *auf Mt. Harqua Hala* (bei Wenden in Arizona). Die Beobachtungen dieser Stationen ABBOTS und seiner Mitarbeiter bilden ein ungemein wertvolles Material zur Untersuchung des Problems der Sonnenstrahlung und ihrer Veränderungen. In Europa ist vor allem das 1909 von KNUT ÅNGSTRÖM errichtete Sonnenobservatorium in *Upsala* hervorzuheben, wo seit 1911 von GRANQUIST[1] den amerikanischen Beobachtungen analoge Messungen ausgeführt werden.

---

[1] GRANQUIST, G.: Solarkonstantens bestämning. Medd. från Vetenskaps akademiens Nobelinstitut Bd. 5, Nr. 13.

Die *Solarkonstante* als Maß der Sonnenstrahlung wird bekanntlich definiert als die Wärmemenge in gr/kal für 1 Minute und 1 cm², die in mittlerer Distanz von der Sonne bei senkrechter Einstrahlung die Erde erreichen würde, wenn keine Atmosphäre vorhanden wäre. Da es sich darum handelt, den Einfluß der Atmosphäre zu eliminieren und die empfangene Strahlung auf den Raum außerhalb der Atmosphäre zu extrapolieren, sind hier große Schwierigkeiten zu überwinden. Abb. 1, die ich einer Arbeit von Dines[1]) entnehme, vermag wohl am eindringlichsten die Kompliziertheit der Vorgänge zwischen Erdoberfläche und Vakuum zu schildern.

Abgesehen von der Frage nach Schwankungen der Sonnenstrahlung sind die in Amerika gewonnenen Mittelwerte von größter Wichtigkeit, da sie u. a. bekanntlich zur Bestimmung der effektiven Sonnentemperatur herangezogen werden. In neuester Zeit z. B. in den Arbeiten von Milne[2]) und Minnaert[3]).

Aus 696 Tagen der Jahre 1902—1912 ergibt sich 1.933 gr/kal,
Aus 1244 Tagen der Jahre 1912—1920 ergibt sich 1.946 gr/kal.

Aus späteren Messungen ergibt sich wieder ein kleinerer Wert. So finde ich aus Januar—September 1922 (dem letzten publizierten Material) 1.924, aus Juni—September allein sogar nur 1.913. Auf dieses Material kommen wir später noch zurück. Es ist bemerkenswert, daß der Mittelwert der zwei letzten Jahrzehnte mit einer der ältesten primitiven Messungen übereinstimmt. 1860 erhielt nämlich Hagen den Wert 1.9.

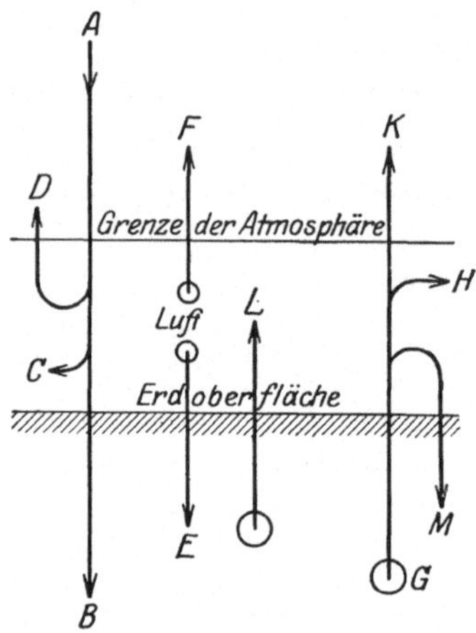

Abb. 1 (nach Dines). *Die Strahlungsvorgänge in der Atmosphäre.*
A Sonnenstrahlung außerhalb der Atmosphäre; davon wird *B* der Erde zugeführt, *C* in der Atmosphäre absorbiert und *D* endlich in Raum reflektiert.
G Strahlung der Erde, davon wird *H* in der Atmosphäre absorbiert, *M* von ihr reflektiert und *K* schließlich in den Raum durchgelassen.
F und *E* Strahlung der Atmosphäre in den leeren Raum bzw. zur Erde; *L* die Wärmeabgabe der Erde an die Luft durch Konvektion.

Abgesehen davon, daß der Mittelwert jedenfalls wegen einer zu geringen Ultraviolettkorrektion (Kron, s. Abschn. 2) um etwa 2—3% zu klein ist — *ein Umstand, der von den Berechnern der Sonnentemperatur bisher nicht berücksichtigt wurde* — kann man die modernen Messungen sicherlich als zutreffend bezeichnen.

[1]) Quart. J. of the R. Met. Soc. Bd. 43. 1917.
[2]) Philos. Trans. 1922. A. 223, 250 — Monthly Not. R. Astr. Soc. 81, 361.
[3]) Bull. of the Astr. Inst. of the Netherlands Bd. 2, Nr. 51.

Die mitunter, z. B. auf Grund eines Einwandes von Bigelow, noch geäußerte Ansicht, daß die S.K. den Wert von 2 gr/kal wesentlich übersteigt, scheint mir nicht stichhaltig zu sein. Der Abbotsche Wert (mit der erwähnten Kronschen Korrektion) erfährt nämlich eine wesentliche Stütze durch Vergleiche der in verschiedenen Höhenschichten der Atmosphäre vorgenommenen Strahlungsmessungen[1]). So fand man in

$$
\begin{array}{lr}
\text{Washington (20 m)} \dots\dots\dots\dots & 1.58 \ \text{gr/kal} \\
\text{Mt. Wilson (1740 m)} \dots\dots\dots\dots & 1.64 \quad ,, \\
\text{Mt. Whitney (4420 m)} \dots\dots\dots\dots & 1.72 \quad ,, \\
\text{Ballonfahrt von Peppler (7500 m)} \dots & 1.76 \quad ,, \\
\text{Ballonfahrt 1914 (22 000 m)} \dots\dots & 1.89 \quad ,,
\end{array}
$$

Man kann daher wohl annehmen, daß außerhalb der Atmosphäre der Wert 2 gr/kal nicht wesentlich überschritten werden wird.

2. Die Langley - Abbotsche Methode zur Bestimmung der Solarkonstante. Die urprünglichen Strahlungsmessungen bestanden darin, mit einem Pyrheliometer (z. B. Pouillet, Ångströms Kompensationspyrheliometer[2]), Abbots Wasserstrompyrheliometer, das als Bezugsinstrument dient, dem Silberscheibenpyrheliometer von Abbot und Aldrich, mit dem gegenwärtig in Amerika beobachtet wird) die auf die Erde einfallende Gesamtstrahlung zu messen und nach dem Gesetz von Bouguer die Extrapolation auf das Vakuum durchzuführen. Es zeigt sich, daß im Mittel von der Gesamtstrahlung der Sonne nur 83—84% die Erde erreichen.

Wie schon aus Rayleighs Gesetz hervorgeht, ist die Extinktion der Strahlung in der Atmosphäre nicht für alle λ gleich, sondern für kleinere λ stärker. Das aus Pyrheliometermessungen gewonnene a ist nur ein Mittelwert (von Abbot scheinbarer Transmissionskoeffizient genannt). Das Gesetz[3]) muß also richtig lauten:

$$ I_\lambda = I_{0\lambda} \, a_\lambda^m $$

In der einfachen Form wäre die Extinktionswirkung zu wenig berücksichtigt, man erhielte zu große a und mithin zu kleine Strahlungsintensitäten I (Langley 1881).

Es bedarf also noch einer Beobachtung der Energieverteilung im ganzen Spektralbereiche. Diese wird mit einem Bolometer durch-

---

[1]) Smithson. Misc. Coll. Bd. 65, Nr. 4. 1915.

[2]) Eine Beschreibung dieses und der übrigen Instrumente ist u. a. zu finden in einem Aufsatz von Dietzius, Naturw. 1923, S. 246.

[3]) I bedeutet die gemessene Intensität, I_0 die Intensität außerhalb der Atmosphäre, a den Transmissionskoeffizienten und m die bei der Beobachtung von der Strahlung durcheilte Luftmasse, proportional der sec z (Zenitdistanz der Sonne). m läßt sich aus Bemporads Tabellen (Mitt. Heidelberg. Sternwarte Bd. 4. 1904) entnehmen.

geführt und die Energiekurve von $0{,}35-2{,}8\,\mu$ photographisch registriert. In der Anordnung von Abbot werden an einem Tage bei verschiedenen Luftmassen (sec z) 6 solche Registrierungen, jede zu sieben Minuten, vorgenommen. Gegenwärtig führt Abbot die Messungen mit einem sog. Vakuumbolometer aus, das zehnmal empfindlicher als der frühere Typ ist. (Da die Platinstreifen hier im Vakuum sind, so gibt es keine Wärmeabgabe an die Luft durch Konvektion.)

Die Bestimmung der Solarkonstante (S.K.) erfolgt, abgesehen von einigen Korrektionen, die ich als nicht wesentlich übergehen möchte, folgendermaßen: Durch die Spitzen des Bologramms wird eine glatte Kurve gelegt, an 40 Punkten der Abszissenachse (λ) die Ordinate ausgewertet (Fläche I_e). Es sei die Intensität an der Erdoberfläche $I'_e = I_e + K_r - K_s$. K_r bedeute die Korrektion für den nichtregistrierten infraroten und ultravioletten Teil, K_s für die Flächen der atmosphärischen Absorption (Einsenkungen gegenüber der glatten Kurve). Man bestimmt nun für ausgewählte λ die a und extrapoliert aus den Bologrammen der verschiedenen Luftmassen auf die Intensität im Vakuum I_0 und erhält schließlich $I'_0 = I_0 + K_r$. Eine negative Korrektur wird nicht angebracht, da angenommen wird, daß die atmosphärischen Banden im Sonnenspektrum nicht vorkommen. Ist P die Pyrheliometerlesung, so ist die S.K.:

$$E = P\,\frac{I'_0}{I'_e}$$

K_r wird nach dem Gesetz von Rayleigh angebracht unter der Annahme, daß in dem betreffenden Teil der Spektren keine stärkeren Absorptionsbanden vorhanden sind. Das trifft, wie Granquist[1]) bemerkt, nicht ganz zu. Schon früher hat Kron[2]) darauf hingewiesen, daß K_r im Ausmaße von $1-2\%$ wegen der Ozonbande im Ultravioletten zu klein ist, demgemäß auch E. Abbot hat dies[3]) anerkannt, aber als nicht nennenswert für das Problem der Schwankungen von E unberücksichtigt gelassen. Die Wasserdampfkorrektur K_s hat Fowle[4]) festgelegt als $\dfrac{\varrho}{\varrho_{sc}}$ ($\varrho =$ Ordinate der tiefsten Einsenkung, ϱ_{sc} der höchsten Erhebung des als $\varrho\,\sigma\,\tau$ bekannten Teiles der Intensitätskurve). Dieser Wert $\dfrac{\varrho}{\varrho_{sc}}$ spielt bei der neuen Methode (Abschn. 4) eine ausschlaggebende Rolle. Bei den Mt. Wilson-Werten zeigt sich merkwürdigerweise nach dieser Korrektion noch immer eine Abhängigkeit des E vom Wasserdampf, so daß eine neuerliche Korrektion des E auf 1 mm

---

[1]) Bestämningar av Solarkonstanten. Upsala 1919.
[2]) Vierteljahrschrift der Astron. Ges. (V. J. S.) Bd. 49, S. 63.
[3]) Annals of the Astroph. Obs. of the Smithsonian Inst. Vol. IV. (Ann. IV.)
[4]) Sm. Ann. III. S. 182.

Dampfdruck sich als notwendig erwies. Von KRON[1]) wird deshalb der Verdacht eines Fehlers in der Methode ausgesprochen. Nun zeigen aber nach ABBOT[2]) die Beobachtungen von Calama diese Abhängigkeit nicht. kch konnte dies aus dem Material der Tab. 47[3]) bestätigen. Desgleichen ɪonnte ich auch bei den Beobachtungen der Station Montezuma Januar—Juni 1922 keine Korrelation zwischen E und $\dfrac{\varrho}{\varrho_{sc}}$ finden.

3. Die definitiven Werte der Solarkonstante in ihrer Beziehung zum Transmissionskoeffizienten. Wenn es also nach dieser Methode gelungen ist, den Einfluß der Atmosphäre zu eliminieren, also exakt auf das Vakuum zu extrapolieren, dürfen die E keinen Zusammenhang mit a besitzen. MÜLLER und KRON[4]) haben zuerst in Teneriffa beobachtet, daß sich die Durchlässigkeit der Atmosphäre im Laufe des Tages ändert. Desgleichen finde ich ähnliche Feststellungen von BEMPORAD[5]). Selbst auf Alta Vista (3260 m) wurde dies von KRON[6]) festgestellt. Daß sich bei den endgültigen Werten der S.K. tatsächlich eine Abhängigkeit des E von a findet, in dem Sinne, daß hohe Werte von E bei kleinem a auftreten und umgekehrt, wurde 1919 zum ersten Male von GRANQUIST[7]) erkannt und 1921 in einer weiteren Arbeit[8]) eingehender untersucht. Ich möchte nicht unterlassen, auf diese Arbeiten des verblichenen Vorsitzenden des Nobelkomitees besonders hinzuweisen, da dieselben, weil schwedisch geschrieben, KNOX-SHAW[9]) und KALITIN[10]) wohl unbekannt geblieben sein dürften. GRANQUIST stellt den Zusammenhang an den Werten von Upsala 1912—1915 und 1919 fest, desgleichen in ABBOTS Mt. Wilson-Resultaten in 1911. Er schließt daraus, daß die beobachteten Schwankungen nicht reell seien. A. ÅNGSTRÖM[11]) bemerkt demgegenüber, daß der Zusammenhang in Upsala wohl durch Durchlässigkeitsschwankungen der Atmosphäre hervorgerufen sein könnte, in Amerika aber nur ein scheinbarer sei, dadurch bedingt, daß sich z. B. ein kleineres a ergibt, wenn E während der Beobachtung steigend war, das gerechnete E aber für das Ende der Beobachtung gilt. Er rechnet unter dieser Annahme die wahrscheinliche Korrektion für a, die aber nicht einmal $^1/_2\%$ der Gesamtintensität

---

[1]) V. J. S. Bd. 49, S. 63.
[2]) Sm. Ann. IV.   [3]) Sm. Ann. IV. S. 75.   [4]) Publ. Astroph. Obs. Nr. 64.
[5]) Neapel, Mem. Bd. 6. 1921.   [6]) V. J. S. Bd. 49, S. 65.
[7]) GRANQUIST, G.: Bestämningar av Solarkonstanten. Upsala 1919.
[8]) GRANQUIST, G.: Är solens strålning variabel? Kosmos, Fysiska Uppsatser utgivna av Svenska Fysika Samfundet. Stockholm 1921.
[9]) Ministry of public works, Egypt: Helwan Obs. Bull. Nr. 23. — KNOX-SHAW, H.: Observ. of solar radiation 1915—21.
[10]) Nachr. d. phys. Hauptobs. 1920, I, Nr. 2. Petrograd. KALITIN: Die Solarkonstante nach Beobachtungen in Pawlowsk. (Russisch.)
[11]) ÅNGSTRÖM, A.: Är solens strålning variabel? Tidskr. f. El. Mat. Fys. och Kemi Bd. 5, Nr. 2.

übersteigt. Ich finde die Korrelation bei Mt. Wilson 1911 auch dann noch bestehend. In seiner letzten Arbeit weist Granquist[1]) den Zusammenhang zwischen a und E für Upsala 1920, für Mt. Wilson 1905—1908, 1909—1911 nach. Immerhin möchte ich in Übereinstimmung mit Ångström bzw. Lundblad[2]), der in einer groß angelegten Arbeit sich auch mit dieser Frage beschäftigt und im wesentlichen zu ähnlichen Ergebnissen wie Granquist gelangt, aus diesem Material allein noch nicht ohne weiteres schon als feststehend betrachten, daß Abbots Schwankungen nicht reell seien. Kalitin[3]) kommt aus seinen Beobachtungen in Pawlowsk zu folgendem Resultat: „Alle Ableitungen führen zur bedauerlichen Feststellung, daß es mit den vorliegenden Methoden unmöglich sei, die zeitlichen Änderungen von E zu bestimmen, weil es unmöglich ist, sich vom Einfluß der Luftdurchlässigkeit zu befreien." Knox-Shaw[4]) kommt in einer Arbeit, die mir leider nur als Referat in „Nature" zugänglich war, auf Grund von Messungen in *Helwan* (Ägypten) zu ähnlichen Ergebnissen. Er findet, offenbar wegen Zunahme der Lufttrübung gegen Mittag, zwischen E und a einen negativen Korrelationsfaktor bis zu $r = -0.6$ und hält die Bestimmung der S.K. durch die Extrapolationsmethode für sehr diskreditiert. Diese Kritik ist nicht unberechtigt. Ich habe daraufhin die bis heute besten Beobachtungen von Montezuma untersucht und finde selbst hier nennenswerte Korrelationen. 1922 Januar—Juli zwar nur:

$$r = -0.295 \pm 0.120$$

Für alle Tage 1922 (33 Fälle) aber $r = -0.387 \pm 0.079 \left(\dfrac{r}{\text{w. F.}} = 4.9 \right)$.

Nimmt man aber nicht Einzelwerte, sondern Monatsmittel, wo ja kurzperiodische Schwankungen eliminiert sind — ich verwendete alle Monatsmittel des Mt. Wilson von 1905—1920, insgesamt 71 Monate —, so ergibt sich auch in diesem Falle eine Bestätigung der Abhängigkeit:

$$r = -0.359 \pm 0.069$$

Die Korrelation ist sehr bemerkenswert, da $\dfrac{r}{\text{w. F.}} = 5.2$ ist[5]). Die Beziehung ist auch in nebenstehender Abb. 2 deutlich ersichtlich. So

---

[1]) Granquist: Om variationerna i den enligt Langleys metod beräknade solarkonstanterna. Tidskr. f. El. Mat. Fys. och Kemi Bd. 5, S. 216.

[2]) Lundblad, R.: Till frågan om solstrålningens variabilitet. Ark. f. Mat. Astr. Fys. Bd. 17, Nr. 14.

[3]) l. c. Ich verdanke Herrn Dr. Kofler die Übersetzung aus dem Russischen.

[4]) l. c. Referiert in Nature 1922, Dezember.

[5]) Eine Beziehung gilt als gesichert, wenn $\dfrac{r}{\text{w. F.}}$ etwa 5 erreicht. Dines stellt in Computers Handbook V London 1915 als Norm auf: Der Korrelationsfaktor r soll etwa gleich viermal dem standarderror (mittl. F.) sein. *Es ist für das Problem wesentlich, daß das neueste Material im Abschn. 6 diese Bedingung keineswegs erreicht.*

erkennt man z. B., daß das höchste Monatsmittel 1.989 Juli 1917 offenbar gar nicht reell ist, sondern nur wegen des außerordentlich tiefen mittleren a von 0.815 so errechnet wurde. Ähnliches kann man von den Monaten mit Mittelwerten der S.K. unter 1.920 behaupten. Hier wurden eben auch auffallend hohe Transmissionskoeffizienten beobachtet.

Überblicken wir die Ergebnisse dieses Abschnittes, so müssen wir wohl sagen, daß die LANGLEYsche Methode zweifellos deshalb zu keinen einwandfreien Ergebnissen kommen kann, weil ihre Voraussetzung, daß die Luftdurchlässigkeit zwischen den einzelnen Beobachtungssätzen un-

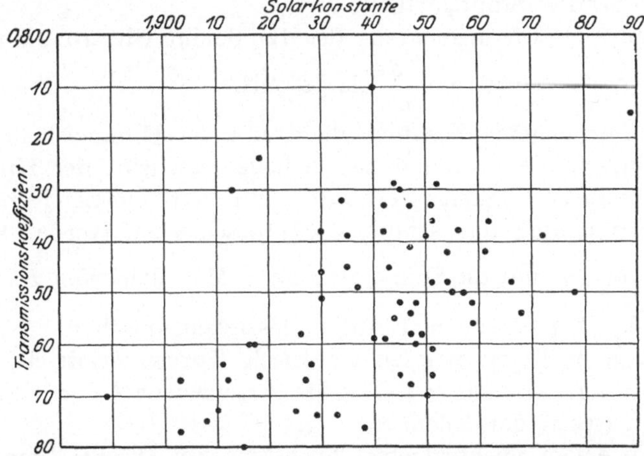

Abb. 2.   Die Beziehung zwischen Monatsmitteln der Solarkonstante und des am Mt. Wilson beobachteten Transmissionskoeffizienten.   1905—1920.

verändert bleibt, auch auf den idealsten Stationen nicht zutrifft und daher nur fehlerhaft auf den Raum außerhalb der Atmosphäre extrapoliert werden kann.

Wir verdanken auch den Arbeiten von LINKE[1]) diesbezüglich wichtige Aufschlüsse. Seine Untersuchungen, zuletzt in Argentinien, bestätigen die Extinktionszunahme in den Mittagsstunden bei klarstem Wetter, und ich halte seinen Vorschlag, den von ihm ermittelten *Trübungsfaktor T*, durch den das Extrapolationsgesetz die Form

$$I = I_0\, a^{T m}$$

annimmt, einzuführen, für höchst beachtenswert. Im übrigen scheint auch ABBOT, nach einigen in seinen Arbeiten eingestreuten Bemerkungen, dieser offenkundige Mangel seiner Methode nicht entgangen zu sein. Die *neue Methode* liefert eine *vollständige* Bestimmung der S.K. (Abschn. 4) schon nach einer Beobachtungszeit von nur ca. 14 Minuten; hier würde diese Abhängigkeit sich also nicht mehr zeigen. KNOX-SHAW[2]) hat

---

[1]) U. a. Meteor. Z. 1922, S. 74 — Astron. Nachr. **221**, S. 182.
[2]) l. c.

diesbezügliche Stichproben durchgeführt und berichtet, daß er tatsächlich keine Korrelation mehr gefunden hat. Daß dies aber leider nicht
voll zutrifft, soll im folgenden gezeigt werden.

4. Die neue Methode mit Pyranometermessungen. Der Transmissionskoeffizient a für verschiedene λ konnte bisher nur aus Beobachtungen bei verschiedener Luftmasse ermittelt werden. Nach ABBOTS
neuer Methode gelingt es, ihn aus einer *einzigen* Beobachtung zu erhalten. Der Grad der Luftdurchlässigkeit wird durch zweierlei bestimmt: 1. durch den Gehalt an Wasserdampf und 2. durch die Menge
von lufttrübenden Staubpartikeln.

1. ergibt sich wie früher aus der Bande im Ultraroten und wird
durch die Ausmessung von $\dfrac{\varrho}{\varrho_{sc}}$ dargestellt.

2. steht im engsten Zusammenhang mit der Himmelshelligkeit; je
stärker nämlich die Trübung, um so heller erscheint der Himmel.

Die Himmelshelligkeit (H) wird nun mit einem *Pyranometer*,
einem Instrumente, das schon 1916[1]) beschrieben wurde, in einem
Umkreise von 15° um die Sonne gemessen. Man kann nun $F = \dfrac{H}{\varrho} \cdot \varrho_{sc}$
als Maß der herrschenden Luftdurchlässigkeit ansehen. Aus einem
Material von 60 Tagen wurden empirisch Kurven ermittelt, die den
Zusammenhang zwischen H und a für verschiedene λ und ausgewählte
Luftmassen (sec z) darstellen.

Da a in einem Spektralgebiet höchstens um 5%, H hingegen um
einige 100 Einheiten variiert, so wird in der Bestimmung des a aus
H nur ein geringer Fehler gemacht. Die Messung von H ist rasch
durchgeführt (10 Minuten); a ist dann sogleich aus den empirischen
Kurven gefunden und der Bestimmung der S.K. M[2]) (mit Zuhilfenahme
der Pyrheliometerlesungen) steht nichts mehr im Wege. Diese Methode
ist Juni 1919 zuerst angewendet worden. Wir wollen hier gleich betonen, daß wegen der empirisch ermittelten Beziehung zwischen H
und a *die neue Methode von der alten nicht unabhängig ist.* Der große
Vorteil der neuen Methode besteht darin, daß man an einem Tage
bei verschiedenen Luftmassen voneinander unabhängige S.K.-Bestimmungen vornehmen kann. Da zeigt es sich aber leider auch, wie stark
die Resultate voneinander abweichen. Dies ist wenig bekannt; in
den Annalen wird nämlich nur der Tagesmittelwert publiziert. Diesen
Werten haftet aber wie allen gewichteten Mittelwerten eine gewisse
persönliche Note an, wenn ich auch gerne zugebe, daß die Gewichte

---

[1]) Sm. Misc. Coll. Bd. 66, Nr. 9 1916.

[2]) Zur Unterscheidung von den nach der alten Methode bestimmten S.K.-
Werten E_0, eind die Resultate nach der neuen Methode übereinstimmend mit
ABBOT mit M bezeichnet. Der Index, z. B. $M_{1,5}$, gibt die Luftmasse (sec z)
während der Beobachtung an.

für die Einzelbeobachtungen, wenigstens was die Luftmassen betrifft, nach einem einheitlichen Schema angebracht werden. Im folgenden einige Beispiele.

$$
\begin{array}{lll}
\text{1921, Juni 9.:} & 1.967\ E_0 & \text{Gew. 2} \\
& 28\ M_3 & 2 \\
& 14\ M_{2,5} & 3 \\
& 32\ M_4 & 4 \\
& 51\ M_{1,5} & 4 \\
\hline
& 1.938\ \text{gew. Mittel.}
\end{array}
\qquad
\begin{array}{l}
\text{Juli 31.:}\ 1.938\ M_{1,88} \\
\phantom{\text{Juli 31.:}\ }1.820\ M_{1,69} \\
\hline
\phantom{\text{Juli 31.:}\ }1.938\ \text{gew. Mittel.}
\end{array}
$$

ABBOT findet[1]) eine sehr gute Übereinstimmung alter und neuer Beobachtungen eines Tages. So seien von 53 Tagen (MOORE und ABBOT) bei 45 Tagen die Abweichungen weniger als 2%, bei 32 weniger als 1%. Ich habe das neue Material darauf näher untersucht und kann eine so gute Übereinstimmung nicht finden. Z. B. obiger Tag 1921, Juni 9., oder 1921, Juni 27.:

$$
\begin{array}{l}
1.964\ E_0 \\
1.938\ M_0 \\
\hline
1.938!\ \text{gew. Mittel.}
\end{array}
$$

Besser ist wieder 1919, Oktober 7.:

$$
\begin{array}{l}
1.887\ E_0 \\
885\ M_3 \\
912\ M_2 \\
\hline
1.891\ \text{gew. Mittel,}
\end{array}
$$

wodurch der auffallend tiefe Wert gesichert zu sein scheint. Am achten ist schon wieder 1.963 gemessen worden. Immerhin sind diese starken Differenzen recht entmutigend. Im letzten mir zugänglichen Material sind Beobachtungen nach alter Methode nur mehr selten; in Harqua Hala wird nur mit der Pyranometermethode gearbeitet.

Im übrigen scheint ABBOT später von der guten Übereinstimmung selbst nicht mehr überzeugt zu sein; 1923 schreibt er[2]), daß der *alten Methode nur mehr das halbe Gewicht* zuerteilt wird. Ich möchte geradezu sagen, daß man Tage mit Beobachtungen nach alter und neuer Methode kaum vergleichen kann. Anfang 1922 war durch einen Unfall das Pyranometer unbrauchbar, statt dessen wurde eine lange Serie nur nach alter Methode ausgeführt. März 1921, wo zum erstenmal wieder beide Methoden auftreten, gibt dann folgendes Bild:

März 23.:

E_0 1.957 M_2 1.928 $M_{1,5}$ 1.944   u. gew. Mittel 1.943!

Hingegen März 25.:

M_2 1.880 M_1 1.916, $M_{1,44}$ 1.934   u. gew. Mittel 1.925!

Man muß hoffen, daß das noch unpublizierte Material eine bessere innere Übereinstimmung zeigen wird.

---

[1]) Monthly Weather Rev. Bd. 47, S. 580. [2]) Monthly Weather Rev. Bd. 51, S. 2.

Für die Zeit 1. Juli 1919 bis 26. Juli 1920 (Calama) mußten an die definitiven Werte Korrektionen angebracht werden, da es sich zeigte, daß die M von $\dfrac{\varrho}{\varrho_{sc}}$ nicht unabhängig waren. Nach Abbot wären die späteren Werte von der Feuchtigkeit unabhängig. *Das trifft aber für Montezuma nicht zu.* Ich habe dieses Material untersucht und finde folgende Korrelationen:

$$1921 \quad \text{Jan.—März} \quad (28 \text{ Tage}) \quad r = +0.377 \pm 0.109$$
$$\text{April—Mai} \quad (30 \text{ Tage}) \quad r = +0.396 \pm 0.104$$
$$\text{Juni—Juli} \quad (34 \text{ Tage}) \quad r = +0.656 \pm 0.091$$

Die Beziehung ist also vorhanden. Bekanntlich ist es auch für die Zuverlässigkeit eines Zusammenhanges maßgebend, wenn ein größeres Material in Gruppen eingeteilt, für jede Gruppe Korrelationen derselben Größenordnung gibt [1]).

Die letzten Beobachtungen des Jahres 1921 sind dagegen einwandfrei:

$$\text{Aug.—Dez. } (41 \text{ Tage}) \quad r = +0,122 \pm 0,104.$$

1922 Montezuma verhielt sich anders. Ich finde hier keine sichere Korrelation zwischen M und $\dfrac{\varrho}{\varrho_{sc}}$. Statt dessen ergibt sich die unerwartete Tatsache, daß diese, nur nach der neuen Methode ausgeführten Beobachtungen eine Abhängigkeit zwischen M und a zeigen. Ich erhalte:

$$\text{Januar—Juli} \quad (45 \text{ Tage}) \quad r = -0.447 \pm 0.080$$
$$\text{Juni—Dezember } (41 \text{ Tage}) \quad r = -0.310 \pm 0.095$$

$$\frac{r}{\text{w. F.}} = 5,6 \text{ bzw. } 3,3$$

Ich muß also gegenüber Knox-Shaw betonen, daß auch bei der neuen Methode bedauerlicherweise eine Abhängigkeit sich zeigt. Während beim Eintreten der Korrelation $M - \dfrac{\varrho}{\varrho_{sc}}$ die Korrelation $M - a$ nicht auftritt, ergibt sich beim Fehlen der Beziehung zur Feuchtigkeit die starke Korrelation zwischen S.K. und Transmissionskoeffizienten. *Auf jeden Fall sind die definitiven Werte der M von den Vorgängen zwischen Erdoberfläche und Vakuum noch nicht völlig befreit.* Die Beobachtungen von Harqua Hala entziehen sich dieser Untersuchung, da weder a noch $\dfrac{\varrho}{\varrho_{sc}}$ in der Monthley Weather Rev. publiziert sind.

5. Die Vorgänge auf der Sonne im Zusammenhange mit den beobachteten Schwankungen der Solarkonstante. Die Realität der

---

[1]) Siehe z. B. Exner, F. M.: Naturwissenschaften, S. 206, u. Naturwiss. Wochenschrift 1913.

Schwankungen der S.K. müßte sich u. a. durch Veränderungen erweisen lassen, die gleichzeitig auf der Sonne beobachtet werden können. So hat Abbot den Helligkeitskontrast zwischen Sonnenzentrum und -rand, der bekanntlich mit zunehmendem λ wächst, beobachtet und Schwankungen desselben untersucht. Das Ergebnis scheint mir wenig günstig. So findet er:

$$1913 \text{ eine Korrelation } r = +0.601 \pm 0.067$$
$$1916 \quad \text{,,} \qquad \text{,,} \qquad r = -0.363 \pm 0.097$$

Da hier sog. Kontrastzahlen, die verkehrt proportional dem Kontraste selbst laufen, verwendet sind, bedeutet das Resultat für 1913 Zunahme des Kontrastes bei abnehmender Sonnenstrahlung, 1916 das Umgekehrte. Spätere Untersuchungen liegen mir nicht mehr vor. Die widersprechenden Ergebnisse sucht Abbot durch drei verschiedene Ursachen, die sich gegenseitig überlagern, zu begründen. Es ist ja kein Zweifel, daß hier sehr komplizierte Verhältnisse herrschen. So müssen wir z. B. auf Grund der wertvollen Untersuchungen Lundblads[1]) annehmen, daß die Strahlung der kleinen λ und ein Teil der ultraroten aus höheren kühleren Schichten kommt, hingegen die sichtbare aus tieferen heißen. [Vgl. auch die Arbeiten von Seeliger[2]), Wilsing[3]), Minnaert[4]).] Jedenfalls vermag man heute aus Kontrastbeobachtungen noch keine Schlüsse auf die Realität der Schwankungen der Sonnenstrahlung zu ziehen.

Moore[5]) hat kurze Zeit in Calama die Absorptionslinien des H, Ca, Mg, Fe im Sonnenspektrum untersucht und geglaubt, eine Verstärkung derselben bei abnehmender S.K. zu beobachten. Nähere Daten sind nicht veröffentlicht, die Beobachtungen später auch nicht mehr fortgesetzt. Doch halte ich diese Untersuchungen für aussichtsreich. In ähnlicher Weise kann man die Intensität der von Fabry und Buisson[6]) gemessenen Ozonbande mit den gleichzeitig beobachteten S.K. vergleichen. Ich finde für 13 gemeinsame Tage die ganz unsichere Korrelation:

$$r = -0.244 \pm 0.178 \, .$$

Abbot hat[7]) drei Korrektionen an den S.K.-Werten vorgenommen, über deren Berechtigung man geteilter Meinung sein kann. Dann ergibt sich $r = -0.574 \pm 0.131$. Dieses kleine Material genügt natürlich für eine Bestätigung noch nicht. Kalitin[8]) glaubte einen Zusammenhang der S.K. in Pawlowsk mit der gleichzeitigen heliographischen

---

[1]) Astroph. J. **58**, 113.   [2]) Sitzungsber. bayr. Akad. d. Wiss. **21**, 247.

[3]) Potsdam, Astroph. Obs. Publ. **23**, Nr. 72.

[4]) Bull. of the Astr. Inst. of the Netherlands. II, Nr. 51. — 1924.

[5]) Sm. Ann. IV. S. 186.

[6]) Astroph. J. **54**, A Study of the ultraviolet End of the Solar Spektrum.

[7]) Sm. Misc. Coll. Vol. 47, Nr. 7. — 1923.

[8]) Astr. Nachr. **215**, S. 17.

Breite der Erde zu finden in dem Sinne, daß niedere S.K. auftreten bei kleiner Breite. Danach wäre die Gegend des Sonnenäquators kühler als die der Pole (EMDEN). LINKE[1]) untersucht daraufhin Beobachtungen aus Calama, die dies nicht bestätigen, im Gegenteil sogar eine schwache Beziehung im umgekehrten Sinne zeigen. Weitergehende Schlüsse sind also auch hier nicht am Platze.

Von CLAYTON[2]) wurden Montezuma-Werte des Jahres 1921 mit in La Plata und am magnetischen Observatorium in Pilar (Cordoba) beobachteten Fackelhäufigkeiten verglichen. Danach scheinen die Maxima der S.K. und der Fackeltätigkeit, die sich nach 3.5 Tagen wiederholten, übereinzustimmen und zugleich eine Beziehung zu argentinischen Wetteränderungen gehabt zu haben. CLAYTON gibt[3]) ein Diagramm: Abszisse S.K., Ordinate eine Indexzahl aus Größe des Feldes und Intensität der Fackeln. Ich vermag aus dieser Figur keine Beziehung zu erkennen.

Ein Zusammenhang zwischen Sonnenstrahlung und Fleckentätigkeit *langperiodischer* Natur wird schon längere Zeit vermutet [z. B. MÜLLER[4])]. A. ÅNGSTRÖM[5]) findet aus den Jahresmitteln 1905—1917 der S.K. von Mt. Wilson und den WOLFERschen Fleckenzahlen eine *positive* Korrelation mit einem Maximum für die Fleckenzahl (N) zwischen 100 und 160. Ich erhalte für diese Jahre $r = +0,627 \pm 0.124$. Besser ist die Übereinstimmung, die ÅNGSTRÖM für die Korrelation S.K. und $\sqrt{N}$ erhält, nämlich: $r = +0.75 \pm 0.09$. ÅNGSTRÖM[6]) findet schließlich als beste Darstellung: S.K. $= 1.903 + 0.011 \sqrt{N} - 0.0006\,N$. ABBOT[7]) glaubt aber, daß man eine solche Beziehung nicht aufstellen kann.

So hohe Korrelationen wie 0.6 erhält man aber nur aus den *Jahresmitteln*. Ich habe aus demselben Material, das schon S. 458 verwendet wurde, die Korrelation zwischen S.K. und WOLFERschen Relativzahlen für 71 *Monatsmittel* und finde nur mehr:

$$r = +0.415 \pm 0.066$$

Die Beziehung ist auch in Abb. 3 dargestellt. Man muß dazu noch bedenken, daß dieselben Werte der S.K. in der gleichgroßen Korrelation zum Transmissionskoeffizienten stehen, also offenbar gar nicht als richtige definitive Monatsmittel der S.K. angegeben werden können! Bei dem neuen Material finde ich für die Monatsmittel:

Harqua Hala Okt. 20—Sept. 22 $r = +0.326 \pm 0.124 \quad \dfrac{r}{\text{w. F.}} = 2.4!$

„   „ Juni—Sept. 22 allein $r = +0.423 \pm 0.175 \quad \dfrac{r}{\text{w. F.}} = 2.7!$

[1]) Astr. Nachr. **221**, S. 182.   [2]) World Weather, New York 1923.
[3]) World Weather S. 291, Abb. 226.   [4]) Potsdam, Publ. Bd. 8.
[5]) Geografiska Annaler 1921, S. 162.   [6]) Astroph. J. **55**, S. 24.
[7]) Ann. IV. S. 215.

Hingegen Calama, 24 Monate, Aug. 18—Juli 20 $r = -0.152 \pm 0.134$
und   Montezuma 1921: $r = -0.209 \pm 0.186$

Wir sehen also zum Teil eine schwache positive Korrelation zwischen den Monatsmitteln der S.K. und den Sonnenflecken-Relativzahlen

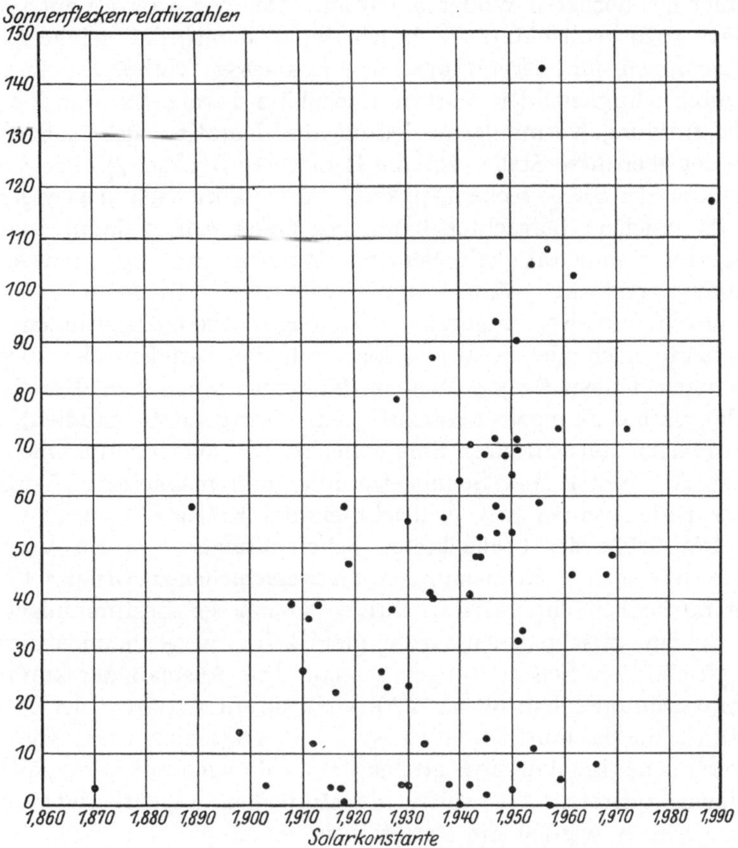

Abb. 3. Die Beziehung zwischen Monatsmitteln der Solarkonstante auf dem Mt. Wilson und Monatsmitteln der Sonnenfleckenhäufigkeit, dargestellt in WOLFER-schen Sonnenflecken-Relativzahlen.

$\left(\text{beachtenswert der auffallend kleine Wert von } \dfrac{r}{\text{w.F.}}\right)$, andererseits *überhaupt keine Beziehung, dem Vorzeichen nach sogar im entgegengesetzten Sinne.* Ebenso ungünstig wird das Ergebnis für *Einzeltage.* ARCTOWSKI[1]) hat dies für alle Tage 1905—1911 untersucht, an denen Greenwicher Fleckenbeobachtungen vorlagen, und konnte keine sichere Entscheidung treffen. Ich habe daraufhin Montezuma-Werte, 1921

---

[1]) Mem. spettr. Ital. Bd. 6, S. 30.

Januar—Mai (47 Fälle), untersucht und auch eine ganz unzulängliche *negative* Korrelation:

$$r = -0.185 \pm 0.095$$

gefunden.

Der 14. März mit der tiefsten S.K. dieses Zeitraums tritt zugleich mit einer der höchsten Wolferzahlen auf. Das hat Abbot[1]) auch schon 22. März 1920 beobachtet. Er begründet es damit, daß solche Einzelfleckengruppen mit hervorstürzenden Gasmassen verbunden sind, die die Durchsichtigkeit der Sonnenatmosphäre herabsetzen und so die Sonnenstrahlung vermindern. Durch die Rotation der Sonne verschwindet aber diese Stelle (vgl. die Hypothese Abschn. 7). Die *positive* Korrelation der S.K.-Fleckentätigkeit, die ich aber nach obigem keineswegs als gesichert betrachten kann, begründet Abbot damit, daß bei gesteigerter Sonnentätigkeit heißeres Material aufsteigt und so die Strahlung verstärkt. Man kann gewiß zwei widersprechende Erscheinungen mit zwei entgegengesetzten Hypothesen begründen. Wir sehen heute noch viel zu wenig klar, um die Tätigkeit der Flecken, dieser nach Russel[2]) gewaltigsten Kältemaschine, zum Beweis der Strahlungsschwankungen heranzuziehen. Fortgesetzte Studien, z. B. über Hales unsichtbare Sonnenflecken[3]), Wiederaufnahme von Bauers Arbeiten[4]) über Sonnenstrahlung und magnetische Vorgänge auf der Erde, werden hier vielleicht einmal Klärung bringen.

6. Die Größe der beobachteten Schwankungen und der Vergleich der gleichzeitigen S.K.-Messungen an verschiedenen Orten. Ist die Sonne tatsächlich ein variabler Stern, so müssen Bestimmungen der S.K., die an verschiedenen Orten gleichzeitig vorgenommen werden, auch gleichzeitige Schwankungen zeigen. Das Ausmaß der kurzperiodischen Variationen beträgt nach Abbot[5]) mitunter bis 10% der Gesamtintensität, durchschnittlich etwa 5%. Es zeigt sich aber, wie auch Marvin[6]) bemerkt, daß diese großen Schwankungen nur in den ältesten Messungen auftreten. Je weiter die Methoden ausgearbeitet werden, um so geringer werden die Variationen. So finde ich z. B. in Montezuma 1920 nur mehr *eine maximale Variation von 1.3% und eine durchschnittliche von 0.3%! 1922: 2.4 resp. 0.8%.* In Harqua Hala, wo weniger günstige klimatische Verhältnisse herrschen, sind sie etwas größer. Analog fand Clayton[7]) aus denselben Gründen die Schwankungen am Mt. Wilson um 14% größer als in Calama. Dies ist ohne weiters be-

---

[1]) Ann. IV. S. 186.
[2]) Contrib. from Mt. Wilson Obs. Nr. 217.
[3]) Z. B. Proc. of the Nat. Academy of Sc. Bd. 8, S. 168.
[4]) Z. B. Zusammenfassung in Popular Astronomy **29**. 555.
[5]) Die letzte Angabe 1920 Proc. of the Nat. Academy of Sc. Bd. 6, S. 670 spricht von $0^{m}1$ zwischen den Extremen und von $0^{m}03$ oft in wenigen Tagen.
[6]) Amerik. meteor. Gesellschaft 1923.
[7]) l. c. S. 218.

greiflich, da ABBOT für Calama den w. F. einer S.K.-Bestimmung zu 0.19%, für Mt. Wilson aber zu 0.31% ermittelt.

Selbst unter der Voraussetzung, daß die Beobachtungsfehler nicht größer sind — nach LINKE[1]) freilich gelangt man, wenn ich nur die für die Pyranometermethode in Betracht kommenden Fehler heranziehe, nahe an 1% —, so sind sie dem Betrag der derzeit konstatierten Schwankungen bedenklich nahegerückt. Es ist daher nicht verwunderlich, wenn ich aus der Untersuchung der letzten Simultanmessungen Argumente gegen die Realität der kurzperiodischen Schwankungen erhalte.

Zum ersten Male hat KRON[2]) den Zusammenhang simultaner S.K.-Messungen untersucht und aus den besten Beobachtungen auf dem Mt. Wilson und in Bassour (Algier) folgende Korrelationen gefunden:

1911 und 1912 (50 Fälle) $r = +0.508 \pm 0.071$

1911 allein (18 ,,) $r = +0.132 \pm 0.156$ $\begin{cases} \text{also keine Beziehung für} \\ \text{dieses Jahr} \end{cases}$

1912 allein (32 ,,) $r = +0.550 \pm 0.083$ $\begin{cases} \text{nach SCHUSTER}[3]) \text{ sogar} \\ r = +0.63 \end{cases}$

Eine verhältnismäßig so hohe Korrelation wie im Jahre 1912 ist in späteren Jahren nie mehr erreicht worden. Es ist naheliegend, den Grund hier nicht in Vorgängen auf der Sonne zu suchen, sondern in den gewaltigen Trübungserscheinungen, die vom Ausbruch des Katmai herrührten und große Gebiete der Erde umfaßten. Diesbezügliche Untersuchungen sind u. a. von GORCZYNSKI[4]) in Warschau und WILSING[5]) in Potsdam vorgenommen worden.

1913 hat A. ÅNGSTRÖM[6]) am *Mt. Whitney* Messungen ausgeführt. Da er 2 Tage ausschalten muß, erhält er nur 4 Tage mit simultanen Beobachtungen am Mt. Whitney und Mt. Wilson, die zu einer Bestätigung der Realität von Schwankungen der Sonnenstrahlung nicht ausreichen. Beobachtungen von 1918, gleichzeitig mit Mt. Wilson auf *Mount Humptain* vorgenommen, ergeben wegen ungünstigen klimatischen Verhältnissen auf letzterer Station nach ABBOT[7]) keine Korrelation. Als Kuriosum möchte ich bemerken — ein größeres Gewicht kann man dieser kleinen Reihe von 10 Tagen nicht geben —, daß ich aus simultanen Beobachtungen in Upsala[8]) und Calama des Jahres 1920 eine außerordentlich hohe Korrelation von $r = -0.795 \pm 0.078$

---

[1]) Meteor. Z. 1924, S. 24.
[2]) V. J. S. der A. G. **49**, S. 68.
[3]) Nach ÅNGSTRÖM: Geogr. Annaler 1951.
[4]) Boll. d. Soc. Met. Ital. 1921, 4—6.
[5]) Potsdam, Astroph. Obs. Publ. **23**, Nr. 72.
[6]) Sm. Misc. Coll. **65**, Nr. 3.
[7]) Ann. IV. S. 177.
[8]) Ich verdanke diese Werte der Freundlichkeit Herrn BÄCKLINS.

erhalte. Der Zusammenhang ist aber entgegengesetzt! Wenn die eine Station eine hohe S.K. mißt, erhält die andere eine tiefe.

Das erste große Material sind die 106 Simultanbestimmungen Mt. Wilson—Calama 1918—1920. Abbot[1] findet daraus:

$$r = +0.491 \pm 0.050$$

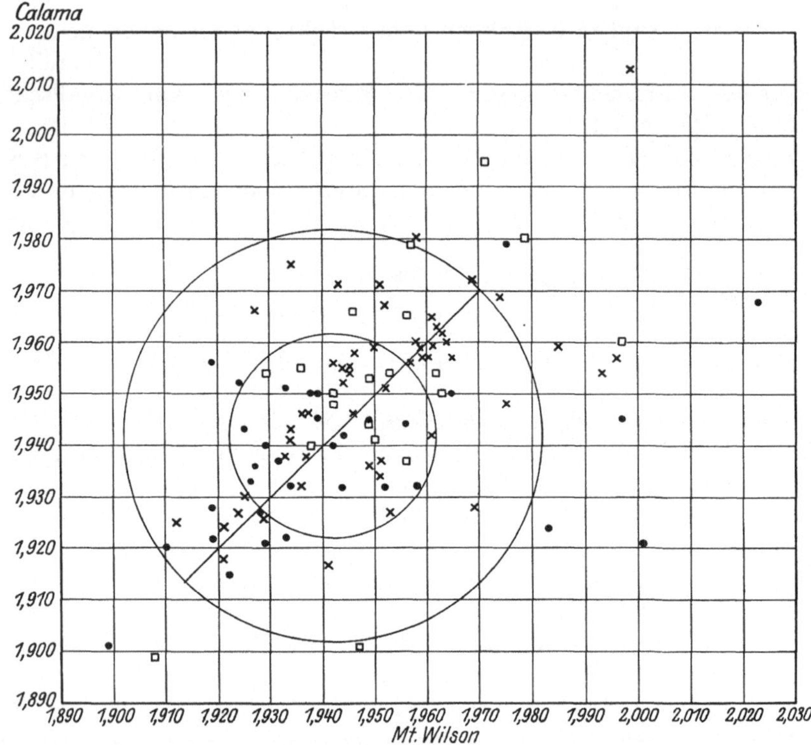

Abb. 4. *Vergleich der simultanen Messungen der Solarkonstante auf dem Mt. Wilson (Cal.) und Calama (Chile)*. Bei Realität der Schwankungen der Sonnenstrahlung müssen die Punkte längs der Linie gruppiert sein, im anderen Falle nach dem Fehlergesetze im Kreise um den Mittelwert. Der kleine Kreis schließt Schwankungen bis 1%, der große bis 2% vom Mittel ein.

□ Werte von 1918 (Juli—Okt.)
× ,, ,, 1919 (Juni—Sept.)
● ,, ,, 1920 (Juli—Sept.)

also eine etwas schwächere Beziehung wie 1912 mit Bassour. Clayton[2] vergleicht 5 Tage-Mittel und erhält dann $r = +0.63 \pm 0.065$. Linke[3] zerlegt das Material und findet 1919 (53 Tage) nur mehr $r = +0.461 \pm 0.084$, für 1920 (33 Tage) sogar nur $r = +0.405 \pm 0.097$.

---

[1] Ann. IV. S. 180.
[2] World Weather, New York 1923, S. 218.
[3] Meteor. Z. 1924, S. 74.

Diese 3 Jahre habe ich in Abb. 4 zusammengestellt. Hier gibt es noch eine nennenswerte Anzahl von Fällen, wo die Schwankungen 2% des Mittelwertes überschreiten. Die Werte innerhalb des kleinen Kreises zeigen aber bereits nahezu die volle Zufallsverteilung. Untersuchen wir das letzte und beste Material der neuen Stationen Montezuma und Harqua Hala, so ergibt sich:

1920 (20 Fälle) $r = + 0{,}144 \pm 0{,}148$　　also überhaupt keine Korrelation!

1921 (55　　,,　) $r = + 0{,}294 \pm 0{,}083$　　　　$\dfrac{r}{\text{w. F.}} = 3{,}5!$

1922 (37　　,,　) $r = + 0{,}330 \pm 0{,}099$[1])　　$\dfrac{r}{\text{w. F.}} = 3{,}3!$

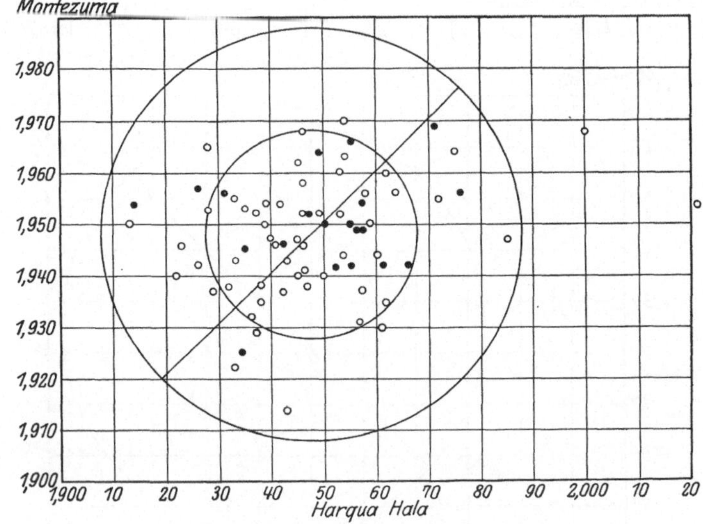

Abb. 5. Vergleich der simultanen Messungen der Solarkonstante auf Harqua Hala (Arizona) und Montezuma (Chile).

● Werte von 1920 (Okt.—Dez.),　　○ Werte von 1921 (Jan.—Dez.), im übrigen vgl. Abb. 4.

Also alle Korrelationen weit unter den früheren Ergebnissen. Dabei sei noch hervorgehoben, daß 1921 das erste vollständige Jahr mit Simultanbeobachtungen aus allen Monaten ist. Wir sehen auch in Abb. 5, wie nur mehr 2 Tage Schwankungen über 2% zeigen, gegenüber 18 Tagen im Material der Abb. 4; andererseits sind die Werte innerhalb des kleinen Kreises im letzten veröffentlichten Zeitraum ganz offenkundig nach einer Zufallsverteilung gruppiert. In der nach-

---

[1]) Schließt man nur 2 Tage aus, so wird die Korrelation bereits wesentlich geringer!

stehenden Tabelle sind zu besserem Vergleiche die Abweichungen vom Mittelwerte in Prozenten der Gesamtzahl der Fälle angeführt.

Abweichungen vom Mittel	unter 1%	1—2%	> 2%
1918	50%	25%	25% der Fälle
1919	40%	49%	11% ,, ,,
1920	51%	30%	19% ,, ,,
alle Wilson-Calama . . .	45%	39%	16% der Fälle
1920	75%	25%	1% der Fälle
1921	71%	25%	4% ,, ,,
1922	57%	20%	13% ,, ,,
Abweichungen vom Mittel der 3 Jahre Montezuma-Harqua Hala	57%	36%	7% der Fälle

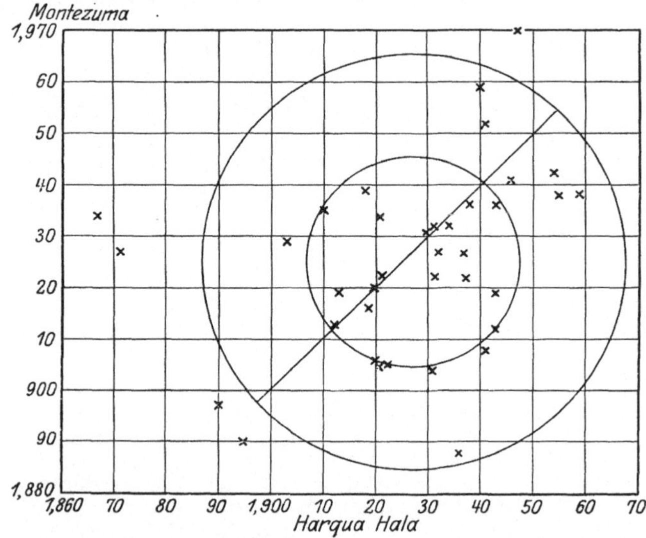

Abb. 6. Harqua Hala-Montezuma 1922, im übrigen vgl. Abb. 4 und 5.

Wir stellen also nicht nur im Laufe der Zeit eine Abnahme der beobachteten Schwankungen fest, sondern ersehen aus den Abbildungen bzw. den errechneten Korrelationen den viel schwerer wiegenden Umstand, daß der Zusammenhang simultaner S. K.-Beobachtungen bei dem neuen, unter bedeutend günstigeren Bedingungen gewonnenen Material, ganz wesentlich unsicherer ist, ein Umstand, den man wohl nur dahin deuten kann, *daß die kurzperiodischen Schwankungen nicht reell sind.*

7. Die Beobachtungen der Sonnenhelligkeit an den Planeten. Ein einwandsfreier Nachweis der Realität der Schwankungen der Sonnenstrahlung könnte an den Planeten erbracht werden, deren Helligkeit diese Schwankungen widerspiegeln müßte. Die ersten derartigen Untersuchungen rühren von Müller[1]) und King[2]) her. Da es sich ja hier nur um sehr kleine Schwankungen handeln kann, sind diese Untersuchungen sehr schwierig und erst durch die Anwendung der Photozelle durch Guthnick und Prager erfolgverheißend geworden. Ein großer Vorzug dieser Methode liegt darin, daß Änderungen der Luftdurchsichtigkeit nur eine verhältnismäßig geringfügige Rolle spielen, da stets nur der *Helligkeitsunterschied* gegen einen nahegelegenen Vergleichsstern gemessen wird. Es wurde von ihnen Saturn in den Oppositionen 1914/15 und 1917 untersucht[3]). Insbesondere den Messungen von 1917 ist hohes Gewicht zuzuteilen. Es zeigte sich, daß nach Berücksichtigung der Phasen- und der schwierigen Ringkorrektion eine Helligkeitsschwankung nicht über $\pm 1\%$ auftrat. Da ja die S.K.-Bestimmungen aus dieser Zeit weitaus größere Schwankungen zeigten, so mußte Guthnick[4]) aus den Saturnmessungen schließen, daß sich die amerikanischen Beobachtungen nur auf Erscheinungen beziehen können, die nicht in der Sonne selbst, sondern in unserer Atmosphäre vor sich gegangen sind. Jupitermessungen[5]) 1917/18 ergaben ebenfalls nur Abweichungen, die meist unter 1% der Gesamtintensität lagen. Die Babelsberger Saturnbeobachtungen des Jahres 1920 wurden Abbot zur Verfügung gestellt und von ihm[6]) publiziert. Schon 1917 hatte sich gezeigt, daß der Müllersche Phasenkoeffizient verkleinert werden mußte, was auch 1920 bestätigt wurde, wobei sich noch die interessante Erscheinung zeigte, daß der Koeffizient vor der Opposition nur $0^{m}\!.025$ und nach $0^{m}\!.037$ betrug. Nachdem alle Korrektionen angebracht wurden, ergab der Vergleich mit den simultanen Calama-Beobachtungen[7]) ganz offenkundig *keinen* Zusammenhang. Da die Beobachtungen kleinerer Phasen als $1°$ gestrichen wurden und ebenso einige weniger sichere Beobachtungen, verblieben im ganzen nur 9 Fälle. Abbot greift nun zu einer Hypothese: Die Hülle der Sonne sei in verschiedenen Richtungen von wechselnder Durchsichtigkeit, infolgedessen die Calama-Werte nicht mit *gleichzeitigen* Planetenbeobachtungen verglichen werden dürfen. Man müsse daher die wegen der Sonnenrotation und der Differenz der heliozentrischen Saturn- und Erdlänge sich ergebende Datumsdifferenz berücksichtigen. Abbot[8]) führt dies durch und findet jetzt

---

1) Potsdam, Publ. Bd. 8. Vgl. auch Very, Témoignage des planètes concernant la radiation solaire. Bull. astronomique. **34**, 129.
2) Harvard Annals, **59**, S. 262 u. **81**, S. 208.
3) Veröff. Sternwarte Berlin-Babelsberg, I, Heft 1 und II, Heft 3.
4) Naturwissenschaften, 1918, S. 136.
5) l. c. Seite 124.
6) Proc. of the Nat. Academy of Sc. Bd. 6, S. 674, auch Ann. IV.
7) Ann. IV. S. 191.
8) Ann. IV. S. 190.

eine schwache Beziehung zwischen Saturnhelligkeiten und den Werten der S.K. Das Material ist aber so klein und die Hypothese viel zu wenig gesichert, daß wir unmöglich daraus die Realität der Variation der Sonnenstrahlung erweisen können. Dies wurde von Bernheimer[1]), Dietzius[2]) und Linke[3]) bereits hervorgehoben, und auch Clayton[4]), der die übrigen Bestätigungen als gegeben ansieht, steht diesen Saturnvergleichen sehr zurückhaltend gegenüber. Prof. Guthnick hat mir in liebenswürdiger Weise, wofür ich ihm auch an dieser Stelle meinen herzlichsten Dank aussprechen möchte, seine unpublizierten Saturnund Jupitermessungen des Jahres 1921 zur Verfügung gestellt. Die Beobachtungen sind, wie er mitteilt, im allgemeinen bei wenig günstigen atmosphärischen Verhältnissen gemacht[5]). Die Saturnopposition ist in der Nähe der Zeit des verschwundenen Ringes. Die Ringkorrektion steigt dabei wesentlich, nämlich von $0^{m}_{.}002$ bis $0^{m}_{.}1$. Ich habe daher nach Reduktion auf mittlere Opposition eine Neubestimmung des Phasenkoeffizienten lieber gleich für die Saturnkugel, also nach Reduktion .auf verschwundenen Ring, gemacht. (Bei den in dieser Opposition vorkommenden kleinen Elevationswinkeln B stimmen die Reduktionsgrößen von Müller und Seeliger überein.) Es bestätigt sich, daß der Phasenkoeffizient vor der Opposition bedeutend geringer ist. Ich erhalte nach Ausgleichungsrechnung für die Saturnkugel: $0^{m}_{.}0162$ vor der Opposition und $0^{m}_{.}0357$ nach der Opposition. Der Vergleich der endgültigen Werte Saturn$-\chi$ Leonis mit Kontrollmessungen $\chi - c$ Leonis zeigt eine deutliche Abhängigkeit, die offenbar in einer Veränderlichkeit des Vergleichssternes gelegen ist. Ich habe dies berücksichtigt und finde schließlich folgende Abweichungen der einzelnen Tage gegen den Mittelwert:

Tag	Phase	$B-R$	Tag	Phase	$B-R$
Februar 23.6	$1^{\circ}_{.}96$	$-0^{m}_{.}002$	März 17.5	$0^{\circ}_{.}70$	$+0^{m}_{.}014$
24.5	1.84	-0.001	April 3.5	2.42	$+0.017$
25.5	1.70	-0.006	9.4	3.04	$+0.015$
März 3.5	1.14	-0.012	10.4	3.10	0.000
7.5	0.64	-0.009	19.4	3.88	-0.011
11.5	0.50	$+0.007$	27.4	4.56	$+0.007$
13.5	0.00	$+0.011$	Mai 2.4	4.94	-0.004
14.5	0.41	$+0.007$	12.4	5.48	-0.009
15.5	0.50	$+0.001$	14.4	5.54	-0.008
16.5	0.58	$+0.007$	16.4	5.66	-0.028

[1]) Om fotoelektr. mätning av himmlakropparnas ljusstyrka. P. Astr. Tidskr. Bd. 3, S. 1. 1922.

[2]) Meteor. Z. 1923. S. 161.

[3]) Meteor. Z. 1924. S. 74.

[4]) World Weather 1923. S. 220.

[5]) Vor der Bearbeitung wurden die ungünstigen Tage Febr. 21. März 8. 12. 22. April 13. ausgeschlossen. Bei Saturn fehlt im endgültigen Ergebnis auch März 30. wegen der Bemerkung des Beobachters: „Bilder außerordentlich schlecht, nicht klar!"

Sie erreichen im Durchschnitt nicht 1%, genau so wie in den früheren Jahren. Freilich zeigen diesmal, wie wir gesehen haben, auch die Schwankungen der S.K. keine größere Amplitude. Die auffallende Anhäufung der positiven Abweichungen in der Mitte des Beobachtungszeitraumes erscheint ebenfalls wie bei den früheren Messungen von Guthnick und Prager. Es dürfte dies, wie Guthnick[1]) bereits hervorgehoben hat, dadurch begründet sein, daß die Seeligersche Theorie der Ringhelligkeit nicht entsprechend berücksichtigt werden konnte. Auf jeden Fall ist natürlich diese Welle etwas ganz anderes als die kurzperiodischen Schwankungen der S.K. Ein Vergleich mit simultanen amerikanischen Messungen (gewichtete Mittel aus Montezuma und Harqua Hala) zeigt wiederum *keine* Übereinstimmung. Ich habe noch versucht, die Abbotsche Hypothese anzuwenden. Der Vergleich ist in Abb. 7 wiedergegeben. Die Linie gibt die theoretische Änderung 0.018 kal der S.K. = 0^{m}.01 des Saturn. Man muß zugeben, daß ein gewisser Zusammenhang zu erkennen ist. Freilich ist das Material wieder außerordentlich gering, insbesondere wenn man be-

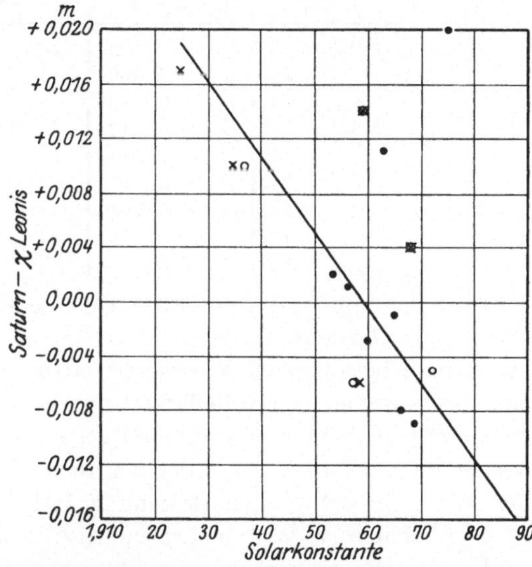

Abb. 7. *Vergleich der Saturnhelligkeit in Berlin-Babelsberg 1921 mit den Werten der Solarkonstante in Amerika.* Die einander entsprechenden Tage sind zufolge einer Hypothese von Abbot ermittelt.

● Zuverlässige Werte.
○ Unsicherheit im Werte der S.K.
× Werte, wo der Phasenwinkel des Saturn unter 1% betrug.
⊗ Zusammentreffen von ○ und ×.

Die Linie entspricht der theoretischen Abhängigkeit.

denkt, daß nur die vollen Punkte in Betracht kommen. Andererseits ist auch bei den voneinander so stark abweichenden amerikanischen Einzelbeobachtungen eines Tages die Wahl der entsprechenden mit Babelsberg korrespondierenden S.K.-Werte von größter Unsicherheit.

In gleicher Weise habe ich die nahezu gleichzeitig mit Saturn vorgenommenen Jupitermessungen Guthnicks bearbeitet. Es bestätigt sich, wie Guthnick und Prager 1917, neuerdings auch King[2]) fest-

---

[1]) Naturwissenschaften 1918, S. 135.
[2]) Harvard Annals. **85**, Nr. 4.

gestellt haben, daß der Phasenkoeffizient den theoretischen Wert
beträchtlich übersteigt. Ich erhalte zwar, wohl mit Rücksicht auf die
schlechteren Beobachtungsbedingungen, einen kleineren Wert als $0^m.0148$,
aber immerhin noch $0^m.0098$. Einen Einfluß der Jupiterrotation auf
die Helligkeit fand ich nur schwach angedeutet und er wurde als nicht
gesichert unberücksichtigt gelassen. Die gefundenen Abweichungen der
einzelnen Beobachtungen vom Mittel:

Tag		$B-R$	Tag		$B-R$
Februar	23.6	$+ 0^m.011$	März	16.5	$- 0^m.002$
	24.5	$- 0.006$		17.5	$+ 0.018$
	25.5	$- 0.013$		30.5	$- 0.005$
März	3.5	$+ 0.013$	April	3.5	$+ 0.012$
	7.5	$+ 0.011$		9.4	$- 0.022$
	11.5	$- 0.005$		10.4	$- 0.006$
	13.5	$+ 0.001$		19.4	$- 0.009$
	14.5	$+ 0.010$		27.4	$- 0.007$
	15.5	$+ 0.010$	Mai	2.4	$- 0.005$

sind wieder nur unbedeutend. Ein Vergleich mit simultanen S.K.-
Werten gibt *keine* Korrelation. Ich habe auch hier wieder die ABBOTsche
Hypothese heranzuziehen versucht, aber auch dann läßt sich keine Beziehung zwischen Jupiterhelligkeit und S.K. aufdecken. Dieses Ergebnis
erscheint wohl bemerkenswert und gegen die Realität der Schwankungen
der Sonnenstrahlung zu sprechen und beeinträchtigt natürlich auch
den Wert der schwachen Beziehung Babelsberg—Amerika für Saturn.

8. Ausblick. Zusammenfassend ergibt sich, daß die Beobachtungen
nach der ursprünglichen LANGLEYschen Methode, in gleicher Weise
aber auch die Beobachtungen nach der neuen Methode, soweit sie
bis heute publiziert sind[1]), *vom Einfluß der Vorgänge in der Erdatmosphäre nicht voll befreit, mithin die Ergebnisse noch nicht als definitiv
zu betrachten sind.* Es ist ferner nicht möglich, die Realität der kurzperiodischen Schwankungen zu beweisen. Im Gegenteil scheinen die
Vergleiche mit Vorgängen auf der Sonne, mit den Planetenhelligkeiten
und vor allem die Untersuchungen der simultanen Beobachtungen der
S.K. selbst, *gegen* die Realität zu sprechen. Das Auftreten größerer
Schwankungen wird mit der Verfeinerung der Beobachtungskunst immer
seltener, die Größe der Schwankungen nähert sich der Größe der Beobachtungsfehler. Die Realität *jährlicher* Schwankungen im Zusammenhang mit der Fleckentätigkeit scheint dagegen tatsächlich vorzuliegen.
Dafür spricht auch die eingangs (Abschn. 1) erwähnte offenkundige Abnahme der Sonnenstrahlung im Jahre 1922. Möglicherweise sind auch
monatliche Schwankungen reell; so finde ich für die Monatsmittel
Montezuma-Harqua Hala von Oktober 1920 bis September 1922 eine

---

[1]) Bis inkl. Februarheft der Monthly Weather Rev. 1924 fehlen Angaben
seit Ende 1922. Früher wurden die S.K. monatlich veröffentlicht.

Korrelation von $r = +0.806 \pm 0.048$. Damit ist freilich noch nicht bewiesen, daß diese monatlichen Schwankungen in der Sonne selbst vor sich gehen. Man müßte z. B. simultane Transmissionskoeffizienten auf Korrelation untersuchen[1]), also nach über weite Gebiete der Erde sich erstreckenden Trübungszonen forschen. Die Zusammenhänge, die CLAYTON[2]) u. a. zwischen Schwankungen der Sonnenstrahlung und meteorologischen Erscheinungen auf der Erde zu finden glaubt (Korrelationen bis zu $r = 0.6$!) näher in Erwägung zu ziehen, erübrigt sich wohl solange, als man nicht einmal in der Lage ist, Veränderungen der Sonnenstrahlung gleichzeitig an zwei Orten gesichert festzustellen. CLAYTONS Schlüsse[3]), die in den 13 Punkten zusammengefaßt werden, sind gewiß sehr interessant, aber heute nur auf schwankendem Grunde aufgebaut.

In Zukunft wird es sich darum handeln, die S.K. überall nur mehr nach der neuen Methode zu messen. Die empirisch ermittelte Beziehung zwischen F und a, da für 1922 schon nicht mehr zutreffend, müßte neu bestimmt werden. Die an einem Tage gefundenen definitiven S.K., errechnet aus verschiedenen Luftmassen, müßten in Zukunft untereinander eine viel bessere Übereinstimmung zeigen. Man kann überzeugt sein, daß ABBOT und seine Mitarbeiter dies auch erreichen werden. Die Simultanbeobachtungen auf Montezuma und Harqua Hala werden vorläufig bis Juli 1925 fortgesetzt. Es wäre zu wünschen, daß dieses bewundernswerte Unternehmen nicht umsonst gemacht wird!

Daneben gilt es in Zukunft rege Kontraststudien vorzunehmen, eine intensive Wiederaufnahme der MOOREschen Arbeiten über die Intensitätsveränderungen der Absorptionslinien, regelmäßige Beobachtungen der Fackelhäufigkeiten und Studium der Fleckentätigkeit. Neue photoelektrische Messungen des Saturn und Jupiter, ebenso auch des Uranus sind geboten, womöglich simultane Beobachtungen, um vom Klima unabhängiger zu sein, also auch mit der Photozelle am Lickobservatorium.

Den vereinten Anstrengungen wird es dann sicherlich gelingen, die Lösung der schwierigen Frage zu erreichen. Sollte es sich dann doch noch erweisen, daß die Schwankungen der Sonnenstrahlung wirklich reell sind, dann erwächst erst das neue große Problem, die Ursachen derselben zu erforschen, zugleich die Suche nach variablen Zwergsternen wieder aufzunehmen und das Problem der Veränderlichen überhaupt mit neuen Kräften anzugeben.

---

[1]) Sind aus dieser Zeit nicht publiziert.
[2]) u. a. World Weather New York. 1923. S. 264.
[3]) l. c. Seite 264.